T0206118

The quantum inverse scattering method is a means of finding exact solutions for two-dimensional models in quantum field theory and statistical physics (such as the sine-Gordon equation or the quantum nonlinear Schrödinger equation). These models are the subject of much attention amongst physicists and mathematicians.

The present text is an introduction to this important and exciting area. It consists of four parts. The first deals with the Bethe Ansatz and calculation of physical quantities. The authors then give a detailed but pedagogical account of the quantum inverse scattering method before applying it in the second half of the book to the calculation of correlation functions. This is one of the most important applications of the method and the authors have made significant contributions to the area. Here they describe some of the most recent and general approaches and include some new results.

The book will be essential reading for all mathematical physicists, at research or graduate level, working in field theory and statistical physics.

CAMBRIDGE MONOGRAPHS ON MATHEMATICAL PHYSICS

General Editors: P.V. Landshoff, D.R. Nelson, D.W. Sciama, S. Weinberg

QUANTUM INVERSE SCATTERING METHOD
AND CORRELATION FUNCTIONS

CAMBRIDGE MONOGRAPHS ON MATHEMATICAL PHYSICS

[†] Issued as a paperback

QUANTUM INVERSE
SCATTERING METHOD
AND CORRELATION FUNCTIONS

V. E. KOREPIN

Institute for Theoretical Physics, State University of New York at Stony Brook

N. M. BOGOLIUBOV

St Petersburg Department of Mathematical Institute of Academy of Sciences of Russia, POMI

A. G. IZERGIN

St Petersburg Department of Mathematical Institute of Academy of Sciences of Russia, POMI

CAMBRIDGE
UNIVERSITY PRESS

PUBLISHED BY THE PRESS SYNDICATE OF THE UNIVERSITY OF CAMBRIDGE
The Pitt Building, Trumpington Street, Cambridge CB2 1RP, United Kingdom

CAMBRIDGE UNIVERSITY PRESS
The Edinburgh Building, Cambridge CB2 2RU, United Kingdom
40 West 20th Street, New York, NY 10011–4211, USA
10 Stamford Road, Oakleigh, Melbourne 3166, Australia

© Cambridge University Press 1993

This book is in copyright. Subject to statutory exception
and to the provisions of relevant collective licensing agreements,
no reproduction of any part may take place without
the written permission of Cambridge University Press.

First published 1993
First paperback edition 1997

Typeset in Times

A catalogue record for this book is available from the British Library

Library of Congress Cataloguing in Publication data available

ISBN 0 521 37320 4 hardback
ISBN 0 521 58646 1 paperback

Transferred to digital printing 2005

TAG

Contents

Preface

This book is devoted to exact solutions of quantum field theory models (in one space dimension plus one time dimension). We also study two-dimensional models of classical statistical physics, which are naturally related to these problems. Complete descriptions of the solvable model are given by the Bethe Ansatz which was discovered by H. Bethe in 1931 [7] while studying the Heisenberg antiferromagnet. The Bethe Ansatz has been very useful for the solution of various problems ([25], [32], [40]–[42], [44], [49], and [57]).

Some of the Bethe Ansatz solvable models have direct physical application. A famous problem solved by the Bethe Ansatz is the Kondo problem (see [56] and [2]). Another model is the Hubbard model [39] which is related to high temperature superconductivity. An important application of the Bethe Ansatz is in nonlinear optics where cooperative spontaneous emission of radiation can be described by an exactly solvable quantum model [48]. The Bethe Ansatz is very useful in modern theoretical physics [46, 47]. Correlation functions provide us with dynamical information about the model. They are described in detail in this book.

Bethe Ansatz solvable models are not free; they generalize free models of quantum field theory in the following sense. Many-body dynamics of free models can be reduced to one-body dynamics. With the Bethe Ansatz, many-body dynamics can be reduced to two-body dynamics. The many-particle scattering matrix is equal to the product of two-particle ones. This leads to the self consistency relation for the two-particle scattering matrix. It is the famous Yang-Baxter equation (a survey of articles can be found in [34]) which is the central concept of exactly solvable models. The role of the Yang-Baxter equation goes beyond the theory of dynamical systems. It is very important in the theory of both knots [58] and quantum groups [34].

Quantum exactly solvable models are closely related to the theory of completely integrable differential equations. The simplest relation is provided by the quasiclassical limit. Quantum correlation functions are described by classical differential equations. The modern way to solve these equations, the inverse scattering method, was founded in 1967 by Gardner, Greene, Kruskal, and Miura [24]. They studied the Korteweg-deVries equation. P. Lax showed that this equation can be represented as a commutativity condition for two linear differential operators [38]. It is interesting to note that the Lax representation is algebraically related to the Yang-Baxter equation [31].

The inverse scattering method has permitted a wide class of nonlinear differential equations to be solved. The Toda lattice is one example (see [21], [22], and [23]). These equations have applications in various areas of physics: plasma physics, nonlinear optics, nonlinear ocean waves, and others. There exists a number of very interesting books on the inverse scattering method ([1], [9], [10], [11], [18], [37], [45], and [61]). One should also mention that there exist multi-dimensional completely integrable differential equations. The most famous is the self-dual Yang-Mills equation [3]. Other examples are the Davey-Stewartson [15] and Kadomtsev-Petviashvilli [35] equations. Completely integrable ordinary differential equations are also extremely important; for example the famous Painlevé transcendent equations (see [29] and references therein). They also arise in two-dimensional gravitation and matrix models (see [8], [16], and [27]) and in the description of quantum correlation functions [4], [33], [43] and [55].

Further development of the inverse scattering method is related to [60] where the Hamiltonian interpretation was understood; L.D. Faddeev and V.E. Zakharov showed that the solution of a model by the inverse scattering method can be considered as a transformation to action-angle variables. This provides an opportunity for quasiclassical quantization. The quantum theory of solitons was constructed in [12]–[14], [20], and [26] where it was shown that after quantization, solitons appear as elementary particles in the spectrum of the Hamiltonian.

The quantum inverse scattering method was discovered in [50]–[52]. It provides a unified point of view of the exact solution of classical and quantum models. It combines the ideas of the Bethe Ansatz and the inverse scattering method. The first model to be solved by means of the quantum inverse scattering method was the nonlinear Schrödinger equation

$$i\partial_t \Psi = -\partial_x^2 \Psi + 2c\Psi^\dagger \Psi \Psi.$$

The Lax representation for this equation was constructed in reference [59]. The Bethe Ansatz for the quantum version of this equation was constructed in [40] and [41]. The quantum inverse scattering method permitted a reproduction of the Bethe Ansatz results starting from the Lax representation. An important development of the quantum inverse scattering method is related to the study of differential equations for quantum correlation functions. It was shown in [28] that differential equations for quantum correlation functions are simply related to the original differential equation which was quantized. The correct language for the description of correlation functions is the language of τ-functions. This is described in our book. We should mention the papers [4] and [55] where differential equations were first obtained for correlation functions of the Ising model.

The quantum inverse scattering method is a well developed method (see reviews [19], [34], [36], [49], and [54]). It has allowed a wide class of nonlinear evolution equations to be solved. It explains the algebraic nature of the Bethe Ansatz. Our book explains this method in detail.

An important example solved by the quantum inverse scattering method [50] is the sine-Gordon equation

$$\partial_t^2 u - \partial_x^2 u + \frac{m^2}{\beta} \sin \beta u = 0.$$

In relation to this model, we would like to mention the new book by Smirnov [53]. The algebraic Bethe Ansatz is related to quantum groups [17]. It is also deeply related to Zamolodchikov's theory of factorized S-matrices [62] and to the theory of exactly solvable lattice models in classical statistical physics (the best review of these models is the book by Baxter [5]) and to conformal field theory ([6] and [30]).

In our book, we try to illustrate general statements for a few simple models. Our main example is the nonlinear Schrödinger equation (in the quantum case this is the model of a one-dimensional Bose gas with δ-repulsion). We consider also the sine-Gordon model, the Heisenberg antiferromagnet, and the Hubbard model.

Our book is divided into four Parts, each of which has an introduction. The first Part explains the coordinate Bethe Ansatz. We evaluate the energy and momentum of excitations and the scattering matrix in the thermodynamic limit. Normally, the ground state of the model is a Fermi sphere (or Dirac sea). The thermodynamics of the model are constructed explicitly.

The second Part explores the quantum inverse scattering method and the algebraic Bethe Ansatz. The classification of exactly solvable models is given there. The important concept of the quantum determinant is introduced (it is related to the antipode in quantum groups). The partition function of the six-vertex model is represented as a determinant for the finite lattice with domain wall boundary conditions. The Pauli principle for one-dimensional interacting bosons is also discussed. Lattice versions of continuous models are constructed in such a way that the most important dynamical characteristics are preserved.

The third and fourth Parts describe the theory of correlation functions. In Part III quantum correlation functions are represented as determinants of integral operators (of a very special form). The third Part starts with an algebraic study of scalar products. For example, it is proved that the square of the norm of the Bethe wave function is equal to the determinant of very simple matrices. It can be obtained by linearization of periodic boundary conditions (in the logarithmic form) near the solution. Correlation functions are represented as determinants of some special integral operators (Fredholm type). In the fourth Part, differential equations for correlation functions are derived. First we represent correlation functions as Fredholm determinants of some integral operator. This integral operator has a very special structure which permits us to consider it as a Gel'fand-Levitan-Marchenko operator of some other differential equation. The last equation drives the correlation functions. The asymptotics of correlation functions are evaluated explicitly; even time- and temperature-dependent correlation functions are evaluated.

Another approach to correlation functions is also discussed. It is related to conformal quantum field theory and permits the evaluation of long distance asymptotics of correlation functions at zero temperature. The models considered here are gapless, and correlation functions decay as powers of the distance; zero temperature is the point of phase transition. These powers are called the critical exponents. They depend on all parameters of the model and we evaluate them explicitly. The method of their evaluation is related to finite size corrections. The central charge of the Virasoro algebra describing the asymptotics of our models is generally equal to one.

The Parts of the book are divided into chapters. The aims and goals of each chapter are explained in an introduction and at the end of each chapter, there is a conclusion which gives a summary and contains bibliographic comments. The references are listed by chapter at the end of the book.

There is double enumeration of the formulæ in the book. The first number corresponds to the section number and the second is the formula number in the section. When referring to formulæ in different chapters, we precede the equation number with the number of the chapter. If we refer to a section from another chapter then the section number will be preceded by the chapter number. Theorems are numbered separately in each chapter. Sections with asterisks can be omitted in the first reading.

This book was started in St Petersburg where we benefitted from discussions with L.D. Faddeev, A.R. Its, N.A. Slavnov, E.K. Sklyanin, N.Yu.

Reshetikhin, L.A. Takhtajan, V.E. Zakharov, P.P. Kulish, V.N. Popov, F.A. Smirnov and A.N. Kirillov. The book was finished in Stony Brook. We greatly appreciate the creative atmosphere of the Institute for Theoretical Physics at Stony Brook. We benefitted from discussions with C.N. Yang, B. McCoy, B. Sutherland, M. Fowler, H. Thacker, H. Flaschka, A. Fokas, A. Newell, M. Ablowitz, J. Palmer, C. Tracy, L.L. Chau, R. Shrock, W. Weisberger, E. Melzer, F. Essler, D. Uglov, H. Frahm, K. Schoutens, F. Figueirido, E. Williams, and S. Ray. The authors wish to thank David A. Coker for proofreading and pedagogical suggestions.

We are grateful to NFS for grant PHY-9107261.

Part I
The Coordinate Bethe Ansatz

Introduction to Part I

A method of solution of a number of quantum field theory and statistical mechanics models in two space-time dimensions is presented in this Part. This method was first suggested by H. Bethe in 1931 [1] and is traditionally called the Bethe Ansatz*. Later on the method was developed by Hulthen, Yang and Yang, Lieb, Sutherland, Baxter, Gaudin and others (see [2], [3], and [4]).

We begin the presentation with the coordinate Bethe Ansatz not only due to historical reasons but also because of its simplicity and clarity. The multi-particle scattering matrix appears to be equal to the product of two-particle matrices for integrable models. This property of two-particle reducibility is of primary importance when constructing the Bethe wave function. The important feature of integrable models is that there is no mass-shell multiple particle production. This property is closely connected to the existence of an infinite number of conservation laws in such models; this will be clear from Part II.

Four main models, namely the one-dimensional Bose gas, the Heisenberg magnet, the massive Thirring model and the Hubbard model, are considered in Part I. Eigenfunctions of the Hamiltonians of these models are constructed. Imposing periodic boundary conditions leads to a system of equations for the permitted values of momenta. These are known as the Bethe equations. This system can also be derived from a certain variational principle, the corresponding action being called the Yang-Yang action. It plays an important role when investigating the models. The Bethe equations are also useful in the thermodynamic limit. The energy of the ground state, the velocity of sound, etc., may be calculated in this limit. Excitations above the ground state i.e., physical

---

* Literally, "start".

1

particles, are also investigated. To define their physical characteristics, the technique of dressing equations is introduced and investigated. The thermodynamics of the model are explained in great detail.

The material in this Part is arranged as follows. The theory of the one-dimensional Bose gas with point-like repulsive interaction between particles is presented in the first chapter. The solution of the XXZ Heisenberg magnet in an external magnetic field is given in the second chapter. The quantum model of the spinor field with four-point self-interaction in two space-time dimensions is solved in the third chapter. This is usually called the massive Thirring model, and is equivalent to the sine-Gordon model (in the zero charge sector). In the last chapter of Part I, the Hubbard model of interacting fermions on the lattice is briefly considered.

I

The One-dimensional Bose Gas

Introduction

The one-dimensional Bose gas with point-like interaction of the particles
(the quantum variant of the nonlinear Schrödinger equation) is one of
the main and most important models which can be solved by means of
the Bethe Ansatz [14], [15]. This model has been thoroughly investigated
([1], [5], [17], [21] and [22]). We shall start with the construction of eigenfunctions of the Hamiltonian in a finite volume. Quantities interesting
from the physical point of view (in the thermodynamic limit at zero temperature) are considered; the thermodynamics at finite temperatures is
investigated in detail as well. A number of essential ideas which will be
applied to other models are introduced.

The construction of eigenfunctions of the Hamiltonian is explained in
section 1. Their explicit form and, in particular, the two-particle reducibility, are common features of the models solvable by means of the
Bethe Ansatz method. Periodic boundary conditions are imposed on
the wave function in section 2; the Bethe equations for the particles' momenta are introduced and analyzed. Taken in the logarithmic form, these
equations realize the extremum condition of a certain functional, the corresponding action being called the Yang-Yang action. The transition to
the thermodynamic limit is considered in section 3. In that same section
the ground state of the gas is constructed. The density of the particle
distribution in momentum space and the energy of the ground state are
calculated. The method of the transition to the thermodynamic limit described in this section is rather general and may be applied to any model
solvable by means of the Bethe Ansatz. In section 4, excitations above
the ground state are constructed and their main characteristics (energy,
momentum and scattering matrix) are determined by means of the dress-

3

ing equations. The ground state of the model is the Dirac sea (also called the Fermi sphere).

The thermodynamics of the model is presented in section 5. The functional integral approach is presented. It allows us to solve various problems at finite temperatures. The basic equations describing the state of thermodynamical equilibrium, the Yang-Yang equation being one of them, are derived in this section. The Yang-Yang equation, which is a nonlinear integral equation, is analyzed in section 6. The theorem showing the existence of solutions is proved. The state of thermodynamic equilibrium with temperature going to zero is considered in section 7. While investigating this limit we can obtain more detailed information about the ground state of the Hamiltonian at zero temperature. The strong coupling limit (in which the model is equivalent to the free fermion model) is discussed. The integral equations are solved exactly in this limit.

Excitations above the state of thermodynamic equilibrium are investigated and their interpretation in terms of particles is given. It is important to note that the formulæ at finite temperature and at zero temperature differ only in the integration measure. The thermodynamics of exactly solvable models is so special that it is possible to construct stable excitations at finite temperature, see section 8. Temperature correlations are also discussed in section 8. For exactly solvable models they are also very special: they can be represented in a form similar to one at zero temperature. Later, in Part IV (Chapters XIII–XVI), this will be used for the explicit evaluation of temperature correlation functions (even if they depend on time). Section 9 is devoted to the evaluation of finite size corrections at zero temperature. Later they will be used for calculation of the long distance asymptotics of correlation functions by means of conformal field theory.

I.1 The coordinate Bethe Ansatz

The one-dimensional Bose gas is described by the canonical quantum Bose fields $\Psi(x,t)$ with canonical equal-time commutation relations:

$$\begin{aligned}
[\Psi(x,t),\, \Psi^\dagger(y,t)] &= \delta(x-y) \\
[\Psi(x,t),\, \Psi(y,t)] &= [\Psi^\dagger(x,t),\, \Psi^\dagger(y,t)] = 0.
\end{aligned} \tag{1.1}$$

Later on, the argument t will be as a rule omitted since all the considerations in this chapter apply to a fixed moment in time.

The Hamiltonian of the model is

$$H = \int dx \left[\partial_x \Psi^\dagger(x)\, \partial_x \Psi(x) + c\, \Psi^\dagger(x)\Psi^\dagger(x)\Psi(x)\Psi(x) \right] \tag{1.2}$$

where c is the coupling constant. The corresponding equation of motion,

$$i\,\partial_t \Psi = -\partial_x^2 \Psi + 2c\, \Psi^\dagger \Psi \Psi, \tag{1.3}$$

is called the nonlinear Schrödinger (NS) equation.

For $c > 0$ the ground state at zero temperature is a Fermi sphere. Only this case will be considered further. The Fock vacuum $|0\rangle$, defined by

$$\Psi(x)|0\rangle = 0, \qquad x \in \mathbb{R}^1 \tag{1.4}$$

is of importance. It will be called the pseudovacuum and it is to be distinguished from the physical vacuum which is the ground state of the Hamiltonian (the Dirac sea). The dual pseudovacuum $\langle 0|$ is defined as $\langle 0| = |0\rangle^\dagger$ and satisfies the relations

$$\langle 0| \Psi^\dagger(x) = 0; \qquad \langle 0|0\rangle = 1 \tag{1.5}$$

where the dagger denotes Hermitian conjugation. The number of particles operator Q and momentum operator P are

$$Q = \int \Psi^\dagger(x)\Psi(x)\, dx \tag{1.6}$$

$$P = -\frac{i}{2} \int \left\{ \Psi^\dagger(x)\partial_x\Psi(x) - \left[\partial_x\Psi^\dagger(x)\right]\Psi(x) \right\} dx. \tag{1.7}$$

These are Hermitian operators and they are integrals of motion

$$[H, Q] = [H, P] = 0. \tag{1.8}$$

It should be noted that in Part II we shall construct an infinite number of integrals of motion. We can now look for the common eigenfunctions $|\psi_N\rangle$ of operators H, P, Q:

$$|\psi_N(\lambda_1,\ldots,\lambda_N)\rangle =$$
$$\frac{1}{\sqrt{N!}} \int d^N z \, \chi_N(z_1,\ldots,z_N \mid \lambda_1,\ldots,\lambda_N)\, \Psi^\dagger(z_1)\ldots\Psi^\dagger(z_N)|0\rangle. \tag{1.9}$$

Here χ_N is a symmetric function of all z_j. The eigenvalue equations,

$$H|\psi_N\rangle = E_N|\psi_N\rangle, \quad P|\psi_N\rangle = p_N|\psi_N\rangle, \quad Q|\psi_N\rangle = N|\psi_N\rangle, \tag{1.10}$$

result in the fact that χ_N is an eigenfunction of both the quantum mechanical Hamiltonian $\mathcal{H}_N$ and the quantum mechanical momentum operator $\mathcal{P}_N$:

$$\mathcal{H}_N = \sum_{j=1}^{N} \left(-\frac{\partial^2}{\partial z_j^2} \right) + 2c \sum_{N \geq j > k \geq 1} \delta(z_j - z_k)$$

$$\mathcal{P}_N = \sum_{j=1}^{N} \left(-i\frac{\partial}{\partial z_j} \right) \tag{1.11}$$

$$\mathcal{H}_N \chi_N = E_N \chi_N. \tag{1.12}$$

This can be explained using the momentum operator P (1.7) as an example. First we integrate (1.7) by parts to represent P in the form

$$P = i \int \left[\partial_x \Psi^\dagger(x) \right] \Psi(x) dx.$$

Acting with this operator on the eigenfunction (1.9) gives

$$P|\psi_N(\lambda_1, \ldots, \lambda_N)\rangle =$$
$$\frac{i}{\sqrt{N!}} \int dx \int d^N z \; \chi_N(z_1, \ldots, z_N \mid \lambda_1, \ldots, \lambda_N) \left[\partial_x \Psi^\dagger(x) \right]$$
$$\times \sum_{k=1}^{N} \Psi^\dagger(z_1) \ldots [\Psi(x), \Psi^\dagger(z_k)] \ldots \Psi^\dagger(z_N)|0\rangle.$$

where (1.4) was used. Formula (1.1) gives a delta function for the commutator which can then be integrated out to give

$$P|\psi_N(\lambda_1, \ldots, \lambda_N)\rangle =$$
$$\frac{i}{\sqrt{N!}} \int d^N z \; \chi_N(z_1, \ldots, z_N \mid \lambda_1, \ldots, \lambda_N)$$
$$\times \sum_{k=1}^{N} \Psi^\dagger(z_1) \ldots \frac{\partial}{\partial z_k} \Psi^\dagger(z_k) \ldots \Psi^\dagger(z_N)|0\rangle.$$

Now we integrate by parts with respect to z_k to get

$$P|\psi_N(\lambda_1, \ldots, \lambda_N)\rangle =$$
$$\frac{1}{\sqrt{N!}} \int d^N z \left\{ -i \sum_{k=1}^{N} \frac{\partial}{\partial z_k} \chi_N(z_1, \ldots, z_N \mid \lambda_1, \ldots, \lambda_N) \right\}$$
$$\times \Psi^\dagger(z_1) \ldots \Psi^\dagger(z_N)|0\rangle.$$

Thus we have proved that the action of (1.7) on (1.9) is equivalent to the action of $\mathcal{P}_N$ on χ_N. The construction of the quantum mechanical Hamiltonian is quite similar.

So the quantum field theory problem is reduced to a quantum mechanical problem. The Hamiltonian $\mathcal{H}_N$ describing N boson particles is repulsive for $c > 0$. Due to the symmetry of χ_N in all the z_j, it is sufficient to consider the following domain T in coordinate space:

$$T \; : \; z_1 < z_2 < \cdots < z_N. \tag{1.13}$$

In this domain the function χ_N is an eigenfunction of the free Hamiltonian

$$\mathcal{H}_N^0 = -\sum_{j=1}^{N} \frac{\partial^2}{\partial z_j^2} \tag{1.14}$$

$$\mathcal{H}_N^0 \chi_N = E_N \chi_N.$$

The following boundary conditions should be satisfied:

$$\left(\frac{\partial}{\partial z_{j+1}} - \frac{\partial}{\partial z_j} - c \right) \chi_N = 0, \qquad z_{j+1} = z_j + 0. \tag{1.15}$$

Equation (1.14) and boundary condition (1.15) are equivalent to equation (1.12). In fact, the potential in (1.11) is equal to zero in the domain T. So integrating equation (1.12) over the variable $(z_{j+1} - z_j)$ in the small vicinity $\mid z_{j+1} - z_j \mid < \epsilon \rightarrow 0$, considering all other z_k $(k \neq j, j+1)$ to be fixed in T, one obtains exactly the condition (1.15).

A function satisfying (1.14) and (1.15) can be constructed as follows. Consider the eigenfunction of the Hamiltonian (1.14) in the domain T given as the determinant of the $N \times N$ matrix $\exp\{i\lambda_j z_k\}$

$$\det \left[\exp \{ i \lambda_j z_k \} \right], \tag{1.16}$$

with arbitrary numbers λ_j. This function is equal to zero on the boundary of the domain T due to its antisymmetry in z_k. It is easily seen then that the function χ_N given by

$$\chi_N = \text{const} \left[\prod_{N \geq j > k \geq 1} \left(\frac{\partial}{\partial z_j} - \frac{\partial}{\partial z_k} + c \right) \right] \det \left[\exp \{ i \lambda_j z_k \} \right] \tag{1.17}$$

satisfies equations (1.14), (1.15). To verify, for instance, the equality

$$\left(\frac{\partial}{\partial z_2} - \frac{\partial}{\partial z_1} - c \right) \chi_N = 0, \qquad z_2 = z_1 + 0 \tag{1.18}$$

let us rewrite χ_N as

$$\chi_N = \left(\frac{\partial}{\partial z_2} - \frac{\partial}{\partial z_1} + c \right) \widetilde{\chi}_N \tag{1.19}$$

where

$$\widetilde{\chi}_N = \text{const} \prod_{j=3}^{N} \left(\frac{\partial}{\partial z_j} - \frac{\partial}{\partial z_1} + c \right) \left(\frac{\partial}{\partial z_j} - \frac{\partial}{\partial z_2} + c \right)$$

$$\times \prod_{N \geq j > k \geq 3} \left(\frac{\partial}{\partial z_j} - \frac{\partial}{\partial z_k} + c \right) \det \left[\exp \{ i \lambda_j z_k \} \right]. \tag{1.20}$$

This function is antisymmetric with respect to $z_1 \leftrightarrow z_2$,

$$\widetilde{\chi}_N(z_1, z_2) = -\widetilde{\chi}_N(z_2, z_1), \tag{1.21}$$

and it is equal to zero when $z_1 = z_2$. Returning to equation (1.18)

$$\left[\left(\frac{\partial}{\partial z_2} - \frac{\partial}{\partial z_1}\right)^2 - c^2\right]\tilde{\chi}_N = 0, \qquad z_2 = z_1 \qquad (1.22)$$

we see that the left hand side changes its sign when $z_1 \leftrightarrow z_2$, and hence, equality (1.22) is correct. We can similarly verify the other boundary conditions. Thus χ_N (1.17) is the desired eigenfunction of the Hamiltonian $\mathcal{H}_N$ (1.11). The determinant in (1.17) can be written as a sum over all the permutations $\mathcal{P}$ of the numbers 1, 2, ..., N:

$$\det\left[\exp\left\{i\,\lambda_j z_k\right\}\right] = \sum_{\mathcal{P}}(-1)^{[\mathcal{P}]}\exp\left\{i\sum_{n=1}^{N} z_n\,\lambda_{\mathcal{P}n}\right\} \qquad (1.23)$$

where $[\mathcal{P}]$ denotes the parity of the permutation. One gets, in domain T (1.13):

$$\chi_N = \left\{N!\prod_{j>k}\left[(\lambda_j - \lambda_k)^2 + c^2\right]\right\}^{-1/2}$$

$$\times \sum_{\mathcal{P}}(-1)^{[\mathcal{P}]}\exp\left\{i\sum_{n=1}^{N} z_n\lambda_{\mathcal{P}n}\right\}\prod_{j>k}(\lambda_{\mathcal{P}j} - \lambda_{\mathcal{P}k} - ic) \qquad (1.24)$$

with the constant specified. Let us now continue χ_N by symmetry to the whole of $\mathbb{R}^N$:

$$\chi_N = \left\{N!\prod_{j>k}\left[(\lambda_j - \lambda_k)^2 + c^2\right]\right\}^{-1/2}$$

$$\times \sum_{\mathcal{P}}(-1)^{[\mathcal{P}]}\exp\left\{i\sum_{n=1}^{N} z_n\lambda_{\mathcal{P}n}\right\}\prod_{j>k}\left[\lambda_{\mathcal{P}j} - \lambda_{\mathcal{P}k} - ic\epsilon(z_j - z_k)\right]$$
$$(1.25)$$

where $\epsilon(z)$ is the sign function. The method stated above was apparently suggested by M. Gaudin [9].

Another useful way of writing χ_N is:

$$\chi_N = \frac{(-i)^{\frac{N(N-1)}{2}}}{\sqrt{N!}}\left\{\prod_{N\geq j>k\geq 1}\epsilon(z_j - z_k)\right\}$$

$$\times \sum_{\mathcal{P}}(-1)^{[\mathcal{P}]}\exp\left\{i\sum_{k=1}^{N} z_k\lambda_{\mathcal{P}k}\right\}$$

$$\times \exp\left\{\frac{i}{2}\sum_{N\geq j>k\geq 1}\epsilon(z_j - z_k)\,\theta(\lambda_{\mathcal{P}j} - \lambda_{\mathcal{P}k})\right\} \qquad (1.26)$$

where

$$\theta(\lambda - \mu) = i \ln \left(\frac{ic + \lambda - \mu}{ic - \lambda + \mu} \right).$$

Formula (1.26) determines the Bethe wave function; this function is two-particle reducible. It must be mentioned that the wave functions of all the models solvable by the Bethe Ansatz are of a form similar to (1.26).

Let us discuss the properties of the wave function χ_N. The function χ_N is a symmetric function of the variables z_j ($j = 1, \ldots, N$) and a continuous function of each z_j. These properties become obvious if one rewrites the representation (1.26) in the following form:

$$\chi_N = \frac{\prod\limits_{j>k} (\lambda_j - \lambda_k)}{\sqrt{N! \prod\limits_{j>k} [(\lambda_j - \lambda_k)^2 + c^2]}} \sum_{\mathcal{P}} \exp \left\{ i \sum_{n=1}^{N} z_n \lambda_{\mathcal{P}n} \right\}$$

$$\times \prod_{j>k} \left[1 - \frac{ic\epsilon(z_j - z_k)}{(\lambda_{\mathcal{P}j} - \lambda_{\mathcal{P}k})} \right]. \qquad (1.27)$$

One can also see from this formula that χ_N is an antisymmetric function of λ_j:

$$\chi_N(z_1, \ldots, z_N \mid \lambda_1, \ldots, \lambda_j, \ldots, \lambda_k, \ldots, \lambda_N)$$

$$= -\chi_N(z_1, \ldots, z_N \mid \lambda_1, \ldots, \lambda_k, \ldots, \lambda_j, \ldots, \lambda_N). \quad (1.28)$$

Hence $\chi_N = 0$ if $\lambda_j = \lambda_k$, $j \neq k$. This is the basis of the Pauli principle for one-dimensional interacting bosons which is of great importance for constructing the ground state, which is the Dirac sea (the complete proof of the Pauli principle is given in section VII.4).

It is known that the theorem connecting spin and statistics is not valid in $1 + 1$ space-time dimensions. Hence some of the boson models are equivalent to fermion ones; for instance, the sine-Gordon model to the massive Thirring model, the one-dimensional Bose gas at $c = \infty$ to the free fermion model.

Above, the common eigenfunctions (1.26), (1.9) of the operators H, P and Q were constructed, the corresponding eigenvalues being equal to

$$E_N = \sum_{j=1}^{N} \lambda_j^2; \; P_N = \sum_{j=1}^{N} \lambda_j; \; Q_N = N. \qquad (1.29)$$

Consider now eigenfunctions in the whole coordinate space $\mathbb{R}^N$: $-\infty <$

$z_j < \infty$ $(j = 1, \ldots, N)$. The normalization in this case was calculated in [9]:

$$\int_{-\infty}^{+\infty} d^N z\, \chi_N^*(z_1, \ldots, z_N \mid \lambda_1, \ldots, \lambda_N)\, \chi_N(z_1, \ldots, z_N \mid \mu_1, \ldots, \mu_N)$$

$$= (2\pi)^N \prod_{j=1}^{N} \delta(\lambda_j - \mu_j). \quad (1.30)$$

The momenta $\{\lambda\}$ and $\{\mu\}$ here are supposed to be ordered:

$$\lambda_1 < \lambda_2 < \cdots < \lambda_N, \qquad \mu_1 < \mu_2 < \cdots < \mu_N. \quad (1.31)$$

In the same book [9] the completeness of the system χ_N is also proved:

$$\int_{-\infty}^{+\infty} d^N \lambda\, \chi_N^*(z_1, \ldots, z_N \mid \lambda_1, \ldots, \lambda_N)\, \chi_N(y_1, \ldots, y_N \mid \lambda_1, \ldots, \lambda_N)$$

$$= (2\pi)^N \prod_{j=1}^{N} \delta(z_j - y_j) \quad (1.32)$$

$$z_1 < z_2 < \cdots < z_N, \qquad y_1 < y_2 < \cdots < y_N. \quad (1.33)$$

I.2 Periodic boundary conditions

To analyze different properties of the Bose gas and especially to construct the thermodynamics of the model, it is convenient to impose periodic boundary conditions on the wave functions. Let us put the system considered into a periodic box of length L. Then the wave function χ_N should be periodic in each z_j with all other z_k $(k \neq j)$ fixed:

$$\chi_N(z_1, \ldots, z_j + L, \ldots, z_N \mid \lambda_1, \ldots, \lambda_N)$$

$$= \chi_N(z_1, \ldots, z_j, \ldots, z_N \mid \lambda_1, \ldots, \lambda_N). \quad (2.1)$$

These requirements result in the following system of equations for the permitted values of momenta λ_j (see (1.26)):

$$\exp\{i\lambda_j L\} = -\prod_{k=1}^{N} \frac{\lambda_j - \lambda_k + ic}{\lambda_j - \lambda_k - ic}, \qquad j = 1, \ldots, N. \quad (2.2)$$

These equations will be called Bethe equations. The Bethe system of equations is of primary importance. Its main properties are discussed in detail below.

Theorem 1. *All the solutions λ_j of (2.2) ($c > 0$) are real numbers.*

Proof: Let us use the following properties of $\exp\{i\,\lambda L\}$ and $(\lambda+ic)/(\lambda-ic)$.

$$
\begin{aligned}
|\exp\{i\,\lambda L\}| &\leq 1 \qquad \text{when Im } \lambda \geq 0;\\
|\exp\{i\,\lambda L\}| &\geq 1 \qquad \text{when Im } \lambda \leq 0;
\end{aligned}
\tag{2.3}
$$

$$
\begin{aligned}
\left|\frac{\lambda+ic}{\lambda-ic}\right| &\geq 1 \qquad \text{when Im } \lambda \geq 0;\\
\left|\frac{\lambda+ic}{\lambda-ic}\right| &\leq 1 \qquad \text{when Im } \lambda \leq 0.
\end{aligned}
\tag{2.4}
$$

Consider the set of complex numbers $\{\lambda_j\}$ which satisfy equation (2.2). Denote the λ_j with maximal imaginary part as $\lambda_{\max}$; then

$$
\begin{aligned}
\text{Im } \lambda_{\max} &\geq \text{Im } \lambda_j, \qquad j = 1, \ldots, N;\\
\lambda_{\max} &\in \{\lambda_j\}.
\end{aligned}
\tag{2.5}
$$

If there are several momenta with this same property then one of them should be taken. Taking the modulus of both sides of the equation for $\lambda_j = \lambda_{\max}$ in (2.2), one makes use of the estimate (2.4) for the right hand side and thus obtains

$$
|\exp\{i\,\lambda_{\max}L\}| = \left|\prod_k \frac{\lambda_{\max} - \lambda_k + ic}{\lambda_{\max} - \lambda_k - ic}\right| \geq 1.
\tag{2.6}
$$

Due to (2.3), this results in Im $\lambda_{\max} \leq 0$ and as consequence of (2.5)

$$
\text{Im } \lambda_j \leq 0, \qquad j = 1, \ldots, N.
\tag{2.7}
$$

Let us define $\lambda_{\min}$: Im $\lambda_{\min} \leq$ Im λ_j. Quite similarly one proves that $0 \leq$ Im $\lambda_{\min} \leq$ Im λ_j. So the only remaining possibility is Im $\lambda_j = 0$, $j = 1, \ldots, N$. Theorem 1 is proved.

Let us now prove the existence of solutions for the system (2.2). To do this, one puts the system into logarithmic form:

$$
\varphi_j = 2\pi\tilde{n}_j, \qquad j = 1, \ldots, N
\tag{2.8}
$$

where $\tilde{n}_j$ is an arbitrary set of integers. The variables φ_j are defined as

$$
\varphi_j = L\lambda_j + \sum_{\substack{k=1\\k\neq j}}^{N} \varphi(\lambda_j - \lambda_k)
\tag{2.9}
$$

where

$$
\varphi(\lambda) = i \ln\left(\frac{\lambda+ic}{\lambda-ic}\right); \qquad -2\pi < \varphi(\lambda) < 0, \quad \text{Im } \lambda = 0.
\tag{2.10}
$$

It is more convenient to use the antisymmetric function $\theta(\lambda)$ instead of $\varphi(\lambda)$:

$$\theta(\lambda) = \varphi(\lambda) + \pi; \qquad \theta(\lambda) = -\theta(-\lambda)$$

$$\theta(\lambda) = i \ln\left(\frac{ic + \lambda}{ic - \lambda}\right). \tag{2.11}$$

This function is monotonically increasing in λ:

$$\theta(\lambda_2) > \theta(\lambda_1), \text{ when } \lambda_2 > \lambda_1; \qquad \theta(\pm\infty) = \pm\pi$$

$$\theta'(\lambda) = \frac{2c}{c^2 + \lambda^2} > 0. \tag{2.12}$$

Let us rewrite (2.8) and (2.9) as

$$L\lambda_j + \sum_{k=1}^{N} \theta(\lambda_j - \lambda_k) = 2\pi n_j \tag{2.13}$$

where the integer or half-integer numbers n_j are defined as

$$n_j = \tilde{n}_j + \frac{N-1}{2}. \tag{2.14}$$

The system (2.13) is equivalent to the system (2.2). Equations (2.13) are also called the Bethe equations.

Theorem 2. *The solutions of the Bethe equations (2.13) exist and can be uniquely parametrized by a set of integer (half-integer) numbers n_j.*

Proof: The proof is based on the fact that the equations (2.13) can be obtained from a variational principle. The corresponding action was introduced by C.N. Yang and C.P. Yang [22]

$$S = \frac{1}{2}L\sum_{j=1}^{N}\lambda_j^2 - 2\pi\sum_{j=1}^{N} n_j\lambda_j + \frac{1}{2}\sum_{j,k} \theta_1(\lambda_j - \lambda_k) \tag{2.15}$$

where $\theta_1(\lambda) = \int_0^\lambda \theta(\mu)\, d\mu$. Equations (2.13) are the extremum conditions for S ($\partial S/\partial \lambda_j = 0$), this extremum being a minimum. To prove this, it is sufficient to establish that the matrix $\partial^2 S/\partial \lambda_j \partial \lambda_k$ of second derivatives is positive definite (all eigenvalues are positive). One has:

$$\frac{\partial^2 S}{\partial \lambda_j \partial \lambda_l} = \frac{\partial \varphi_j}{\partial \lambda_l} = \varphi'_{jl} = \delta_{jl}\left[L + \sum_{m=1}^{N} K(\lambda_j, \lambda_m)\right] - K(\lambda_j, \lambda_l), \tag{2.16}$$

where

$$K(\lambda, \mu) = \varphi'(\lambda - \mu) = \theta'(\lambda - \mu) = \frac{2c}{c^2 + (\lambda - \mu)^2}, \tag{2.17}$$

and so

$$\sum_{j,l} \frac{\partial^2 S}{\partial \lambda_j \partial \lambda_l} v_j v_l = \sum_{j=1}^{N} L v_j^2 + \sum_{j>l=1}^{N} K(\lambda_j, \lambda_l)(v_j - v_l)^2 \geq L \sum_{j=1}^{N} v_j^2 > 0$$

(2.18)

for any vector v_j with real components. So the matrix of second derivatives is positive definite: the action is indeed convex. Thus S has a unique minimum which defines the solutions of the Bethe equations. Theorem 2 is proved.

In Chapter X we shall see that the squared norm of the wave function in the periodic box is equal to the determinant of the second derivative of the Yang-Yang action evaluated on the Bethe equation solutions:

$$\int_0^L d^N z \, |\chi_N|^2 = \det\left(\frac{\partial^2 S}{\partial \lambda_j \partial \lambda_l}\right).$$

(2.19)

The antisymmetry of $\theta(\lambda)$ leads to the following equality

$$P_N = \sum_{j=1}^{N} \lambda_j = \frac{2\pi}{L} \sum_{j=1}^{N} n_j,$$

(2.20)

where P_N is the momentum (1.29) of the system.

The solution of the system (2.13) possesses the following important property:

Theorem 3. *If $n_j > n_k$, then $\lambda_j > \lambda_k$. If $n_j = n_k$ then $\lambda_j = \lambda_k$.*

Proof: Let us substract the k-th equation of the system from the j-th equation:

$$L(\lambda_j - \lambda_k) + \sum_{l=1}^{N} [\theta(\lambda_j - \lambda_l) - \theta(\lambda_k - \lambda_l)] = 2\pi(n_j - n_k).$$

(2.21)

Due to the monotonic increase of the function $\theta(\lambda)$ in λ, the left hand side is of the same sign as its first term. The theorem is proved.

In particular, if $n_j = n_k$, then $\lambda_j = \lambda_k$, which results in the fact that the wave function becomes equal to zero due to the Pauli principle. So only $n_j \neq n_k$ $(j \neq k)$ has to be taken into account. Since the wave function is antisymmetric in λ, we can always put $\lambda_j > \lambda_k$ when $j > k$, hence:

$$n_j > n_k \quad \text{when } j > k.$$

(2.22)

To pass to the thermodynamic limit we shall need the following properties:

(1) The different solutions λ_j of the system (2.13) are separated by some interval

$$\frac{2\pi(n_j - n_k)}{L} \geq |\lambda_j - \lambda_k| \geq \frac{2\pi(n_j - n_k)}{L(1 + \frac{2D}{c})} \geq \frac{2\pi}{L(1 + \frac{2D}{c})}; j \neq k \quad (2.23)$$

where $D = N/L$ is the density of the Bose gas. This estimate follows from (2.21) if one uses the inequalities

$$0 < K(\lambda, \mu) < \frac{2}{c}; \quad \text{Im}\,\lambda = \text{Im}\,\mu = 0 \quad (2.24)$$

and

$$\theta(\lambda) - \theta(\mu) = \int_{\mu}^{\lambda} K(\nu, 0)\, d\nu \leq \frac{2}{c}(\lambda - \mu), \quad \lambda > \mu. \quad (2.25)$$

(2) The energy functional $\sum_{j=1}^{N} \lambda_j^2$ in the sector with a fixed number of particles, N, under the condition that $\{\lambda_j\}$ are solutions of the Bethe equations (2.13) is minimized by the following set of numbers n_j that are integer (N odd) or half-integer (N even) [9]:

$$n_j = -\left(\frac{N-1}{2}\right) + j - 1, \quad j = 1, \ldots, N. \quad (2.26)$$

(This is obvious when $c = \infty$.)

(3) Let us define a function $\lambda(x)$ ($x \in \mathbb{R}^1$) closely connected to the solution $\{\lambda_j\}$ of the Bethe equations (2.13). This function is defined by the following relation:

$$L\lambda(x) + \sum_{k=1}^{N} \theta(\lambda(x) - \lambda_k) = 2\pi L x. \quad (2.27)$$

Introducing the action S similar to (2.15)

$$S = \frac{1}{2}L\lambda^2(x) + \sum_{k=1}^{N} \theta_1(\lambda(x) - \lambda_k) - 2\pi L x \lambda(x) \quad (2.28)$$

one can easily prove that a unique value $\lambda(x)$ does exist for every real x, and that $\lambda(x)$ is a monotonically increasing and one-to-one function of x. The value of the function $\lambda(x)$ at $x = n_j/L$ is just the corresponding λ_j from the solution of (2.13):

$$\lambda\left(\frac{n_j}{L}\right) = \lambda_j, \quad n_j \in \{n_k\}. \quad (2.29)$$

The values of λ_j will be called the momenta of the particles present in the state χ_N (1.26). Taking now the number $m \notin \{n_j\}$ (m is integer for odd N and half-integer for even N) one can call the corresponding value of $\lambda(x)$

$$\lambda_m = \lambda\left(\frac{m}{L}\right)$$

the momentum λ_m of the hole. So each number n (integer or half-integer), $n \pmod 1 = [(N-1)/2]$, defines a "vacancy". A filled vacancy is a particle, and a free vacancy is a hole. The total number of particles and holes gives the complete number of vacancies. The quantity $\rho_t(\lambda)$,

$$\rho_t(\lambda(x)) = \frac{dx(\lambda)}{d\lambda}, \tag{2.30}$$

is the density of vacancies. Differentiating equation (2.27) with respect to λ one has

$$1 + \frac{1}{L}\sum_{k=1}^{N} K(\lambda(x), \lambda_k) = 2\pi\rho_t(\lambda(x)). \tag{2.31}$$

It is sometimes useful to consider antiperiodic boundary conditions

$$\chi_N(z_j + L) = -\chi_N(z_j) \tag{2.32}$$

instead of the periodic conditions (2.1). This condition leads to elimination of the minus sign on the right hand side of the Bethe equations (2.2). The further considerations are just the same.

Let us make some remarks about the case $c = \infty$. It is easy to see that the Bethe equations (2.2) become

$$e^{iL\lambda_j} = (-1)^{N+1} \tag{2.33}$$

or in the logarithmic form

$$L\lambda_j = 2\pi\tilde{\tilde{n}}_j. \tag{2.34}$$

This equation describes noninteracting particles and, due to the Pauli principle (VII.4), the model in this limit is equivalent to the free fermion model ([11], [15], and [22]; see Appendix 1).

I.3 The thermodynamic limit at zero temperature

In the thermodynamic limit, the number of particles N and the volume (the length of the box) L tend to infinity so that their ratio $D = N/L$ remains finite:

$$N \to \infty; \qquad L \to \infty; \qquad D = \frac{N}{L} = \text{const.} \tag{3.1}$$

Let us consider the system at zero temperature. We recall that the state with the lowest energy in the sector with a fixed number of particles corresponds to the solutions λ_j of the following Bethe equations:

$$L\lambda_j + \sum_{k=1}^{N} \theta(\lambda_j - \lambda_k) = 2\pi \left[j - \left(\frac{N+1}{2} \right) \right], \quad j = 1, \ldots, N \qquad (3.2)$$

where the numbers n_j are chosen according to (2.26). In the thermodynamic limit, the values of λ_j are condensed ($\lambda_{j+1} - \lambda_j = O(1/L)$, see (2.23)), and fill the symmetric interval $[-q, q]$. In quantum field theory this state is called the Dirac sea and in solid state physics the Fermi sphere, where

$$q = \lim \lambda_N. \qquad (3.3)$$

Let us denote by $\rho(\lambda)$ the density of particles in momentum space (see (2.22), (2.23) and theorem 3) in the following way:

$$\rho(\lambda_k) = \lim \frac{1}{L(\lambda_{k+1} - \lambda_k)} > 0. \qquad (3.4)$$

As all the vacancies inside the interval $[-q, q]$ are occupied, one has that (see (2.31)):

$$\rho(\lambda) = \rho_t(\lambda) = \frac{dx}{d\lambda} > 0, \quad -q \leq \lambda \leq q. \qquad (3.5)$$

By definition $\rho(\lambda)$ is positive. The quantity $L \int \rho(\lambda) d\lambda$ is equal to the number of particles in the interval $[-q, q]$.

Turning now to (2.31), one changes the sum to an integral:

$$\frac{1}{L} \sum_l K(\lambda(x), \lambda_l) = \int_{-N/2L}^{N/2L} K(\lambda(x), \lambda(y)) \, dy$$

$$= \int_{-q}^{q} K(\lambda(x), \mu) \rho(\mu) \, d\mu, \qquad (3.6)$$

which leads to the linear integral equation for $\rho(\lambda)$

$$\rho(\lambda) - \frac{1}{2\pi} \int_{-q}^{q} K(\lambda, \mu) \, \rho(\mu) \, d\mu = \frac{1}{2\pi}. \qquad (3.7)$$

This equation was obtained for the first time in [14]; we shall call it the Lieb equation. In the quantum inverse scattering method λ is the additive spectral parameter and q is the value of the spectral parameter on the boundary of the Fermi sphere. (A more complete definition of q is given in section 9.) From the definition of $\rho(\lambda)$ one has:

$$D = \frac{N}{L} = \int_{-q}^{q} \rho(\lambda)\, d\lambda. \tag{3.8}$$

With the help of this equation and (3.7) we can calculate the Fermi momentum as a unique function of D.

It is convenient to introduce the linear operator $\widehat{K}$ with positive kernel $K(\lambda, \mu)$. This operator acts on the function $\rho(\lambda)$ as follows:

$$\left(\widehat{K}\rho\right)(\lambda) = \int_{-q}^{q} K(\lambda, \mu)\, \rho(\mu)\, d\mu$$

$$K(\lambda, \mu) = \frac{2c}{c^2 + (\lambda - \mu)^2}. \tag{3.9}$$

That equation (3.7) has a unique solution follows from the nondegeneracy of the operator $1 - \widehat{K}/2\pi$. Taking the thermodynamic limit in (2.18) (see (2.16), (3.7)), one has

$$\int_{-q}^{q} d\lambda\, v(\lambda)^2 - \frac{1}{2\pi} \int_{-q}^{q} d\lambda \int_{-q}^{q} d\mu\, K(\lambda, \mu) v(\lambda) v(\mu) \geq \int_{-q}^{q} \frac{v(\lambda)^2}{2\pi \rho(\lambda)} \geq 0 \tag{3.10}$$

for any real function $v(\lambda)$. So the operator $1 - \widehat{K}/2\pi$ is indeed nondegenerate and its eigenvalues are positive, being separated from zero by a gap $(2\pi \rho_{\max})^{-1}$. Here $\rho_{\max}$ is the maximum value of $\rho(\lambda)$ on the interval $-q \leq \lambda \leq q$. One obtains from (2.23) that

$$\frac{1}{2\pi}\left(1 + 2\frac{D}{c}\right) \geq \rho_{\max} \geq \rho(\lambda) \geq \frac{1}{2\pi}$$

$$0 < \frac{1}{2\pi} K \leq \frac{2D}{2D + c} < 1 \tag{3.11}$$

where K is an eigenvalue of the operator $\widehat{K}$. The positivity of $\widehat{K}$ is proved in [14]. When $c \to \infty$, the kernel $K(\lambda, \mu) \to 0$ and all the equations can be solved exactly. The model is then equivalent to the free fermion model at the point $c = \infty$ (see Appendix 1) with

$$\rho(\lambda) = \frac{1}{2\pi}, \qquad \text{when } |\lambda| \leq q;$$

$$\rho(\lambda) = 0, \qquad \text{when } |\lambda| > q. \tag{3.12}$$

So the ground state $|\Omega\rangle$ of the system at $T = 0$ is constructed. It is described by equations (3.7), (3.8). We considered the microcanonical ensemble since the constructed ground state is an eigenfunction of the

Hamiltonian. The energy of this state is

$$\frac{\langle\Omega|H|\Omega\rangle}{\langle\Omega|\Omega\rangle} = E_L = L \int\limits_{-q}^{q} \lambda^2 \rho(\lambda)\, d\lambda. \tag{3.13}$$

Let us now discuss another approach, which is the grand canonical ensemble. One changes the Hamiltonian H to

$$H_h = H - hQ \tag{3.14}$$

where Q (1.6) is the operator for the number of particles and h is the chemical potential. Positive chemical potential ($h > 0$) corresponds to positive density of gas at zero temperature, negative chemical potential corresponds to zero density at zero temperature (see section 7). The number of particles N is now not fixed; it depends on the value of the chemical potential, as well as the energy of the ground state. We recall that $[H, Q] = 0$, and in section 1 the common eigenfunctions of H and Q were constructed. The eigenvalues of H_h are

$$E_N^h = \sum_{j=1}^{N} (\lambda_j^2 - h). \tag{3.15}$$

In the framework of the grand canonical ensemble we can consider excitations with the number of particles different from the ground state. In this way we shall construct one-particle excitations. Since the energy of the particles with small momenta λ_j is negative, the ground state of the Hamiltonian corresponds to the same set of numbers, (2.26) and (3.2). In the thermodynamic limit, one obtains again the Lieb equation (3.7), but the density D and parameter q are now defined by the value of the chemical potential. We define the function $\varepsilon(\lambda)$ as the solution of the linear integral equation

$$\varepsilon(\lambda) - \frac{1}{2\pi} \int\limits_{-q}^{q} K(\lambda, \mu)\varepsilon(\mu)\, d\mu = \lambda^2 - h \equiv \varepsilon_0(\lambda), \tag{3.16}$$

demanding that

$$\varepsilon(q) = \varepsilon(-q) = 0. \tag{3.17}$$

This condition uniquely defines the dependence of q on h. The density D, also defined by h, is given by (3.8).

In the following sections we shall consider the thermodynamics of the model at nonzero temperature. Equations (3.16) and (3.17) will be obtained in a natural way in the zero temperature limit (see section 7), thus proving the existence and uniqueness of both (3.16) and (3.17). It will also be shown that the function $\varepsilon(\lambda)$ possesses the following properties:

$$\varepsilon'(\lambda) > 0 \quad \text{if} \quad \lambda > 0;$$
$$\varepsilon(\lambda) = \varepsilon(-\lambda); \tag{3.18}$$

$$\varepsilon(\lambda) < 0 \quad \text{if} \quad -q < \lambda < q; \tag{3.19}$$

$$\varepsilon(\lambda) > 0 \quad \text{if} \quad \lambda > q, \lambda < -q. \tag{3.20}$$

The meaning of the function $\varepsilon(\lambda)$ will be clarified. It is the energy of the particle excitation above the ground state energy. Equation (3.17) follows from equilibrium and shows that there is no gap in the energy spectrum.

The density is a monotonically increasing function (hence one-to-one) of the chemical potential on the semi-axis $h > 0$ (see section 7):

$$\frac{\partial D}{\partial h} > 0, \quad D\big|_{h=0} = 0, \quad D\big|_{h=\infty} = \infty. \tag{3.21}$$

For negative chemical potentials ($h < 0$) the density is zero ($D = 0$). We shall see that even the thermodynamics at positive and negative values of the chemical potential h will differ in an essential way. In Part IV temperature correlation functions are evaluated and they depend qualitatively on the sign of h. The inequality $\partial D / \partial h > 0$ corresponds to the condition of thermodynamic stability at zero temperature.

We have constructed the ground state $|\Omega\rangle$ (the physical vacuum). The ground state energy of the Hamiltonian (3.14) is:

$$\frac{\langle\Omega|H_h|\Omega\rangle}{\langle\Omega|\Omega\rangle} = E_L^h = L \int_{-q}^{q} (\lambda^2 - h)\rho(\lambda)\,d\lambda = \frac{L}{2\pi} \int_{-q}^{q} \varepsilon(\lambda)\,d\lambda. \tag{3.22}$$

The derivation of this last equality is given in Appendix 4, see formulæ (A.4.40)–(A.4.42).

We have constructed the ground state of the Bose gas at zero temperature with positive density D. How do we imagine it? The particles in the ground state move in a Brownian way. What is the probability that on some space interval there will be no particles at all? This important characteristic shows the relation between the bare vacuum (zero density) and the dressed vacuum (true ground state at positive density). It is calculated in [12] (see formula (9.11) in that paper).

I.4 Excitations at zero temperature

We will first consider excitations over the physical vacuum in the sector with zero physical charge (i.e., excitations with the number of particles

N in the excited state the same as the number of bare particles in the ground state). We shall start with periodic boundary conditions (2.13)

$$L\lambda_j + \sum_{k=1}^{N} \theta(\lambda_j - \lambda_k) = 2\pi n_j.$$

The ground state is described by the special set of integers n_j, see (2.26) and (3.2). All other sets of $\{n_j\}$ (they should be different, $n_j \neq n_k$) give excited states. This is the complete description of all excited states. These excitations are obtained by removing some number of particles with momenta $-q < \lambda_h < q$ from the vacuum distribution of particles (i.e. making holes with momenta λ_h) and by adding an equal number of particles with momenta $|\lambda_p| > q$. First we will construct the state where one particle having momentum $\lambda_p > q$ scatters with a hole having momentum $-q < \lambda_h < q$. The particle and the hole being now present, the permitted values of the vacuum particle momenta are changed: $\lambda_j \to \tilde{\lambda}_j$, so that the Bethe equations for the vacuum particles are now rewritten as

$$L\tilde{\lambda}_j + \sum_k \theta(\tilde{\lambda}_j - \tilde{\lambda}_k) + \theta(\tilde{\lambda}_j - \lambda_p) - \theta(\tilde{\lambda}_j - \lambda_h) = 2\pi \left(j - \frac{N+1}{2} \right). \quad (4.1)$$

Subtracting this from the vacuum distribution (3.2) and taking into account that $\lambda_j - \tilde{\lambda}_j = O(L^{-1})$, $\theta(\lambda + \Delta) - \theta(\lambda) = O(\Delta)$, one obtains

$$L(\lambda_j - \tilde{\lambda}_j) - \theta(\lambda_j - \lambda_p) + \theta(\lambda_j - \lambda_h)$$
$$+ (\lambda_j - \tilde{\lambda}_j) \sum_k K(\lambda_j, \lambda_k) - \sum_k K(\lambda_j, \lambda_k)(\lambda_k - \tilde{\lambda}_k) = 0. \quad (4.2)$$

Using equations (2.31), (3.5), and (3.7), one has

$$2\pi \rho(\lambda_j) L(\lambda_j - \tilde{\lambda}_j) - \theta(\lambda_j - \lambda_p) + \theta(\lambda_j - \lambda_h)$$
$$- \sum_k K(\lambda_j, \lambda_k)(\lambda_k - \tilde{\lambda}_k) \frac{(\lambda_{k+1} - \lambda_k)}{(\lambda_{k+1} - \lambda_k)} = 0. \quad (4.3)$$

Now one introduces the "shift function" F:

$$F(\lambda_j | \lambda_p, \lambda_h) \equiv \frac{(\lambda_j - \tilde{\lambda}_j)}{(\lambda_{j+1} - \lambda_j)}. \quad (4.4)$$

In the thermodynamic limit, one can change the sum in (4.3) to an integral, which gives the following:

$$F(\mu | \lambda_p, \lambda_h) - \int_{-q}^{q} \frac{d\nu}{2\pi} K(\mu, \nu) F(\nu | \lambda_p, \lambda_h) = \frac{\theta(\mu - \lambda_p) - \theta(\mu - \lambda_h)}{2\pi}.$$

$$(4.5)$$

So we are able to describe the vacuum polarization caused by a particle and a hole. It allows the calculation of observables (energy, momentum, and scattering matrix) for excitations above the ground state. These observable quantities are obtained by adding contributions from vacuum polarization to the corresponding "bare" quantities. First we calculate the observable energy ΔE which is equal to the energy of the excited state minus the energy of the ground state:

$$\Delta E(\lambda_p, \lambda_h) = \varepsilon_0(\lambda_p) - \varepsilon_0(\lambda_h) + \sum_j [\varepsilon_0(\tilde{\lambda}_j) - \varepsilon_0(\lambda_j)]$$

$$= \varepsilon_0(\lambda_p) - \varepsilon_0(\lambda_h) - \int_{-q}^{q} \varepsilon_0'(\mu) F(\mu|\lambda_p, \lambda_h) \, d\mu \qquad (4.6)$$

where $\varepsilon_0(\lambda) = \lambda^2 - h$. Similarly, one has for the observable momentum (the bare momentum is simply equal to λ):

$$\Delta P(\lambda_p, \lambda_h) = \lambda_p - \lambda_h - \int_{-q}^{q} F(\mu|\lambda_p, \lambda_h) \, d\mu. \qquad (4.7)$$

Let us now prove that

$$\Delta E(\lambda_p, \lambda_h) = \varepsilon(\lambda_p) - \varepsilon(\lambda_h), \qquad (4.8)$$

where the function $\varepsilon(\lambda)$ is defined by

$$\varepsilon(\lambda) - \frac{1}{2\pi} \int_{-q}^{q} K(\lambda, \mu)\varepsilon(\mu) \, d\mu = \lambda^2 - h = \varepsilon_0(\lambda) \qquad (4.9)$$

$$\varepsilon(q) = \varepsilon(-q) = 0.$$

In order to prove (4.8), it is useful to differentiate (4.9) with respect to λ. The result can be written as follows:

$$[(1 - (2\pi)^{-1}\widehat{K})\varepsilon'](\lambda) = \varepsilon_0'(\lambda), \qquad (4.10)$$

where the operator $\widehat{K}$ is defined in (3.9). It is convenient also to define the linear integral operator $\widehat{L}$ as follows:

$$(1 + \widehat{L})(1 - (2\pi)^{-1}\widehat{K}) = 1$$
$$(2\pi)^{-1}(1 + \widehat{L})\widehat{K} = \widehat{L}. \qquad (4.11)$$

The kernel of the operator $\widehat{L}$ is the symmetric function $L(\lambda, \mu)$. Equation (4.10) can be now rewritten as

$$\varepsilon'(\lambda) - \varepsilon_0'(\lambda) = \int_{-q}^{q} L(\lambda, \mu)\varepsilon_0'(\mu) \, d\mu. \qquad (4.12)$$

Equation (4.5) for F can be transformed as follows:

$$\left[\left(1 - \frac{\widehat{K}}{2\pi}\right) F\right](\mu) = \frac{1}{2\pi} \int\limits_{\lambda_p}^{\lambda_h} \theta'(\mu - \nu) d\nu = \frac{1}{2\pi} \int\limits_{\lambda_p}^{\lambda_h} K(\mu, \nu)\, d\nu. \quad (4.13)$$

Acting with the operator $(1 + \widehat{L})$ on both sides and using (4.11), one has:

$$F(\mu | \lambda_p, \lambda_h) = \int\limits_{\lambda_p}^{\lambda_h} L(\mu, \nu)\, d\nu. \quad (4.14)$$

Putting this representation into (4.6), one obtains

$$\Delta E(\lambda_p, \lambda_h) = \varepsilon_0(\lambda_p) - \varepsilon_0(\lambda_h) - \int\limits_{\lambda_p}^{\lambda_h} d\nu \int\limits_{-q}^{q} d\mu\, L(\nu, \mu)\varepsilon_0'(\mu)$$

and after using (4.12) we have

$$\Delta E(\lambda_p, \lambda_h) = \varepsilon_0(\lambda_p) - \varepsilon_0(\lambda_h) - \int\limits_{\lambda_p}^{\lambda_h} d\nu\, [\varepsilon'(\nu) - \varepsilon_0'(\nu)] = \varepsilon(\lambda_p) - \varepsilon(\lambda_h)$$

$$(4.15)$$

thus proving (4.8).

Let us now prove that $\Delta P(\lambda_p, \lambda_h)$ from (4.7) can be written as

$$\Delta P(\lambda_p, \lambda_h) = \lambda_p - \lambda_h + \int\limits_{-q}^{q} \left[\theta(\lambda_p - \mu) - \theta(\lambda_h - \mu)\right]\rho(\mu)\, d\mu. \quad (4.16)$$

The vacuum particle density $\rho(\lambda)$ satisfies equation (3.7) which can be rewritten as

$$2\pi\rho(\mu) - 1 = \int\limits_{-q}^{q} L(\mu, \nu)\, d\nu = \int\limits_{-q}^{q} K(\mu, \nu)\rho(\nu)\, d\nu. \quad (4.17)$$

Using this equation, one obtains from (4.7) and (4.14)

$$\Delta P(\lambda_p, \lambda_h) = \lambda_p - \lambda_h - \int\limits_{-q}^{q} d\mu \int\limits_{\lambda_p}^{\lambda_h} d\nu\, L(\mu, \nu)$$

$$= \lambda_p - \lambda_h - \int\limits_{\lambda_p}^{\lambda_h} d\mu \int\limits_{-q}^{q} d\nu\, K(\mu, \nu)\rho(\nu) \quad (4.18)$$

which results in (4.16).

All the excitations in the zero charge sector can be constructed as a scattering state of equal numbers of particles and holes. The energy and momentum of these excitations are equal to the sum of the energies and momenta of the individual particles and holes. The one-particle and one-hole excitation constructed above is a two-body state. In the grand canonical ensemble we can change the number of particles.

Let us construct a one-particle excitation with energy

$$\varepsilon(\lambda) - \frac{1}{2\pi} \int\limits_{-q}^{q} K(\lambda,\mu)\varepsilon(\mu)\,d\mu = \lambda^2 - h = \varepsilon_0(\lambda)$$

and momentum $k(\lambda_p)$ equal to

$$k(\lambda_p) = \lambda_p + \int\limits_{-q}^{q} \theta(\lambda_p - \mu)\rho(\mu)\,d\mu \qquad (4.19)$$

(see (3.7)). This is a topological excitation (we should change the boundary conditions to antiperiodic). The value λ_p should be outside the Fermi sphere, $|\lambda_p| > q$, $\operatorname{Im} \lambda_p = 0$. One can also construct another topological (elementary hole) excitation with energy equal to $-\varepsilon(\lambda_h)$ and momentum equal to

$$k_h(\lambda_h) = -\lambda_h - \int\limits_{-q}^{q} \theta(\lambda_h - \mu)\rho(\mu)\,d\mu \qquad (4.20)$$

where $-q < \lambda_h < q$. This shows that excited states in the neutral sector constructed above consist of two elementary excitations (compare formulæ (4.19) and (4.20) with (4.16)). To construct these topological excitations, one should change the boundary conditions to be antiperiodic when introducing an excitation

$$\chi_N(z_1 + L, \ldots, z_N) = -\chi_N(z_1, \ldots, z_N). \qquad (4.21)$$

This topological excitation has a fermionic nature. For impenetrable bosons, this is explicitly shown in Appendix 1. Thus, we shall introduce another particle in the ground state and change boundary conditions.

In the excited state, there are $N + 1$ particles with momenta $\tilde{\lambda}_j$; $j = 1, \ldots, N + 1$. The corresponding Bethe equations are

$$L\tilde{\lambda}_j + \sum_{k}^{N} \theta(\tilde{\lambda}_j - \tilde{\lambda}_k) + \theta(\tilde{\lambda}_j - \tilde{\lambda}_{N+1}) = 2\pi \left(j - \frac{N+1}{2} \right) \qquad (4.22)$$

and

$$L\tilde{\lambda}_{N+1} + \sum_{k}^{N} \theta(\tilde{\lambda}_{N+1} - \tilde{\lambda}_k) = 2\pi n_{N+1}, \quad n_{N+1} > \frac{N+1}{2}. \qquad (4.23)$$

The excited state is characterized by the set of $N+1$ numbers $\{n_j, j = 1, \ldots, N+1\}$; the first N numbers are those corresponding to the ground state (2.26). Let us denote λ_{N+1} by λ_p. It is convenient to introduce a shift function F similar to that in (4.4):

$$F(\lambda_j | \lambda_p) = \frac{(\lambda_j - \tilde{\lambda}_j)}{(\lambda_{j+1} - \lambda_j)}. \qquad (4.24)$$

In the thermodynamic limit, the shift function satisfies the following integral equation:

$$F(\mu | \lambda_p) - \frac{1}{2\pi} \int_{-q}^{q} K(\mu, \nu) F(\nu | \lambda_p) \, d\nu = \frac{1}{2\pi} \theta(\mu - \lambda_p) \qquad (4.25)$$

where $|\lambda_p| > q$. The function $F(\mu | \lambda)$ is defined for $|\mu| < q$; however, by means of equation (4.25), F can be analytically continued to the whole real axis.

With the help of the shift function, we can calculate observable quantities in the thermodynamic limit (energy, momentum, and scattering matrix). The shift function describes the cloud of virtual particles which surrounds the particle λ_p or, in other words, the vacuum polarization due to the bare particle λ_p. Calculations similar to those at the beginning of this section show that the energy of a particle is $\varepsilon(\lambda_p)$ (see (4.9)) and the momentum is given by (4.19).

A "hole" excitation can be treated in a similar fashion. The number of particles in this excitation is equal to $N - 1$ and the observable charge is equal to -1. The wave function χ_N should again be antiperiodic. This state is characterized by the set of integer numbers $\{n_j\}$ obtained by eliminating one number from the vacuum set. The shift function satisfies the equation

$$F_h(\mu | \lambda_h) - \frac{1}{2\pi} \int_{-q}^{q} K(\mu, \nu) F_h(\nu | \lambda_h) \, d\nu = -\frac{1}{2\pi} \theta(\mu - \lambda_h) \qquad (4.26)$$

where $|\lambda_h| < q$ is the bare momentum of the hole. With the help of this function, the energy and momentum can be obtained:

$$\Delta E(\lambda_h) = -\varepsilon(\lambda_h) > 0, \quad |\lambda_h| < q \qquad (4.27)$$

and

$$k_h(\lambda_h) = -\lambda_h - \int_{-q}^{q} \theta(\lambda_h - \mu)\rho(\mu)\,d\mu, \quad |\lambda_h| < q. \tag{4.28}$$

This function can be replaced by $-k(\lambda_h)$ found from (4.19).

Arbitrary excitations are constructed from several particles and holes. The energy and momentum of such excitations are simply the sum of the contributions from the individual elementary excitations. Thus, the energy and momentum are given by

$$\begin{aligned}
\Delta E &= \sum_{\text{particles}} \varepsilon(\lambda_p) - \sum_{\text{holes}} \varepsilon(\lambda_h) \\
\Delta P &= \sum_{\text{particles}} k(\lambda_p) + \sum_{\text{holes}} k_h(\lambda_h) = \sum_{\text{particles}} k(\lambda_p) - \sum_{\text{holes}} k(\lambda_h)
\end{aligned} \tag{4.29}$$

The many-body scattering matrix is simply the product of two-body scattering matrices. First, the scattering matrix for two particles will be evaluated. Adding two particles ($\lambda_2 > \lambda_1 > q$) to the vacuum shifts the values of the vacuum particle momenta: $\lambda_j \to \tilde{\tilde{\lambda}}_j$ ($j = 1,\ldots,N$). The shift function

$$F(\lambda|\lambda_1, \lambda_2) = \frac{(\lambda_j - \tilde{\tilde{\lambda}}_j)}{(\lambda_{j+1} - \lambda_j)} \tag{4.30}$$

is equal to the sum of the single-particle shift functions (4.25):

$$F(\lambda|\lambda_1, \lambda_2) = F(\lambda|\lambda_1) + F(\lambda|\lambda_2). \tag{4.31}$$

Now consider the scattering matrix of two particles possessing momenta λ_1 and λ_2 with $\lambda_2 > \lambda_1 > q$ as above. In this case, the scattering matrix is simply a numerical factor with modulus unity; hence, it can be written as

$$S = \exp\{i\delta(\lambda_2, \lambda_1)\}. \tag{4.32}$$

The phase δ is real and is given by

$$\delta(\lambda_2, \lambda_1) = \varphi_{21} - \varphi_2 \tag{4.33}$$

where φ_2 is the complete phase which the second particle gains when moving through the whole box in the case when the first particle is absent:

$$\varphi_2 = L\lambda_2 + \sum_{k=1}^{N} \theta(\lambda_2 - \tilde{\lambda}_k). \tag{4.34}$$

The phase φ_{21} is the complete phase which the second particle gains when moving through the whole box in the presence of the first one:

$$\varphi_{21} = L\lambda_2 + \sum_{k=1}^{N} \theta(\lambda_2 - \tilde{\tilde{\lambda}}_k) + \theta(\lambda_2 - \lambda_1). \tag{4.35}$$

Using (4.34) and (4.35), the scattering phase is given by

$$\delta(\lambda_2, \lambda_1) = \theta(\lambda_2 - \lambda_1) + \sum_{k=1}^{N} \left[\theta(\lambda_2 - \tilde{\tilde{\lambda}}_k) - \theta(\lambda_2 - \tilde{\lambda}_k) \right]$$

$$= \theta(\lambda_2 - \lambda_1) + \sum_{k=1}^{N} K(\lambda_2, \lambda_k) \Big[F(\lambda_k | \lambda_1, \lambda_2)$$

$$- F(\lambda_k | \lambda_2) \Big] (\lambda_{k+1} - \lambda_k). \tag{4.36}$$

Changing the sum to an integral in the thermodynamic limit and using (4.31), one obtains

$$\delta(\lambda_2, \lambda_1) = \theta(\lambda_2 - \lambda_1) + \int_{-q}^{q} K(\lambda_2, \mu) F(\mu | \lambda_1) \, d\mu \tag{4.37}$$

and from (4.25)

$$\delta(\lambda_2, \lambda_1) = 2\pi F(\lambda_2 | \lambda_1), \quad \lambda_2 > \lambda_1. \tag{4.38}$$

Thus, the scattering phase satisfies the following integral eqation

$$\delta(\lambda, \mu) - \frac{1}{2\pi} \int_{-q}^{q} K(\lambda, \nu) \delta(\nu, \mu) \, d\nu = \theta(\lambda - \mu). \tag{4.39}$$

We have demonstrated that physical excitations over the Dirac sea are obtained from "bare" excitations over the Fock vacuum $|0\rangle$ (which are described by the Bethe wave functions) by means of the dressing equations. These linear integral dressing equations are universal ones. To see this, it is sufficient to compare the dressing equations for the energy (4.9), scattering phase (4.39), and density (3.7). The dressing equations are also very useful for the investigation of the correlation functions and finite size corrections as we shall see in later sections.

The scattering of two holes having bare momenta λ_1 and λ_2, with $-q \leq \lambda_{1,2} \leq q$, is also equal to $\exp\{i\delta(\lambda_2, \lambda_1)\}$ ($\lambda_2 > \lambda_1$) where δ is defined by (4.39). The scattering matrix of the particle λ_p with the hole λ_h is equal to

$$S = \exp\{-i\delta(\lambda_p, \lambda_h)\}. \tag{4.40}$$

The scattering matrix of several particles is equal to the product of the two-particle scattering matrices.

It should also be mentioned that the excitation structures of other models (XXZ Heisenberg antiferromagnet and sine-Gordon) are very similar. The energy, momentum, and scattering matrix are evaluated in the same way.

I.5 Thermodynamics of the model

Let us consider the canonical ensemble and calculate the partition function Z of the model

$$Z = \text{tr}(e^{-H/T}) = e^{-F/T}. \tag{5.1}$$

Here H is the Hamiltonian given by (1.2) and T is the temperature. The free energy F is given by (5.1). Let us recall that we are studying the thermodynamic limit ($L \to \infty$, $N \to \infty$) with the density of the gas remaining fixed:

$$D = \frac{N}{L} = \text{const.} \tag{5.2}$$

In the thermodynamic limit, vacancies, particles and holes (those defined in section 2 (see (2.29), (2.30)) have finite distribution densities $\rho_t(\lambda)$, $\rho_p(\lambda)$ and $\rho_h(\lambda)$ in momentum space which are defined as follows:

$$L\rho_p(\lambda)\, d\lambda \quad \text{number of particles in} \quad [\lambda, \lambda + d\lambda] \tag{5.3}$$
$$L\rho_h(\lambda)\, d\lambda \quad \text{number of holes in} \quad [\lambda, \lambda + d\lambda] \tag{5.4}$$
$$L\rho_t(\lambda)\, d\lambda \quad \text{number of vacancies in} \quad [\lambda, \lambda + d\lambda] \tag{5.5}$$

the number of vacancies being simply the sum of the numbers of particles and holes:

$$\rho_t(\lambda) = \rho_p(\lambda) + \rho_h(\lambda). \tag{5.6}$$

By vacancies we mean potential positions, in momentum space, which can be occupied by particles or holes. In the thermodynamic limit, the sum in equation (2.31) turns into an integral involving the density $\rho_p(\lambda)$ and one has:

$$2\pi\rho_t(\lambda) = 1 + \int_{-\infty}^{+\infty} K(\lambda, \mu)\rho_p(\mu)\, d\mu. \tag{5.7}$$

It should be emphasized that we now pass from the microscopic description of the model (with the help of the set n_j) to the macroscopic description in terms of densities ρ_p, ρ_h and ρ_t. The given macroscopic situation described by fixed ρ_p, ρ_h and ρ_t corresponds to many sets of microscopic states (the sets of numbers n_j). Indeed, there exist many ways of putting

$L\rho_p(\lambda)\,d\lambda$ particles into $L\rho_t(\lambda)\,d\lambda$ vacancies. The number of possibilities is

$$\exp\{dS\} = \frac{[L\rho_t(\lambda)\,d\lambda]!}{[L\rho_p(\lambda)\,d\lambda]!\,[L\rho_h(\lambda)\,d\lambda]!}. \tag{5.8}$$

This number is large in the thermodynamic limit. Using Stirling's formula for the factorial asymptotics, one has

$$dS = L\,d\lambda\,[\rho_t(\lambda)\ln\rho_t(\lambda) - \rho_p(\lambda)\ln\rho_p(\lambda) - \rho_h(\lambda)\ln\rho_h(\lambda)]. \tag{5.9}$$

Later on we shall see that the energy and other observables of the system depend only on the macroscopic variables ρ_p and ρ_h. The quantity

$$S = \int dS \tag{5.10}$$

is the entropy. We turn now to the partition function (5.1). It can be represented in the form

$$Z_N = \frac{1}{N!}\sum_{n_1,\dots,n_N}\exp\left\{-\frac{E_N}{T}\right\} = \sum_{n_1 < n_2 < \dots < n_N}\exp\left\{-\frac{E_N}{T}\right\} \tag{5.11}$$

where $E_N = \sum\limits_{j=1}^{N}\lambda_j^2$ and the momenta λ_j are the solutions of the Bethe equations (2.13). We shall consider the thermodynamics of the gas which is at rest as a whole, hence the total momentum P_N (1.29) is equal to zero; this means that $\sum\limits_{j=1}^{N} n_j = 0$. Introducing new variables $n_{j+1,j} = n_{j+1} - n_j$, we can rewrite (5.11):

$$Z_N = \sum_{n_{21}=1}^{\infty}\sum_{n_{32}=1}^{\infty}\cdots\sum_{n_{N,N-1}=1}^{\infty} e^{-E_N/T}. \tag{5.12}$$

In the thermodynamic limit, the energy of a state

$$E_N = L\int\limits_{-\infty}^{+\infty}\rho_p(\lambda)\lambda^2\,d\lambda \tag{5.13}$$

depends only on the macroscopic variable $\rho_p(\lambda)$; this allows one to pass from the summation over the microscopic variables in (5.12) to integration with respect to macroscopic variables. To do that, let us make the substitution

$$n_{j+1,j} \longrightarrow \frac{\rho_t(\lambda)}{\rho_p(\lambda)}.$$

In fact, we can calculate the ratio of the number of vacancies to the number of particles (in $d\lambda$) in terms of micro- and macrovariables and then compare the answers. The variable $n_{j+1,j}$ can be considered to be the number of vacancies for the j-th particle (i.e. only one vacancy from $n_{j+1,j}$ is occupied). Macroscopically this ratio is given by the ratio of the corresponding densities (5.3), (5.5), i.e. by $\rho_t(\lambda)/\rho_p(\lambda)$. So the sum in (5.12) can be changed to a functional integral. While doing this one has to remember, however, that a large number ((5.8) and (5.9)) of sets of microscopic variables correspond to given macroscopic variables. Thus one obtains the functional integral representation for the partition function Z_N:

$$Z_N = \text{const} \int \mathcal{D}\left(\frac{\rho_t(\lambda)}{\rho_p(\lambda)}\right) \delta\left(L\int \rho_p(\lambda)\,d\lambda - DL\right) e^{S-E_N/T} \quad (5.14)$$

$$S - \frac{E_N}{T} = -\frac{L}{T}\int d\lambda \left\{\lambda^2\rho_p(\lambda) - T[\rho_t(\lambda)\ln\rho_t(\lambda) - \rho_p(\lambda)\ln\rho_p(\lambda)\right.$$

$$\left. - \rho_h(\lambda)\ln\rho_h(\lambda)]\right\}. \quad (5.15)$$

The fixed number of particles N in the canonical ensemble leads to the appearence of the δ-function in (5.14). Using the representation

$$\delta(x) = \frac{1}{2\pi i}\int\limits_{-i\infty}^{i\infty} e^{hx}\,dh$$

we can rewrite (5.14) as

$$Z_N = \text{const}\int dh \int \mathcal{D}\left(\frac{\rho_t(\lambda)}{\rho_p(\lambda)}\right)$$
$$\times \exp\left\{S - \frac{E_N}{T} + \frac{L}{T}h\left(\int\rho_p(\lambda)\,d\lambda - D\right)\right\}. \quad (5.16)$$

As $L \to \infty$, the steepest descent method can be used to calculate the functional integral (5.16). The variation of the exponent in (5.16), using (5.7), is

$$-\frac{L}{T}\int d\lambda\,\delta\rho_p(\lambda)\left\{\lambda^2 - h - \varepsilon(\lambda) - \frac{T}{2\pi}\int K(\lambda,\mu)\ln\left(1 + e^{-\varepsilon(\mu)/T}\right)d\mu\right\}$$
$$+\frac{L}{T}\delta h\left(\int\rho_p(\lambda)\,d\lambda - D\right) \quad (5.17)$$

where we have used the following convenient notation

$$\frac{\rho_h(\lambda)}{\rho_p(\lambda)} = e^{\varepsilon(\lambda)/T}. \quad (5.18)$$

Thus the steepest descent method leads to the following equations:

$$\varepsilon(\lambda) = \lambda^2 - h - \frac{T}{2\pi} \int\limits_{-\infty}^{+\infty} K(\lambda, \mu) \ln(1 + e^{-\varepsilon(\mu)/T}) \, d\mu \qquad (5.19)$$

$$D = \int\limits_{-\infty}^{+\infty} \rho_p(\lambda) \, d\lambda. \qquad (5.20)$$

It can be proved that the value of h at the stationary point is real; the matrix of the second derivatives of the exponent in (5.16) has the correct sign (Appendix 2). These statements justify the applicability of the steepest descent method.

Thus we calculate the partition function and the free energy (5.1):

$$Z = e^{-F/T}; \quad F = Nh - \frac{TL}{2\pi} \int\limits_{-\infty}^{+\infty} d\lambda \ln(1 + e^{-\varepsilon(\lambda)/T}). \qquad (5.21)$$

By definition, the pressure is the derivative of the free energy with respect to the volume at fixed temperature:

$$\mathcal{P} = -\left(\frac{\partial F}{\partial L}\right)_T = \frac{T}{2\pi} \int\limits_{-\infty}^{+\infty} d\lambda \ln(1 + e^{-\varepsilon(\lambda)/T}). \qquad (5.22)$$

Direct calculation shows that the basic thermodynamic identity is fulfilled:

$$d\mathcal{P} = \frac{S}{L} dT + D \, dh. \qquad (5.23)$$

Thus S (5.10) is the entropy. Let us list the equations which determine the state of thermodynamic equilibrium:

$$2\pi \rho_t(\lambda) = 1 + \int\limits_{-\infty}^{+\infty} K(\lambda, \mu) \rho_p(\mu) \, d\mu;$$

$$\qquad (5.24)$$

$$D = \int\limits_{-\infty}^{+\infty} \rho_p(\lambda) \, d\lambda;$$

$$\varepsilon(\lambda) = \lambda^2 - h - \frac{T}{2\pi} \int K(\lambda, \mu) \ln\left(1 + e^{-\varepsilon(\mu)/T}\right) d\mu; \qquad (5.25)$$

$$\frac{\rho_h(\lambda)}{\rho_p(\lambda)} = e^{\varepsilon(\lambda)/T}; \quad \frac{\rho_p(\lambda)}{\rho_t(\lambda)} = \frac{1}{1 + e^{\varepsilon(\lambda)/T}} \equiv \vartheta(\lambda); \qquad (5.26)$$

$$\rho_t(\lambda) = \rho_p(\lambda) + \rho_h(\lambda); \quad \rho_p(\lambda) \geq 0; \quad \rho_h(\lambda) \geq 0. \qquad (5.27)$$

The density $D > 0$, the temperature $T > 0$, and the coupling constant $c > 0$ are free parameters here. These equations define the dependence of pressure on density and temperature. The equilibrium state is by no means a pure quantum mechanical state: it is not described by a single eigenfunction of the Hamiltonian (as for $T = 0$). It is a mixture of eigenfunctions. The number of eigenstates present in the state of thermal equilibrium is $\exp S$ (see (5.9), (5.10)). Here S is the entropy. Thus $\exp S$ sets $\{n_j\}$ in (2.13) correspond to a given $\rho_p(\lambda)$ and $\rho_h(\lambda)$. Let us pick one of the eigenstates present in the state of thermal equilibrium and denote it by $|\Omega_T\rangle$. The equation (5.25) for $\varepsilon(\lambda)$ plays a most important role. It was obtained in [22] and will be called the Yang-Yang equation. It will be investigated in the next section. The function $\varepsilon(\lambda)$ has a clear physical interpretation as the energy of elementary excitations above the equilibrium state. This will be shown directly in section 8. One can, however, understand this without doing much calculation. Consider the ratio $\vartheta(\lambda)$ of the number of occupied vacancies to the total number of vacancies (from (5.18)):

$$\frac{\rho_p(\lambda)}{\rho_t(\lambda)} = \vartheta(\lambda) = \frac{1}{1 + e^{\varepsilon(\lambda)/T}}. \tag{5.28}$$

Comparing this with the well known Fermi distribution, one concludes that $\varepsilon(\lambda)$ should be just the energy mentioned above. The function $\vartheta(\lambda)$ will be called the Fermi weight.

So we have constructed the state of thermal equilibrium for the Bose gas at finite temperature and finite density. What is it like? In the state of thermal equilibrium particles move in a Brownian way. An interesting question is: "What is the probability that in some space interval there will be no particles at all?" The reader can find the answer in section 2 of Chapter XVI ((XVI.2.28) for the impenetrable case). It shows the relation between the true state of thermal equilibrium and the bare vacuum (no particles). At large distance this probability can be expressed in terms of the pressure (see section XVII.5).

In conclusion, we give some useful estimates which are similar to (3.10) and (3.11). Using (2.23) we obtain in the thermodynamic limit

$$\frac{1}{2\pi} \leq \rho_t(\lambda) \leq \frac{1}{2\pi}\left(1 + \frac{2D}{c}\right). \tag{5.29}$$

It is also convenient to introduce the integral operator $\widehat{K_T}$ acting on an arbitrary function $v(\lambda)$ as follows:

$$\left(\widehat{K_T}v\right)(\lambda) = \int\limits_{-\infty}^{+\infty} K(\lambda, \mu)\vartheta(\mu)v(\mu)\,d\mu. \tag{5.30}$$

One has for the eigenvalues K_T of this operator

$$0 < K_T \leq \frac{4\pi D}{2D + c} < 2\pi. \tag{5.31}$$

The proof is similar to that of (3.11). Equations (5.24)–(5.27) will be studied in the next section and in Appendix 2. Let us mention here that h is a one-to-one function of D, $\partial D/\partial h > 0$.

It is worth mentioning that the thermodynamics of the grand canonical ensemble are described by the same equations (5.24)–(5.27). In this case, the chemical potential is an arbitrary parameter while the density D is defined by the corresponding equations.

I.6 The Yang-Yang equation

Our first task here is to prove the existence of a solution of the Yang-Yang equation

$$\varepsilon(\lambda) = \lambda^2 - h - \frac{T}{2\pi} \int\limits_{-\infty}^{+\infty} K(\lambda, \mu) \ln\left(1 + e^{-\varepsilon(\mu)/T}\right) d\mu. \tag{6.1}$$

This is done following [22].

Theorem 4. *A solution to the Yang-Yang equation does exist.*

Proof: Let us construct the following sequence of functions:

$$\varepsilon_0(\lambda) = \lambda^2 - h; \qquad \varepsilon_{n+1}(\lambda) = \lambda^2 - h + A_n \tag{6.2}$$

$$A_n = -\frac{T}{2\pi} \int\limits_{-\infty}^{+\infty} K(\lambda, \mu) \ln\left(1 + e^{-\varepsilon_n(\mu)/T}\right) d\mu. \tag{6.3}$$

This sequence decreases at every point λ:

$$\varepsilon_0(\lambda) > \varepsilon_1(\lambda) > \varepsilon_2(\lambda) > \cdots > \varepsilon_n(\lambda) > \varepsilon_{n+1}(\lambda) > \cdots \tag{6.4}$$

and is bounded from below:

$$\varepsilon_n(\lambda) \geq \lambda^2 + x_0 \tag{6.5}$$

where x_0 is some constant. These properties proved below mean that the limit

$$\varepsilon(\lambda) = \lim_{n \to \infty} \varepsilon_n(\lambda) \tag{6.6}$$

does exist and is a solution of the Yang-Yang equation. Property (6.4) is obvious from the fact that $A_n < 0$ and $\delta A_n / \delta \varepsilon_n > 0$:

$$\delta A_n = \frac{1}{2\pi} \int\limits_{-\infty}^{+\infty} K(\lambda, \mu)(1 + e^{\varepsilon_n(\mu)/T})^{-1} \, \delta \varepsilon_n(\mu) \, d\mu. \tag{6.7}$$

The proof of (6.5) is not so straightforward. First we establish some important properties of $\varepsilon_n(\lambda)$, namely that

$$\varepsilon_n(\lambda) = \varepsilon_n(-\lambda) \tag{6.8}$$

and

$$\varepsilon_n(\lambda_2) > \varepsilon_n(\lambda_1), \qquad \text{if } \lambda_2 > \lambda_1 \geq 0. \tag{6.9}$$

The first property is quite obvious. The second property can be proved by induction in n. Supposing that (6.9) is true for some ε_n, one establishes that it is also true for ε_{n+1}. Indeed, one has from (6.2)

$$
\begin{aligned}
\varepsilon'_{n+1}(\lambda) &= 2\lambda + \frac{1}{2\pi} \int\limits_{-\infty}^{+\infty} K(\lambda, \mu) \frac{\varepsilon'_n(\mu)}{1 + e^{\varepsilon_n(\mu)/T}} \, d\mu \\
&= 2\lambda + \frac{1}{2\pi} \int\limits_{0}^{\infty} [K(\lambda, \mu) - K(\lambda, -\mu)] \frac{\varepsilon'_n(\mu)}{1 + e^{\varepsilon_n(\mu)/T}} \, d\mu > 0
\end{aligned}
\tag{6.10}
$$

as $K(\lambda, \mu) > K(-\lambda, \mu)$ at $\lambda > 0$ and $\mu > 0$ (see (3.9)). Thus we have proved (6.9).

As the term A_n in (6.3) is also an even function which increases monotonically on the positive semi-axis, one has

$$\varepsilon_n(\lambda) \geq \lambda^2 + \varepsilon_n(0). \tag{6.11}$$

Taking into account (6.2), (6.7) and (6.11) one can write the following inequality

$$\varepsilon_{n+1}(0) \geq -h - \frac{T}{2\pi} \int\limits_{-\infty}^{+\infty} K(0, \mu) \ln(1 + e^{-[\varepsilon_n(0) + \mu^2]/T}) \, d\mu. \tag{6.12}$$

Defining the function

$$
\begin{aligned}
f(x) &\equiv -h - \frac{T}{2\pi} \int\limits_{-\infty}^{+\infty} K(0, \mu) \ln(1 + e^{-[\mu^2 + x]/T}) \, d\mu \\
&= -h + x - \frac{T}{2\pi} \int\limits_{-\infty}^{+\infty} K(0, \mu) \ln(e^{x/T} + e^{-\mu^2/T}) \, d\mu
\end{aligned}
\tag{6.13}
$$

one rewrites (6.12) as follows

$$\varepsilon_{n+1}(0) \geq f(\varepsilon_n(0)).$$ (6.14)

The function $f(x)$ increases monotonically, and $f(x) < -h$. The function $f(x) - x$ decreases monotonically taking values in the interval $[-\infty, +\infty]$. Thus the equation

$$f(x_0) - x_0 = 0$$ (6.15)

possesses a unique solution. At fixed temperature, the constant x_0 is uniquely defined by h, and, vice versa, a given x_0 defines h:

$$h = -\frac{T}{2\pi} \int\limits_{-\infty}^{+\infty} K(0, \mu) \ln(e^{x_0/T} + e^{-\mu^2/T}) \, d\mu.$$ (6.16)

So h is a monotonically decreasing function of x_0 taking values in the interval $[-\infty, +\infty]$. One can now prove that

$$\varepsilon_n(0) \geq x_0, \qquad n \geq 0.$$ (6.17)

First one notes that

$$\varepsilon_0(0) = -h > f(x_0) = x_0.$$ (6.18)

Then one uses induction in n. Supposing $\varepsilon_n(0) \geq x_0$, one has from (6.14) and from the monotonicity of the function $f(x)$ that

$$\varepsilon_{n+1}(0) \geq f(\varepsilon_n(0)) \geq f(x_0) = x_0.$$ (6.19)

So inequality (6.17) is proved. Combining this with (6.11) one arrives at (6.5), which finishes the proof of the existence of the solution to the Yang-Yang equation.

Simultaneously, the following properties of this solution are in fact also established:

$$\varepsilon(\lambda) = \varepsilon(-\lambda);$$ (6.20)

$$\lambda^2 - h \geq \varepsilon(\lambda) \geq \lambda^2 + x_0;$$ (6.21)

$$\frac{\partial \varepsilon(\lambda)}{\partial \lambda} > 0, \qquad \text{if } \lambda > 0;$$ (6.22)

$$\varepsilon(\lambda) \to \lambda^2 - h + O(1/\lambda^2); \qquad \lambda \to \pm\infty.$$ (6.23)

It follows that the function $\varepsilon(\lambda)$ either has no zeros on the real axis (if $x_0 \geq 0$) or possesses only two zeros (if $h > 0$).

Considering the limits of zero and infinite density we obtain

$$\varepsilon(\lambda) \to -\infty; \ \rho_h \to 0; \ D \to \infty \text{ as } h \to \infty;$$ (6.24)

$$\varepsilon(\lambda) \to +\infty; \ x_0 \to +\infty; \ \rho_p \to 0; \ D \to 0 \text{ as } h \to -\infty.$$ (6.25)

So the density is positive for any real value of the chemical potential h. (This is not the case at zero temperature, where $D = 0$ for $h < 0$.) But the thermodynamics still depends qualitatively on the sign of h. This will be especially evident from the asymptotics of temperature correlations, evaluated in Part IV. Let us also emphasize that the analytic properties of $\varepsilon(\lambda)$ are studied in the Appendix to Chapter XVII.

I.7 Limiting cases

Let us first discuss the zero temperature limit, $T \to 0$. Consider what happens with the estimate (6.21) to the solution of the Yang-Yang equation, $\varepsilon(\lambda)$, at $T = 0$:

$$\lambda^2 - h \geq \varepsilon(\lambda) \geq \lambda^2 + x_0. \tag{7.1}$$

The constant x_0 is a one-to-one monotonically decreasing function of h (see (6.16)). When $T = 0$, the point $h = 0$ corresponds to $x_0 = 0$. The semi-axes are mapped in the following way:

$$\text{when } h > 0 \quad \text{then } x_0 < 0; \tag{7.2}$$

$$\text{when } h < 0 \quad \text{then } x_0 > 0. \tag{7.3}$$

In the last case ($x_0 > 0$ and $h < 0$), it is easy to see that $x_0 = -h$. The function $\varepsilon(\lambda)$ is then equal to

$$\varepsilon(\lambda) = \lambda^2 - h > 0, \qquad h < 0 \tag{7.4}$$

and has no zeros on the real axis. This corresponds to the case of zero density, $D = 0$.

When $h > 0$ then $x_0 < 0$ ($x_0 = -y$) and

$$y = h + \frac{1}{2\pi} \int\limits_{-\sqrt{y}}^{\sqrt{y}} d\mu \, \frac{2c}{c^2 + \mu^2}(y - \mu^2) > h > 0. \tag{7.5}$$

This is the case when $\varepsilon(\lambda)$ possesses two zeros on the real axis

$$\varepsilon(\pm q) = 0, \qquad h > 0. \tag{7.6}$$

The function $\varepsilon(\lambda)$ is an even function of λ monotonic on the semi-axis $\lambda > 0$ and also

$$\varepsilon(\lambda) > 0 \quad \text{when } |\lambda| > q; \tag{7.7}$$

$$\varepsilon(\lambda) < 0 \quad \text{when } |\lambda| < q. \tag{7.8}$$

These properties are obtained directly from (6.20)–(6.22) in the limit $T \to 0$. So the Yang-Yang equation becomes linear when $T = 0$:

$$\varepsilon(\lambda) - \frac{1}{2\pi} \int_{-q}^{q} K(\lambda, \mu)\varepsilon(\mu)\, d\mu = \lambda^2 - h \tag{7.9}$$

$$\varepsilon(q) = \varepsilon(-q) = 0.$$

The existence of the solution to this equation and other properties follow from the existence theorem for the Yang-Yang equation for $T > 0$. Thus we have justified equations (3.16), (3.17), (3.19) and (3.20). In treating the $T \to 0$ case, we have used the fact that at $T = 0$:

$$e^{\varepsilon(\lambda)/T} = \frac{\rho_h(\lambda)}{\rho_p(\lambda)} = \begin{cases} \infty, & |\lambda| > q; \\ 0, & |\lambda| < q \end{cases} \tag{7.10}$$

which means that

$$\rho_p(\lambda) = 0; \quad \rho_h(\lambda) = \rho_t(\lambda), \ |\lambda| > q \ (T = 0) \tag{7.11}$$

$$\rho_h(\lambda) = 0; \quad \rho_p(\lambda) = \rho_t(\lambda), \ |\lambda| < q \ (T = 0). \tag{7.12}$$

The equation (5.24) for densities turns into

$$2\pi\rho_p(\lambda) - \int_{-q}^{q} K(\lambda, \mu)\rho_p(\mu)\, d\mu = 1; \quad -q \leq \lambda \leq q \tag{7.13}$$

$$\int_{-q}^{q} \rho_p(\lambda)\, d\lambda = D. \tag{7.14}$$

It follows from (7.4) that if $h < 0$ then $D = 0$, if $h > 0$ then $D > 0$ and $q > 0$. Since $\partial D/\partial h > 0$ (see Appendix 2) and h takes values in the interval $[0, \infty]$ for $0 < D < \infty$, $D(h)$ is a one-to-one monotonically increasing function of h on the positive semi-axis. Thus we have justified (3.21).

The equilibrium state becomes a pure quantum mechanical state at $T = 0$. The entropy S tends to zero at $T \to 0$ due to (7.11) and (7.12). So we have justified all the results of section 3. At zero temperature, the pressure (5.22) is equal to

$$\mathcal{P} = -\frac{1}{2\pi} \int_{-q}^{q} \varepsilon(\lambda)\, d\lambda. \tag{7.15}$$

Thus, the pressure is the density of the ground state energy (3.22).

Let us also mention the sound velocity. There are two definitions for this quantity—the macroscopic and the microscopic ones. The macroscopic velocity is defined as the derivative of the pressure $\mathcal{P}$ with respect to the density D:

$$v^2 = 2\frac{\partial \mathcal{P}}{\partial D}. \tag{7.16}$$

The microscopic sound velocity, known as the Fermi velocity, is defined as the derivative of the dressed energy (3.16) with respect to the dressed momentum (4.19) on the Fermi boundary:

$$v_F = v = \left.\frac{\partial \varepsilon(\lambda)}{\partial k(\lambda)}\right|_{\lambda=q}. \tag{7.17}$$

The equivalence of these velocities for the nonrelativistic Bose gas is proved in Appendix 3, following [15].

Now consider another limiting case, supposing that T is arbitrary but the coupling is strong ($c \to \infty$). As $c \to \infty$, the kernel $K(\lambda, \mu)$ of the operator $\widehat{K}$ tends to zero and all the integral equations can be solved explicitly:

$$\varepsilon(\lambda) = \lambda^2 - h - \frac{2}{c}\mathcal{P} + O\left(\frac{1}{c^3}\right);$$

$$\rho_p(\lambda) = \frac{1}{2\pi}\left(\frac{1 + \frac{2D}{c}}{1 + e^{\varepsilon(\lambda)/T}}\right); \quad \rho_h(\lambda) = \frac{1}{2\pi}\left(\frac{1 + \frac{2D}{c}}{1 + e^{-\varepsilon(\lambda)/T}}\right);$$

$$\rho_t(\lambda) = \frac{1}{2\pi}\left(1 + \frac{2D}{c}\right); \tag{7.18}$$

$$D = \int\limits_{-\infty}^{+\infty} \rho_p(\lambda)\, d\lambda.$$

It is worth mentioning once more that the strong coupling limit ($c \to \infty$) corresponds to free fermions.

I.8 Excitations and correlations at nonzero temperature

For exactly solvable models excitations at finite temperature are stable; they do not decay (due to the infinite set of conservation laws). Let us construct them for the Bose gas. We have already described the thermodynamic equilibrium state in section 5. This equilibrium state is represented as a mixture of different eigenfunctions of the Hamiltonian. However, to obtain excitations over the equilibrium state, we take one of these eigenfunctions $|\Omega_T\rangle$ and then construct excitations over this eigenfunction using the same method as for zero temperature in section 4. It turns out that the observable quantities of the excitations calculated in this way (energy, momentum and scattering matrix) depend on the macroscopic characteristics of the equilibrium state only and do not depend on the particular eigenfunction chosen. One may say that the macroscopic quantities, ρ_h and ρ_p, (5.3)–(5.6), are "gauge invariant" while $\exp S$ is the volume of the gauge group. The transition from one set of microscopic variables $\{n_j\}$ (2.13) to another one corresponding to the same values of ρ_p and ρ_h is similar to a gauge transformation.

Turning to the Bethe equations, it is convenient to introduce the function $\lambda(x)$ as was done in (2.27):

$$L\lambda(x) + \sum_{k=1}^{N} \theta(\lambda(x) - \lambda_k) = 2\pi L x \tag{8.1}$$

where λ_k are the permitted values of the particle momenta in the equilibrium state. The simplest excited state in the canonical ensemble (the number of particles being fixed) is obtained by making a hole and by simultaneously adding a particle. Taking into account that $\lambda(x)$ will slightly change, $\lambda(x) \to \tilde{\lambda}(x)$, one obtains

$$L\tilde{\lambda}(x) + \sum_{k=1}^{N} \theta(\tilde{\lambda}(x) - \tilde{\lambda}_k) + \theta(\tilde{\lambda}(x) - \lambda_p) - \theta(\tilde{\lambda}(x) - \lambda_h) = 2\pi L x \tag{8.2}$$

where $\tilde{\lambda}_k$ are the new permitted values of particle momenta and λ_p, λ_h are the bare momenta of the added particle and hole. Introducing a shift function analogous to (4.5):

$$F\left(\lambda\left(\frac{j}{L}\right) \middle| \lambda_p, \lambda_h\right) = \left\{ \frac{\lambda\left(\frac{j}{L}\right) - \tilde{\lambda}\left(\frac{j}{L}\right)}{\lambda\left(\frac{j+1}{L}\right) - \lambda\left(\frac{j}{L}\right)} \right\} \tag{8.3}$$

one derives the following integral equation for this function:

$$2\pi F\left(\lambda|\lambda_p, \lambda_h\right) - \int_{-\infty}^{+\infty} K(\lambda, \mu)\vartheta(\mu)F\left(\mu|\lambda_p, \lambda_h\right) d\mu$$
$$= \theta(\lambda - \lambda_p) - \theta(\lambda - \lambda_h). \tag{8.4}$$

This equation and the equation at zero temperature, (4.5), differ only in the integration measure:

$$\int_{-q}^{q} d\lambda \longrightarrow \int_{-\infty}^{+\infty} d\lambda\, \vartheta(\lambda) \tag{8.5}$$

where $\vartheta(\lambda)$ is the Fermi weight (5.26), (5.28):

$$\vartheta(\lambda) = \frac{\rho_p(\lambda)}{\rho_t(\lambda)} = \frac{1}{1 + e^{\varepsilon(\lambda)/T}}. \tag{8.6}$$

The energy and momentum of the excited state are obtained similarly to those in section 4:

$$\Delta E(\lambda_p, \lambda_h) = \lambda_p^2 - \lambda_h^2 - \int_{-\infty}^{+\infty} \varepsilon_0'(\mu) F(\mu|\lambda_p, \lambda_h) \vartheta(\mu) \, d\mu; \qquad (8.7)$$

$$k(\lambda_p, \lambda_h) = \lambda_p - \lambda_h - \int_{-\infty}^{+\infty} F(\mu|\lambda_p, \lambda_h) \vartheta(\mu) \, d\mu. \qquad (8.8)$$

It is not difficult to prove that

$$\Delta E(\lambda_p, \lambda_h) = \varepsilon(\lambda_p) - \varepsilon(\lambda_h); \qquad (8.9)$$

$$k(\lambda_p, \lambda_h) = k(\lambda_p) - k(\lambda_h)$$

$$= \lambda_p - \lambda_h + \int_{-\infty}^{+\infty} d\mu \, \rho_p(\mu) \left[\theta(\lambda_p - \mu) - \theta(\lambda_h - \mu) \right]. \qquad (8.10)$$

The function $\varepsilon(\lambda)$ is just the solution of the Yang-Yang equation (6.1).

The scattering matrix can also be computed:

$$S(\lambda_p, \lambda_h) = \exp\left\{ -i\delta(\lambda_p, \lambda_h) \right\}, \qquad \lambda_p > \lambda_h \qquad (8.11)$$

with the scattering phase δ satisfying the following integral equation:

$$\delta(\lambda_p, \lambda_h) - \frac{1}{2\pi} \int_{-\infty}^{+\infty} K(\lambda_p, \mu) \vartheta(\mu) \delta(\mu, \lambda_h) \, d\mu = \theta(\lambda_p - \lambda_h). \qquad (8.12)$$

So indeed all the observables depend only on the macroscopic variables. All the expressions for observable quantities at nonzero temperature are obtained from the corresponding formulæ for the zero temperature case by substitution of the integration measure (8.5). So following the ideas of C.N. Yang and C.P. Yang we have constructed stable excitations at finite temperatures.

Let us now discuss quantum correlation functions at finite temperature. The state of thermal equilibrium was constructed in section 5. The construction shows that not all the eigenfunctions of the Hamiltonian are present in the state of thermal equilibrium (i.e., contribute to the free energy and partition function); only those eigenfunctions which satisfy equations (5.24)–(5.26) are present. Let us take one of these eigenfunctions and denote it by $|\Omega_T\rangle$. We know that the number of these eigenfunctions is equal to $\exp S$. Here the entropy S is equal to

$$S = L \int_{-\infty}^{+\infty} d\lambda \left[\rho_t(\lambda) \ln \rho_t(\lambda) - \rho_p(\lambda) \ln \rho_p(\lambda) - \rho_h(\lambda) \ln \rho_h(\lambda) \right]. \qquad (8.13)$$

From calculations in parts III and IV (see sections X.4, XI.5, XII.5, XVII.3 and XIII.6) it will be clear that the thermodynamic limit of the mean value

$$\frac{\langle \Omega_T | \Psi^\dagger(x) \Psi(0) | \Omega_T \rangle}{\langle \Omega_T | \Omega_T \rangle} \tag{8.14}$$

exists and depends only on the macroscopic characteristics of $|\Omega_T\rangle$ ($\rho_t(\lambda)$, $\rho_h(\lambda)$, $\rho_p(\lambda)$ and $\varepsilon(\lambda)$). All the eigenfunctions $|\Omega_T\rangle$ (present in the state of thermal equilibrium) have the same macroscopic characteristics, they differ only in their distributions of the integer numbers n_j (see (2.13) and (5.8)). So the mean value is the same for all $|\Omega_T\rangle$. It is independent of the particular choice of eigenfunction $|\Omega_T\rangle$ present in the state of thermal equilibrium. This is enough to show that the mean value (8.14) is equal to the temperature correlation function

$$\frac{\mathrm{tr}\left(e^{-H/T} \Psi^\dagger(x) \Psi(0)\right)}{\mathrm{tr}\, e^{-H/T}} = \frac{\langle \Omega_T | \Psi^\dagger(x) \Psi(0) | \Omega_T \rangle}{\langle \Omega_T | \Omega_T \rangle}. \tag{8.15}$$

Let us show this.

We shall write down a functional integral representation for this correlation function, similar to (5.14). First, let us write

$$\mathrm{tr}\left[e^{-H/T} \Psi^\dagger(x) \Psi(0)\right] = \frac{1}{N!} \sum_{n_1, \ldots, n_N} e^{-E_n/T} \frac{\langle \Psi_N | \Psi^\dagger(x) \Psi(0) | \Psi_N \rangle}{\langle \Psi_N | \Psi_N \rangle}. \tag{8.16}$$

Here $|\Psi_N\rangle$ is the complete set of all eigenfunctions of the Hamiltonian H. This representation is similar to the representation (5.11) for $\mathrm{tr}\, e^{-H/T}$. Now one can write a functional integral representation for (8.16), similar to (5.14):

$$\mathrm{tr}\left[e^{-H/T} \Psi^\dagger(x) \Psi(0)\right] =$$
$$\mathrm{const} \int D\left(\frac{\rho_t(\lambda)}{\rho_p(\lambda)}\right) \delta\left(L \int \rho_p(\lambda)\, d\lambda - D\right) \frac{\langle \Psi_N | \Psi^\dagger(x) \Psi(0) | \Psi_N \rangle}{\langle \Psi_N | \Psi_N \rangle} e^{S - E/T}. \tag{8.17}$$

The mean value

$$\frac{\langle \Psi_N | \Psi^\dagger(x) \Psi(0) | \Psi_N \rangle}{\langle \Psi_N | \Psi_N \rangle}$$

has a finite thermodynamic limit. On the contrary, the entropy S and energy E are both proportional to the length of the box, L, in the thermodynamic limit ($L \to \infty$). This shows that the stationary point of (8.17) coincides with the stationary point of (5.14) and is given by equations (5.24)–(5.26) (this is the state of thermodynamic equilibrium). So $|\Psi_N\rangle$

in (8.17) should be replaced by $|\Omega_T\rangle$. But the mean values

$$\frac{\langle \Omega_T|\Psi^\dagger(x)\Psi(0)|\Omega_T\rangle}{\langle \Omega_T|\Omega_T\rangle} \tag{8.18}$$

are the same for all eigenfunctions of the Hamiltonian that are present in the state of thermodynamic equilibrium. So, finally, we have

$$\langle \Psi^\dagger(x)\Psi(0)\rangle_T = \frac{\text{tr}\left(e^{-H/T}\,\Psi^\dagger(x)\Psi(0)\right)}{\text{tr}\,e^{-H/T}}$$

$$= \frac{\langle \Omega_T|\Psi^\dagger(x)\Psi(0)|\Omega_T\rangle}{\langle \Omega_T|\Omega_T\rangle}. \tag{8.19}$$

Here $|\Omega_T\rangle$ is one of the eigenfunctions which are present in the state of thermodynamic equilibrium and the right hand side of (8.19) does not depend on the particular choice of $|\Omega_T\rangle$. This is very similar to what has happened with excitations at finite temperature. We are considering correlation functions of fields. But formulæ like (8.15) are also correct for any other correlation functions. Take the current correlation function

$$\langle j(x,t)j(0,0)\rangle_T = \frac{\text{tr}\left(e^{-H/T}\,j(x,t)j(0,0)\right)}{\text{tr}\,e^{-H/T}} = \frac{\langle \Omega_T|j(x,t)j(0,0)|\Omega_T\rangle}{\langle \Omega_T|\Omega_T\rangle} \tag{8.20}$$

or that for $\exp\{\alpha Q_1(x)\}$,

$$\frac{\text{tr}\left(e^{-H/T}\,e^{\alpha Q_1(x)}\right)}{\text{tr}\,e^{-H/T}} = \frac{\langle \Omega_T|e^{\alpha Q_1(x)}|\Omega_T\rangle}{\langle \Omega_T|\Omega_T\rangle}. \tag{8.21}$$

Here

$$Q_1(x) = \int\limits_0^x \Psi^\dagger(y)\Psi(y)\,dy. \tag{8.22}$$

Formula (8.15) remains valid for a time-dependent field $\Psi^\dagger(x,t)$.

Finally, let us mention that in parts III and IV we shall obtain a different integral representation for correlation functions. If we compare positive and zero temperature correlation functions they will always differ only by the change of the integration measure (8.5)

$$\int\limits_{-q}^{q} d\lambda \longrightarrow \int\limits_{-\infty}^{+\infty} d\lambda\,\vartheta(\lambda) \tag{8.23}$$

$$\vartheta(\lambda) = \frac{1}{1 + e^{\varepsilon(\lambda)/T}}.$$

In Part IV, the asymptotics of temperature correlation functions will be explicitly evaluated even for time-dependent correlation functions.

I.9 Finite size corrections

In this section, we shall come back to zero temperature. We shall also include a chemical potential, h, in the Hamiltonian

$$H_h = \int dx \, \{\partial_x \Psi^\dagger \partial_x \Psi + c \Psi^\dagger \Psi^\dagger \Psi \Psi - h \Psi^\dagger \Psi\}. \tag{9.1}$$

In sections 1 and 2, the model was solved in a finite periodic box for a finite number of particles. The eigenvalues of (9.1) will now be called the energy which is given by

$$E = \sum_j \varepsilon_0(\lambda_j); \quad \varepsilon_0(\lambda_j) = \lambda_j^2 - h \tag{9.2}$$

and the corresponding momentum is given by

$$P = \sum_j p_0(\lambda_j); \quad p_0(\lambda_j) = \lambda_j \tag{9.3}$$

where the $\{\lambda_j\}$ satisfy the Bethe equations (2.13)

$$L p_o(\lambda_j) + \sum_{k=1}^N \theta(\lambda_j - \lambda_k) = 2\pi n_j, \quad j = 1, \ldots, N. \tag{9.4}$$

For the ground state, we have (3.2)

$$L p_0(\lambda_j) + \sum_{k=1}^N \theta(\lambda_j - \lambda_k) = 2\pi \left(j - \frac{N+1}{2} \right). \tag{9.5}$$

In section 3, the thermodynamic limit was studied for $L \to \infty$, $N \to \infty$, such that $N/L = D = $ constant. Basically, sums were replaced by integrals.

In Chapter XVIII we shall see that finite size corrections define the long distance asymptotics of correlation functions, which conformal field theory makes clear.

Finite size corrections to the ground state energy (3.22) will be evaluated first. The function $\lambda(x)$ introduced in (2.27) will be used. The ground state energy can be rewritten using (9.2) and (9.5) as a summation with respect to an integer n_j:

$$\frac{E}{L} = \frac{1}{L} \sum_{n_j} \varepsilon_0 \left(\lambda \left(\frac{n_j}{L} \right) \right). \tag{9.6}$$

Using the Euler-Maclaurin formula for approximating sums by integrals, the following is obtained:

$$\frac{E}{L} = \int_{-N/2L}^{N/2L} \varepsilon_0(\lambda(x))\, dx - \frac{1}{24L^2}\frac{\partial \varepsilon_0}{\partial x}\bigg|_{x=N/2L} + \frac{1}{24L^2}\frac{\partial \varepsilon_0}{\partial x}\bigg|_{x=-N/2L} + \text{h.o.c.}$$

(9.7)

Here, h.o.c. means higher order corrections.

Similar corrections should be taken into account when deriving equation (3.7) from (2.31):

$$\rho_L(\lambda) - \frac{1}{2\pi}\int_{-q}^{q} K(\lambda,\mu)\rho_L(\mu)\, d\mu$$

$$= \frac{1}{2\pi}\left\{ p_0'(\lambda) + \frac{1}{24L^2\rho(q)}\left[K'(\lambda,q) - K'(\lambda,-q) \right] \right\}$$

(9.8)

where equation (2.30) was also used. The function $\rho_L(\lambda)$ should be used when changing the summation to an integration in (9.7):

$$E = L\int_{-q}^{q} \varepsilon_0(\lambda)\rho_L(\lambda)\, d\lambda - \frac{\pi}{6L}\frac{\varepsilon_0'(q)}{2\pi\rho(q)} + \text{h.o.c.}$$

(9.9)

Here, the zeroth order in L is absorbed in the definition of q,

$$q = \lambda_N + \frac{1}{2L\rho(\lambda_N)}.$$

(9.10)

Indeed, the limit of integration in (9.9) should be $q = \lambda(x = N/2L)$ where the function $\lambda(x)$ was defined by formula (2.27). However, λ_N stands for $\lambda_N = \lambda(x = (N-1)/2L)$. Substituting $\rho_L(\lambda)$ from (9.8) into (9.9) gives

$$E = L\int_{-q}^{q} \varepsilon_0(\lambda)\rho(\lambda)\, d\lambda$$

$$+ \frac{1}{48\pi L\rho(q)}\int_{-q}^{q} d\lambda\, \varepsilon(\lambda)\left[K'(\lambda,q) - K'(\lambda,-q) \right] - \frac{\varepsilon_0'(q)}{12L\rho(q)}.$$

(9.11)

Using (4.9) we obtain

$$\varepsilon'(q) = \varepsilon_0'(q) + \frac{1}{2\pi}\int_{-q}^{q} K'(q,\mu)\varepsilon(\mu)\, d\mu.$$

(9.12)

This allows (9.11) to be rewritten as

$$E = L \int_{-q}^{q} \varepsilon_0(\lambda)\rho(\lambda)\, d\lambda - \frac{\pi}{6L} v_F + \text{h.o.c.} \qquad (9.13)$$

where v_F is the Fermi velocity,

$$v_F = \frac{1}{2\pi\rho(q)} \frac{\partial \varepsilon}{\partial \lambda}\bigg|_{\lambda=q}. \qquad (9.14)$$

We shall show that this definition coincides with (7.17),

$$v = \frac{\partial \varepsilon}{\partial k}\bigg|_{\lambda=q} = \frac{\partial \varepsilon}{\partial \lambda}\bigg|_{\lambda=q} \left(\frac{\partial k}{\partial \lambda}\right)^{-1}\bigg|_{\lambda=q}. \qquad (9.15)$$

From (4.19), it immediately follows that

$$\frac{\partial k}{\partial \lambda} = 1 + \int_{-q}^{q} K(\lambda_p, \mu)\rho(\mu)\, d\mu. \qquad (9.16)$$

However, equation (3.7) for $\rho(\lambda)$ shows that

$$\frac{\partial k}{\partial \lambda} = 2\pi\rho(\lambda). \qquad (9.17)$$

This proves the equivalence of (9.14) and (9.15). The Fermi velocity of the Bose gas can also be expressed in terms of the derivative of the pressure with respect to the density (see (7.16) and Appendix 3.).

Now let us consider excited states. In the thermodynamic limit, there is no gap in the spectrum of the model. The energy on the Fermi sphere is equal to zero, $\varepsilon(q) = \varepsilon(-q) = 0$, see (4.9). However, in a finite box, a gap arises, but it is small (of order $1/L$).

Now consider low lying excitations which are related to the following three physical processes. (1) The integer value n_j corresponding to the particle on the Fermi surface will be changed by a finite amount N^+ (at q) and N^- (at $-q$). (2) The number of particles can be changed by ΔN. (3) Some particles (let us say d) can undergo a backscattering process under which these particles can jump from one Fermi boundary $(-q)$ to another Fermi boundary $(+q)$. The changes in energy and momentum corresponding to these processes are

$$\delta E = \frac{2\pi v}{L} \left[\left(\frac{\Delta N}{2\mathcal{Z}}\right)^2 + (\mathcal{Z}d)^2 + N^+ + N^- \right]$$

$$\delta P = 2k_F d + \frac{2\pi}{L}(N^+ - N^- + \Delta N d) \qquad (9.18)$$

where the Fermi momentum k_F is

$$k_F = \pi D = \pi \frac{N}{L} \tag{9.19}$$

and $\mathcal{Z}$ is the value of the function $Z(\lambda)$ at the Fermi boundary. The function $Z(\lambda)$ is defined by the integral equation

$$Z(\lambda) - \frac{1}{2\pi} \int\limits_{-q}^{q} K(\lambda, \mu) Z(\mu)\, d\mu = 1 \tag{9.20}$$

$$\mathcal{Z} = Z(q) = Z(-q).$$

In the model under consideration, the function $Z(\lambda)$ (which we shall refer to as the dressed charge) is proportional to the ground state distribution function (3.7), $Z(\lambda) = 2\pi\rho(\lambda)$. However, for other models solvable by the Bethe Ansatz, this is not true. The derivation of (9.18) is given in Appendix 4. It should be emphasized once more that finite size corrections are very useful for conformal field theories (see Chapter XVIII). For example, the critical exponents which drive the long distance asymptotics of correlation functions are defined by the dressed charge $Z(\lambda)$. In the Bose gas, the critical exponent θ which drives the long distance asymptotics of local density correlations (and fields) can be expressed in terms of $Z(\lambda)$ (9.20) in the following way:

$$\theta = 2\mathcal{Z}^2. \tag{9.21}$$

Using the identities of Appendices 3 and 4, one can express the same θ in terms of the density D (3.8) and the Fermi velocity (7.17):

$$\theta = \frac{4\pi D}{v}. \tag{9.22}$$

Conclusion

The one-dimensional Bose gas has been solved. The problem of quantum field theory was first reduced to a many-body problem of quantum mechanics (see section 1) which was possible due to particle number conservation. Then, the many-body problem was solved; this was straightforward since many-body dynamics are reducible to two-body dynamics. Next, periodic boundary conditions for the wave function were studied, which led to determination of the spectral parameters via the Bethe equations. These equations have the very nice property that they can be obtained from the Yang-Yang variational principle; they are used to derive the thermodynamic limit. The structure of excitations over the ground state in the thermodynamic limit can be treated in terms of particles and holes. Also, the energy, momentum and scattering matrix were evaluated. In the last section, we studied finite size corrections, which will be used later in the description of correlation functions. Finally, the model was studied at finite temperature where the equations differ from the zero temperature ones only by the measure of integration. Following the ideas of C.N. Yang and C.P. Yang [22], we have constructed stable excitations at finite temperature. We have also shown how to calculate temperature correlation functions. This will be used in sections X.4, XI.5, XII.5, XIII.2, XIII.3, XIII.6 and XVII.3.

The one-dimensional Bose gas was solved by E.H. Lieb and W. Liniger; see [14] and [15]. The way of constructing eigenfunctions which we have presented was proposed by M. Gaudin in [9]. The papers [1], [5], [7], [8], [17], [18], [19] and [21] have played an important role in the development and generalization of the theory of one-dimensional particles interacting via a δ potential. One may find a review of works on the subject in [16]. C.N. Yang and C.P. Yang were the first to construct the thermodynamics of the model; see [20] and [22]. We have followed the ideas of these papers. The calculation method of the scattering matrix by means of the Bethe Ansatz was developed in [13]. The thermodynamics of the model integer were investigated using functional integration in [3]. The finite size corrections for the Bose gas were first calculated in [4], see also [2].

Appendix I.1: Fermionization

The Fermi-Bose correspondence, proposed by H. Thacker for impenetrable bosons [6], is very useful:

$$\Psi_B^\dagger(x) = \Psi_F^\dagger(x) \exp\left\{ i\pi \int\limits_{-\infty}^{x} \Psi_F^\dagger(z)\Psi_F(z)dz \right\}. \qquad (A.1.1)$$

Here $\Psi_B(x)$ is the canonical Bose field in section 1 of this chapter and $\Psi_F(x)$ is the canonical Fermi field with standard anticommutation relations. Bose eigenfunctions of the Hamiltonian (1.9) can be recalculated with a Fermi basis:

$$\int \chi_N^B(z_1,\ldots,z_N)\Psi_B^\dagger(z_1)\cdots\Psi_B^\dagger(z_N)|0\rangle$$

$$\longrightarrow \int \chi_N^F(z_1,\ldots,z_N)\Psi_F^\dagger(z_1)\cdots\Psi_F^\dagger(z_N)|0\rangle. \quad (A.1.2)$$

Here the Fermi wave function χ_N^F is given by

$$\chi_N^F(z_1,\ldots,z_N) = \prod_{1 \le i < j \le N} \epsilon(z_j - z_i)\chi_N^B(z_1,\ldots,z_N) \qquad (A.1.3)$$

which is known as the Girardeau formula [12]. The Fermi wavefunction is antisymmetric as it should be. The fermionic two-body scattering matrix is different from the bosonic one by a factor -1. Comparison with (2.1) makes it clear that the periodic boundary conditions for fermions imply the following boundary conditions for bosons:

$$\chi_N^B(x_1 + L,\ldots,x_N) = (-1)^{N-1}\chi_N^B(x_1,\ldots,x_N). \qquad (A.1.4)$$

In other models, it is also possible to construct fermi-like excitations by changing the boundary conditions. For the Heisenberg antiferromagnet, the Jordan-Wigner transformation allows the local Hamiltonian to be rewritten in terms of fermions at any value of the coupling constant. In

47

the sine-Gordon model, the Mandelstam-Coleman correspondence plays a similar role.

Formula (A.1.1) can be inverted because

$$\Psi_B^\dagger(x)\Psi_B(x) = \Psi_F^\dagger(x)\Psi_F(x)$$

so we have

$$\Psi_F^\dagger(x) = \Psi_B^\dagger(x)\exp\left\{i\pi\int\limits_{-\infty}^{x}\Psi_B^\dagger(z)\Psi_B(z)dz\right\}. \qquad (A.1.5)$$

Here we have used the fact that the spectrum of the operator

$$\int\limits_{-\infty}^{x}\Psi_B^\dagger(z)\Psi_B(z)dz$$

consists of integer numbers.

Appendix I.2: Stability
of thermal equilibrium

The calculations in section 5 using the steepest descent method are valid if the matrix of the second derivatives in the exponent (5.16) is negative definite. To prove this we shall integrate (5.16) first over ρ_p and then over h. Straightforward calculations following [22] yield

$$
\delta^2 \left(S - \frac{E}{T} + \frac{L}{T} h \left[\int \rho_p(\lambda) \, d\lambda - D \right] \right)
$$
$$
= -\frac{L}{T^2} \int d\lambda \, (\delta\varepsilon(\lambda))^2 \, \rho_p(\lambda) \left(1 + e^{-\varepsilon(\lambda)/T} \right)^{-1} < 0.
$$

(A.2.1)

Let us remember that in section 5 we represented the partition function Z as a functional integral with respect to the function $\varepsilon(\lambda)$.

Using the steepest descent method one obtains

$$
Z = \text{const} \int_{-i\infty}^{+i\infty} dh \, \exp\left\{ -\frac{L}{T} f \right\};
$$

$$
f = Dh - \frac{T}{2\pi} \int_{-\infty}^{+\infty} d\lambda \ln\left(1 + e^{-\varepsilon(\lambda)/T} \right).
$$

(A.2.2)

Equations (5.24)–(5.26) define the dependence of f on h; its first derivative is

$$
\frac{\partial f}{\partial h} = D - \int_{-\infty}^{+\infty} \rho_p(\lambda) \, d\lambda
$$

(A.2.3)

and the second one is

$$\frac{\partial^2 f}{\partial h^2} = -\frac{\partial}{\partial h} \int\limits_{-\infty}^{+\infty} \rho_p(\lambda)\, d\lambda. \tag{A.2.4}$$

If the right-hand side of this expression is negative then the integration over h along the imaginary axis corresponds to the correct passage of the stationary point (see (5.16)).

It follows from (5.19) and (5.7) that

$$\frac{\partial \varepsilon(\lambda)}{\partial h} = -2\pi \rho_t(\lambda). \tag{A.2.5}$$

Here one should use the equality $\rho_t(\lambda) = \rho_p(\lambda)[1 + \exp\{\varepsilon(\lambda)/T\}]$. It can also be proved that the function $\partial \rho_p(\lambda)/\partial h$ satisfies the equation

$$\left(1 + e^{\varepsilon(\lambda)/T}\right) \frac{\partial \rho_p(\lambda)}{\partial h} - \frac{1}{2\pi} \int\limits_{-\infty}^{+\infty} K(\lambda,\mu) \frac{\partial \rho_p(\mu)}{\partial h}\, d\mu$$

$$= \frac{2\pi}{T} e^{\varepsilon(\lambda)/T} \rho_p(\lambda) \rho_t(\lambda) \tag{A.2.6}$$

and hence

$$-\frac{\partial^2 f}{\partial h^2} = \frac{\partial}{\partial h} \int \rho_p(\lambda)\, d\lambda$$

$$= \frac{(2\pi)^2}{T} \int d\lambda\, e^{\varepsilon(\lambda)/T} \left(1 + e^{\varepsilon(\lambda)/T}\right) \rho_p(\lambda)^3 > 0. \tag{A.2.7}$$

This inequality means that

$$\frac{\partial D}{\partial h} = \frac{\partial}{\partial h} \int \rho_p(\lambda)\, d\lambda > 0, \qquad T > 0. \tag{A.2.8}$$

At zero temperature:

$$\frac{\partial D}{\partial h} = \frac{8\pi^2 \rho(q)^3}{\varepsilon'(q)}, \qquad T = 0. \tag{A.2.9}$$

Let us also mention the alternative way of writing the partition function (5.21)

$$e^{Nh/T} \operatorname{tr} e^{-H/T} = \prod_{\lambda_j = 2\pi j/L} \left(\frac{\rho_t(\lambda_j)}{\rho_h(\lambda_j)}\right). \tag{A.2.10}$$

This is the product with respect to all the integers j $(-\infty < j < +\infty)$.

Using (5.26) we can rewrite the right-hand side of (A.2.10) as

$$\prod_{j=-\infty}^{+\infty} \left(1 + e^{-\varepsilon(\lambda_j)/T}\right) = \exp\left\{\sum_{j=-\infty}^{+\infty} \ln\left(1 + e^{-\varepsilon(\lambda_j)/T}\right)\right\}$$

$$= \exp\left\{\frac{L}{2\pi} \int_{-\infty}^{+\infty} d\lambda \, \ln\left(1 + e^{-\varepsilon(\lambda_j)/T}\right)\right\}.$$

$$(A.2.11)$$

We are talking about the thermodynamic limit $L \to \infty$.

Appendix I.3: Fermi velocity

Here we prove the equivalence between the two definitions (7.16) and (7.17) for the velocity of sound in the model considered:

$$v = \frac{\partial \varepsilon(\lambda)}{\partial k(\lambda)}\Big|_{\lambda=q} = \left(\frac{\partial \varepsilon(\lambda)}{\partial \lambda}\right) \Big/ \left(\frac{\partial k(\lambda)}{\partial \lambda}\right)\Big|_{\lambda=q} \tag{A.3.1}$$

$$v^2 = 2\frac{\partial P}{\partial D} = 2\left(\frac{\partial P}{\partial h}\right) \Big/ \left(\frac{\partial D}{\partial h}\right). \tag{A.3.2}$$

Comparing (3.7) and (3.16) one arrives at

$$\frac{\partial \varepsilon(\lambda)}{\partial h} = -2\pi\rho(\lambda). \tag{A.3.3}$$

It follows from (7.15) that

$$\frac{\partial P}{\partial h} = -\frac{1}{2\pi}\int\limits_{-q}^{q}\frac{\partial \varepsilon(\lambda)}{\partial h}\,d\lambda = D \tag{A.3.4}$$

while (4.19) and (3.7) give

$$\frac{\partial k(\lambda)}{\partial \lambda} = 2\pi\rho(\lambda). \tag{A.3.5}$$

The proof of the equivalence of (A.3.1) and (A.3.2) is easily reduced to proving the equality

$$\varepsilon'(q)\rho(q) = D \tag{A.3.6}$$

by virtue of (A.2.9). In fact, it follows from (4.5) and (4.6) that (see (4.8) and (4.9)):

$$\frac{\partial \varepsilon(\lambda)}{\partial \lambda} = 2\lambda + 2\int\limits_{-q}^{q}\mu\dot{F}(\mu|\lambda)\,d\mu. \tag{A.3.7}$$

52

Here $\dot{F}$ is defined by

$$\dot{F}(\lambda|\mu) - \frac{1}{2\pi} \int\limits_{-q}^{q} K(\lambda,\nu)\dot{F}(\nu|\mu)d\nu = \frac{1}{2\pi}K(\lambda|\mu).$$

Differentiating (3.7) with respect to λ we obtain

$$\frac{\partial\rho(\lambda)}{\partial\lambda} = -\rho(q)\left[\dot{F}(\lambda|q) - \dot{F}(\lambda|-q)\right].$$

Substituting this equality into the following identity

$$\int\limits_{-q}^{q} \lambda\frac{\partial\rho(\lambda)}{\partial\lambda}\,d\lambda = 2q\rho(q) - D$$

one obtains (A.3.6) and thus completes the proof. Let us mention also that the following equality is true:

$$v = \frac{\partial h}{\partial q}. \tag{A.3.8}$$

One can also prove that $\partial D/\partial q = 2D/v$ and $\theta = 2\pi\partial D/\partial q$; see (9.21).

Appendix I.4: Finite size corrections

Equations (9.18) will be derived in this appendix. The three processes affecting low lying excitations (described before in section 9, see (9.18)) are additive; thus, contributions to the change in the ground state energy from each process can be considered separately.

(1) First, change the maximal n_j (in the ground state it is equal to $(N-1)/2$) by an integer N^+. Then, the complete momentum is given by

$$P = \frac{2\pi}{L} \sum_j n_j = \frac{2\pi}{L} N^+. \tag{A.4.1}$$

On the other hand, this excitation can be considered as a particle–hole excitation, as in section 4, with momentum

$$P = k(\lambda_p) - k(\lambda_h) = \left.\frac{\partial k}{\partial \lambda}\right|_{\lambda=q} (\lambda_p - \lambda_h) \tag{A.4.2}$$

and energy

$$\delta E = \varepsilon(\lambda_p) - \varepsilon(\lambda_h) = \left.\frac{\partial \varepsilon}{\partial \lambda}\right|_{\lambda=q} (\lambda_p - \lambda_h). \tag{A.4.3}$$

From (A.4.1) and (A.4.2), we get

$$\lambda_p - \lambda_h = \frac{2\pi N^+}{Lk'(q)}. \tag{A.4.4}$$

Substituting this into (A.4.3) gives

$$\delta E = \frac{2\pi}{L} \frac{\varepsilon'}{k'} N^+ = \frac{2\pi}{L} v N^+$$

where the definition (9.15) of the Fermi velocity is used. This reproduces the contribution of N^+ in (9.18). Local excitations at the other Fermi boundary, $(-q)$, corresponding to a moderate change of the minimal integer number $n_j = -(N-1)/2$, give the contribution of N^- in (9.18).

(2) Now consider a second type of excitation: a change in the number of particles. The energy of the ground state is changed when the number of particles is changed by a small number ΔN. The density of the ground state energy is, from (9.2) (see (3.22)),

$$e = \int_{-q}^{q} \varepsilon_0(\lambda)\rho(\lambda)\,d\lambda, \tag{A.4.5}$$

$$\varepsilon_0(\lambda) = \lambda^2 - h$$

(where $e = E/L$) and can be separated into contributions from the microcanonical ensemble ($e_0(D) = L\int_{-q}^{q} \lambda^2 \rho(\lambda)\,d\lambda$) and from the chemical potential,

$$e(D) = e_0(D) - hD. \tag{A.4.6}$$

In the case under consideration, the excitation density D is changed by a small amount, $\Delta N/L$. Thus, the change in the ground state energy is given by

$$\delta E = L\left\{e_0\left(D + \frac{\Delta N}{L}\right) - e_0(D) - h\frac{\Delta N}{L}\right\}. \tag{A.4.7}$$

Now decompose $e_0(D+\Delta N/L)$ into a Taylor series expansion with respect to $\Delta N/L$. Thus,

$$\delta E = L\left\{\frac{\Delta N}{L}\left(\frac{\partial e_0(D)}{\partial D} - h\right) + \frac{1}{2}\left(\frac{\Delta N}{L}\right)^2 \frac{\partial^2 e_0(D)}{\partial D^2} + \text{h.o.c.}\right\}. \tag{A.4.8}$$

From the definition of the grand canonical ensemble, it follows that the density minimizes the free energy at a fixed value of the chemical potential $\partial e_0(D)/\partial D = h$. This permits us to rewrite (A.4.8) in the following form:

$$\delta E = \frac{(\Delta N)^2}{2L}\left(\frac{\partial D}{\partial h}\right)^{-1} + \text{h.o.c.} \tag{A.4.9}$$

or more conveniently

$$\delta E = \frac{(\Delta N)^2}{2L}\left(\frac{\partial D}{\partial q}\right)^{-1}\left(\frac{\partial h}{\partial q}\right). \tag{A.4.10}$$

Below, it will be shown that

$$\frac{\partial D}{\partial q} = 2\rho(q)Z(q) \tag{A.4.11}$$

and

$$\frac{\partial h}{\partial q} = \frac{\varepsilon'(q)}{Z(q)}. \tag{A.4.12}$$

Substitution of these into (A.4.10) gives the correct contribution in (9.18) from a change in the number of particles:

$$\delta E = \frac{2\pi v}{L} \left(\frac{\Delta N}{2Z(q)} \right)^2 . \tag{A.4.13}$$

Here equation (9.14) was used, $v = \varepsilon'/[2\pi\rho(q)]$.

Before continuing, equations (A.4.12) and (A.4.11) should be proved. First, it will be shown that (A.4.12) is correct. Comparing the equations for $\varepsilon(\lambda)$

$$\varepsilon(\lambda) - \frac{1}{2\pi} \int_{-q}^{q} K(\lambda, \mu)\varepsilon(\mu)\, d\mu = \varepsilon_0(\lambda) = \lambda^2 - h \tag{A.4.14}$$

and $Z(\lambda)$

$$Z(\lambda) - \frac{1}{2\pi} \int_{-q}^{q} K(\lambda, \mu)Z(\mu)\, d\mu = 1 \tag{A.4.15}$$

immediately gives

$$\frac{\partial \varepsilon}{\partial h} = -Z(\lambda). \tag{A.4.16}$$

The function $\varepsilon(\lambda)$ actually depends on two variables, λ and q; thus we write $\varepsilon(\lambda|q)$. Changing q in the equation $\varepsilon(\lambda|q)\Big|_{\lambda=q} = 0$ also changes λ because in this case they are equal. Differentiating the latter equation with respect to q gives

$$\frac{\partial \varepsilon}{\partial \lambda}\Big|_{\lambda=q} + \frac{\partial \varepsilon}{\partial q}\Big|_{\lambda=q} = 0. \tag{A.4.17}$$

Putting $\lambda = q$ and rewriting (A.4.16) using derivatives with respect to q gives

$$\frac{\partial q}{\partial h} \left(\frac{\partial \varepsilon}{\partial q}\Big|_{\lambda=q} \right) = -Z(q). \tag{A.4.18}$$

Using (A.4.17), we see that

$$\frac{\partial q}{\partial h} \left(\frac{\partial \varepsilon}{\partial \lambda}\Big|_{\lambda=q} \right) = Z(q).$$

In this way, we have proved (A.4.12).

Next, we wish to prove (A.4.11). The density D can be represented in the form

$$D = \int_{-q}^{q} \rho(\lambda)\, d\lambda \qquad (A.4.19)$$

where the distribution function is defined by the integral equation

$$\rho(\lambda) - \frac{1}{2\pi} \int_{-q}^{q} K(\lambda, \mu)\rho(\mu)\, d\mu = \frac{1}{2\pi} p_0'(\lambda). \qquad (A.4.20)$$

In the Bose gas case, $p_0' = 1$, but in the XXZ Heisenberg model, the Hubbard model and other models, this is not the case. Thus, the general situation will be studied, where $p_0(\lambda)$ is the bare momentum of a particle, $p_0(-\lambda) = -p_0(\lambda)$. Differentiating (A.4.19) with respect to q gives

$$\frac{\partial D}{\partial q} = \rho(q) + \rho(-q) + \int_{-q}^{q} \frac{\partial \rho(\lambda)}{\partial q}\, d\lambda. \qquad (A.4.21)$$

We shall make use of the fact that $\rho(\lambda) = \rho(-\lambda)$. In order to represent $\partial\rho(\lambda)/\partial q$, equation (A.4.20) can be differentiated to give

$$\frac{\partial \rho(\lambda)}{\partial q} - \frac{1}{2\pi} \int_{-q}^{q} K(\lambda, \mu) \frac{\partial \rho(\mu)}{\partial q}\, d\mu = \frac{1}{2\pi} \rho(q)\left[K(\lambda, q) + K(\lambda, -q)\right].$$

$$(A.4.22)$$

The formal solution of this equation is

$$\frac{\partial \rho(\lambda)}{\partial q} = \frac{1}{2\pi}\rho(q) \int_{-q}^{q} d\mu \left(1 - \frac{1}{2\pi}K\right)^{-1}(\lambda, \mu)\left[K(\mu, q) + K(\mu, -q)\right].$$

$$(A.4.23)$$

An integral of this function is essential to determine (A.4.21):

$$\int_{-q}^{q} d\lambda\, \frac{\partial \rho(\lambda)}{\partial q}$$

$$= \frac{1}{2\pi}\rho(q) \int_{-q}^{q} d\mu \int_{-q}^{q} d\lambda \left(1 - \frac{1}{2\pi}K\right)^{-1}(\lambda, \mu)\left[K(\mu, q) + K(\mu, -q)\right].$$

$$(A.4.24)$$

From (A.4.15), it follows that

$$\int_{-q}^{q} d\lambda \left(1 - \frac{1}{2\pi}K\right)^{-1}(\lambda, \mu) = Z(\mu) \qquad (A.4.25)$$

where the symmetry of the kernel is used, $K(\lambda, \mu) = K(\mu, \lambda)$. Substituting equation (A.4.25) into (A.4.24) gives

$$\int_{-q}^{q} d\lambda \, \frac{\partial \rho(\lambda)}{\partial q} = \frac{1}{2\pi} \rho(q) \int_{-q}^{q} d\mu \, Z(\mu) \left[K(\mu, q) + K(\mu, -q) \right]. \qquad \text{(A.4.26)}$$

Substituting this into (A.4.21) gives

$$\frac{\partial D}{\partial q} = \rho(q) \left\{ 2 + \frac{1}{2\pi} \int_{-q}^{q} d\mu \, K(q, \mu) Z(\mu) + \frac{1}{2\pi} \int_{-q}^{q} d\mu \, K(-q, \mu) Z(\mu) \right\}$$

$$\text{(A.4.27)}$$

which can be transformed using equation (A.4.15) at $\lambda = q$ and $\lambda = -q$ to the following equations:

$$Z(q) = 1 + \frac{1}{2\pi} \int_{-q}^{q} K(q, \mu) Z(\mu) \, d\mu \qquad \text{(A.4.28)}$$

and

$$Z(-q) = 1 + \frac{1}{2\pi} \int_{-q}^{q} K(-q, \mu) Z(\mu) \, d\mu. \qquad \text{(A.4.29)}$$

It should be remembered that $Z(\lambda)$ is an even function of its argument. Substituting equations (A.4.28) and (A.4.29) into equation (A.4.27) and using the symmetry property of K gives the following result:

$$\frac{\partial D}{\partial q} = 2\rho(q) Z(q) \qquad \text{(A.4.30)}$$

which is the same as equation (A.4.11). This finalizes the proof of formula (A.4.13) which gives the contribution of ΔN in (9.13).

(3) It is now time to consider the effect of the backscattering process which shifts all the n_j in the ground state (9.5) by an integer number d. This describes the state where d particles have jumped over the sea. The Bethe equations for the excited state are as follows:

$$L\tilde{\lambda}_j + \sum_{k} \theta(\tilde{\lambda}_j - \tilde{\lambda}_k) = 2\pi \left(j - \frac{N+1}{2} \right) + 2\pi d. \qquad \text{(A.4.31)}$$

As was done before, the ground state equation (9.5) is subtracted from (A.4.31). The calculations are similar to those in section 4 (formulæ (4.1)–(4.5)) Again, we shall use the shift function

$$F(\lambda_j) = \frac{\lambda_j - \tilde{\lambda}_j}{\lambda_{j+1} - \lambda_j} \qquad \text{(A.4.32)}$$

which is similar to (4.4). This satisfies the following integral equation

$$F(\lambda) - \frac{1}{2\pi} \int_{-q}^{q} K(\lambda, \mu) F(\mu) \, d\mu = -d. \qquad (A.4.33)$$

Comparison with the integral equation for $Z(\lambda)$, (A.4.15), shows that

$$F(\lambda) = -Z(\lambda) d \qquad (A.4.34)$$

which is useful in evaluating the shift δ of the Fermi boundary q.

Now let us consider the last equation for $\lambda = q$, $F(q) = -Z(q)d$. Using the equation for $F(\lambda)$ in (A.4.32), F can be rewritten as

$$F(q) = \frac{-L\delta}{L(\lambda_{j+1} - \lambda_j)}\bigg|_{\lambda_j = q} = -L\rho(q)\delta \qquad (A.4.35)$$

which gives:

$$\delta = d\frac{Z(q)}{L\rho(q)}. \qquad (A.4.36)$$

Both Fermi boundaries have shifted in the same direction by the same amount, $q \to q + \delta$ and $-q \to -q + \delta$. Thus, we shall treat this excited state as a ground state with new Fermi boundaries, $-q+\delta$ and $q+\delta$. The energy of this excited state can thus be represented in a similar fashion to (A.4.5)

$$E = L \int_{-q+\delta}^{q+\delta} \varepsilon_0(\lambda) \tilde{\rho}(\lambda) \, d\lambda \qquad (A.4.37)$$

where $\tilde{\rho}(\lambda)$ is the new distribution function defined by

$$\tilde{\rho}(\lambda) - \frac{1}{2\pi} \int_{-q+\delta}^{q+\delta} K(\lambda, \mu) \tilde{\rho}(\mu) \, d\mu = \frac{1}{2\pi} p_0'(\lambda) \qquad (A.4.38)$$

which should be compared with (A.4.20). It is now possible to rewrite the ground state energy using the dressed energy from equation (A.4.14). Let us derive a formula similar to (3.22); first, we shall consider the ground state. Formally solving equation (A.4.20) gives

$$\rho(\lambda) = \frac{1}{2\pi} \int_{-q}^{q} \left(1 - \frac{1}{2\pi}K\right)^{-1}(\lambda, \mu) \, p_0'(\mu) \, d\mu \qquad (A.4.39)$$

which can be substituted into (A.4.5) to give

$$
E = L \int_{-q}^{q} \varepsilon_0(\lambda)\rho(\lambda) \, d\lambda
$$

$$
= \frac{1}{2\pi} L \int_{-q}^{q} d\lambda \int_{-q}^{q} d\mu \varepsilon_0(\lambda) \left(1 - \frac{1}{2\pi}K\right)^{-1}(\lambda, \mu) \, p_0'(\mu). \quad \text{(A.4.40)}
$$

Using the symmetry of the kernel and performing the integral with respect to λ gives

$$
\int_{-q}^{q} d\lambda \, \varepsilon_0(\lambda) \left(1 - \frac{1}{2\pi}K\right)^{-1}(\lambda, \mu) = \varepsilon(\mu) \quad \text{(A.4.41)}
$$

which is the solution of (A.4.14). Finally, the ground state energy is given by

$$
E = \frac{1}{2\pi} L \int_{-q}^{q} d\mu \, \varepsilon(\mu) p_0'(\mu). \quad \text{(A.4.42)}
$$

In a similar way, the energy of the backscattered excited state is given by

$$
\widetilde{E}(\delta) = \frac{1}{2\pi} L \int_{-q+\delta}^{q+\delta} d\mu \, \widetilde{\varepsilon}(\mu) p_0'(\mu) \quad \text{(A.4.43)}
$$

where $\widetilde{\varepsilon}(\mu)$ is defined by the equation

$$
\widetilde{\varepsilon}(\lambda) - \frac{1}{2\pi} \int_{-q+\delta}^{q+\delta} K(\lambda, \mu)\widetilde{\varepsilon}(\mu) \, d\mu = \varepsilon_0(\lambda). \quad \text{(A.4.44)}
$$

At $\delta = 0$, the excited state coincides with the ground state. Decomposing the excited state energy $\widetilde{E}(\delta)$ through a Taylor series expansion in δ gives

$$
\widetilde{E}(\delta) = E + \left.\frac{\partial \widetilde{E}}{\partial \delta}\right|_{\delta=0} \delta + \frac{1}{2} \left.\frac{\partial^2 \widetilde{E}}{\partial \delta^2}\right|_{\delta=0} \delta^2 + O(\delta^3). \quad \text{(A.4.45)}
$$

It should be noted that at $\delta = 0$, $\widetilde{\varepsilon}(\lambda) = \varepsilon(\lambda)$. Thus $\widetilde{\varepsilon}(\lambda)$ can also be decomposed through a Taylor series in δ:

$$
\widetilde{\varepsilon}(\lambda) = \varepsilon(\lambda) + \left.\frac{\partial \widetilde{\varepsilon}(\lambda)}{\partial \delta}\right|_{\delta=0} \delta + \frac{1}{2} \left.\frac{\partial^2 \widetilde{\varepsilon}(\lambda)}{\partial \delta^2}\right|_{\delta=0} \delta^2 + O(\delta^3). \quad \text{(A.4.46)}
$$

Differentiating (A.4.44) with respect to δ gives

$$\frac{\partial \widetilde{\varepsilon}(\lambda)}{\partial \delta} - \frac{1}{2\pi} \int\limits_{-q+\delta}^{q+\delta} K(\lambda, \mu) \frac{\partial \widetilde{\varepsilon}(\mu)}{\partial \delta} \, d\mu$$

$$= \frac{1}{2\pi} K(\lambda, q+\delta) \widetilde{\varepsilon}(q+\delta) - \frac{1}{2\pi} K(\lambda, -q+\delta) \widetilde{\varepsilon}(-q+\delta). \quad \text{(A.4.47)}$$

We have

$$\widetilde{\varepsilon}(q) = \varepsilon(q) = 0 \quad \text{and} \quad \widetilde{\varepsilon}(-q) = \varepsilon(-q) = 0 \quad \text{for} \quad \delta = 0 \quad \text{(A.4.48)}$$

which makes the right hand side of (A.4.47) equal to zero so that

$$\left. \frac{\partial \widetilde{\varepsilon}(\lambda)}{\partial \delta} \right|_{\delta=0} = 0. \quad \text{(A.4.49)}$$

Differentiating (A.4.47) with respect to δ once more while setting $\delta = 0$ and using equations (A.4.48) and (A.4.49) gives

$$\frac{\partial^2 \widetilde{\varepsilon}(\lambda)}{\partial \delta^2} - \frac{1}{2\pi} \int\limits_{-q}^{q} K(\lambda, \mu) \frac{\partial^2 \widetilde{\varepsilon}(\mu)}{\partial \delta^2} \, d\mu = \frac{1}{2\pi} \varepsilon'(q) \left[K(\lambda, q) + K(\lambda, -q) \right]$$

$$\text{(A.4.50)}$$

where $\varepsilon'(q) = \partial \varepsilon / \partial \lambda$ at $\lambda = q$. Now we are ready to evaluate the Taylor expansion (A.4.45). Differentiating (A.4.43) with respect to δ gives

$$\frac{\partial \widetilde{E}}{\partial \delta} = \frac{1}{2\pi} L \left[p_0'(q+\delta) \widetilde{\varepsilon}(q+\delta) - p_0'(-q+\delta) \widetilde{\varepsilon}(-q+\delta) \right]$$

$$+ \int\limits_{-q+\delta}^{q+\delta} p_0'(\mu) \frac{\partial \widetilde{\varepsilon}(\mu)}{\partial \delta} \, d\mu. \quad \text{(A.4.51)}$$

At $\delta = 0$, the right hand side of this equation is zero. This is easily seen by using (A.4.48) and (A.4.49); thus,

$$\left. \frac{\partial \widetilde{E}}{\partial \delta} \right|_{\delta=0} = 0. \quad \text{(A.4.52)}$$

Differentiating (A.4.51) with respect to δ once more and putting $\delta = 0$ gives

$$\left. \frac{\partial^2 \widetilde{E}}{\partial \delta^2} \right|_{\delta=0} = \frac{1}{2\pi} L \left[2 p_0'(q) \varepsilon'(q) + \int\limits_{-q}^{q} p_0'(\lambda) \frac{\partial^2 \varepsilon(\lambda)}{\partial \delta^2} \, d\lambda \right] \quad \text{(A.4.53)}$$

where $\varepsilon'(-\lambda) = -\varepsilon'(\lambda)$ and $p_0'(\lambda) = p_0'(-\lambda)$ were used. This equation can also be solved in a formal way with

$$\frac{\partial^2 \varepsilon(\lambda)}{\partial \delta^2} = \frac{1}{2\pi} \varepsilon'(q) \int_{-q}^{q} d\mu \left(1 - \frac{1}{2\pi}K\right)^{-1}(\lambda, \mu) \left[K(\mu, q) + K(\mu, -q)\right].$$

(A.4.54)

Substituting this into (A.4.53) gives

$$\frac{\partial^2 \tilde{E}}{\partial \delta^2}\bigg|_{\delta=0} = \frac{1}{2\pi} L\varepsilon'(q) \left[2p_0'(q) + \frac{1}{2\pi} \int_{-q}^{q} d\lambda \int_{-q}^{q} d\mu \, p_0'(\lambda) \left(1 - \frac{1}{2\pi}K\right)^{-1}(\lambda, \mu)\right.$$

$$\left. \times \left[K(\mu, q) + K(\mu, -q)\right]\right]. \quad \text{(A.4.55)}$$

Using (A.4.39), the integral with respect to λ can be evaluated to give

$$\frac{\partial^2 \tilde{E}}{\partial \delta^2}\bigg|_{\delta=0} = \frac{1}{2\pi} L\varepsilon'(q) \left[2p_0'(q) + \int_{-q}^{q} d\mu \, \rho(\mu) \left[K(\mu, q) + K(\mu, -q)\right]\right].$$

(A.4.56)

Using equation (A.4.20) for $\lambda = q$ and $\lambda = -q$ gives

$$2p_0'(q) + \int_{-q}^{q} d\mu \, \rho(\mu) \left[K(\mu, q) + K(\mu, -q)\right]$$

$$= 2\pi \rho(q) + 2\pi \rho(-q) = 4\pi \rho(q). \quad \text{(A.4.57)}$$

Combining the above leads to

$$\frac{\partial^2 \tilde{E}}{\partial \delta^2}\bigg|_{\delta=0} = 2L\varepsilon'(q)\rho(q). \quad \text{(A.4.58)}$$

Thus, the change in energy due to the backscattering process can be evaluated. Using (A.4.45), (A.4.52), (A.4.58) and (A.4.36), we obtain

$$\tilde{E}(\delta) - E = \frac{1}{2} \frac{\partial^2 \tilde{E}}{\partial \delta^2}\bigg|_{\delta=0} \delta^2$$

$$= L\varepsilon'(q)\rho(q) \left(\frac{Z(q)d}{L\rho(q)}\right)^2 \quad \text{(A.4.59)}$$

$$= \frac{2\pi}{L} v Z^2(q) d^2$$

where $v = \varepsilon'(q)/[2\pi \rho(q)]$. Thus (9.18) has been shown and it is seen that finite size corrections to the energy can be accounted for by the above three processes.

II

The One-dimensional
Heisenberg Magnet

Introduction

The Heisenberg model, providing a non-trivial example of an interacting many-body system, was introduced in [13]. We shall consider a chain of interacting atoms of spin 1/2. The spin operator is $S = \sigma/2$ with components proportional to the Pauli matrices

$$\sigma_x = \begin{pmatrix} 0 & 1 \\ 1 & 0 \end{pmatrix}; \; \sigma_y = \begin{pmatrix} 0 & -i \\ i & 0 \end{pmatrix}; \; \sigma_z = \begin{pmatrix} 1 & 0 \\ 0 & -1 \end{pmatrix} \qquad (0.1)$$

with the usual commutation relations $[\sigma_p, \sigma_q] = 2i\epsilon_{pqr}\sigma_r$ (p, q, $r = x$, y, z and ϵ_{pqr} is the completely antisymmetric tensor, $\epsilon_{xyz} = 1$). One introduces the local spin operator $\sigma^{(m)}$ associated with the m-th lattice site. These operators commute at different sites, so that the commutation relations are

$$[\sigma_p^{(m)}, \sigma_q^{(n)}] = 2i\epsilon_{pqr}\delta_{mn}\sigma_r^{(m)}. \qquad (0.2)$$

Let us consider a Hamiltonian with nearest-neighbor interactions on a lattice with M sites:

$$H = -\sum_{m=1}^{M} \left\{ J_x\sigma_x^{(m)}\sigma_x^{(m+1)} + J_y\sigma_y^{(m)}\sigma_y^{(m+1)} + J_z\sigma_z^{(m)}\sigma_z^{(m+1)} \right\}. \qquad (0.3)$$

Periodic boundary conditions are imposed, $\sigma^{(M+1)} = \sigma^{(1)}$.

When all three real constants J_x, J_y, J_z are different the model is named the XYZ Heisenberg model. The important special cases, $J_x = J_y \neq J_z$ and $J_x = J_y = J_z$, are called the XXZ and XXX models, respectively.

The Heisenberg model was first solved by means of the Bethe Ansatz. The eigenfunctions of the isotropic model were constructed by Hans Bethe

[3] in 1931. The explicit form of these eigenfunctions is called the Bethe Ansatz.

It is of special interest to investigate the interaction of the Heisenberg magnet with a magnetic field. The model in an external magnetic field can be solved in the case of partial isotropy (the XXZ magnet) when the magnetic field is directed along the z-axis. Therefore, the results obtained by means of the Bethe Ansatz for the XXZ model in an external field will be considered further. Since the method of solution of this model is analogous to that of the one-dimensional Bose gas, the presentation will be shorter in this chapter than in the previous one. Let us note that the Bethe equations for the XXZ magnet are quite similar to those for the Thirring model, which will be studied in the next chapter. It is in that chapter that the role of bound states will be considered.

The eigenfunctions of the XXZ Heisenberg Hamiltonian in an external field are constructed in section 1, where the Bethe equations are derived. The bound states are discussed in section 2. The ground state of the model is also constructed (the Dirac sea must be filled within the Bethe Ansatz). A number of physical characteristics such as magnetic suscepti- bility and Fermi velocity are calculated in section 3 along with the model's reaction to a magnetic field.

The isotropic case of the Heisenberg magnet is considered in section 4. In section 5 it is shown that fractional charge (dressed charge) defines the finite size corrections. In section 6, the fractional charge is investigated in detail as a function of the magnetic field (magnetization of the ground state). This will help to calculate correlation functions in Chapter XVIII.

II.1 The Bethe equations

The Hamiltonian H of the Heisenberg magnet in an external magnetic field directed along the z-axis is given by

$$H = H_0 - 2hS_z \tag{1.1}$$

$$H_0 = -\sum_{m=1}^{M} \left\{ \sigma_x^{(m)} \sigma_x^{(m+1)} + \sigma_y^{(m)} \sigma_y^{(m+1)} + \Delta \left(\sigma_z^{(m)} \sigma_z^{(m+1)} - 1 \right) \right\}. \tag{1.2}$$

We shall consider the periodic chain ($\sigma^{(M+1)} = \sigma^{(1)}$) with an even num- ber of sites M. The operator of the third component of total spin is

$$S_z = \frac{1}{2} \sum_{m=1}^{M} \sigma_z^{(m)}. \tag{1.3}$$

The change in the sign of H_0 leads to a change in the sign of the constant

Δ due to the existence of the similarity transformation

$$-H_0(-\Delta) = \mathcal{U}H_0(\Delta)\mathcal{U}^{-1},$$

with $\mathcal{U} = \prod_{m=1}^{M/2} \sigma_z^{(2m)}$. This changes the boundary conditions. Let us also note that it is sufficient to consider the case $h > 0$ since the change of the field sign corresponds to the similarity transformation $H \to \mathcal{V}H\mathcal{V}^{-1}$, $\mathcal{V} = \prod_{m=1}^{M} \sigma_x^{(m)}$. The anisotropy parameter Δ $(-\infty < \Delta < \infty)$ defines the physical nature of the model. When $\Delta \geq 1$, the ground state of the Hamiltonian (state of minimum energy) in zero magnetic field is ferromagnetic ($\Delta = 1$ corresponds to the isotropic ferromagnet). When $\Delta < 1$, the magnetization of the ground state is equal to zero (for zero magnetic field); $\Delta = -1$ corresponds to the isotropic antiferromagnet.

We shall focus our attention mainly on the domain $-1 \leq \Delta < 1$. It is this domain which is of interest from the point of view of the investigation of the correlation functions since the Hamiltonian spectrum is gapless; thus there is a power law decrease of correlations at zero temperature. In this domain, it is convenient to introduce the coupling constant η instead of Δ:

$$\Delta = \cos 2\eta; \qquad -1 \leq \Delta < 1; \qquad \pi \geq 2\eta > 0. \tag{1.4}$$

Let us now construct eigenfunctions of the Hamiltonian by means of the Bethe Ansatz. The eigenfunctions are common for both H_0 (1.2) and S_z (1.3) which is possible since H_0 commutes with S_z.

Define the basis vectors $|\uparrow\rangle_m$, $|\downarrow\rangle_m$ on the m-th lattice site by

$$\sigma_z^{(m)}|\uparrow\rangle_m = |\uparrow\rangle_m; \qquad \sigma_z^{(m)}|\downarrow\rangle_m = -|\downarrow\rangle_m$$
$$_m\langle\uparrow|\uparrow\rangle_n = {}_m\langle\downarrow|\downarrow\rangle_n = \delta_{mn}; \qquad _n\langle\uparrow|\downarrow\rangle_n = {}_m\langle\downarrow|\uparrow\rangle_m = 0. \tag{1.5}$$

The ferromagnetic eigenfunction with all spins up is obvious:

$$|0\rangle = \bigotimes_{m=1}^{M} |\uparrow\rangle_m; \qquad H|0\rangle = -hM|0\rangle; \qquad S_z|0\rangle = \frac{M}{2}|0\rangle. \tag{1.6}$$

There is another ferromagnetic state with all spins down:

$$|0'\rangle = \bigotimes_{m=1}^{M} |\downarrow\rangle_m; \qquad S_z|0'\rangle = -\frac{M}{2}|0'\rangle. \tag{1.7}$$

The next step is to define the spin raising and lowering operators $\sigma_\pm$:

$$\sigma_\pm^{(m)} = \frac{1}{2}\left(\sigma_x^{(m)} \pm i\sigma_y^{(m)}\right); \qquad [\sigma_z^{(m)}, \sigma_\pm^{(n)}] = \pm 2\delta_{mn}\sigma_\pm^{(n)};$$
$$[\sigma_+^{(m)}, \sigma_+^{(n)}] = [\sigma_-^{(m)}, \sigma_-^{(n)}] = 0; \qquad \left(\sigma_+^{(m)}\right)^2 = \left(\sigma_-^{(m)}\right)^2 = 0. \tag{1.8}$$

Acting with $\sigma_-^{(m)}$ on the ferromagnetic state, we obtain the eigenfunctions of S_z:

$$S_z \left(\prod_{j=1}^{N} \sigma_-^{(m_j)} \right) |0\rangle = \left\{ \frac{M}{2} - N \right\} |0\rangle, \tag{1.9}$$

$$1 \le m_j \le M; \quad m_j \neq m_k \quad (j \neq k) \quad N \le M.$$

We shall parametrize the eigenfunctions $|\psi_N\rangle$ of the Hamiltonian by the "spectral parameters" (generally complex-valued), λ_α ($\alpha = 1, \ldots, N$):

$$H|\psi_N(\lambda_1, \ldots, \lambda_N)\rangle = E_N(\lambda_1, \ldots, \lambda_N)|\psi_N(\lambda_1, \ldots, \lambda_N)\rangle. \tag{1.10}$$

The Bethe Ansatz scheme for $|\psi_N\rangle$ is analogous to that of the one-dimensional Bose gas (I.1.9):

$$|\psi_N(\lambda_1, \ldots, \lambda_N)\rangle =$$
$$\frac{1}{N!} \sum_{m_1, \ldots, m_N = 1}^{M} \chi_N(m_1, \ldots, m_N | \lambda_1, \ldots, \lambda_N) \sigma_-^{(m_1)} \cdots \sigma_-^{(m_N)} |0\rangle. \tag{1.11}$$

The wave function χ_N is

$$\chi_N = \left\{ \prod_{N \ge b > a \ge 1} \epsilon(m_b - m_a) \right\} \sum_{\mathcal{Q}} (-1)^{[\mathcal{Q}]} \exp \left\{ -i \sum_{a=1}^{N} m_a p_0(\lambda_{\mathcal{Q}a}) \right\}$$
$$\times \exp \left\{ \frac{-i}{2} \sum_{N \ge b > a \ge 1} \theta(\lambda_{\mathcal{Q}b} - \lambda_{\mathcal{Q}a}) \epsilon(m_b - m_a) \right\} \tag{1.12}$$

and the summation is performed over all $N!$ permutations $\mathcal{Q} = \{q_\alpha; \alpha = 1, \ldots, N\}$ of the integers 1, 2, ..., N. Here $\epsilon(x) = x/|x|$, and the "momentum" $p_0(\lambda)$ and the two-particle scattering phase $\theta(\lambda - \mu)$ are:

$$p_0(\lambda) = i \ln \left[\cosh(\lambda - i\eta) / \cosh(\lambda + i\eta) \right]; \tag{1.13}$$

$$p_0(0) = 0; \tag{1.14}$$

$$\theta(\lambda_1 - \lambda_2) = i \ln \frac{\sinh(2i\eta + \lambda_1 - \lambda_2)}{\sinh(2i\eta - \lambda_1 + \lambda_2)}. \tag{1.15}$$

with $\theta(\lambda)$ antisymmetric. When $\lambda \to \infty$

$$\theta(\lambda) \xrightarrow[\lambda \to \infty]{} \pi - 4\eta. \tag{1.16}$$

The periodic boundary conditions result in $|\psi_N\rangle$ being an eigenfunction only if the spectral parameters λ_α ($\alpha = 1, \ldots, N$) satisfy the Bethe Ansatz equations (see (I.2.2)):

$$\exp\left\{-ip_0(\lambda_\alpha)M\right\} = \exp\left\{i\sum_{\substack{\beta=1\\\beta\neq\alpha}}^{N}\theta(\lambda_\alpha - \lambda_\beta)\right\}(-1)^{N-1} \qquad (1.17)$$

or, in the logarithmic form (see (I.2.13)):

$$Mp_0(\lambda_\alpha) + \sum_{\beta=1}^{N}\theta(\lambda_\alpha - \lambda_\beta) = 2\pi n_\alpha, \quad \alpha = 1,\ldots,N \qquad (1.18)$$

where $\{n_\alpha\}$ is a set of integers (half-integers) when N is odd (even).

These equations, like in the one-dimensional Bose gas (I.2.15), are generated by the Yang-Yang action:

$$S = \sum_{\alpha=1}^{N}\left\{M\widehat{p}(\lambda_\alpha) - 2\pi n_\alpha\lambda_\alpha\right\} + \frac{1}{2}\sum_{\alpha,\beta}\widehat{\theta}(\lambda_\alpha - \lambda_\beta) \qquad (1.19)$$

where

$$\widehat{p}(\lambda) = \int_0^\lambda p_0(\mu)\,d\mu; \quad \widehat{\theta}(\lambda) = \int_0^\lambda \theta(\mu)\,d\mu. \qquad (1.20)$$

To study the convexity of the Yang-Yang action for real λ, let us calculate the following quadratic form:

$$\sum_{\alpha,\beta}v_\alpha v_\beta \frac{\partial^2 S}{\partial\lambda_\alpha\partial\lambda_\beta} = M\sum_\alpha p_0'(\lambda_\alpha)v_\alpha^2 + \sum_{\alpha>\beta}K(\lambda_\alpha,\lambda_b)(v_\alpha - v_\beta)^2 \qquad (1.21)$$

where v is an arbitrary real vector and

$$K(\lambda,\mu) = \theta'(\lambda - \mu) = \frac{\sin 4\eta}{\sinh(\lambda - \mu + 2i\eta)\sinh(\lambda - \mu - 2i\eta)} \qquad (1.22)$$

$$p_0'(\lambda) = \frac{\sin 2\eta}{\cosh(\lambda - i\eta)\cosh(\lambda + i\eta)}. \qquad (1.23)$$

Remember that the domain $0 < 2\eta \le \pi$ is considered. From the formulæ cited above, one concludes that the Yang-Yang action is convex for real λ only if

$$0 < 2\eta \le \frac{\pi}{2}. \qquad (1.24)$$

It is in this domain where the theorem on the existence and uniqueness of the solutions of the Bethe equations is valid and the Pauli principle (see VII.4) may be proved.

Formulæ (1.10)–(1.17) present the Bethe Ansatz for the XXZ model. They are correct both for $-1 \le \Delta \le 1$ and for $\Delta > 1$, $\Delta < -1$ (in the latter case η (1.4) is purely imaginary).

The eigenvalue of energy E_N (1.10) is

$$E_N = \sum_{\alpha=1}^{N} \varepsilon_0(\lambda_\alpha) - hM \qquad (1.25)$$

where the one-particle dispersion $\varepsilon_0(\lambda)$ is a function of the momentum (1.13):

$$\varepsilon_0(\lambda) = -2\sin 2\eta \frac{dp_0(\lambda)}{d\lambda} + 2h$$

$$= -2\sin^2 2\eta \left[\cosh(\lambda + i\eta)\cosh(\lambda - i\eta)\right]^{-1} + 2h. \qquad (1.26)$$

Thus the state $|\psi_N\rangle$ may be interpreted as containing N particles with momenta $p_0(\lambda_\alpha)$ $(\alpha = 1, \ldots, N)$ and energies $\varepsilon_0(\lambda_\alpha)$ over the ferromagnetic state $|0\rangle$ (1.6). These excitations may be called spin-waves.

One should also mention that the normalization of the eigenfunctions (1.11) and (1.12) is given by the determinant of the matrix $\partial^2 S / \partial \lambda_\alpha \partial \lambda_\beta$ (see (1.21)) as was the case for the Bose gas (I.2.19) [17].

II.2 The ground state

The Bethe equations define the allowed spectrum of excitations over the ferromagnetic vacuum $|0\rangle$ (1.6). The structure of the Bethe equations solutions is now more complicated than in the Bose gas case since complex-valued solutions λ_α can also be present. Since the momentum p_0 (1.13) and scattering phase θ (1.15) are periodic functions with period $i\pi$, we identify points λ and $\lambda + in\pi$ on the complex plane (n integer, $-\infty < n < \infty$). The set of complex-valued solutions is self-conjugate: if $\{\lambda_\alpha; \alpha = 1, \ldots, N\}$ are solutions of (1.17), then $\{\lambda_\alpha\} = \{\lambda_\alpha^*\}$ are also solutions (where the asterisk means complex conjugation) [23].

Each λ is interpreted as the rapidity (spectral parameter) of a spin-wave. The structure of the Bethe equations in the massive Thirring model is very similar to the XXZ ones (which are under consideration). In the massive Thirring model (Chapter III) the spectral parameter λ will be called the rapidity and the word "spin-wave" will be replaced by "particle". If all the λ are real or $\text{Im}\,\lambda = \pi/2$ then the wave function (1.12) describes the scattering of spin-waves (particles). They can also form bound states which are often called strings. This means (as usual) that the wave function (1.12) will decay exponentially with respect to the difference in coordinates. Now consider the bound state of n particles, the corresponding λ_α being situated equidistantly along the imaginary axis with spacing equal to η:

$$\lambda_\alpha^{(n)} = s + \frac{i\pi}{4}(1 - v_n) + i\eta(n - 2\alpha - 1), \qquad (2.1)$$

$$\text{Im}\,s = 0, \quad v_n = \pm 1, \quad \alpha = 0, 1, \ldots, n - 1$$

where v_n is the parity of the string.

The number n of particles in the bound state is not arbitrary. It is determined by the condition that the product ([11], [19] and [20])

$$\sin[2\eta(n - m)] \sin 2\eta m$$

should have the same sign for each $m = 1, \ldots, n - 1$. In other words, all the ratios

$$\frac{\sin[2\eta(n - m)] \sin 2\eta m}{\sin[2\eta(n - 1)] \sin 2\eta} > 0, \quad m = 1, 2, \ldots, n - 1. \tag{2.2}$$

should be positive. This is explained in [19] and [20].

Now we can describe the ground state of the Hamiltonian (1.1). When $\Delta \geq 1$ and $h = 0$, the state with minimum energy is the ferromagnetic state (1.6). The state with all spins down $|0'\rangle$ (1.7) possesses the same energy, but in the physically interesting case $M \to \infty$, transitions between $|0\rangle$ and $|0'\rangle$ are absent. Hence when $\Delta \geq 1$, the ground state is the ferromagnetic one and the theory is especially simple.

The ferromagnetic state is not the ground state for $-1 \leq \Delta < 1$ and small field h (when $h = 0$, the ground state is the antiferromagnetic one). It is proved in [25]–[28] that the ground states of the XXZ Heisenberg model for $0 < 2\eta \leq \pi$ and of the Bose gas model are constructed in essentially the same way (see I.3). Integers (half-integers) n_α in the Bethe equation (1.18) fill all possible vacancies $n_{\alpha+1} - n_\alpha = 1$ in the symmetric interval $-(N - 1)/2 \leq n_\alpha \leq (N - 1)/2$.

Let us describe the ground state in the thermodynamic limit for the case $-1 \leq \Delta < 1$. Remember that this limit corresponds to $N \to \infty$, $M \to \infty$ with fixed magnetization

$$\langle \sigma_z^{(m)} \rangle = \frac{2}{M} \langle S_z \rangle = \sigma = 1 - 2D.$$

If we call each overturned spin a particle, then D is the particle density

$$D = \frac{N}{M}. \tag{2.3}$$

Consider the Bethe equations for real λ with (see (2.1)):

$$\lambda^{(1)} = \lambda = s; \quad \text{Im}\,\lambda = 0; \quad v_n = 1; \tag{2.4}$$

and define the local density in "λ-space"

$$\rho(\lambda_\alpha) = \frac{1}{M(\lambda_{\alpha+1} - \lambda_\alpha)}. \tag{2.5}$$

In the thermodynamic limit, one obtains the following integral equation for the density, analogous to (I.3.7)

$$2\pi\rho(\lambda) - \int_{-\Lambda}^{\Lambda} K(\lambda, \mu)\rho(\mu)\, d\mu = \frac{dp_0(\lambda)}{d\lambda};$$

$$D = \int_{-\Lambda}^{\Lambda} \rho(\lambda)\, d\lambda = \frac{N}{M}.$$

(2.6)

$p_0(\lambda)$ is the momentum (1.13). The kernel $K(\lambda, \mu)$ is defined through the two-particle phase $\theta(\lambda - \mu)$ (1.15) as

$$K(\lambda, \mu) = \frac{\partial\theta(\lambda - \mu)}{\partial\lambda} = \frac{\sin 4\eta}{\sinh(\lambda - \mu + 2i\eta)\sinh(\lambda - \mu - 2i\eta)}. \quad (2.7)$$

The Fermi spectral parameter $\Lambda > 0$ can be defined as in the Bose gas case by the requirement that the energy of the hole should be equal to zero on the Fermi boundary (i.e., $\lambda = \pm\Lambda$):

$$\varepsilon(\pm\Lambda) = 0. \quad (2.8)$$

The excitation energy satisfies the integral equation (see (I.3.16))

$$\varepsilon(\lambda) - \frac{1}{2\pi} \int_{-\Lambda}^{\Lambda} K(\lambda, \mu)\varepsilon(\mu)\, d\mu = \varepsilon_0(\lambda),$$

(2.9)

$$\varepsilon_0(\lambda) = 2h - 2(\sin^2 2\eta)\left[\cosh(\lambda + i\eta)\cosh(\lambda - i\eta)\right]^{-1},$$

where $\varepsilon_0(\lambda)$ is the one-particle dispersion (1.26); $\varepsilon(\lambda) = \varepsilon(-\lambda)$. It can be shown that when $-1 \leq \Delta < 1$ this equation has a unique solution.

Equation (2.8) has a solution if

$$0 < h < h_c; \quad h_c = 4\sin^2\eta = 2(1 - \Delta) \quad (2.10)$$

when $h \to h_c$, $\Lambda \to 0$. For $h > h_c$ the ground state is ferromagnetic. The Fermi boundary tends to infinity when the external field goes to zero:

$$\Lambda \to \infty, \quad h \to 0 \quad (-1 \leq \Delta < 1). \quad (2.11)$$

The integral equations are easily solved when $h = 0$ by Fourier transforms; for instance

$$\rho(\lambda) = \left\{2(\pi - 2\eta)\cosh\left[\frac{\pi\lambda}{(\pi - 2\eta)}\right]\right\}^{-1} \quad \text{at } h = 0. \quad (2.12)$$

Thus we have constructed the ground state in the thermodynamic limit. Note that the magnetic field in the XXZ model is analogous to the chemical potential of the Bose gas.

II.3 Interaction with a magnetic field

We shall consider only the domain $-1 \leq \Delta < 1$. The density of the ground state energy is $\epsilon = E/M$, where E is the ground state energy and M is the number of lattice sites. From (1.25) and (2.5) it follows that

$$\epsilon = -h + \int_{-\Lambda}^{\Lambda} \varepsilon_0(\lambda)\rho(\lambda)\, d\lambda \qquad (3.1)$$

where the one-particle dispersion ε_0 is defined in (1.26) and the density $\rho(\lambda)$ satisfies equation (2.6). On the other hand, using (2.6) and (2.9) the expression (3.1) can be rewritten in terms of the dressed energy $\varepsilon(\lambda)$:

$$\epsilon = -h + \frac{1}{2\pi} \int_{-\Lambda}^{\Lambda} \varepsilon(\lambda)p_0'(\lambda)\, d\lambda \qquad (3.2)$$

where $p_0(\lambda)$ is the bare momentum (1.13).

The important characteristics are the magnetization σ and the magnetic susceptibility χ:

$$\sigma = \langle \sigma_z^{(n)} \rangle = \langle \sigma_z^{(1)} \rangle; \quad \chi = \frac{\partial \sigma}{\partial h}. \qquad (3.3)$$

The mean value here is taken with respect to the ground state of the magnet $|\Omega\rangle$. For an arbitrary operator A, $\langle A \rangle = \langle \Omega|A|\Omega\rangle/\langle \Omega|\Omega\rangle$. The magnetization is connected with the density D (2.3) and the density of the ground state energy ϵ (3.1) through

$$\sigma = 1 - 2D = -\frac{\partial \epsilon}{\partial h}. \qquad (3.4)$$

Remember that D is expressed through $\rho(\lambda)$ (2.5) as

$$0 \leq D = \int_{-\Lambda}^{\Lambda} \rho(\lambda)\, d\lambda \leq \frac{1}{2}. \qquad (3.5)$$

Moreover, $D = 0$ if $h \geq h_c$ (2.10), hence $\sigma = 1$, and $D = 1/2$ only if $h = 0$, hence $\sigma = 0$. With the Magnetic Field approaching the critical value from below, it is easy to verify that

$$\sigma = 1 - \frac{2}{\pi}\sqrt{h_c - h} \qquad (h \to h_c; \quad h \leq h_c). \qquad (3.6)$$

For a weak magnetic field

$$\sigma = \chi_0 h \qquad (h \to 0) \qquad (3.7)$$

where χ_0 is the susceptibility in zero field:

$$\chi_0 = \frac{\pi - 2\eta}{2\pi\eta\sin 2\eta}.\tag{3.8}$$

Let us now discuss the dependence of the Fermi spectral parameter Λ on the magnetic field. This can be defined by (2.8) and (2.9). As discussed in section 2, $\Lambda \to 0$ as $h \to h_c$, so

$$\Lambda = (2\tan\eta)^{-1}\sqrt{h_c - h} \to 0 \qquad (h \to h_c; \quad h \le h_c).\tag{3.9}$$

In a weak magnetic field, one gets the following by applying the Wiener–Hopf method:

$$\Lambda = \left\{\frac{\pi - 2\eta}{\pi}\right\}\ln\left(\frac{h_0}{h}\right) \qquad (h \to 0)\tag{3.10}$$

where

$$h_0 = \frac{8\eta\sqrt{\pi}\sin 2\eta}{(\pi - 2\eta)}\Gamma\left(\frac{3\pi - 4\eta}{2(\pi - 2\eta)}\right)$$

$$\times \Gamma^{-1}\left(\frac{\pi - \eta}{\pi - 2\eta}\right)\exp\left\{\frac{\pi\beta}{2(\pi - 2\eta)}\right\}\tag{3.11}$$

$$\beta = \frac{\pi - 2\eta}{\pi}\ln\left(\frac{\pi - 2\eta}{\pi}\right) + \frac{2\eta}{\pi}\ln\left(\frac{2\eta}{\pi}\right).\tag{3.12}$$

The Wiener-Hopf method is explained in detail in paper [26].

The Fermi velocity v_F defined as

$$v_F = \frac{\partial\varepsilon(\lambda)}{\partial k(\lambda)}\bigg|_{\lambda=\Lambda}\tag{3.13}$$

is of importance. Here $\varepsilon(\lambda)$ is the dressed energy (2.9) and $k(\lambda)$ is the dressed momentum of the excitation:

$$k(\lambda) = p_0(\lambda) - \int_{-\Lambda}^{\Lambda} p_0'(\mu)F(\lambda|\mu)\,d\mu$$

$$= p_0(\lambda) + \int_{-\Lambda}^{\Lambda} \theta(\lambda - \mu)\rho(\mu)\,d\mu\tag{3.14}$$

with bare momentum $p_0(\lambda)$ from (1.13). The shift function $F(\mu|\lambda)$ satisfies an equation similar to (2.6) with $\theta/2\pi$ (1.15) on the right hand side:

$$F(\lambda|\mu) - \frac{1}{2\pi}\int_{-\Lambda}^{\Lambda} K(\lambda, \nu)F(\nu|\mu)\,d\nu = \frac{1}{2\pi}\theta(\lambda - \mu).\tag{3.15}$$

As for the Bose gas (I.9.17):

$$k'(\lambda) = 2\pi\rho(\lambda). \tag{3.16}$$

Another representation for the Fermi velocity (3.13) is

$$v_F = -\frac{2\sin 2\eta}{h}\left(\frac{\partial \Lambda}{\partial h}\right)^{-1}. \tag{3.17}$$

To derive this relation, one uses (see (2.8))

$$\left.\frac{\partial \varepsilon(\lambda)}{\partial \lambda}\right|_{\lambda=\Lambda} = -\left.\frac{\partial \varepsilon(\lambda)}{\partial \Lambda}\right|_{\lambda=\Lambda} \tag{3.18}$$

and the following connection between equations (2.6) and (2.9):

$$\varepsilon(\lambda) = h\frac{\partial \varepsilon(\lambda)}{\partial h} - 4\pi\sin(2\eta)\rho(\lambda). \tag{3.19}$$

It is now easy to calculate the Fermi velocity for fields near the critical field $h \to h_c$, $h \leq h_c$:

$$v_F = 16\sin^2\eta\frac{\sqrt{h_c - h}}{h} \tag{3.20}$$

and for weak fields ($h \to 0$):

$$v_F = \frac{2\pi\sin 2\eta}{\pi - 2\eta}. \tag{3.21}$$

II.4 The *XXX* magnet

Putting $\Delta = \pm 1$ in H (1.1), one comes to the interesting particular case of the isotropic Heisenberg magnet. When $\Delta = 1$ the model describes the ferromagnet, while $\Delta = -1$ ($\eta = \pi/2$ (1.4)) corresponds to the isotropic antiferromagnet. Eigenfunctions for the model are constructed below by means of the algebraic Bethe Ansatz, see (VII.1.10) and (VII.3.13).

The bare momentum (1.13) and energy (1.26) must be changed in (2.6) and (2.9) to

$$p_0(\lambda) = i\ln\left(\frac{\frac{i}{2} + \lambda}{\frac{i}{2} - \lambda}\right) + \pi \tag{4.1}$$

$$\theta(\lambda_1 - \lambda_2) = i\ln\left(\frac{i - \lambda_1 + \lambda_2}{i + \lambda_1 - \lambda_2}\right) \tag{4.2}$$

$$\varepsilon_0(\lambda) = -\frac{2}{\lambda^2 + \frac{1}{4}} + 2h. \tag{4.3}$$

The kernel $K(\lambda, \mu)$ (now a rational function) and the scattering phase $\theta(\lambda, \mu)$ are

$$K(\lambda, \mu) = -\frac{2}{(\lambda - \mu)^2 + 1} \qquad \theta(\lambda, \mu) = i\ln\left(\frac{i - \lambda + \mu}{i + \lambda - \mu}\right). \tag{4.4}$$

The ground state in the thermodynamic limit is the Fermi sphere filled with the following states: $-\Lambda \leq \lambda \leq \Lambda$. The Fermi boundary Λ is defined by (2.8) ($\varepsilon(\pm\Lambda) = 0$). The critical value of the magnetic field obtained from (4.1) and (2.10) is

$$h_c = 4. \tag{4.5}$$

Instead of (3.10), one obtains for Λ in a weak magnetic field (due to the rationality of the kernel (4.4)):

$$\Lambda = \frac{1}{2\pi} \ln \left[\frac{(2\pi)^3}{eh^2} \right] \tag{4.6}$$

where e is the base of natural logarithms. The magnetic susceptibility and Fermi velocity (at zero magnetic field) are given by

$$\chi = \frac{1}{\pi^2} \quad \text{and} \quad v_F = 2\pi. \tag{4.7}$$

The Heisenberg magnet is a well investigated model. The asymptotic completeness of the Bethe wave functions is established in [1], [3], [10] and [21], and the spins of the physical excitations over the antiferromagnetic ground state are calculated in [9] and [10].

II.5 Finite size corrections

Evaluation of finite size corrections in the Heisenberg model is performed as in the Bose gas case (see section 9 of Chapter I). The formal difference between the models depends on certain commensuration conditions of the lattice models [24]. In Chapter XVIII we shall evaluate long distance asymptotics of the correlation functions using finite size corrections and ideas of conformal field theory. Thus, finite size corrections to the ground state energy (3.1) are equal to

$$E = M\epsilon - \frac{\pi}{6M} v + \text{h.o.c.} \tag{5.1}$$

where M is the size of our periodic lattice, ϵ is the ground state energy density (3.1), and v_F is the Fermi velocity (3.13). v_F can also be represented in the form

$$v_F = \frac{1}{2\pi\rho(\lambda)} \frac{\partial \varepsilon(\lambda)}{\partial \lambda} \bigg|_{\lambda=\Lambda} \tag{5.2}$$

where $\varepsilon(\lambda)$ is the dressed energy (2.9), $\rho(\lambda)$ is the distribution in the ground state (2.6), and Λ is the Fermi boundary. In the infinite lattice there is no gap in the spectrum of the Hamiltonian but in a finite (however large) lattice, a gap arises. The expression for the gap is similar to that for the Bose gas (I.9.18):

$$\delta E = \frac{2\pi v}{M} \left[\left(\frac{\Delta N}{2\mathcal{Z}} \right)^2 + (\mathcal{Z}d)^2 + N^+ + N^- \right] + \text{h.o.c.} \qquad (5.3)$$

The change in momentum is also similar to that for the Bose gas:

$$\delta P = 2k_F d + \frac{2\pi}{M} \left(N^+ - N^- + d\,\Delta N \right) + \text{h.o.c.} \qquad (5.4)$$

Let us now explain the nature of these corrections.

We need to consider the Bethe equation (1.18) for the ground state $(n_{\alpha+1} - n_\alpha = 1)$. The three following processes are essential for the classification of the low lying excitations (these are the same as for the Bose gas):

(1) The integer n_α corresponding to the particle on the Fermi surface Λ (maximal λ) can be changed by a finite amount N^+. A similar change on the other Fermi surface $-\Lambda$ is denoted by N^-.

(2) One can change the number of particles (spin down sites) by ΔN.

(3) Backscattering processes. Some particles (let us say d) can jump over the sea (from one Fermi surface to another). The Fermi momentum k_F is equal to

$$k_F = \pi D = \pi \frac{N}{M}. \qquad (5.5)$$

As with the Bose gas, $\mathcal{Z}$ in (5.3) is the value of some function $Z(\lambda)$ (which we call the dressed charge) on the Fermi boundary. The function $Z(\lambda)$ is defined by the following integral equation:

$$Z(\lambda) - \frac{1}{2\pi} \int_{-\Lambda}^{\Lambda} K(\lambda, \mu) Z(\mu)\, d\mu = 1. \qquad (5.6)$$

Thus the quantities entering into formula (5.3) are explained. The derivation of formula (5.3) is similar to that for the Bose gas (see Appendix 4 of Chapter I). Equation (5.6) for the dressed charge is studied in detail in the next section.

Later we shall see that the whole spectrum of conformal dimensions (which define the long distance asymptotics of the correlation functions) can be expressed in terms of $\mathcal{Z}$.

II.6 Fractional charge

We have already introduced the density of vacuum particles and the dressed energy of one-particle excitations. These quantities are defined by the "dressing equations" (2.6) and (2.9) respectively. It is remarkable that these equations, as well as (3.15), are of the same type (with different right hand sides only). The "fractional charge" $Z(\lambda)$ (see (5.6)) is defined by the simplest "dressing equation", with the right hand side equal to 1:

$$Z(\lambda) - \frac{1}{2\pi} \int\limits_{-\Lambda}^{\Lambda} K(\lambda, \mu) Z(\mu) \, d\mu = 1. \tag{6.1}$$

The origin of the name "fractional charge" will become clear from the consideration of the charge of the elementary excitations in the massive Thirring model (section III.3). The function $Z(\lambda)$ appears also in the expression for finite size corrections.

We now turn to the physical meaning of the "fractional charge" in the Heisenberg model. Differentiating (2.9) with respect to h and recalling (2.8) we find that

$$Z(\lambda) = \frac{1}{2} \frac{\partial \varepsilon(\lambda)}{\partial h}. \tag{6.2}$$

It then follows from (3.4) and (3.2) that the magnetization σ is equal to

$$\sigma = 1 - \frac{1}{\pi} \int\limits_{-\Lambda}^{\Lambda} Z(\lambda) p_0'(\lambda) \, d\lambda \tag{6.3}$$

and hence $Z(\lambda)$ is the intrinsic magnetic moment of the elementary excitations. Using F (3.15) we can write $Z(\lambda)$ as

$$Z(\lambda) = 1 + \int\limits_{-\Lambda}^{\Lambda} \dot{F}(\mu|\lambda) \, d\mu$$

$$\dot{F}(\mu|\lambda) = -\frac{\partial F(\mu|\lambda)}{\partial \lambda}. \tag{6.4}$$

However, this is not the whole story. As the reader may have already noticed, the values of the microscopic quantities on the Fermi boundary are of importance (see, for instance, the expression for the Fermi velocity (3.13)). These values can be related to the macroscopic characteristics of the model; thus, the dressed momentum (3.14) is a function of the density D (2.3) and hence of the magnetization σ (3.3):

$$k(\Lambda) = \pi D = \frac{\pi}{2}(1 - \sigma); \quad k'(\Lambda) = 2\pi\rho(\Lambda). \tag{6.5}$$

Define now

$$\theta = 2Z^2(\Lambda) \tag{6.6}$$

which is the critical exponent since this function defines the power law decrease of the long distance asymptotics of the correlation functions at zero temperature (see Chapter XVIII). We shall now show that for the

Heisenberg magnet, θ is a simple function of the susceptibility χ (3.3) and of the Fermi velocity v_F (3.13), namely

$$\theta = \frac{\pi}{2} v_F \chi. \tag{6.7}$$

Differentiating (2.6) with respect to Λ and comparing the result with (3.15), one obtains

$$\frac{\partial \rho(\lambda)}{\partial \Lambda} = \rho(\Lambda) \left[\dot{F}(\Lambda|\lambda) + \dot{F}(-\Lambda|\lambda) \right]. \tag{6.8}$$

Hence, for D ((2.3), (3.5)) one has from (6.4)

$$\frac{\partial D}{\partial \Lambda} = 2\rho(\Lambda) Z(\Lambda). \tag{6.9}$$

As in the Bose gas case, it is easy to calculate the derivative of Λ with respect to the external field h. Equation (2.9) and conditions (2.8) and (6.2) result in:

$$\frac{\partial \Lambda}{\partial h} = -\frac{2Z(\Lambda)}{\varepsilon'(\Lambda)}. \tag{6.10}$$

Since $k'(\Lambda) = 2\pi\rho(\Lambda)$ (6.5) the Fermi velocity (3.13) is:

$$v_F = \frac{\varepsilon'(\Lambda)}{k'(\Lambda)} = \frac{\varepsilon'(\Lambda)}{2\pi\rho(\Lambda)}. \tag{6.11}$$

The combination of (6.9)–(6.11) results in

$$\theta = 2Z^2(\Lambda) = -\pi v_F \frac{\partial D}{\partial h}. \tag{6.12}$$

As $\partial D/\partial h = -(1/2)\, \partial\sigma/\partial h = -\chi/2$ (see (3.4)), one obtains (6.7). Another representation is also valid:

$$\theta = \frac{\partial}{\partial \Lambda} \int_{-\Lambda}^{\Lambda} Z(\lambda)\, d\lambda. \tag{6.13}$$

Let us now discuss the dependence of the critical exponent θ on the magnetic field. Explicit analytical answers may be obtained only in two limiting cases: for a weak magnetic field by means of the Wiener-Hopf method (see paper [26]) and for a field close to the critical value h_c by perturbation theory. As $h \to h_c$, $h \le h_c$

$$\theta = 2 + 4 \left(\pi \tan \eta \tan 2\eta \right)^{-1} \sqrt{h_c - h} \tag{6.14}$$

(one makes use of (3.19)).

Solving equation (6.1) by the Wiener-Hopf method, we obtain for the weak field [4]:

$$\theta = (\pi/2\eta) \left[1 + \alpha_1 h^2 \right], \quad h \to 0, \quad 0 < 2\eta < \frac{2\pi}{3} \tag{6.15}$$

$$\theta = (\pi/2\eta)\left[1 + \alpha_2 h^{(2\pi/\eta)-4}\right], \quad h \to 0, \quad \frac{2\pi}{3} < 2\eta < \pi. \quad (6.16)$$

The behavior of the critical exponent in h appears to depend on the coupling constant η since it is essentially different below and above the point $\eta = 2\pi/3$. The coefficients α_1 and α_2 are functions of η:

$$\alpha_1 = \left(\frac{\pi - 2\eta}{\pi}\right)^2 \frac{\cot[\pi\eta/(\pi - 2\eta)]}{8\eta \sin^2 2\eta}$$

$$\alpha_2 = \frac{4\eta}{\pi} e^{\pi\beta/\eta} \tan\left(\frac{\pi^2}{2\eta}\right) \frac{\Gamma^2\left(1 + \frac{\pi}{2\eta}\right)}{\Gamma^2\left(\frac{1}{2} + \frac{\pi}{2\eta}\right)} h_0^{-(2\pi/\eta)+4}$$

where h_0 is defined by (3.11) and β by (3.12).

For the XXX model in the case of a weak magnetic field, one obtains from (4.6)

$$\theta = 1 + [2\ln(h_0/h)]^{-1}; \quad h_0 = \left(\frac{8\pi^3}{e}\right)^{1/2} \quad (6.17)$$

and in the case $h \to h_c$, $h \le h_c$, one gets

$$\theta = 2\left(1 - \frac{1}{\pi}\sqrt{h_c - h}\right). \quad (6.18)$$

The fractional charge at $h = 0$ ($\Lambda = \infty$) is easily calculated: $Z(\lambda) = Z(0) = \pi/4\eta$. This is the fractional charge of the elementary excitations in the massive Thirring model, obtained in Chapter III. However, at large but finite Λ, $Z(\lambda) \approx \pi/4\eta$ only if $|\lambda| \ll \Lambda$. Using the Wiener-Hopf method, one obtains the following relation:

$$\lim_{\Lambda \to \infty} Z(\Lambda) = \sqrt{\lim_{\Lambda \to \infty} Z(0)} = \sqrt{\frac{\pi}{4\eta}}. \quad (6.19)$$

It follows from (6.15), (6.16) and (6.19) that for zero magnetic field

$$\theta = \frac{\pi}{2\eta} \quad (6.20)$$

for the XXZ model, and

$$\theta = 1 \quad (6.21)$$

for the XXX model.

Conclusion

This chapter was intended as an introduction to the Heisenberg magnet. We hope that we have succeeded in giving the reader an idea of the model's perfection. The results of this chapter will be applied to the calculation of the asymptotics of correlation functions in Chapter XVIII. We shall return to the Heisenberg chain in Part II with the quantum inverse scattering method, which offers new possibilities for the investigation of this model.

Due to shortage of space, a number of results on the theory of the Heisenberg magnet were not mentioned in this chapter. They may be found in [2], [7], [9], [10], [12], [17], [22] and in references therein. The Heisenberg chain has been extensively investigated after Bethe's solution of its isotropic variant [3]. One should mention the work of Hulthén [14] in which the ground state of the model in the thermodynamic limit was constructed. The Bethe Ansatz for the XXZ model was constructed by Yang and Yang [25]–[28] and for the XYZ model by Baxter [2]. The model's thermodynamics were investigated in the papers of Takhahashi ([19], [20] and [21]) the bound states playing an important role in this analysis. The papers in which the completeness of the eigenfunction of the XXX model is investigated should be mentioned; see [1], [10], [15], and [16]. The Heisenberg chain has been analyzed not only with the help of the exact solution but also by numerical methods [6]. Finite size corrections were studied in [5], [8], and [18].

III

The Massive Thirring Model

Introduction

The Thirring model, introduced in [25], was the first model of a relativistic quantum field theory solved by means of the Bethe Ansatz. The theory of the massive Thirring model is presented in this chapter. The Bethe Ansatz is constructed in section 1. The ground state of the Hamiltonian (the physical vacuum) as a Dirac sea is constructed in section 2, the ultraviolet cutoff being introduced and mass renormalization being discussed. Excitations above the physical vacuum are described in section 3, and the energy, momentum and scattering matrix are calculated. The fractional charge (dressed excitation charge) is introduced. The "repulsion" of vacuum particles beyond the cutoff is the phenomenon due to which the dressed charge is different from the bare (integer) one.

It should be mentioned that the massive Thirring model in the zero charge sector is equivalent to the sine-Gordon model (see sections V.3, VI.4, VII.3, VIII.2 and VIII.5).

III.1 The Bethe Ansatz

The massive Thirring model is a relativistic quantum field theory model in $1 + 1$ space-time dimensions. It is defined by a two-component quantum fermion field $\Psi_k(x, t)$ with the following equal-time anticommutation relations:

$$
\begin{aligned}
\{\Psi_k(x),\ \Psi_m^\dagger(y)\} &\equiv \Psi_k(x)\Psi_m^\dagger(y) + \Psi_m^\dagger(y)\Psi_k(x) = \delta_{km}\delta(x - y); \\
\{\Psi_k(x),\ \Psi_m(y)\} &= \{\Psi_k^\dagger(x),\ \Psi_m^\dagger(y)\} = 0; \quad k, m = 1, 2
\end{aligned} \tag{1.1}
$$

where the dagger denotes Hermitian conjugation. The Lagrangian of the model is

$$\mathcal{L} = \int \left[i\bar{\Psi}\gamma^\mu \partial_\mu \Psi - m_0\bar{\Psi}\Psi - \frac{1}{2}g : (\bar{\Psi}\gamma_\mu \Psi)(\bar{\Psi}\gamma^\mu \Psi) : \right] dx \qquad (1.2)$$

where

$$\gamma_0 \equiv \sigma_x, \quad \gamma_1 \equiv i\sigma_y \qquad (1.3)$$

and

$$\sigma_x = \begin{pmatrix} 0 & 1 \\ 1 & 0 \end{pmatrix}, \quad \sigma_y = \begin{pmatrix} 0 & -i \\ i & 0 \end{pmatrix}, \quad \sigma_z = \begin{pmatrix} 1 & 0 \\ 0 & -1 \end{pmatrix} \qquad (1.4)$$

are the Pauli matrices. Normal ordering means that Ψ stands to the right and $\Psi^\dagger$ to the left. The Dirac conjugation is defined as usual: $\bar{\Psi} = \Psi^\dagger \gamma^0$. The positive constant m_0 is the fermion bare mass and g is the bare coupling constant. The Hamiltonian H, the momentum operator P and the operator Q of the number of particles are:

$$H = \int dx \left[i\Psi^\dagger \sigma_z \partial_x \Psi + m_0\Psi^\dagger \sigma_x \Psi + 2g\Psi_1^\dagger \Psi_2^\dagger \Psi_2 \Psi_1 \right] \qquad (1.5)$$

$$P = -i \int dx \, \Psi^\dagger \partial_x \Psi; \quad Q = \int dx \, \Psi^\dagger \Psi. \qquad (1.6)$$

When constructing eigenfunctions, the Fock vacuum $|0\rangle$ and the dual vacuum $\langle 0|$ are of importance:

$$\Psi(x)|0\rangle = 0; \quad \langle 0|\Psi^\dagger(x) = 0; \quad \langle 0| = (|0\rangle)^\dagger; \quad \langle 0|0\rangle = 1. \qquad (1.7)$$

We shall construct common eigenfunctions of the operators H, P and Q in a way similar to section I.1 for the one-dimensional Bose gas:

$$|\psi_N(\beta_1, \ldots, \beta_N)\rangle =$$
$$\int d^N x \, \chi^{j_1 \cdots j_N}(x_1, \ldots, x_N | \beta_1, \ldots, \beta_N) \Psi_{j_1}^\dagger(x_1) \cdots \Psi_{j_N}^\dagger(x_N)|0\rangle.$$
$$(1.8)$$

For $|\psi_N\rangle$ to be the an eigenfunction of the Hamiltonian (1.5), it is necessary and sufficient that χ should be an eigenfunction (with the same eigenvalue) of the following N-particle quantum mechanical Hamiltonian $\mathcal{H}_N$:

$$\mathcal{H}_N = \sum_{k=1}^N \left[i\sigma_z^{(k)} \frac{\partial}{\partial x_k} + m_0\sigma_x^{(k)} \right] + 2g \sum_{k<n} \delta(x_k - x_n). \qquad (1.9)$$

In the one-particle sector ($N = 1$), the eigenfunction χ^j is the two-component spinor

$$\chi(x|\beta) = \begin{pmatrix} e^{-\beta/2} \\ e^{\beta/2} \end{pmatrix} e^{ixm_0 \sinh \beta}. \qquad (1.10)$$

The energy and momentum of this relativistic particle are

$$E = m_0 \cosh \beta, \quad P = m_0 \sinh \beta, \tag{1.11}$$

hence β is the rapidity. The N-particle wave-function is of the form:

$$\chi^{j_1 \cdots j_N}(x_1, \ldots, x_N | \beta_1, \ldots, \beta_N) = \sum_{\mathcal{P}} (-1)^{[\mathcal{P}]} \left\{ \prod_{k=1}^{N} \chi^{j_k}(x_k | \beta_{\mathcal{P}k}) \right\}$$

$$\times \prod_{k<l} \exp \left\{ \frac{i}{2} \epsilon(x_l - x_k) \Phi(\beta_{\mathcal{P}l} - \beta_{\mathcal{P}k}) \right\}. \tag{1.12}$$

Here the summation should be held with respect to all permutations $\mathcal{P}$ of the numbers $1, \ldots, N$ ($\mathcal{P} : (1, \ldots, N) \to (\mathcal{P}1, \ldots, \mathcal{P}N)$). The value $\epsilon(x) = \text{sign } x = x/|x|$ is the sign function. The two-body scattering matrix is equal to

$$S = e^{i\Phi(\beta)} = e^{-ig} \frac{e^\beta + e^{ig}}{e^\beta + e^{-ig}};$$

$$\Phi(\beta) = i \ln \frac{\cosh(\beta/2 + ig/2)}{\cosh(\beta/2 - ig/2)} = i \ln \frac{\sinh(i\gamma - \beta/2)}{\sinh(i\gamma + \beta/2)}; \tag{1.13}$$

$$\Phi(\beta) = -\Phi(-\beta); \quad \Phi(\beta) \to \mp g \quad (\beta \to \pm\infty).$$

The coupling constant γ is given by (1.16). The total energy and momentum are sums of the energies (momenta) of the individual particles

$$E_N = m_0 \sum_k \cosh \beta_k;$$

$$P_N = m_0 \sum_k \sinh \beta_k. \tag{1.14}$$

The wavefunction is an antisymmetric function of all momenta. This is the Pauli principle. An important difference between this case and the Bose gas is that the eigenfunctions are not continuous in the coordinates x_j. There exist different regularizations of the operator (1.9). This results in the two-particle scattering matrix having a different dependence on the coupling constant g (g may be replaced by some function of g). This situation is usual for quantum field theories where only the dependence on the renormalized coupling constant is essential. The regularization scheme changes the dependence of the physical quantities on the bare coupling constant. In spite of this, the dependence of these quantities on the observable (dressed) coupling constant should not change.

There are elementary particles with positive and negative energies in the model if $\text{Im } \beta = 0$ and $\text{Im } \beta = \pi$, respectively. Bound states are also present. A bound state may be formed by n particles if the products

$$\sin \gamma p \sin[\gamma(n - p)] \quad (p = 1, \ldots, n - 1) \tag{1.15}$$

are all of the same sign [17]; the same is true (II.2.2) for the XXZ model, where

$$\gamma = \frac{\pi - g}{2}. \tag{1.16}$$

The wave function describing the n-particle bound state is obtained from (1.12) in the following way. Putting $N = n$ and taking the rapidities β_k as

$$\beta_k = B + i\gamma(n - 1 - 2k), \quad k = 0, 1, \ldots, n - 1, \tag{1.17}$$

with $\operatorname{Im} B = 0$ if all expressions in (1.15) are positive and $\operatorname{Im} B = \pi$ if they are all negative, one obtains a wave function decreasing along the differences $|x_j - x_k|$. The important set of solutions of (1.15) may be presented immediately:

$$n = 1, 2, \ldots, \left[\frac{\pi}{\gamma}\right] + 1. \tag{1.18}$$

The square brackets here are the integer part of π/γ.

Let us consider the wave function (1.12) and put the system into a periodic box of length L. This leads to the Bethe equations:

$$\exp\{im_0 L \sinh \beta_l\} = \prod_{k=1}^{N} \exp\{-i\Phi(\beta_l - \beta_k)\}. \tag{1.19}$$

Since the eigenfunction (1.12) is a periodic function of g, it is sufficient to consider the region

$$-\pi < g < \pi \; ; \; 0 < \gamma < \pi. \tag{1.20}$$

III.2 The ground state

Let us consider elementary particles with negative energies and rapidities written as

$$\beta = \alpha + i\pi, \quad \operatorname{Im} \alpha = 0. \tag{2.1}$$

The logarithmic form of the Bethe equations (1.19) for the massive Thirring model are:

$$m_0 L \sinh \alpha_l = 2\pi n_l + \sum_{k=1}^{N} \Phi(\alpha_l - \alpha_k). \tag{2.2}$$

The energy and momentum are

$$E_N = -m_0 \sum_k \cosh \alpha_k, \quad P_N = -m_0 \sum_k \sinh \alpha_k. \tag{2.3}$$

As in the Bose gas case (I.2.15), equations (2.2) are generated by a Yang-Yang action. This action is convex only if

$$0 < g < \pi, \quad 0 < \gamma < \frac{\pi}{2} \tag{2.4}$$

and only this domain will be considered in further detail.

The ground state is constructed in the following way. All states with negative energies must be filled. This means that for integers n_l one has

$$n_{l+1} - n_l = 1. \tag{2.5}$$

They should also fill a symmetric interval.

Since the massive Thirring model is a quantum field theory model, it possesses ultraviolet divergences. The problem of ultraviolet regularization will be solved here in the standard way for perturbation theory. We shall take into account only the particles with rapidities α:

$$-\Lambda < \alpha < \Lambda \tag{2.6}$$

where Λ is the ultraviolet cut-off. The most rigorous regularization performed by the lattice sine-Gordon model (see Part II, Chapter VIII, sections 2 and 5) leads to the same physical answers.

The permitted values of the rapidities are condensed ($\alpha_{l+1} - \alpha_l = O(1/L)$) in the $L \to \infty$ limit. Let us introduce the distribution function

$$\rho(\alpha) = \frac{1}{L(\alpha_{l+1} - \alpha_l)}. \tag{2.7}$$

A linear integral equation for $\rho(\alpha)$ is obtained similar to (I.3.7):

$$\rho(\alpha) - \frac{1}{2\pi} \int\limits_{-\Lambda}^{\Lambda} K(\alpha, \beta)\rho(\beta) \, d\beta = \frac{m_0}{2\pi} \cosh \alpha \tag{2.8}$$

where

$$K(\alpha, \beta) = \frac{\sin 2\gamma}{2 \sinh\left(\frac{\alpha - \beta}{2} + i\gamma\right) \sinh\left(\frac{\alpha - \beta}{2} - i\gamma\right)}. \tag{2.9}$$

Thus, the ground state is the Dirac sea filled with elementary particles having negative energies.

In the limit $\Lambda \to \infty$, the bare mass m_0 must depend on Λ in such a way that the observable mass of the physical excitation remains finite:

$$m_0 = a \exp\left\{-\frac{g\Lambda}{\pi + g}\right\} \tag{2.10}$$

where a is a constant. The corresponding calculations will be given in the next section. Using (2.10) one obtains the asymptotic solution ($\Lambda \to \infty$) of (2.8)

$$\rho(\alpha) = \frac{ag}{2(\pi + g)\sin g}\cosh\left(\frac{\pi\alpha}{\pi + g}\right); \quad |\alpha| \ll \Lambda. \tag{2.11}$$

The distribution of particles in the Dirac sea appears to differ from the free distribution ($g = 0$) by a renormalization of the rapidity:

$$\alpha \to \frac{\pi}{\pi + g}\alpha. \tag{2.12}$$

We may obtain formula (2.10) from equation (2.8) directly by choosing the dependence of m_0 on Λ in such a way as to ensure the finiteness of the solution $\rho(\alpha)$ (2.8) for $|\alpha| \ll \Lambda$, $\Lambda \to \infty$.

The case $-\pi < g < 0$ was investigated in papers [4], [12] and [18]. It was shown there that the ground state in this case is the Dirac sea filled with both free particles and bound states.

III.3 Fractional charge and repulsion beyond the cutoff

Let us consider excitations over the ground state (i.e., over the physical vacuum). We begin with an excitation describing an elementary fermion, that is, with a hole in the Dirac sea. The absence of the particle with rapidity α_h in the sea leads to a shift in the rapidities of the remaining particles. As in section I.4, one introduces the shift function F:

$$F(\alpha_j|\alpha_h) = \frac{\alpha_j - \tilde{\alpha}_j}{\alpha_{j+1} - \alpha_j} \tag{3.1}$$

which satisfies the linear integral equation

$$F(\alpha|\alpha_h) - \frac{1}{2\pi}\int\limits_{-\Lambda}^{\Lambda} K(\alpha, \beta)F(\beta|\alpha_h)\,d\beta = \frac{1}{2\pi}\Phi(\alpha - \alpha_h). \tag{3.2}$$

According to the ultraviolet regularization scheme, one has to take into account only those particles with rapidities $|\alpha| < \Lambda$. In the excited state, some of the particles are repulsed beyond the cutoff. These are particles with $\tilde{\alpha}_j < -\Lambda$ and $\tilde{\alpha}_j > \Lambda$. The change in the number of particles remaining in the interval $[-\Lambda, \Lambda]$ can be calculated:

$$\Delta N = z = -1 + F(\Lambda|\alpha_h) - F(-\Lambda|\alpha_h). \tag{3.3}$$

The quantity z is the fractional (dressed) charge of the excitation. The functions F and z are finite when $\Lambda \to \infty$. The derivative of F with respect to α is of importance:

$$f(\alpha|\alpha_h) = \frac{\partial}{\partial\alpha}F(\alpha|\alpha_h). \tag{3.4}$$

At $\Lambda = \infty$, equation (3.2) can be solved by means of the Fourier transform. The Fourier transform of f is

$$\tilde{f}(\omega) = \int\limits_{-\infty}^{+\infty} e^{i\alpha\omega} f(\alpha)\, d\alpha \tag{3.5}$$

$$= -\frac{\sinh[\omega(\pi - 2\gamma)]}{2\sinh(\omega\gamma)\cosh[\omega(\pi - \gamma)]}.$$

Hence, the fractional charge is equal to

$$z = -1 + \int\limits_{-\infty}^{+\infty} f(\alpha)\, d\alpha = -1 + \tilde{f}(0) = -\frac{\pi}{2\gamma}. \tag{3.6}$$

The fractional charge can be obtained from the bare charge (which is equal to -1) by means of the dressing equation similar to (I.9.20) and (II.5.6):

$$z(\alpha) - \frac{1}{2\pi} \int\limits_{-\infty}^{+\infty} K(\alpha, \beta) z(\beta)\, d\beta = -1. \tag{3.7}$$

Let us now calculate the observable (dressed) energy and momentum of the hole-type excitation. Doing everything as explained in section I.4, one should take into account the energy (momentum) of the Dirac sea particles which are not repulsed beyond the cutoff ($-\Lambda < \tilde{\alpha}_j < \Lambda$):

$$\varepsilon_h = m_0 \cosh\alpha_h - m_0 \int\limits_{-\Lambda}^{\Lambda} \cosh\alpha\, f(\alpha|\alpha_h)\, d\alpha;$$

$$\tag{3.8}$$

$$k_h = m_0 \sinh\alpha_h - m_0 \int\limits_{-\Lambda}^{\Lambda} \sinh\alpha\, f(\alpha|\alpha_h)\, d\alpha.$$

As $\Lambda \to \infty$, the integrals at the right hand side go to infinity. To obtain a finite answer one lets the bare mass m_0 go to zero, as $\Lambda \to \infty$, in the following way (see (1.16), (2.10)):

$$m_0 = a \exp\left\{ -\frac{\pi - 2\gamma}{2(\pi - \gamma)}\Lambda \right\}. \tag{3.9}$$

In terms of the dressed mass

$$M_h = \frac{a}{\pi - 2\gamma} \cot \frac{\gamma\pi}{2(\pi - \gamma)} \tag{3.10}$$

the answers are the following:

$$\varepsilon_h = M_h \cosh \frac{\pi}{2(\pi - \gamma)}\alpha_h; \quad k_h = M_h \sinh \frac{\pi}{2(\pi - \gamma)}\alpha_h. \tag{3.11}$$

This excitation corresponds to an elementary fermion. Excitations corresponding to the bound state of the fermion and antifermion also exist. Let us put the bound state of n particles into the vacuum, where

$$n = 1, \ldots, \left[\frac{\pi}{\gamma}\right] - 1 \tag{3.12}$$

(see (1.18)). The observable energy, momentum and charge of these excitations are:

$$\varepsilon_n = M_n \cosh \frac{\pi}{2(\pi - \gamma)} \alpha_n; \quad k_n = M_n \sinh \frac{\pi}{2(\pi - \gamma)} \alpha_n$$

$$M_n = 2M_h \sin \left[\frac{\pi \gamma}{2(\pi - \gamma)} n\right]; \quad z = 0 \tag{3.13}$$

(the charge is equal to zero).

We will now discuss the fermion-antifermion scattering matrix. This scattering matrix depends on the spacial parity of the fermion-antifermion state. One should consider the state obtained by putting in the vacuum two holes with rapidities α_1 and α_2 together with the bound state with rapidity $(\alpha_1 + \alpha_2)/2$. The number of particles in this bound state is equal to

$$m = \left[\frac{\pi}{\gamma}\right] \quad \text{or} \quad \left[\frac{\pi}{\gamma}\right] + 1. \tag{3.14}$$

The spatial parity of the state thus obtained is $(-1)^m$ and the charge is zero. The dressed charges of the bound states (1.18) were calculated in [17]. It was shown there that the charge of the bound state (3.14) does indeed compensate the charge (3.6) of the two holes, and the energies and momenta of the bound states are equal to zero. The rapidity of the bound state is chosen in such a way that the wave function of the system $(\alpha_1 + \alpha_2 = 0)$ is of the needed parity. The fermion-antifermion scattering matrix is calculated using the rules given in section I.4. The corresponding calculations were made in paper [17]; the answer is Zamolodchikov's scattering matrix ([26] and [15]):

$$S^{\pm}(\theta) = U_{\pm}(\theta)S(\theta); \quad \theta = \frac{\pi}{2(\pi - \gamma)}(\alpha_1 - \alpha_2); \quad \gamma' = \frac{\pi\gamma}{\pi - \gamma};$$

$$S(\theta) = \exp\left\{-\int_0^\infty \frac{dx}{x} \frac{\sinh[2i\theta x/\gamma'] \sinh[x(\pi - \gamma')/\gamma']}{\sinh(x) \cosh(\pi x/\gamma')}\right\};$$

$$U_+(\theta) = \frac{\sinh[\pi(i\pi + \theta)/2\gamma']}{\sinh[\pi(i\pi - \theta)/2\gamma']};$$

$$U_-(\theta) = -\frac{\cosh[\pi(i\pi + \theta)/2\gamma']}{\cosh[\pi(i\pi - \theta)/2\gamma']}. \tag{3.15}$$

The scattering matrix of the bound states is obtained from this by taking the residue at the poles ([11], [15] and [27]) or directly by the methods of section I.4 (see formulæ (I.4.32) and (I.4.39)). Direct calculation was done in paper [17] where its explicit form can be found. The scattering matrix of several particles is equal to the product of the two-particle ones.

Conclusion

We have presented the solution of the massive Thirring model by means of the Bethe Ansatz. The calculations of the observable quantities carried out in the present chapter are a good example of the calculation of observables in any integrable model of quantum field theory. The massive Thirring model was solved in papers [2], [3] and [17]. It should be mentioned that the massive Thirring model was obtained as the limit of the XYZ Heisenberg model in [21], [22].

The massive Thirring model is equivalent to the quantum sine-Gordon model (in the sector with zero charge: [8], [23] and [24]). The fermion of the massive Thirring model is related to the soliton in the sine-Gordon model. The soliton scattering matrix in the sine-Gordon model was calculated in [15] and [26].

In the massive Thirring model, the S-matrix was calculated dynamically in paper [17], this method later being used for other models (see paper [20]). It will be shown in Part II (see Chapter VII, formulæ (VII.3.26), (VII.3.27) and section 5 of Chapter VIII) that the Bethe equations for the quantum sine-Gordon model essentially coincide with those for the massive Thirring model (the same may be said about the expressions for energy and momentum); therefore, all the observable quantities coincide. The sine-Gordon model has also been investigated by means of perturbation theory. The scattering matrix of the basic particles in this model (corresponding to the bound states of the massive Thirring model) was calculated by means of Feynman diagram techniques in [1] where the absence of mass-shell particle production was also demonstrated.

The fermions of the Thirring model correspond to the solitons of the sine-Gordon model (solitons are particle-like solutions of the classical equation). The quasiclassical quantization of solitons was performed in [9], [10] and [19]. The mass spectrum (3.13) was predicted and the quasiclassical approximation of the S-matrix up to contributions exponentially small in Planck's constant were calculated in paper [16]. The thermodynamics of the massive Thirring model (and of the sine-Gordon model) are also very interesting; see [5], [6], [7], [13] and [14].

IV

The Hubbard Model

Introduction

The Hubbard model is a one-dimensional lattice model for interacting spin 1/2 fermions. It describes electrons that can hop between nearest-neighbor sites of the chain and interact if two of them on the same site have opposite spins. This is an example of a model with more than one type of interaction which can be solved by the Bethe Ansatz. The Hubbard model is very important in condensed matter as it describes strongly correlated electronic systems. The model was solved by E.H. Lieb and F.Y. Wu [6]. Recently there has been considerable interest in the Hubbard model in relation to superconductivity.

In section 1 we describe the Bethe Ansatz for this model and in section 2 we calculate the finite size corrections.

IV.1 Bethe Ansatz solution for the Hubbard model

This model describes electrons with spin 1/2 on a lattice. The electrons are described in terms of canonical Fermi fields $\psi_{j,\sigma}$ where j is the number of the lattice site ($j = 1, \ldots, N$) and $\sigma = \uparrow, \downarrow$ is the spin. The $\psi_{j,\sigma}$ obey standard anticommutation relations:

$$\{\psi_{i,\sigma}^\dagger, \psi_{j,\sigma'}^\dagger\} = \{\psi_{i,\sigma}, \psi_{j,\sigma'}\} = 0; \quad \{\psi_{i,\sigma}, \psi_{j,\sigma'}^\dagger\} = \delta_{ij}\delta_{\sigma\sigma'};$$
$$\psi_{i,\sigma}|0\rangle = \langle 0|\psi_{i,\sigma}^\dagger = 0. \tag{1.1}$$

Here $|0\rangle$ and $\langle 0|$ are the Fock vacuum and the dual Fock vacuum, respectively.

The Hamiltonian for the one-dimensional Hubbard model is given by the following expression:

$$\mathcal{H} = -\sum_{j=1}^{N}\sum_{\sigma}(\psi^{\dagger}_{j+1,\sigma}\psi_{j,\sigma} + \psi^{\dagger}_{j,\sigma}\psi_{j+1,\sigma}) + 4u\sum_{j=1}^{N} n_{j\uparrow}n_{j\downarrow}$$

$$+ \mu\sum_{j=1}^{N}(n_{j\uparrow} + n_{j\downarrow}) - \frac{h}{2}\sum_{j=1}^{N}(n_{j\uparrow} - n_{j\downarrow}), \qquad (1.2)$$

where $n_{j,\sigma} = \psi^{\dagger}_{j,\sigma}\psi_{j,\sigma}$ is the number of spin σ electrons at the site j, $4u$ ($u > 0$) is the on-site Coulomb repulsion, μ is the chemical potential and h is the external magnetic field. E.H. Lieb and F.Y. Wu [6] first constructed the Bethe Ansatz for this model. We shall denote the number of spin up electrons by $N_{\uparrow}$ and the number of spin down electrons by $N_{\downarrow}$. The total number of electrons is denoted by $N_c = N_{\uparrow} + N_{\downarrow}$, ($0 \leq N_c \leq 2N$), and the number of spin down electrons is also denoted by $N_s = N_{\downarrow}$. The eigenfunctions are characterized by N_c momenta k_j of charged spinless excitations (recently called holons) and N_s rapidities λ of spin waves (spinons). Explicit expression for the eigenfunctions can be found for example in [4] where the complete set of eigenfunctions of the model were constructed. Imposing periodic boundary conditions on the wave functions leads to the Bethe Ansatz equations

$$Nk_j = 2\pi I_j - \sum_{\beta=1}^{N_s} 2\arctan\left(\frac{\sin k_j - \lambda_{\beta}}{u}\right),$$

$$\sum_{j=1}^{N_c} 2\arctan\left(\frac{\lambda_{\alpha} - \sin k_j}{u}\right) = 2\pi J_{\alpha} + \sum_{\beta=1}^{N_s} 2\arctan\left(\frac{\lambda_{\alpha} - \lambda_{\beta}}{2u}\right). \qquad (1.3)$$

where N is the length of the system in units of lattice spacing. The quantum numbers I_j and J_{α} are integer or half-integer, depending on the parity of the numbers of down and up spins, respectively:

$$I_j = \frac{N_s}{2} \pmod{1}, \quad J_{\alpha} = \frac{N_c - N_s + 1}{2} \pmod{1}. \qquad (1.4)$$

The energy and momentum of the system in a state corresponding to a solution of (1.3) are

$$E = -2\sum_{j=1}^{N_c} \cos k_j + \mu N_c + h\left(N_s - \frac{N_c}{2}\right),$$

$$P = \sum_{j=1}^{N_c} k_j = \frac{2\pi}{N}\left(\sum_{j}^{N_c} I_j + \sum_{\alpha}^{N_s} J_{\alpha}\right). \qquad (1.5)$$

In the thermodynamic limit ($N \to \infty$, with N_c/N, N_s/N kept constant) the ground state of the model is two Fermi seas, characterized by two distribution functions: $\rho_c(k)$, the distribution function of charges with momentum k and $\rho_s(\lambda)$, the distribution of down spins with spin wave rapidity λ. The functions $\rho_c(k)$ and $\rho_s(\lambda)$ satisfy the following integral equations:

$$\rho_c(k) = \frac{1}{2\pi} + \frac{\cos k}{2\pi} \int_{-\lambda_0}^{\lambda_0} d\lambda \, K_1(\sin k - \lambda)\rho_s(\lambda);$$

$$\rho_s(\lambda) = \frac{1}{2\pi} \int_{-k_0}^{k_0} dk \, K_1(\lambda - \sin k)\rho_c(k)$$
$$- \frac{1}{2\pi} \int_{-\lambda_0}^{\lambda_0} d\mu \, K_2(\lambda - \mu)\rho_s(\mu).$$

$$(1.6)$$

This is a matrix generalization of equations (I.3.7) and (II.2.6). The kernels K_1, K_2 of these integral equations are given by

$$K_1(x) = \frac{2u}{u^2 + x^2}; \quad K_2(x) = \frac{4u}{4u^2 + x^2}. \tag{1.7}$$

The values of λ_0 and k_0 are fixed by the following equations:

$$\int_{-k_0}^{k_0} dk \, \rho_c(k) = \frac{N_c}{N} \equiv n_c;$$

$$\int_{-\lambda_0}^{\lambda_0} d\lambda \, \rho_s(\lambda) = \frac{N_s}{N} \equiv \frac{1}{2}n_c - M$$

$$(1.8)$$

where n_c is the total density of electrons and $M = (N_\uparrow - N_\downarrow)/2N$ is the magnetization per lattice site. The ground state energy per lattice site is

$$\epsilon_0 = \int_{-k_0}^{k_0} dk \left(\mu - \frac{h}{2} - 2\cos k\right) \rho_c(k) + h \int_{-\lambda_0}^{\lambda_0} d\lambda \, \rho_s(\lambda). \tag{1.9}$$

Alternatively, ϵ_0 may be given in terms of the dressed energy

$$\epsilon_0 = \frac{1}{2\pi} \int_{-k_0}^{k_0} dk \, \varepsilon_c(k) \tag{1.10}$$

where $\varepsilon_c(k)$ is a solution of the following system of coupled integral equations:

$$\varepsilon_c(k) = \varepsilon_c^{(0)}(k) + \frac{1}{2\pi} \int\limits_{-\lambda_0}^{\lambda_0} d\lambda\, K_1(\sin k - \lambda)\varepsilon_s(\lambda);$$

$$\varepsilon_s(\lambda) = \varepsilon_s^{(0)}(\lambda) + \frac{1}{2\pi} \int\limits_{-k_0}^{k_0} dk\, \cos k K_1(\lambda - \sin k)\, \varepsilon_c(k)$$

$$- \frac{1}{2\pi} \int\limits_{-\lambda_0}^{\lambda_0} d\mu\, K_2(\lambda - \mu)\, \varepsilon_s(\mu).$$

(1.11)

The bare energies $\varepsilon_{c,s}^{(0)}$ are obtained from (1.5) as

$$\varepsilon_c^{(0)}(k) = \mu - \frac{h}{2} - 2\cos k; \quad \varepsilon_s^{(0)}(\lambda) = h. \tag{1.12}$$

The solutions of equations (1.11) define the energy bands. The ground state configuration corresponds to the filling of all states with $\varepsilon_c(k) < 0$ and $\varepsilon_s(\lambda) < 0$. Consequently,

$$\varepsilon_c(k_0) = 0 \quad \text{and} \quad \varepsilon_s(\lambda_0) = 0 \tag{1.13}$$

provide another way to define k_0 and λ_0 for a given magnetic field h and chemical potential μ. Here we consider a less than half-filled band ($n_c < 1$), a repulsive value of the coupling constant and a moderate magnetic field. In this case all excitations are gapless. Equations (1.11) are matrix generalizations of equations (I.3.16) and (II.2.9).

IV.2 Finite size corrections

The finite size corrections can be evaluated as for the Bose gas (section 9 of Chapter 1). The finite size correction formula is similar to those for the Bose gas (I.9.13) and the Heisenberg antiferromagnet (II.5.1).

The finite size corrections to the ground state energy are equal to

$$E = N\epsilon_0 - \frac{\pi v_c}{6N} - \frac{\pi v_s}{6N} + \text{h.o.c.} \tag{2.1}$$

where ϵ_0 is the density of the ground state energy (see (1.9)). The values v_c and v_s are the Fermi velocities of charge and spin density waves:

$$v_c = \frac{\varepsilon_c'(k_0)}{2\pi\rho_c(k_0)}, \quad v_s = \frac{\varepsilon_s'(\lambda_0)}{2\pi\rho_s(\lambda_0)}. \tag{2.2}$$

In the infinite volume limit, there is no gap in the spectrum of the Hamiltonian, but in a finite volume a gap arises. To describe it we need to introduce some notation. First let us form a diagonal matrix

$$V = \begin{pmatrix} v_c & 0 \\ 0 & v_s \end{pmatrix} \tag{2.3}$$

and introduce a dressed charge matrix

$$\begin{pmatrix} Z_{cc}(k) & Z_{cs}(\lambda) \\ Z_{sc}(k) & Z_{ss}(\lambda) \end{pmatrix} \tag{2.4}$$

which is defined by the following system of integral equations:

$$Z_{cc}(k) = 1 + \frac{1}{2\pi} \int_{-\lambda_0}^{\lambda_0} d\lambda \, Z_{cs}(\lambda) K_1(\lambda - \sin k);$$

$$Z_{cs}(\lambda) = \frac{1}{2\pi} \int_{-k_0}^{k_0} dk \, \cos k \, Z_{cc}(k) K_1(\sin k - \lambda)$$

$$- \frac{1}{2\pi} \int_{-\lambda_0}^{\lambda_0} d\mu \, Z_{cs}(\mu) K_2(\mu - \lambda);$$

$$\tag{2.5}$$

$$Z_{sc}(k) = \frac{1}{2\pi} \int_{-\lambda_0}^{\lambda_0} d\lambda \, Z_{ss}(\lambda) K_1(\lambda - \sin k);$$

$$Z_{ss}(\lambda) = 1 + \frac{1}{2\pi} \int_{-k_0}^{k_0} dk \, \cos k \, Z_{sc}(k) K_1(\sin k - \lambda)$$

$$- \frac{1}{2\pi} \int_{-\lambda_0}^{\lambda_0} d\mu \, Z_{ss}(\mu) K_2(\mu - \lambda).$$

The similarity of these equations to (1.6) and (1.11) is evident. They are the correct matrix generalization of (I.9.20).

The values of these functions at the Fermi boundaries (λ_0 and k_0) will be important for determining finite size corrections: let us introduce the matrix $\mathcal{Z}$

$$\mathcal{Z} = \begin{pmatrix} \mathcal{Z}_{cc} & \mathcal{Z}_{cs} \\ \mathcal{Z}_{sc} & \mathcal{Z}_{ss} \end{pmatrix} = \begin{pmatrix} Z_{cc}(k_0) & Z_{cs}(\lambda_0) \\ Z_{sc}(k_0) & Z_{ss}(\lambda_0) \end{pmatrix}. \tag{2.6}$$

The gap is given by the matrix generalization of formulæ (I.9.18) and (II.5.3),

$$
\delta E = \frac{2\pi}{N} \left\{ \frac{1}{4} \Delta \vec{N}^T (\mathcal{Z}^{-1})^T V \mathcal{Z}^{-1} \Delta \vec{N} + \vec{d}^{\,T} \mathcal{Z} V \mathcal{Z}^T \vec{d} \right.
$$
$$
\left. + v_c (N_c^+ + N_c^-) + v_s (N_s^+ + N_s^-) \right\} + \text{h.o.c.},
$$
(2.7)

where $\Delta \vec{N}$ and $\vec{d}$ are two-dimensional vectors

$$
\Delta \vec{N} = \begin{pmatrix} \Delta N_c \\ \Delta N_s \end{pmatrix} ; \quad \vec{d} = \begin{pmatrix} d_c \\ d_s \end{pmatrix} .
$$
(2.8)

ΔN_c is an integer which describes the change in the total number of electrons and ΔN_s is an integer which describes the change in the total spin of the electrons. The integers $N_c^{\pm}$ describe the possible change of maximal and minimal I_j in (1.3) and the integers $N_s^{\pm}$ describe the possible change of maximal and minimal J_α in (1.3). The values d_c and d_s describe the backscattering process (over sea jumps) and can be integer or half-integer as defined by the equalities

$$
d_c = (\Delta N_c + \Delta N_s)/2 \quad (\text{mod } 1);
$$
$$
d_s = \Delta N_c/2 \quad (\text{mod } 1)
$$
(2.9)

which follow from (1.4).

Now all quantities in formula (2.7) are defined. The change in momentum is given by

$$
\Delta P = \frac{2\pi}{N} \left(\Delta \vec{N} \vec{d} + N_c^+ - N_c^- + N_s^+ - N_s^- \right.
$$
$$
\left. + 2 d_c P_{F,\uparrow} + 2 (d_c + d_s) P_{F,\downarrow} \right) + \text{h.o.c.}
$$
(2.10)

where $P_{F,\uparrow}$ and $P_{F,\downarrow}$ are the Fermi momenta for electrons with spin up and spin down respectively, i.e.

$$
P_{F,\uparrow(\downarrow)} = \frac{1}{2} (\pi n_c \pm 2\pi \mathcal{M}).
$$
(2.11)

In Chapter XVIII we shall see that finite size corrections drive the long distance asymptotics of the correlation functions. We would also like to emphasize that formulæ (2.7) and (2.10) give the correct finite size corrections for arbitrary multicomponent Bethe Ansatz solvable models (the matrix $\mathcal{Z}$ can have other dimensions).

Conclusion

The one-dimensional Hubbard model is an example of a system with different types of excitation (spin-density waves and charge-density waves) which can be solved by the Bethe Ansatz. Sometimes the Bethe Ansatz for this type of model is called the "nested Bethe Ansatz". It was discovered in paper [14]. The Yang-Baxter equation appears as the consistency condition for the dynamics of the model to be two-body reducible. The one-dimensional Hubbard model was solved by E.H. Lieb and F.Y. Wu in [6]. The complete set of eigenfunctions was constructed in [4]. The zero-temperature case for both attractive and repulsive coupling was treated in [7], [8], [10] and [13] in the thermodynamic limit. The thermodynamics of the model were discussed in [9].

Finite size corrections for the Hubbard model (and more general systems) were studied in [1], [2], [3], [11] and [12]. In paper [5], the general formula for finite size corrections in Bethe Ansatz solvable models was obtained. Attention was paid mainly to the nested Bethe Ansatz. Formulæ (2.7) and (2.10) for finite size corrections in the Hubbard model are a special case of the general formula presented in [5].

Part II
The Quantum Inverse Scattering Method

Introduction to Part II

The quantum inverse scattering method relates the Bethe Ansatz to the theory of classical completely integrable differential equations. These are sometimes called soliton equations. The modern way to solve them is called the classical inverse scattering method. In a sense this is a nonlinear generalization of the Fourier transform.

In this Part the quantum inverse scattering method is expounded. The main statements of the classical inverse scattering method necessary for quantization are given in Chapter V where the Lax representation is introduced. The Hamiltonian structure of integrable models is also discussed along with the infinite number of integrals of motion. The most convenient method of analyzing the Hamiltonian structure relies on the classical r-matrix. Some concrete models will be considered. Chapter VI is devoted in particular to the quantum inverse scattering method. The R-matrix, which is the main object of this method, is introduced. The Yang-Baxter equation for the R-matrix is discussed. The main statements of the method are given and a number of examples are presented. The algebraic formulation of the Bethe Ansatz, one of the main achievements of the quantum inverse scattering method, is presented in Chapter VII. The notion of the determinant of the transition matrix in the quantum case is introduced in this chapter. (This is closely related to the concept of the antipode in quantum groups.)

Integrable models of quantum field theory on the lattice are presented in Chapter VIII. The quantum inverse scattering method provides a mechanism for transferring continuous models of quantum field theory to the lattice while preserving the R-matrix. For classical models this means that the structure of the action-angle variables remains the same [2]. In the quantum case, this leads to conservation of the scattering matrix and critical exponents that drive the long distance asymptotics of the correlation functions. For relativistic models of field theory (such as the sine-Gordon model), the lattice variant provides a rigorous solution to the problem of ultraviolet divergences. We can study continuous models as condensed lattice models (i.e., the lattice spacing $\Delta \to 0$). Our construction guarantees that there will be no phase transition. The explicit form of the L-operator allows the classification of all integrable models having a given R-matrix.

The close relationship of the quantum inverse scattering method to other methods of contemporary mathematical physics should be mentioned. Firstly, it is connected with quantum groups [1] and the theory of knots [5]. It is also related to the method used in classical statistical physics for solving two-dimensional lattice models. We shall abbreviate the name "quantum inverse scattering method" to QISM. We must note that the physical characteristics of particles—the dressed energy, momentum and S-matrix—are all calculated within QISM exactly as for the coordinate Bethe Ansatz.

It is interesting to mention that later in the book (see Chapter XIV) we shall obtain classical completely integrable differential equations for quantum correlation functions. We shall study them from a different point of view. The most important problem in this case is determining the long distance asymptotics of the correlation functions. We shall do this by means of the Riemann-Hilbert problem; see Chapters XV and XVI and, for example, [3], [4].

V

The Classical r-matrix

Introduction

The modern way to solve partial differential equations is called the classical inverse scattering method. (One can think of it as a nonlinear generalization of the Fourier transform.)

Nowadays, the classical inverse scattering method (CISM) is a well developed branch of mathematical physics (see Preface references [1], [9], [10], [11], [18], [21]–[24], [29], [37], [45]). In this chapter, we shall give only the information necessary for the quantization which will be performed in the next chapter. The concepts of the Lax representation, the transition matrix and the trace identities are stated in section 1. Classical completely integrable partial differential equations will appear once more in this book. In Chapters XIV and XV we shall derive them for quantum correlation functions. In those chapters we shall study completely integrable differential equations from a different point of view. We shall apply the Riemann-Hilbert problem in order to evaluate the asymptotics. The classical r-matrix, which enables calculation of the Poisson brackets between matrix elements of the transition matrix (and also construction of the action-angle variables) is introduced in section 2. As explained there, the existence of the r-matrix guarantees the existence of the Lax representation. The r-matrix satisfies a certain bilinear relation (the classical Yang-Baxter relation). The existence of the r-matrix also guarantees the existence of an infinite number of conservation laws which restrict in an essential way the dynamics of the system. In the next chapter, the notion of the r-matrix will be generalized to the quantum case. In the first two sections of this chapter, general statements are demonstrated by example using the nonlinear Schrödinger equation which is the simplest dynamical model (it should be mentioned that in the classical case this name is more natural than the one-dimensional Bose gas). Other mod-

els (the sine-Gordon equation, the Mikhailov-Shabat-Zhiber model) are
considered in section 3. Tensor notation, the application of which sub-
stantially simplifies the calculations both in the classical and quantum
cases, is discussed in the appendix to this chapter.

If the reader finds this chapter too brief, we recommend the excellent book by L.D. Faddeev and L.A. Takhtajan (see [18] in the Preface
references), which describes CISM in explicit detail.

V.1 The Lax representation

Let us consider a classical nonlinear Hamiltonian evolutionary equation
in two-dimensional space-time. The corresponding Hamiltonian will be
denoted by H. We study the system on a periodic interval of length L
($0 \leq x \leq L$). The traditional base for application of the inverse scattering
method (ISM) to this equation is that it can be represented in the Lax
form:

$$[\partial_t - U(x|\lambda), \ \partial_x + V(x|\lambda)] = 0 \qquad (1.1)$$

which is valid for all λ. Here U and V are $k \times k$ matrices (the integer k
depends on the equation under consideration) which depend on a complex
spectral parameter λ and on the dynamical variables of the problem. The
matrix $V(x|\lambda)$ is called the potential and $U(x|\lambda)$ is the time evolution
operator. Condition (1.1) must be valid at any λ and can be regarded as
the consistency condition for the following differential equations:

$$\begin{aligned}
\partial_t \Phi(x,t) &= U(x|\lambda)\Phi(x,t), \\
\partial_x \Phi(x,t) &= -V(x|\lambda)\Phi(x,t).
\end{aligned} \qquad (1.2)$$

Here $\Phi(x,t)$ is an unknown vector function which also depends on λ. In
the book [5], the similarity to the Yang-Mills fields is explained. Condition (1.1) plays the role of the zero-curvature condition, with U and V
playing the role of gauge fields. In Chapters XIV and XV we shall derive
nonlinear partial differential equations for quantum correlation functions,
starting from the Lax representation. These will give interesting examples
of completely integrable differential equations.

It is useful to consider the translation of the solution of the system
(1.2), Φ, along the x-direction (at fixed time t):

$$\Phi(x) = \mathsf{T}(x,y|\lambda)\Phi(y). \qquad (1.3)$$

(Here we have suppressed the time argument.) The matrix $\mathsf{T}(x,y|\lambda)$ is
called the transition matrix. Below we shall discuss this $k \times k$ matrix in
detail. But first, let us discuss the Lax representation for lattice models
(with continuous time). We shall use lattice versions of quantum field
theory models to solve the problem of ultraviolet divergences. For quantization, we also need the Lax representation on the periodic lattice with

M sites and lattice spacing Δ:

$$\partial_t \mathsf{L}(n|\lambda) = U(n+1|\lambda)\mathsf{L}(n|\lambda) - \mathsf{L}(n|\lambda)U(n|\lambda). \tag{1.4}$$

Here L and U are $k \times k$ matrices which depend on the spectral parameter and dynamical variables. The equality (1.4) is a consequence of the consistency condition for the following problem on the lattice:

$$\partial_t \Phi(n,t) = U(n|\lambda)\Phi(n,t),$$
$$\Phi(n+1,t) = \mathsf{L}(n,\lambda)\Phi(n,t),$$

where n is the lattice site number. To study continuous models, it is convenient to consider the infinitesimal lattice ($\Delta \to 0$). The coordinate of the n-th site of the lattice thus obtained is $x_n = n\Delta$ ($n = 1, \ldots, M$, and $M = L/\Delta$). For such a lattice

$$\mathsf{L}(n|\lambda) = I - V(x_n|\lambda)\Delta + O\left(\Delta^2\right), \tag{1.5}$$

where I is the $k \times k$ unit matrix.

Let us now study the transition matrix $\mathsf{T}(x,y|\lambda)$ (1.3) which plays an important role in the ISM. For the continuous case, this $k \times k$ matrix is defined on the interval $[y,x]$ ($x \geq y$), by the following requirements:

$$[\partial_x + V(x|\lambda)]\,\mathsf{T}(x,y|\lambda) = 0;$$
$$\mathsf{T}(y,y|\lambda) = I. \tag{1.6}$$

Sometimes it is useful to write down a formal solution of this equation:

$$\mathsf{T}(x,y|\lambda) = \mathsf{P}\exp\left\{ -\int_y^x V(z|\lambda)\,dz \right\}, \tag{1.7}$$

where P denotes path ordering of noncommutative factors.

The transition matrix possesses the following group-like property: if z is any interior point in the interval $[y,x]$ then

$$\mathsf{T}(x,z|\lambda)\mathsf{T}(z,y|\lambda) = \mathsf{T}(x,y|\lambda) \qquad (x \geq z \geq y).$$

The left hand side here is the product of two $k \times k$ matrices. The transition matrix for the whole periodic interval $[0,L]$ is called the monodromy matrix $\mathsf{T}(L,0|\lambda)$.

The lattice transition matrix from the m-th site to the $(n+1)$-th site can be represented as the product of $(n-m+1)$ matrices:

$$\mathsf{T}(n,m|\lambda) = \mathsf{L}(n|\lambda)\mathsf{L}(n-1|\lambda)\cdots\mathsf{L}(m|\lambda), \quad n \geq m \tag{1.8}$$

where $\mathsf{L}(k,\lambda) \equiv \mathsf{T}(k,k|\lambda)$ is the elementary transition matrix for one lattice site. The transition matrix for the whole lattice length, $\mathsf{T}(M,1|\lambda)$, is called the monodromy matrix. The matrix $\mathsf{L}(k|\lambda)$ is called the L-operator.

The trace of the monodromy matrix in both the continuous and lattice cases plays a particularly important role:

$$\tau(\lambda) = \mathrm{tr}\ \mathsf{T}(L,0|\lambda); \quad \tau(\lambda) = \mathrm{tr}\ \mathsf{T}(M,1|\lambda). \tag{1.9}$$

In the next section we shall see that $\tau(\lambda)$ is time-independent. The Hamiltonian of the initial evolutionary equation is expressed in terms of logarithmic derivatives of $\tau(\lambda)$ by means of trace identities.

As an example, we shall consider the nonlinear Schrödinger equation

$$i\partial_t \Psi = -\partial_x{}^2 \Psi + 2c\Psi^*\Psi\Psi \tag{1.10}$$

with the Hamiltonian

$$H = \int_0^L dx\ (\partial_x\Psi^*\,\partial_x\Psi + c\Psi^*\Psi^*\Psi\Psi) \tag{1.11}$$

and the Poisson brackets of the fields Ψ, Ψ^* given as

$$\{\Psi(x),\ \Psi^*(y)\} = i\delta(x-y). \tag{1.12}$$

The charge (number of particles) Q and momentum P

$$Q = \int_0^L \Psi^*\Psi\,dx; \qquad P = -i\int_0^L \Psi^*\partial_x\Psi\,dx \tag{1.13}$$

commute with H: $\{H,Q\} = \{H,P\} = 0$. This model is the classical limit of the quantum nonlinear Schrödinger equation studied in detail in Chapter I. The nonlinear Schrödinger equation can be represented in the Lax form (1.1); the 2×2 matrices V and U are given by

$$V(x|\lambda) = i\frac{\lambda}{2}\sigma_z + \Omega(x), \tag{1.14}$$

$$U(x|\lambda) = i\frac{\lambda^2}{2}\sigma_z + \lambda\Omega(x) + i\sigma_z\left(\partial_x\Omega + c\Psi^*\Psi\right), \tag{1.15}$$

where σ_z is the Pauli matrix ($\sigma_z = \mathrm{diag}(1,-1)$) and the matrix Ω is

$$\Omega(x) = \begin{pmatrix} 0 & i\sqrt{c}\Psi^*(x) \\ -i\sqrt{c}\Psi(x) & 0 \end{pmatrix}. \tag{1.16}$$

The transition matrix has the following properties:

$$\det \mathsf{T}(x,y|\lambda) = 1; \tag{1.17}$$

$$\sigma_x\mathsf{T}^*(x,y|\lambda^*)\sigma_x = \mathsf{T}(x,y|\lambda); \qquad \sigma_x = \begin{pmatrix} 0 & 1 \\ 1 & 0 \end{pmatrix}. \tag{1.18}$$

The corresponding L-operator on the infinitesimal lattice is:

$$\mathsf{L}(n|\lambda) = \begin{pmatrix} 1 - i\frac{\lambda\Delta}{2} & -i\sqrt{c}\Psi_n^*\Delta \\ i\sqrt{c}\Psi_n\Delta & 1 + i\frac{\lambda\Delta}{2} \end{pmatrix} + O(\Delta^2), \tag{1.19}$$

$$\Psi_n = \frac{1}{\Delta} \int\limits_{x_{n-1}}^{x_n} \Psi(x)\, dx; \qquad \{\Psi_n, \Psi_m^*\} = \frac{i}{\Delta}\delta_{nm}. \qquad (1.20)$$

The trace identities for this model are as follows (see (1.9), (1.11), (1.13)):

$$\ln\left[e^{i\lambda L/2}\tau(\lambda)\right] \xrightarrow[\lambda\to i\infty]{} ic\left[\lambda^{-1}Q + \lambda^{-2}P + \lambda^{-3}H + O\left(\lambda^{-4}\right)\right]. \qquad (1.21)$$

Let us derive this formula. Taking $\lambda \to i\infty$ the potential V in (1.14) becomes close to diagonal and one can represent the transition matrix as follows:

$$\mathsf{T}(x, y|\lambda) = G(x|\lambda)D(x, y|\lambda)G^{-1}(y|\lambda). \qquad (1.22)$$

Here D is a diagonal matrix, and the matrix G is chosen have the following form:

$$G(x|\lambda) = I + \sum_{k=1}^{\infty}\lambda^{-k}G_k(x), \qquad (1.23)$$

where I is the unit matrix and the G_k are antidiagonal matrices. The meaning of representation (1.22) is that the transition matrix can be diagonalized by means of a gauge transformation. The differential equation (1.6) results in the following equation for D:

$$[\partial_x + W(x|\lambda)]\, D(x, y|\lambda) = 0; \qquad D(y, y|\lambda) = I \qquad (1.24)$$

where the potential W is equal to

$$W(x|\lambda) = G^{-1}(x)\, \partial_x G(x) + i\frac{\lambda}{2}G^{-1}(x)\sigma_z G(x) + G^{-1}(x)\Omega(x)G(x). \qquad (1.25)$$

The matrices G_k are defined by the requirement that the potential W be a diagonal matrix. It is easy to show that

$$G_1 = i\sigma_z\Omega; \qquad G_2 = -\partial_x\Omega; \qquad G_3 = i\sigma_z\left(-\partial_x^2\Omega + \Omega^3\right). \qquad (1.26)$$

Thus, the potential W is

$$W = i\frac{\lambda}{2}\sigma_z + \lambda^{-1}W_1 + \lambda^{-2}W_2 + \lambda^{-3}W_3 + O\left(\lambda^{-4}\right) \qquad (1.27)$$

where

$$W_1 = -i\sigma_z\Omega^2; \qquad W_2 = -\Omega\partial_x\Omega;$$
$$W_3 = i\sigma_z\left[\Omega\,\partial_x^2\Omega - \Omega^4\right]. \qquad (1.28)$$

Due to the diagonality of the matrix W, equation (1.24) can be solved explicitly:

$$D(x, y|\lambda) = \exp\left\{-\int_y^x W(z|\lambda)\, dz\right\}.$$
(1.29)

Now take $y = 0$ and $x = L$. The periodic boundary conditions imply that $G(L) = G(0)$. Using (1.17), (1.22) one has that $\det D(L, 0) = 1$. Hence

$$D(L, 0|\lambda) = \exp\{\sigma_z Z(\lambda)\}$$
(1.30)

with a scalar function $Z(\lambda)$. It is easily shown from (1.27)–(1.29) that

$$Z(\lambda) = -i\frac{\lambda L}{2} + ic\left[\lambda^{-1}Q + \lambda^{-2}P + \lambda^{-3}H + O\left(\lambda^{-4}\right)\right].$$
(1.31)

Due to the periodic boundary conditions, we have

$$\tau(\lambda) = \operatorname{tr} \mathsf{T}(L, 0|\lambda) = \operatorname{tr} D(L, 0|\lambda)$$
(1.32)

and, for $\lambda \to i\infty$, $D_{11}(L, 0|\lambda) \gg D_{22}(L, 0|\lambda)$. Using (1.30)–(1.32) we can calculate $\ln \tau(\lambda)$ and obtain (1.21). Actually, higher order terms in (1.21) are also interesting:

$$\ln\left[e^{i\lambda L/2}\tau(\lambda)\right] \xrightarrow[\lambda \to i\infty]{} ic \sum_{n=1}^\infty \lambda^{-n} I_n.$$
(1.33)

In the next section, we shall see that each I_n is time-independent. Thus, the $\{I_n\}$ comprise the infinite set of conservation laws that exactly solvable models possess.

Let us write down the first nontrivial conservation law:

$$\int dx \left\{\Psi^*\Psi_{xxx} - \frac{3c}{2}\Psi^{*2}\left(\Psi^2\right)_x\right\}.$$
(1.34)

V.2 The classical r-matrix

To construct the action-angle variables, it is necessary to calculate the Poisson brackets (PB) between matrix elements of the transition matrix (see [5]). There exists an effective method for performing such calculations, based on the classical r-matrix. We shall use the following notation for tensor products. The tensor product of two $k \times k$ matrices A and B will be denoted by $A \otimes B$ (a $k^2 \times k^2$ matrix). The $k^2 \times k^2$ permutation matrix $\sqcap$ has the following property:

$$\sqcap (A \otimes B) \sqcap = B \otimes A.$$
(2.1)

This equality is valid for all numerical matrices A and B. The minimum dimension of Π is 4×4; in this case it can be written as

$$\Pi = \begin{pmatrix} 1 & 0 & 0 & 0 \\ 0 & 0 & 1 & 0 \\ 0 & 1 & 0 & 0 \\ 0 & 0 & 0 & 1 \end{pmatrix}. \tag{2.2}$$

Definition: The Poisson brackets of the tensor product $\{A \overset{\otimes}{,} B\}$ is a $k^2 \times k^2$ matrix, its matrix elements being equal to the PB of some matrix element of A with some matrix element of B. The labeling of the elements of the matrix $\{A \overset{\otimes}{,} B\}$ is the same as for the matrix $A \otimes B$. (Tensor notation is discussed in detail in Appendix 1.)

Theorem 1. *If the PB between matrix elements of $V(x|\lambda)$ can be represented in the form*

$$\{V(x|\lambda) \overset{\otimes}{,} V(y|\mu)\} = \delta(x - y) \left[r(\lambda, \mu), V(x|\lambda) \otimes I + I \otimes V(x|\mu) \right] \tag{2.3}$$

then the PB between matrix elements of the transition matrix is given by

$$\{T(x, y|\lambda) \overset{\otimes}{,} T(x, y|\mu)\} = [T(x, y|\lambda) \otimes T(x, y|\mu), r(\lambda, \mu)]. \tag{2.4}$$

The square brackets on the right hand side denote the matrix commutator of two $k^2 \times k^2$ matrices. The matrix r, a $k^2 \times k^2$ matrix with elements depending on λ and μ, acts in the tensor product of the two spaces (r does not depend on space variables).

Proof: The transition matrix $T(x, y|\lambda)$ depends on the dynamical variables only through the potential $V(x|\lambda)$. Thus the PB of its matrix elements can be expressed in the form

$$\{T_{ij}(x, y|\lambda), T_{kl}(x, y|\mu)\} = \int_y^x dz_\lambda \int_y^x dz_\mu \left(\frac{\delta T_{ij}(x, y|\lambda)}{\delta V_{ab}(z_\lambda|\lambda)} \right) \left(\frac{\delta T_{kl}(x, y|\mu)}{\delta V_{cd}(z_\mu|\mu)} \right)$$

$$\times \{V_{ab}(z_\lambda|\lambda), V_{cd}(z_\mu|\mu)\}. \tag{2.5}$$

The variational derivative of $T(x, y|\lambda)$ with respect to $V(z|\lambda)$ is obtained from

$$\delta T(x, y|\lambda) = - \int_y^x T(x, z|\lambda) \, \delta V(z|\lambda) T(z, y|\lambda) \, dz. \tag{2.6}$$

Substituting this expression into (2.5), one obtains in tensor notation

$$\{T(x,y|\lambda) \overset{\otimes}{,} T(x,y|\mu)\} = \int_y^x dz_\lambda \int_y^x dz_\mu \; (T(x,z_\lambda|\lambda) \otimes T(x,z_\mu|\mu))$$

$$\times \{V(z_\lambda|\lambda) \overset{\otimes}{,} V(z_\mu|\mu)\} \, (T(z_\lambda,y|\lambda) \otimes T(z_\mu,y|\mu)). \quad (2.7)$$

Using (2.3) one writes

$$\{T(x,y|\lambda) \overset{\otimes}{,} T(x,y|\mu)\} = \int_y^x dz \; (T(x,z|\lambda) \otimes T(x,z|\mu))$$

$$\times [r(\lambda,\mu), V(z|\lambda) \otimes I + I \otimes V(z|\mu)] \, (T(z,y|\lambda) \otimes T(z,y|\mu)) \quad (2.8)$$

It follows from (1.7) that

$$\partial_y T(x,y|\lambda) = T(x,y|\lambda)V(y|\lambda) \qquad (2.9)$$

which removes $V(z)$ from (2.8) to give

$$\{T(x,y|\lambda) \overset{\otimes}{,} T(x,y|\mu)\} = - \int_y^x dz \, \frac{d}{dz} \Big[(T(x,z|\lambda) \otimes T(x,z|\mu))$$

$$\times r(\lambda,\mu) \, (T(z,y|\lambda) \otimes T(z,y|\mu)) \Big]. \quad (2.10)$$

Evaluating the integral over the total derivative and using $T(x,x|\lambda) = I$ one arrives at formula (2.4). The theorem is proved. (In [6] and [7], this theorem was generalized to the case where r depends on x.)

With the help of the r-matrix, the PB of the transition matrix in the discrete case can also be calculated.

Theorem 2. *If the PB between the elements of* $L(\lambda)$ *can be expressed in the form*

$$\{L(k|\lambda) \overset{\otimes}{,} L(l|\mu)\} = \delta_{kl}[L(k|\lambda) \otimes L(l|\mu), \, r(\lambda,\mu)] \qquad (2.11)$$

then the PB of $T(\lambda)$ *is given by*

$$\{T(n,m|\lambda) \overset{\otimes}{,} T(n,m|\mu)\} = [T(n,m|\lambda) \otimes T(n,m|\mu), \, r(\lambda,\mu)]. \quad (2.12)$$

Proof: The theorem is proved by induction using expression (2.11) as the basis of the induction. Using (1.8), now consider

$$T(n+1,m|\lambda) = L(n+1,m|\lambda)T(n,m|\lambda). \qquad (2.13)$$

It follows from (2.11) that $\{\mathsf{L}(n+1|\lambda) \overset{\otimes}{,} \mathsf{T}(n,m|\mu)\} = 0$. Supposing that (2.12) is valid, one can easily obtain

$$\{\mathsf{T}(n+1,m|\lambda) \overset{\otimes}{,} \mathsf{T}(n+1,m|\mu)\} = [\mathsf{T}(n+1,m|\lambda) \otimes \mathsf{T}(n+1,m|\mu),\, \mathsf{r}(\lambda,\mu)]. \tag{2.14}$$

The proof is thus finished.

The r-matrix in (2.4), (2.12) must satisfy some identities. It acts on the tensor product of two k-dimensional vector spaces. Consider the triple tensor product of k-dimensional vector spaces. Then

$$[\mathsf{r}_{13}(\lambda,\nu),\, \mathsf{r}_{23}(\mu,\nu)] + [\mathsf{r}_{12}(\lambda,\mu),\, \mathsf{r}_{13}(\lambda,\nu) + \mathsf{r}_{23}(\mu,\nu)] = 0. \tag{2.15}$$

The spaces on which the r-matrix acts non-trivially are denoted by subscripts (see Appendix 1). The square brackets denote the commutator, in (2.15).

As follows from (2.4), the r-matrix is defined up to a scalar term $a(\lambda,\mu)E$ (E is the $k^2 \times k^2$ unit matrix, a is an arbitrary c-number function). The r-matrix can be chosen as an antisymmetric matrix: $\mathsf{r}(\lambda,\mu) = -\Pi \mathsf{r}(\mu,\lambda)\Pi$ (Π is the permutation matrix). The relation (2.15) is called the classical Yang-Baxter relation. It is merely a restatement of the Jacobi identity for the PB (2.12). The point $\lambda = \mu = \nu$ is a pole of the r-matrix; thus the relation (2.15) should be defined more precisely at this point [12].

The simplest r-matrix is

$$\mathsf{r}(\lambda,\mu) = \frac{c}{\lambda - \mu}\,\Pi, \tag{2.16}$$

which is the r-matrix for the nonlinear Schrödinger equation.

The existence of the r-matrix for a given potential V (2.3) or L-operator (2.11) is not obvious in advance. However, r-matrices exist for the majority of models solved by ISM.

The existence of the r-matrix results in the presence of infinitely many conservation laws in the model. This follows from the fact that the traces of the monodromy matrix, defined in (1.9), commute for different values of the spectral parameter:

$$\{\tau(\lambda),\, \tau(\mu)\} = 0. \tag{2.17}$$

To obtain this relation, one calculates the traces of both sides of (2.4) (or (2.12)) in the $k^2 \times k^2$-matrix space and then uses the fact that the trace of the tensor product of two matrices is equal to the product of the traces of each matrix. From (2.17) it follows that $\{\ln \tau(\lambda),\, \tau(\mu)\} = 0$. If we expand this in inverse powers of λ and use the trace identities (1.21), we shall arrive at $\{H,\, \tau(\mu)\} = 0$. This means that the trace of the monodromy matrix, $\tau(\mu)$, is time-independent: $\partial_t \tau(\mu) = 0$. Thus, all the I_n in (1.33) give rise to conservation laws. This means that $\tau(\mu)$ is the generating functional for the integrals of motion. It should be mentioned that the r-matrix replaces, in a sense, the time evolution operator $U(x|\lambda)$ (1.2). It can be proved that if the Hamiltonian and Poisson brackets are given, and if the potential V (or L-operator) and the r-matrix exist and the trace identities hold, then the initial nonlinear equation can be expressed in the Lax form (1.1).

Let us explain the construction of the corresponding time evolution operator $U(x|\lambda)$ in the continuous case. Changing $\partial_t V$ in (1.1) to $\partial_t V = \{H, V\}$ and using the fact that the Hamiltonian is expressed by means of the trace identities (1.21) in terms of $\tau(\lambda)$, one concludes that it is sufficient to consider the Poisson brackets between the monodromy matrix and the potential.

Theorem 3. *The following Lax representation is valid:*

$$\{\tau(\mu), V(x|\lambda)\} = \partial_x U(x|\lambda, \mu) + [V(x|\lambda), U(x|\lambda, \mu)]. \tag{2.18}$$

The proof is analogous to that of Theorem 1. Note that on the left hand side of (2.18) τ is a scalar and V is a matrix. Let us indicate by a subscript 2 the space in which the matrix V acts and introduce an auxiliary space with subscript 1. Rewriting (2.18) in the form

$$\{\tau(\mu), V_2(x|\lambda)\} = \partial_x U_2(x|\lambda, \mu) + [V_2(x|\lambda), U_2(x|\lambda, \mu)], \tag{2.19}$$

we can introduce U_2 as a trace in the first space:

$$U_2(x|\lambda, \mu) = \text{tr}_1 \left(\mathsf{T}_1(L, x|\mu) \mathsf{r}_{12}(\mu, \lambda) \mathsf{T}_1(x, 0|\mu) \right). \tag{2.20}$$

The matrices act in the spaces indicated by the subscripts, and T is the transition matrix. It is natural to call $U_2(x|\lambda, \mu)$ (2.20) the generating functional of time evolution operators. An analogous theorem is valid for lattice models. Due to Theorem 3, we shall not (as a rule) use the time evolution operator further.

To get (1.1) from (2.18) one should first rewrite (2.18) in the form

$$\{\ln \tau(\mu), V(x|\lambda)\} = \partial_x \left(\tau^{-1}(\mu) U(x|\lambda, \mu) \right)$$
$$+ [V(x|\lambda), \left(\tau^{-1}(\mu) \right) U(x|\lambda, \mu)]. \tag{2.21}$$

Then one should use the trace identities. In the case of the nonlinear Schrödinger equation, one should expand both sides of equation (2.21) in inverse powers of μ (see (1.21)). Then one should consider the coefficient of μ^{-3} in (2.21). On the left hand side, one will obtain $ic\{H, V(x|\lambda)\}$. This is proportional to $\partial_t V(x|\lambda)$. This means that $\tau^{-1}(\mu) U(x|\lambda, \mu)$ when expanded in inverse powers of μ gives $U(x|\lambda)$ (from (1.1)) as the coefficient of μ^{-3}. This example shows the relation between (2.18) and (1.1). In other words, $\tau^{-1}(\mu)U(x|\lambda,\mu)$ is the generating functional for time evolution operators $U(x|\lambda)$.

The method of constructing action-angle variables for completely integrable Hamiltonian equations is explained in [5]. Basically, one considers the scattering of plane-waves on the potential in the equation $\partial_x\Phi + V(x|\lambda)\Phi = 0$. The scattering matrix provides an opportunity for the explicit construction of action-angle variables.

The classical r-matrix fixes the structure of the action-angle variables. This suggests the construction of lattice variants of continuous field theory models. The explicit form of the r-matrix should be unchanged when changing the continuous Lax operator $(\partial_x + V)$ to the discrete operator $\mathsf{L}(n|\lambda)$. This will guarantee the conservation of the structure of the action-angle variables in the transition from the continuum to the lattice. Thus, the continuum limit $\Delta \to 0$ (condensation of the lattice) will proceed very smoothly.

V.3 Examples

In this section examples of completely integrable Lorentz-invariant equations are given. .

(1) The sine-Gordon (SG) model. The equation of motion is given by

$$\left(\partial_t^2 - \partial_x^2\right) u + \frac{m^2}{\beta} \sin \beta u = 0. \tag{3.1}$$

The Hamiltonian, momentum and charge are

$$H = \int dx \left(\frac{1}{2}\pi^2 + \frac{1}{2}\left(\partial_x u\right)^2 + \frac{m^2}{\beta^2}\left(1 - \cos \beta u\right) \right); \tag{3.2}$$

$$P = -\int dx\, \pi\, \partial_x u; \qquad Q = \frac{\beta}{2\pi}\int dx\, \partial_x u, \tag{3.3}$$

where $\pi(x) = \partial_t u(x)$, $\{\pi(x),\, u(y)\} = \delta(x - y)$. This model has many physical applications. It is of interest as a nontrivial model of a relativistic scalar field. The equation of motion possesses a zero curvature representation with the potential

$$V(x|\lambda) = \begin{pmatrix} i\beta\pi(x)/4 & (m/2)\sinh\{-\lambda + [i\beta u(x)/2]\} \\ (m/2)\sinh\{\lambda + [i\beta u(x)/2]\} & -i\beta\pi(x)/4 \end{pmatrix}.$$

$$\tag{3.4}$$

The L-operator on the infinitesimal lattice is

$$L(n|\lambda) = \begin{pmatrix} 1 - i\beta p_n/4 & (m\Delta/2)\sinh\left[\lambda - (i\beta u_n/2)\right] \\ -(m\Delta/2)\sinh[\lambda + (i\beta u_n/2)] & 1 + i\beta p_n/4 \end{pmatrix}$$

$$(3.5)$$

where

$$u_n = \frac{1}{\Delta}\int\limits_{x_{n-1}}^{x_n} u(x)\,dx; \qquad p_n = \int\limits_{x_{n-1}}^{x_n}\pi(x)\,dx; \qquad \{p_n,\, u_m\} = \delta_{nm}.$$

$$(3.6)$$

The classical r-matrix (2.3) is

$$r(\lambda,\mu) = \begin{pmatrix} 0 & 0 & 0 & 0 \\ 0 & r_{22}^{11} & r_{21}^{12} & 0 \\ 0 & r_{12}^{21} & r_{11}^{22} & 0 \\ 0 & 0 & 0 & 0 \end{pmatrix}$$

$$(3.7)$$

with the nonzero elements $(\gamma \equiv \beta^2/8)$

$$r_{22}^{11} = r_{11}^{22} = \gamma\coth(\lambda - \mu);$$
$$r_{21}^{12} = r_{12}^{21} = -\frac{\gamma}{\sinh(\lambda - \mu)}.$$

$$(3.8)$$

The trace identities can be found in [5]. For quantization, we must put the model on the lattice. The trace identities for the lattice model, which differ from the continuous case, are constructed in section 2 of Chapter VIII. The transition matrix possesses the following symmetry properties:

$$\sigma_y T^*(x, y|\lambda^*)\sigma_y = T(x, y|\lambda); \tag{3.9}$$
$$\det T(x, y|\lambda) = 1. \tag{3.10}$$

(2) The Zhiber-Shabat-Mikhailov model. The Hamiltonian and the PB of this model are

$$H = \int dx \left\{ \frac{\gamma}{2}\pi^2 + \frac{1}{2\gamma}\left(\partial_x u\right)^2 \right.$$
$$\left. + \frac{m^2}{\gamma}\left[\exp(u) + \frac{1}{2}\exp(-2u) - \frac{3}{2}\right] \right\};$$
$$\pi(x) = \frac{1}{\gamma}\partial_t u(x), \qquad \{\pi(x),\, u(y)\} = \delta(x - y). \tag{3.11}$$

The equation of motion

$$\left(\partial_t^2 - \partial_x^2\right) u + m^2\left[\exp(u) - \exp(-2u)\right] = 0 \tag{3.12}$$

can be expressed in the Lax form (1.1) with potential

$$V(x|\lambda) = \frac{1}{2}\begin{pmatrix} 0 & im\exp(u/2-\lambda) & -im\exp(u/2+\lambda) \\ -im\exp(u/2+\lambda) & -\partial_t u & im\exp(-u-\lambda) \\ im\exp(u/2-\lambda) & -im\exp(-u+\lambda) & \partial_t u \end{pmatrix}$$

$$(3.13)$$

The classical r-matrix has the following nonzero elements ($\kappa \equiv \exp(\lambda - \mu)$):

$$r_{11}^{11} = \gamma\frac{2-\kappa^3-\kappa^{-3}}{4(\kappa^3-\kappa^{-3})};$$

$$r_{22}^{11} = r_{33}^{11} = r_{11}^{22} = r_{11}^{33} = -\gamma\frac{2+\kappa^3+\kappa^{-3}}{4(\kappa^3-\kappa^{-3})};$$

$$r_{33}^{22} = r_{22}^{33} = -\gamma\frac{\kappa^3+\kappa^{-3}}{2(\kappa^3-\kappa^{-3})};$$

$$r_{21}^{12} = r_{13}^{31} = \gamma\frac{\kappa^2+\kappa^{-1}}{2(\kappa^3-\kappa^{-3})};$$

$$r_{13}^{12} = r_{21}^{31} = \gamma\frac{\kappa^2-\kappa^{-1}}{2(\kappa^3-\kappa^{-3})};$$

$$r_{31}^{13} = r_{12}^{21} = \gamma\frac{\kappa+\kappa^{-2}}{2(\kappa^3-\kappa^{-3})};$$

$$r_{12}^{13} = r_{31}^{21} = -\gamma\frac{\kappa-\kappa^{-2}}{2(\kappa^3-\kappa^{-3})};$$

$$r_{32}^{23} = \gamma\frac{\kappa^2}{(\kappa^3-\kappa^{-3})};$$

$$r_{23}^{32} = \gamma\frac{\kappa^{-2}}{(\kappa^3-\kappa^{-3})}.$$

$$(3.14)$$

One should mention finally that there are very few completely integrable systems where one relativistic scalar (Hermitian) field, u, satisfies an equation of the form

$$\partial_t^2 u - \partial_x^2 u + f(u) = 0.$$

Here $f(u)$ is some arbitrary function of the field. Besides the sine-Gordon (sinh-Gordon) and Zhiber-Shabat-Mikhailov equations, the Liouville ($f(u) = e^u$) and Klein-Gordon ($f(u) = m^2 u$) equations are the only other forms that are completely integrable.

More examples of classical completely integrable differential equations can be found in Chapter XIV. They will appear in relation to quantum correlation functions.

Conclusion

In this chapter we have learned that completely integrable evolution equations are very special. They have an infinite set of conservation laws. The trace of the monodromy matrix $\tau(\lambda)$ is the generating functional of these conservation laws. The Lax representation (1.1) introduces an auxiliary matrix structure (and spectral parameter λ), which helps in solving the model. Theorem 3 in section 2 shows that the existence of an r-matrix guarantees the existence of a Lax representation for the given equation. This is the reason why we shall mainly pay attention to the spatial shift operator $(\partial_x + V)$ rather than the time-evolution operator $(\partial_t - U)$.

The Hamiltonian aspects of the classical inverse scattering method which we have presented above will be used for quantization in the next chapter. The method of calculating the PB based on the classical r-matrix was introduced in [13]. The proof of Theorem 1 which we have given appeared in [6], and the proof of its discrete analogue (Theorem 2) in [9]. Solutions of equation (2.15) are completely investigated in [2], [4], [12] and [10]. Let us mention that the nonlinear Schrödinger equation was solved with the help of CISM in [19] and its r-matrix (2.16) found in [13]. The trace identities were obtained in [18]. The sine-Gordon model was solved in [1], [8], [15], [16], [17] and [20] and the r-matrix was found in [14]. The trace identities can be found, for example, in [5]. The Zhiber-Shabat-Mikhailov model was introduced in [3] and solved in [11] and [21]. Its r-matrix was computed in [6]. It should be noted that all integrable equations for one interacting relativistic scalar field in two dimensions were enumerated in [11] and [21]. These are the sine-Gordon model, the hyperbolic sine-Gordon model (obtained by analytic continuation in the coupling constant) and the Zhiber-Shabat-Mikhailov model (the Liouville model is a limiting case of the latter).

In order to find out how the Lax representation helps in actually solving differential equations one should look through chapters XIV–XVI.

Appendix V.1: Tensor notation

We use the following notation. The tensor product of two $k \times k$ matrices A and B (which is a $k^2 \times k^2$ matrix $A \otimes B$) is defined as

$$(A \otimes B)^{ij}_{kl} = A_{ij} B_{kl}, \qquad (i, j, k, l = 1, \ldots, k). \tag{A.1.1}$$

The matrix elements of the $k^2 \times k^2$ matrices are labeled by the block indices i, j (i is the number of the block row and j that of the block column) and by the intrinsic indices k, l (k is the number of the row and l that of the column). As an example, we shall write out a 4×4 matrix R which acts in the tensor product of two 2×2 spaces:

$$R = \begin{pmatrix} R^{11}_{11} & R^{11}_{12} & R^{12}_{11} & R^{12}_{12} \\ R^{11}_{21} & R^{11}_{22} & R^{12}_{21} & R^{12}_{22} \\ R^{21}_{11} & R^{21}_{12} & R^{22}_{11} & R^{22}_{12} \\ R^{21}_{21} & R^{21}_{22} & R^{22}_{21} & R^{22}_{22} \end{pmatrix}. \tag{A.1.2}$$

In this notation, the matrix product of two $k^2 \times k^2$ matrices is written in the following way:

$$(RS)^{ij}_{kl} = R^{im}_{kn} S^{mj}_{nl} \tag{A.1.3}$$

(summation over repeated indices is assumed). The unit matrix E and permutation matrix Π (both $k^2 \times k^2$ matrices) are:

$$E = I \otimes I, \qquad E^{ij}_{kl} = \delta_{ij} \delta_{kl}; \tag{A.1.4}$$

$$\Pi^{ij}_{kl} = \delta_{il} \delta_{kj}, \qquad \Pi^2 = E. \tag{A.1.5}$$

Let us define the PB tensor product of two matrices A and B (both of which depend on the dynamical variables) by

$$\{A \overset{\otimes}{,} B\}^{ij}_{kl} = \{A_{ij}, B_{kl}\}. \tag{A.1.6}$$

The right hand side is the usual PB of the corresponding matrix elements of the matrices A and B.

Sometimes a different notation is used. Denoting the space where a matrix acts by a subscript, we can rewrite the relation

$$\{\mathsf{T}(\lambda) \overset{\otimes}{,} \mathsf{T}(\mu)\} = [\mathsf{T}(\lambda) \otimes \mathsf{T}(\mu),\, \mathsf{r}(\lambda, \mu)] \qquad \text{(A.1.7)}$$

in the form

$$\{\mathsf{T}_1(\lambda),\, \mathsf{T}_2(\mu)\} = [\mathsf{T}_1(\lambda)\mathsf{T}_2(\mu),\, \mathsf{r}_{12}(\lambda, \mu)]. \qquad \text{(A.1.8)}$$

This notation is used in (2.15), (2.19).

VI

The Quantum Inverse Scattering Method

Introduction

The quantum inverse scattering method (QSIM) appears as the quantized form of the classical inverse scattering method. It allows us to reproduce the results of the Bethe Ansatz and to move ahead. QISM is now a well developed branch of mathematical physics. In this chapter the fundamentals of QISM are given and illustrated by concrete examples.

In section 1 the general scheme of QISM, which allows the calculation of commutation relations between elements of the transfer matrix (necessary to construct eigenfunctions of the Hamiltonian in Chapter VII) is presented, and the quantum R-matrix is introduced. As in the classical case, the existence of an R-matrix and trace identities ensures that a Lax representation for the model exists. Thus, there are infinitely many conservation laws.

The Yang-Baxter equation, which is satisfied by the R-matrix, is discussed in section 2. Some important features of the R-matrix are also mentioned. The trace identities for the quantum nonlinear Schrödinger equation are proved in section 3. The general scheme of QISM is applied to the quantum sine-Gordon and Zhiber-Shabat-Mikhailov models in section 4. Spin models of quantum statistical physics are discussed in section 5. It is shown that a fundamental spin model can be constructed with the help of any given R-matrix.

The connection between classical statistical models on a two-dimensional lattice and QISM is established in section 6. QISM is the generalization of the classical inverse scattering method. In section 6 the partition function of the six-vertex model with domain wall boundary conditions is introduced. Later it will be studied extensively, and finally it will be represented as the determinant of a matrix. This idea will be developed in Part III. Finally this will permit us to write down a determinant representation for quantum correlation functions, and in Part IV this will lead to the complete solution of the problem of correlation functions. Hence, the main formalism of Chapter V will be used here again, but in the quantum version.

VI.1 General scheme

The Hamiltonian H and the matrix elements of the matrices $\mathsf{L}(n|\lambda)$ and $V(n|\lambda)$ (see section 1 of Chapter V) for a given model are now quantum operators depending on the dynamical variables of the system. For relativistic systems, regularization is necessary due to ultraviolet divergences. Thus, we shall consider the lattice formulation where the inverse lattice spacing, Δ^{-1}, has the meaning of an ultraviolet cut-off. Continuous models can also be considered within QISM (see section 3, where the quantum continuous nonlinear Schrödinger equation is considered). We shall consider models on a periodic lattice with M sites.

As in the classical case, the transition matrix T is defined by the following relation:

$$\mathsf{T}(n, m|\lambda) = \mathsf{L}(n|\lambda) \cdots \mathsf{L}(m|\lambda), \qquad (n \geq m). \tag{1.1}$$

Here, $\mathsf{L}(m|\lambda)$ can be obtained by direct quantization of the corresponding classical expression. This definition coincides with (V.1.8), but the matrix elements of T and L are now quantum operators which, generally speaking, do not commute with each other. The monodromy matrix, $\mathsf{T}(\lambda)$, is the transition matrix through the entire lattice:

$$\mathsf{T}(\lambda) \equiv \mathsf{T}(M, 1|\lambda). \tag{1.2}$$

The transfer matrix, $\tau(\lambda)$, defined as the trace of the monodromy matrix (exactly as in (V.1.9)),

$$\tau(\lambda) = \operatorname{tr} \mathsf{T}(\lambda), \tag{1.3}$$

plays an important role. Eigenfunctions of the quantum Hamiltonian will be constructed with the help of the monodromy matrix in Chapter VII. Therefore, it is important to find the commutators between all its matrix elements. This problem is solved within QISM by means of the R-matrix method, which is applicable if the commutation relations between the elements of the L-operator can be represented in the form

$$\mathsf{R}(\lambda, \mu) \left(\mathsf{L}(k|\lambda) \bigotimes \mathsf{L}(k|\mu) \right) = \left(\mathsf{L}(k|\mu) \bigotimes \mathsf{L}(k|\lambda) \right) \mathsf{R}(\lambda, \mu) \tag{1.4}$$

where the matrix elements of the L-operator commute at different sites:

$$[\mathsf{L}_{ij}(p|\lambda), \mathsf{L}_{kl}(q|\mu)] = 0, \qquad \text{when } p \neq q. \tag{1.5}$$

The relation (1.5) is usually referred to as the ultralocality property.

Theorem 1. *If the relations (1.4) and (1.5) are valid, then the commutation relations between matrix elements of the transition matrix $T(n,m|\lambda)$ (1.1) are given by a formula similar to (1.4):*

$$R(\lambda,\mu)\Big(T(n,m|\lambda)\bigotimes T(n,m|\mu)\Big) = \Big(T(n,m|\mu)\bigotimes T(n,m|\lambda)\Big)R(\lambda,\mu).$$

$$(1.6)$$

Here $R(\lambda,\mu)$ is a c-number $k^2 \times k^2$ matrix depending on the spectral parameters λ and μ only.

Proof: The proof can be easily obtained by induction. The relation (1.4) is the basis of the induction. Fixing m in (1.6) and assuming that this relation is valid for $n = j$, one then easily proves that (1.6) is valid for $n = j + 1$. Indeed,

$$R(\lambda,\mu)\Big(T(j+1,m|\lambda)\bigotimes T(j+1,m|\mu)\Big)$$
$$= R(\lambda,\mu)\Big(L(j+1|\lambda)\bigotimes L(j+1|\mu)\Big)$$
$$\times R^{-1}(\lambda,\mu)R(\lambda,\mu)\Big(T(j,m|\lambda)\bigotimes T(j,m|\mu)\Big)R^{-1}(\lambda,\mu)R(\lambda,\mu)$$
$$= \Big(L(j+1|\mu)\bigotimes L(j+1|\lambda)\Big)\Big(T(j,m|\mu)\bigotimes T(j,m|\lambda)\Big)R(\lambda,\mu)$$
$$= \Big(T(j+1,m|\mu)\bigotimes T(j+1,m|\lambda)\Big)R(\lambda,\mu). \qquad (1.7)$$

So Theorem 1 is proved.

The relations (1.4) and (1.6) are usually called bilinear relations. Relation (1.4) means that the L-operator is "intertwined" by the R-matrix. The transition matrix is "intertwined" by the same R-matrix (1.6).

Sometimes it is preferable to rewrite the bilinear relation in a different notation. Multiplying (1.6) from the left by the permutation matrix Π (V.2.2), one obtains

$$\tilde{R}(\lambda,\mu)\Big(T(\lambda)\bigotimes I\Big)\Big(I\bigotimes T(\mu)\Big) = \Big(I\bigotimes T(\mu)\Big)\Big(T(\lambda)\bigotimes I\Big)\tilde{R}(\lambda,\mu)$$

$$(1.8)$$

where

$$\tilde{R}(\lambda,\mu) = \Pi R(\lambda,\mu). \qquad (1.9)$$

In (1.8) we have suppressed the space coordinates of $T(n,m|\lambda)$. Equation (1.8) is also valid for the monodromy matrix (1.2). The $k \times k$ matrix $T(\lambda)$ acts only on the first vector space, and the matrix $T(\mu)$ acts only on the second vector space. Thus we can rewrite (1.8) in the form

$$\tilde{R}_{12}(\lambda,\mu)T_1(\lambda)T_2(\mu) = T_2(\mu)T_1(\lambda)\tilde{R}_{12}(\lambda,\mu), \qquad (1.10)$$

where the subindices denote the vector space(s) on which the matrix acts. The matrix $\tilde{\mathsf{R}}$ acts on the tensor product of the two spaces 1 and 2. (Tensor notation is explained in the Appendix to Chapter V.)

To construct integrable lattice models, we have to consider tensor products of many vector spaces. Then expression (1.10) will be generalized as

$$\tilde{\mathsf{R}}_{\alpha\beta}(\lambda,\mu)\mathsf{T}_{\alpha}(\lambda)\mathsf{T}_{\beta}(\mu) = \mathsf{T}_{\beta}(\mu)\mathsf{T}_{\alpha}(\lambda)\tilde{\mathsf{R}}_{\alpha\beta}(\lambda,\mu) \qquad (1.11)$$

where α and β denote (different) vector spaces.

A consequence of (1.6) is the existence of a commuting family of transfer matrices $\tau(\lambda)$, for which

$$[\tau(\lambda),\,\tau(\mu)] = 0. \qquad (1.12)$$

To prove this, one rewrites the relation (1.6) as

$$\mathsf{R}(\lambda,\mu)\,\mathsf{T}(\lambda)\bigotimes\mathsf{T}(\mu)\,\mathsf{R}^{-1}(\lambda,\mu) = \mathsf{T}(\mu)\bigotimes\mathsf{T}(\lambda) \qquad (1.13)$$

and takes the trace in the space of $k^2 \times k^2$ matrices to obtain $\tau(\lambda)\tau(\mu) = \tau(\mu)\tau(\lambda)$ (using the fact that the trace of the tensor product of matrices is equal to the product of their traces).

In the quantum case, as in the classical one, the Hamiltonian of a model can be represented as a linear combination of logarithmic derivatives of the transfer matrix $\tau(\lambda)$ (1.3) at some points ν_a:

$$H = \sum_{k}\sum_{a} c_{ka}\frac{d^k}{d\lambda^k}\ln\tau(\lambda)\Big|_{\lambda=\nu_a} \qquad (1.14)$$

(here the c_{ka} are coefficients). The points ν_a are chosen as to make the Hamiltonian local. These formulæ are known as the trace identities. Examples are given in sections 3 and 5. It follows from (1.12) that (as in the classical case) an infinite number of conservation laws exist in the infinite lattice limit $M \to \infty$. Indeed, from (1.12) it follows immediately that $[\ln\tau(\lambda),\,\tau(\mu)] = 0$. Differentiating this with respect to λ, and using (1.14), we get $[H,\,\tau(\mu)] = 0$. Now we can expand this in a Taylor series in μ and obtain an infinite set of conservation laws, similar to the classical case (section 2, Chapter V).

The Lax representation for the nonlinear equation also follows from the existence of the R-matrix. First, let us construct the generating functional $U(n|\lambda,\mu)$ of the time evolution operators (in a way similar to that of the classical case (V.2.19)–(V.2.20)).

Theorem 2. *The following Lax representation exists in the quantum case:*

$$i\left[\frac{d}{d\mu}\ln\tau(\mu), \mathsf{L}(n|\lambda)\right] = U(n+1|\lambda,\mu)\mathsf{L}(n|\lambda) - \mathsf{L}(n|\lambda)U(n|\lambda,\mu). \quad (1.15)$$

The generating functional of the time evolution operators is equal to

$$U(n|\lambda,\mu) = i\frac{d}{d\mu}\ln\tau(\mu)I - iq^{-1}(n|\lambda,\mu)\frac{d}{d\mu}q(n|\lambda,\mu) \quad (1.16)$$

where $q(n|\lambda,\mu)$ is the $k \times k$ matrix

$$q_2(n|\lambda,\mu) = \mathrm{tr}_1\left[\mathsf{T}_1(M,n|\mu)\tilde{\mathsf{R}}_{12}(\mu,\lambda)\mathsf{T}_1(n-1,1|\mu)\right]. \quad (1.17)$$

The subscripts 1, 2 label different vector spaces. (Note that the subindex 2 is only the label of the auxiliary vector space and can be dropped.)

Proof: The commutation relation between $\mathsf{L}(n|\lambda)$ and $q(n|\lambda,\mu)$ is easily obtained from (1.4), (1.6) and (1.10):

$$q(n+1|\lambda,\mu)\mathsf{L}(n|\lambda) = \mathsf{L}(n|\lambda)q(n|\lambda,\mu) \quad (1.18)$$

where these are matrix products of L and q on both sides. Differentiating this expression with respect to μ and multiplying it by $q^{-1}(n+1|\lambda,\mu)$ on the left gives

$$q^{-1}(n+1|\lambda,\mu)\frac{d}{d\mu}q(n+1|\lambda,\mu)\mathsf{L}(n|\lambda) =$$

$$q^{-1}(n+1|\lambda,\mu)\mathsf{L}(n|\lambda)\frac{d}{d\mu}q(n|\lambda,\mu). \quad (1.19)$$

To transform the right hand side, let us rewrite the commutation relation (1.18) between q and L as

$$q^{-1}(n+1|\lambda,\mu)\mathsf{L}(n|\lambda) = \mathsf{L}(n|\lambda)q^{-1}(n|\lambda,\mu) \quad (1.20)$$

so that

$$q^{-1}(n+1|\lambda,\mu)\frac{d}{d\mu}q(n+1|\lambda,\mu)\mathsf{L}(n|\lambda)$$

$$= \mathsf{L}(n|\lambda)q^{-1}(n|\lambda,\mu)\frac{d}{d\mu}q(n|\lambda,\mu). \quad (1.21)$$

Now we see that (1.15) follows from (1.16) and (1.21). The theorem is thus proved. The first proof was published in [30].

With the help of the trace identities, it is possible to change the left hand side in (1.15) to $\partial_t\mathsf{L}(n|\lambda) = i[H, \mathsf{L}(n|\lambda)]$ and the time evolution operator $U(n|\lambda)$ for the Hamiltonian H is then equal to a linear combination of $d^l U(n|\lambda,\mu)/d\mu^l$ at $\mu = \nu_a$. For the Hamiltonian (1.14), the zero curvature condition exists:

$$\partial_t \mathsf{L}(n|\lambda) = i[H, \mathsf{L}(n|\lambda)] = U(n+1|\lambda)\mathsf{L}(n|\lambda) - \mathsf{L}(n|\lambda)U(n|\lambda). \quad (1.22)$$

Moreover, the evolution operator is constructed from the generating functional as follows:

$$U(n|\lambda) = \sum_k \sum_a c_{ka} \frac{d^{(k-1)}}{d\mu^{(k-1)}} U(n|\lambda, \mu)\Big|_{\mu=\nu_a}. \quad (1.23)$$

This operator possesses the correct quasiclassical limit and is local (for Hamiltonians with local interactions).

All considerations in this section also remain valid for quantum continuum models (they can be obtained, for example, as a formal continuous limit of lattice models). The relation between lattice and continuous models was shown in Chapter V (quantization does not change this).

VI.2 The Yang-Baxter relation

Not any $k^2 \times k^2$ matrix $\mathsf{R}(\lambda, \mu)$ can be an R-matrix for some integrable model. There are some restrictions which can be derived as the compatibility conditions for the relations (1.6). Namely, consider the tensor product $\mathsf{T}(\lambda) \otimes \mathsf{T}(\mu) \otimes \mathsf{T}(\nu)$. This expression can be reduced to $\mathsf{T}(\nu) \otimes \mathsf{T}(\mu) \otimes \mathsf{T}(\lambda)$ in two different ways. Using (1.6) one has

$$
\begin{aligned}
\mathsf{T}(\lambda) &\otimes \mathsf{T}(\mu) \otimes \mathsf{T}(\nu) \\
&= \left(\mathsf{R}^{-1}(\lambda, \mu) \otimes I\right) \left(I \otimes \mathsf{R}^{-1}(\lambda, \nu)\right) \left(\mathsf{R}^{-1}(\mu, \nu) \otimes I\right) \\
&\quad \times \left(\mathsf{T}(\nu) \otimes \mathsf{T}(\mu) \otimes \mathsf{T}(\lambda)\right) \\
&\quad \times \left(\mathsf{R}(\mu, \nu) \otimes I\right) \left(I \otimes \mathsf{R}(\lambda, \nu)\right) \left(\mathsf{R}(\lambda, \mu) \otimes I\right) \\
&= \left(I \otimes \mathsf{R}^{-1}(\mu, \nu)\right) \left(\mathsf{R}^{-1}(\lambda, \nu) \otimes I\right) \left(I \otimes \mathsf{R}^{-1}(\lambda, \mu)\right) \\
&\quad \times \left(\mathsf{T}(\nu) \otimes \mathsf{T}(\mu) \otimes \mathsf{T}(\lambda)\right) \\
&\quad \times \left(I \otimes \mathsf{R}(\lambda, \mu)\right) \left(\mathsf{R}(\lambda, \nu) \otimes I\right) \left(I \otimes \mathsf{R}(\mu, \nu)\right). \quad (2.1)
\end{aligned}
$$

The sufficient condition for the validity of this equality is the famous Yang-Baxter relation

$$
\begin{aligned}
\left(I \otimes \mathsf{R}(\lambda, \mu)\right) \left(\mathsf{R}(\lambda, \nu) \otimes I\right) \left(I \otimes \mathsf{R}(\mu, \nu)\right) = \\
= \left(\mathsf{R}(\mu, \nu) \otimes I\right) \left(I \otimes \mathsf{R}(\lambda, \nu)\right) \left(\mathsf{R}(\lambda, \mu) \otimes I\right). \quad (2.2)
\end{aligned}
$$

Later we shall discover that exactly solvable models which have the same R-matrix but different L-operators are closely related to each other. One can compare this with different representations of the same Lie algebra. The R-matrix should be compared with the set of structure constants and the Yang-Baxter equation can be viewed as the analogue of the Jacobi identity. In section 6 of Chapter VII, we shall enumerate all the monodromy matrices whose commutation relations are given by a fixed R-matrix. In section 4 of Chapter VIII we shall enumerate all the L-operators which give rise to the same R-matrix. This is similar to the enumeration of all representations of a fixed Lie algebra.

The simplest example of an R-matrix is the rational R-matrix for the XXX model:

$$R(\lambda, \mu) = i\hbar E + (\mu - \lambda)\Pi. \tag{2.3}$$

An example of a nontrivial R-matrix is presented in section 4. Before continuing, a few comments must be made about some properties of the R-matrix. Usually the spectral parameter is a complex number which can be chosen so that the R-matrix depends on the difference of its arguments:

$$R(\lambda, \mu) \equiv R(\lambda - \mu). \tag{2.4}$$

Thus, we can shift the spectral parameter of the L-operator. If some operator $L(\lambda)$ satisfies the bilinear relation (1.4), then $L(\lambda - \nu)$ satisfies this relation also:

$$R(\lambda - \mu)\left(L(\lambda - \nu)\bigotimes L(\mu - \nu)\right) = \left(L(\mu - \nu)\bigotimes L(\lambda - \nu)\right)R(\lambda - \mu) \tag{2.5}$$

for any complex number ν.

It follows from (1.6) that the R-matrix can be multiplied by an arbitrary complex-valued function, which can be chosen such that

$$R(\lambda, \mu) = R^{-1}(\mu, \lambda), \qquad R(\lambda, \lambda) = E. \tag{2.6}$$

The quasiclassical limit for such R-matrices is especially simple:

$$R \xrightarrow[\hbar \to 0]{} \Pi\left(E - i\hbar r\right)$$

where r is the corresponding classical r-matrix. As a rule, this normalization condition (2.6) will be used below.

At the end of section 4 we include, as an example, an R-matrix which depends on a three-dimensional spectral parameter $\widetilde{\lambda}$ (actually, $\widetilde{\lambda}$ can be realized as a 2×2 matrix belonging to $SL(2, \mathbb{C})$).

VI.3 Trace identities

We now discuss the trace identities for the quantum nonlinear Schrödinger equation which describes a one-dimensional Bose gas. The Hamiltonian is:

$$H = \int dx \left[\partial_x \Psi^\dagger \, \partial_x \Psi + c\Psi^\dagger \Psi^\dagger \Psi \Psi\right]. \tag{3.1}$$

Here Ψ, $\Psi^\dagger$ are Hermitian-conjugate bosonic operators with the following canonical commutation relations:

$$[\Psi(x), \Psi^\dagger(y)] = \delta(x - y);$$
$$[\Psi(x), \Psi(y)] = [\Psi^\dagger(x), \Psi^\dagger(y)] = 0. \tag{3.2}$$

The operators for momentum, P, and number of particles, Q, are given by (V.1.13). The quantum transition matrix is constructed from the classical one by normal ordering (the operators $\Psi^\dagger$ must stand to the left of the operators Ψ):

$$\mathsf{T}_{\mathrm{q}}(x, y|\lambda) = \; : \mathsf{T}_{\mathrm{cl}}(x, y|\lambda) :$$

In the quantum case, one can act almost as in the classical case. Let us consider the quantum operator $\partial_x + V(x|\lambda)$ (see (V.1.14)). One obtains for the transfer matrix analogously to (V.1.6):

$$\left(\partial_x + i\frac{\lambda}{2}\sigma_z + \Omega(x)\right)\mathsf{T}(x, y|\lambda) = 0. \tag{3.3}$$

Periodic boundary conditions in x are imposed on $\Psi(x)$ (the period being equal to L). It should be mentioned that the matrix elements of the 2×2 matrix T are now functionals of the quantum field operators $\Psi(z)$, $\Psi^\dagger(z)$ ($x \geq z \geq y$). The transition matrix possesses the following property:

$$\sigma_x \mathsf{T}^*(x, y|\lambda^*)\sigma_x = \mathsf{T}(x, y|\lambda). \tag{3.4}$$

Here the asterisk following T means Hermitian conjugation of its matrix elements (but no transposition is performed). The monodromy matrix is defined as in the classical case:

$$\mathsf{T}(\lambda) = \mathsf{T}(L, 0|\lambda). \tag{3.5}$$

As before, the matrix trace of the monodromy matrix, $\tau(\lambda) = \mathrm{tr}\,\mathsf{T}(\lambda)$, is the transfer matrix. One can express the integrals of motion H, P and Q with the help of the trace identities:

$$\ln\left(e^{i\lambda L/2}\tau(\lambda)\right) \xrightarrow[\lambda \to i\infty]{} ic\Big\{\lambda^{-1}Q + \lambda^{-2}[P - (ic/2)Q]$$
$$+ \lambda^{-3}[H - icP - (c^2/3)Q] + O\left(\lambda^{-4}\right)\Big\}. \tag{3.6}$$

The proof of this formula is similar to that in the classical case (see section 1 of Chapter V). Making a gauge transformation (V.1.22)–(V.1.28) which diagonalizes $\mathsf{T}(x, y|\lambda)$, one looks for $D(L, 0|\lambda)$ of the following form:

$$D(L, 0|\lambda) = \exp\left\{-i\frac{\lambda L}{2}\sigma_z\right\}\left[1 + A_1\lambda^{-1} + A_2\lambda^{-2} + A_3\lambda^{-3} + O\left(\lambda^{-4}\right)\right] \tag{3.7}$$

and, using (V.1.27) one finds that

$$A_1 = - : \int_0^L W_1(z)\,dz : \tag{3.8}$$

$$A_2 = - : \int_0^L W_2(z)\,dz : + : \int_0^L W_1(z)\,dz \int_0^z W_1(y)\,dy : \tag{3.9}$$

$$A_3 = - : \int_0^L W_3(z)\,dz : + : \int_0^L W_2(z)\,dz \int_0^z W_1(y)\,dy :$$

$$+ : \int_0^L W_1(z)\,dz \int_0^z W_2(y)\,dy :$$

$$- : \int_0^L W_1(z)\,dz \int_0^z W_1(y)\,dy \int_0^y W_1(t)\,dt : \tag{3.10}$$

Here, the W_i are the same as in the classical case (see V.1.28). Due to periodic boundary conditions, we have $G(0) = G(L)$ and $\tau(\lambda) = \operatorname{tr} D(L,0|\lambda)$. As $D_{22} \ll D_{11}$ for $\lambda \to i\infty$, one has that $\tau(\lambda) = D_{11}(L,0|\lambda)$ $(\lambda \to i\infty)$. Taking the logarithm one gets:

$$\ln\left(e^{i\lambda L/2}\tau(\lambda)\right) \xrightarrow[\lambda \to i\infty]{} \ln\left(1 + a_1\lambda^{-1} + a_2\lambda^{-2} + a_3\lambda^{-3}\right)$$
$$= \lambda^{-1}b_1 + \lambda^{-2}b_2 + \lambda^{-3}b_3. \tag{3.11}$$

Here, by a_i we denote the element in position 11 of the corresponding matrix A_i (see (3.7)–(3.10)), and

$$b_1 = a_1; \qquad b_2 = a_2 - \frac{a_1^2}{2};$$
$$b_3 = a_3 - \frac{a_1a_2 + a_2a_1}{2} + \frac{a_1^3}{3}. \tag{3.12}$$

The quantities a_1, a_2, a_3 and b_1 are normal ordered, but the quantities b_2 and b_3 are not. Reducing b_2 and b_3 to the normal ordered form, one obtains "quantum corrections" (differences between the quantum and the classical trace identities):

$$b_1 = icQ; \qquad b_2 = icP + \frac{c^2}{2}Q;$$
$$b_3 = icH + c^2 P - i\frac{c^3}{3}Q. \tag{3.13}$$

Thus we obtain the quantum trace identities (3.6). When constructing the Hamiltonian of the lattice nonlinear Schrödinger model (see Chapter VIII, section 3), we shall return to the quantum trace identities.

Let us now calculate the R-matrix, which gives the commutation relations between matrix elements of the monodromy matrix. The calculation will be obvious if we use the infinitesimal lattice formulation instead of the continuous one. The L-operator is given by the same classical formula ((V.1.19) and (V.1.20)):

$$L(n|\lambda) = \begin{pmatrix} 1 - i\frac{\lambda\Delta}{2} & -i\sqrt{c}\Psi_n^\dagger\Delta \\ i\sqrt{c}\Psi_n\Delta & 1 + i\frac{\lambda\Delta}{2} \end{pmatrix} + O\left(\Delta^2\right) \tag{3.14}$$

where the quantum operators Ψ_n, $\Psi_n^\dagger$ satisfy

$$[\Psi_n, \Psi_m^\dagger] = \frac{1}{\Delta}\delta_{nm}. \tag{3.15}$$

Elementary calculations show that the bilinear relation (1.4) is valid and the R-matrix is the same as (2.3):

$$R(\lambda, \mu) = \Pi - i\frac{c}{\lambda - \mu}E \tag{3.16}$$

or, in explicit form,

$$R(\lambda, \mu) = \begin{pmatrix} f(\mu, \lambda) & 0 & 0 & 0 \\ 0 & g(\mu, \lambda) & 1 & 0 \\ 0 & 1 & g(\mu, \lambda) & 0 \\ 0 & 0 & 0 & f(\mu, \lambda) \end{pmatrix} \tag{3.17}$$

where

$$f(\mu, \lambda) = 1 + \frac{ic}{\mu - \lambda}, \qquad g(\mu, \lambda) = \frac{ic}{\mu - \lambda}. \tag{3.18}$$

This R-matrix will be called the R-matrix of the *XXX* model (the *XXX* R-matrix).

VI.4 Quantum field theory models

In this section some examples of quantum field theory models solvable by QISM will be presented.

(1) The quantum sine-Gordon (SG) model is given by the same Hamiltonian (V.3.2) as in the classical case (see section 3 of Chapter V). The fields $\pi(x)$ and $u(x)$ are now quantum bosonic operators with canonical commutation relations $[u(x), \pi(y)] = i\delta(x - y)$. The lattice quantities u_n and p_n are also canonical operators:

$$[u_n, p_m] = i\delta_{mn}. \tag{4.1}$$

The correct version of the infinitesimal quantum L-operator of the model,

$$\mathsf{L}(n|\lambda) = \begin{pmatrix} \exp(-i\beta p_n/4) & (m\Delta/2)\sinh(\lambda - i\beta u_n/2) \\ (-m\Delta/2)\sinh(\lambda + i\beta u_n/2) & \exp(i\beta p_n/4) \end{pmatrix}$$
$$+ O\left(\Delta^2\right) \qquad (4.2)$$

was introduced in [54]. This formula will be clarified in Chapter VIII, section 2. It is interesting to note that in the classical case this L-operator can also be used. The difference between the quantum and classical cases is the following: in the classical case p_n can be considered as bounded and $\exp\{i\beta p_n/4\}$ can be replaced by $1 + i\beta p_n/4$. This is impossible in the quantum case, because p_n is an unbounded operator.

The quantum R-matrix is

$$\mathsf{R}(\lambda, \mu) = \begin{pmatrix} f(\mu, \lambda) & 0 & 0 & 0 \\ 0 & g(\mu, \lambda) & 1 & 0 \\ 0 & 1 & g(\mu, \lambda) & 0 \\ 0 & 0 & 0 & f(\mu, \lambda) \end{pmatrix} \qquad (4.3)$$

where the functions f and g are

$$f(\mu, \lambda) = \frac{\sinh(\mu - \lambda - i\gamma)}{\sinh(\mu - \lambda)}; \qquad g(\mu, \lambda) = -i\frac{\sin\gamma}{\sinh(\mu - \lambda)} \qquad (4.4)$$

and

$$0 < \gamma \equiv \beta^2/8 < \pi. \qquad (4.5)$$

This R-matrix will be called the R-matrix of the XXZ model (XXZ R-matrix).

The monodromy matrix possesses the involution

$$\sigma_y \mathsf{T}^*(\lambda^*)\sigma_y = \mathsf{T}(\lambda). \qquad (4.6)$$

Later on (see Chapter VIII), we shall put the model on a lattice while preserving the property of complete integrability; that is, we will construct the exact L-operator intertwined by the R-matrix (4.3), (4.4). This will be the direct solution of the ultraviolet divergence problem.

(2) The Zhiber-Shabat-Mikhailov model in the quantum case is given by the Hamiltonian (V.3.11) with

$$[u(y), \pi(x)] = i\delta(x - y). \qquad (4.7)$$

The nonzero elements of the 9×9 R-matrix are given by (setting $\varphi = \gamma/8$, $\beta = \lambda - \mu$ and $s = e^{-3\beta + 5i\varphi} - e^{3\beta - 5i\varphi} + e^{-i\varphi} - e^{i\varphi}$):

$$s\mathsf{R}_{11}^{11} = -e^{3\beta - 3i\varphi} + e^{-3\beta + 3i\varphi} + e^{5i\varphi} - e^{-5i\varphi} - e^{3i\varphi} + e^{-3i\varphi} - e^{i\varphi} + e^{-i\varphi};$$

$$s\mathsf{R}_{22}^{11} = s\mathsf{R}_{11}^{33} = e^{-2\beta + 5i\varphi} - e^{-2\beta + i\varphi} + e^{\beta - i\varphi} - e^{\beta - 5i\varphi};$$

$R_{22}^{22} = R_{33}^{33} = 1;$

$sR_{33}^{11} = sR_{11}^{22} = e^{2\beta - i\varphi} - e^{2\beta - 5i\varphi} + e^{-\beta + 5i\varphi} - e^{-\beta + i\varphi};$

$sR_{33}^{22} = e^{-2\beta + 5i\varphi} - e^{-2\beta - 3i\varphi} - e^{\beta + i\varphi} + e^{\beta - i\varphi} + e^{\beta - 3i\varphi} - e^{\beta - 5i\varphi};$

$sR_{22}^{33} = e^{2\beta + 3i\varphi} - e^{2\beta - 5i\varphi} + e^{-\beta + 5i\varphi} - e^{-\beta + 3i\varphi} + e^{-\beta - i\varphi} - e^{-\beta + i\varphi};$

$sR_{21}^{12} = sR_{12}^{21} = sR_{31}^{13} = sR_{13}^{31} = -e^{3\beta - 3i\varphi} + e^{-3\beta + 3i\varphi} - e^{3i\varphi} + e^{-3i\varphi};$

$sR_{13}^{12} = sR_{31}^{21} = e^{2\beta} - e^{2\beta - 4i\varphi} - e^{-\beta} + e^{-\beta - 4i\varphi};$

$sR_{12}^{13} = sR_{21}^{31} = e^{-2\beta + 4i\varphi} - e^{-2\beta} - e^{\beta + 4i\varphi} + e^{\beta};$

$sR_{32}^{23} = sR_{23}^{32} = -e^{3\beta - i\varphi} + e^{-3\beta + i\varphi} + e^{-i\varphi} - e^{i\varphi}.$

$$(4.8)$$

This R-matrix is an example of a nontrivial solution of the Yang-Baxter equation.

Let us also mention another curious solution of the Yang-Baxter equation ([8] and [31]). This is an R-matrix which depends on a three-dimensional spectral parameter. In this case, we can represent the spectral parameter as a 2×2 matrix with unit determinant

$$\widehat{\lambda} = \begin{pmatrix} \lambda_{11} & \lambda_{12} \\ \lambda_{21} & \lambda_{22} \end{pmatrix}, \qquad \det \widehat{\lambda} = 1.$$

In other words, $\widehat{\lambda} \in SL(2, \mathbb{C})$. The R-matrix depends on the ordered ratio of two spectral parameters:

$$R(\widehat{\lambda}, \widehat{\mu}) = R(\widehat{\lambda} \widehat{\mu}^{-1}).$$

Let us define $\widehat{g} = \widehat{\lambda} \widehat{\mu}^{-1}$. The nonzero matrix elements of R can now be written in terms of the matrix elements of

$$\widehat{g} = \begin{pmatrix} g_{11} & g_{12} \\ g_{21} & g_{22} \end{pmatrix} :$$

$$R_{11}^{11} = g_{11}; \qquad R_{22}^{22} = g_{22};$$
$$R_{21}^{12} = i g_{12}; \qquad R_{12}^{21} = i g_{21};$$
$$R_{22}^{11} = R_{11}^{22} = 1 :$$

The most interesting example is quantization of the nonabelian Toda lattice, see [40].

VI.5 Fundamental spin models

Spin models present another set of examples solved by QISM. Their L-operators are constructed directly from R-matrices satisfying the Yang-Baxter equation. First, we shall construct the L-operator and then the monodromy and transfer matrices. Then, by means of trace identities, we shall construct the Hamiltonian.

Let us consider a one-dimensional lattice with M sites, and let n number the sites ($1 \le n \le M$). One begins the construction of the spin model connected with a given $k^2 \times k^2$ R-matrix by constructing the corresponding L-operator. This L-operator is a $k \times k$ matrix with matrix elements $\mathsf{L}_{\alpha\beta}(n|\lambda)$ ($\alpha, \beta = 1, \ldots, k$) which are themselves $k \times k$ matrices (quantum operators):

$$\mathsf{L}_{\alpha\beta}^{ij}(n|\lambda) = \mathsf{R}_{\alpha j}^{i\beta}(\lambda, \nu)\varphi(\lambda, \nu) = \tilde{\mathsf{R}}_{ij}^{\alpha\beta}(\lambda, \nu)\varphi(\lambda, \nu),$$
$$(\alpha, \beta, i, j = 1, \ldots, k). \tag{5.1}$$

Here $\varphi(\lambda, \nu)$ is an arbitrary c-number function. It is possible to include φ because all the relations are linear. Sometimes it is convenient to choose φ in a nontrivial way. This L-operator is essentially the R-matrix (1.9) at some fixed point $\mu = \nu$. The indices α, β are called matrix indices, while i, j are quantum indices. The matrix indices originate from the classical L-operator. The quantum space number is labeled by the index n. The Yang-Baxter equation can be rewritten with the help of (5.1) as a bilinear relation:

$$\mathsf{R}(\lambda, \mu)\Big(\mathsf{L}(n|\lambda) \bigotimes \mathsf{L}(n|\mu)\Big) = \Big(\mathsf{L}(n|\mu) \bigotimes \mathsf{L}(n|\lambda)\Big) \mathsf{R}(\lambda, \mu). \tag{5.2}$$

Thus, the L-operator is constructed directly from the R-matrix. The equality of dimensions of the quantum space and the matrix space is a distinctive property of these models. The models defined by these L-operators are called fundamental. Relation (5.2) can be generalized by multiplying the R-matrix by an arbitrary complex-valued function.

The further investigation of spin models is analogous to that of section 1. The monodromy matrix $\mathsf{T}(\lambda)$ and the transfer matrix $\tau(\lambda)$ are constructed as explained there. In the present case, $\tau(\lambda)$ is a quantum operator acting in the tensor product of M different quantum spaces. The Hamiltonian is defined by the trace identities:

$$H = \text{const} \left.\frac{d}{d\lambda} \ln \tau(\lambda)\right|_{\lambda=\nu}. \tag{5.3}$$

The point $\lambda = \nu$ is chosen in such a way that $\mathsf{R}(\nu, \nu) = E$; thus, the Hamiltonian describes the interaction of nearest neighbors on a lattice and is easily calculated:

$$H = \text{const} \sum_{n=1}^{M} H_{n-1,n}. \tag{5.4}$$

The operator $H_{n-1,n}$ acts nontrivially only at the two neighboring sites (n and $n-1$):

$$(H_{n-1,n})_{\{k\}}^{\{i\}} = \left.\left(\frac{d}{d\lambda}\mathsf{R}_{i_n k_n}^{i_{n-1} k_{n-1}}(\lambda, \nu)\right)\right|_{\lambda=\nu} + \left.\frac{d}{d\lambda}\ln\varphi(\lambda, \nu)\right|_{\lambda=\nu}, \tag{5.5}$$

while at sites $j \neq n - 1$, n it acts trivially. The second term in (5.5) is evident. Let us explain that the first term in (5.5) is presented correctly. The proof of these formulæ is based on the fact that $\tau(\nu)$ is the shift operator on one lattice site (provided that $R(\nu, \nu) = E$):

$$(\tau(\nu))_{\{k\}}^{\{i\}} = \prod_{j=1}^{M} \delta_{i_j, k_{j-1}}, \qquad (M + 1 \equiv 1). \qquad (5.6)$$

Let us discuss the Lax representation of the equations of motion which are generated by the above Hamiltonian. This representation exists and the explicit form of the time evolution operator can be obtained from the general formulæ (1.15), (1.16), (1.17) and (1.23):

$$[U_{\alpha\beta}(n|\lambda, \nu)]_{\{k\}}^{\{i\}} = i \frac{d}{d\mu} \left\{ \delta_{\alpha\beta} \, [R(\mu, \nu)]_{i_n k_n}^{i_{n-1} k_{n-1}} \right.$$

$$\left. - \left[R^{-1}(\nu, \lambda) \right]_{\alpha b}^{i_{n-1} a} [R(\mu, \nu)]_{i_n k_n}^{bc} \, [R(\mu, \lambda)]_{c\beta}^{a k_{n-1}} \right\} \Bigg|_{\mu=\nu}. \qquad (5.7)$$

As a quantum operator, it acts nontrivially on two neighboring sites.

The function $\varphi(\lambda, \nu)$ controls the dependence of the above construction on the normalisation of the R-matrix. The starting point above was $R(\nu, \nu) = E$. We shall use different normalizations, φ, in the examples which follow.

Let us consider the R-matrix of the *XXX* model (3.16)–(3.18) as an example. It is convenient, instead, to use the matrix $\tilde{R}$ (1.9):

$$\tilde{R}_{12}(\lambda, \mu) = E_{12} - \frac{ic}{\lambda - \mu} \Pi_{12} \qquad (5.8)$$

(the subscripts 1, 2 denote the vector spaces). We can rewrite this R-matrix as

$$\tilde{R}_{12}(\lambda, \mu) = \left[1 - \frac{ic}{2(\lambda - \mu)} \right] E_{12} - \frac{ic}{2(\lambda - \mu)} \sum_{i=x,y,z} \sigma_i^1 \sigma_i^2 \qquad (5.9)$$

where we have used the identity

$$2\Pi_{12} = I^1 I^2 + \sigma_x^1 \sigma_x^2 + \sigma_y^1 \sigma_y^2 + \sigma_z^1 \sigma_z^2. \qquad (5.10)$$

(Here σ are the Pauli matrices, with superscripts denoting the space upon which they act. In both (5.9) and (5.10), we have a tensor product of σ matrices.)

The corresponding L-operator (5.1) is

$$L(n|\lambda) = \lambda - \frac{ic}{2} \begin{pmatrix} \sigma_z^n & 2\sigma_-^n \\ 2\sigma_+^n & -\sigma_z^n \end{pmatrix} \qquad (5.11)$$

where $2\sigma_\pm = \sigma_x \pm i\sigma_y$. This can be rewritten as

$$\mathsf{L}(n|\lambda) = \lambda - \frac{ic}{2}\sigma_z\sigma_z^n - ic\left(\sigma_-\sigma_+^n + \sigma_+\sigma_-^n\right). \tag{5.12}$$

Usually, when discussing magnets, we shall assume that $c = 1$, which can always be achieved by rescaling the spectral parameter.

To obtain this L-operator, we must multiply the R-matrix (5.8) by $\varphi(\lambda, \nu) = (\lambda - \mu)$, use (5.1) and put $\nu = -i/2$. It should be noted that if $\mathsf{R}(\lambda, \mu)$ satisfies the Yang-Baxter equation, then $(\lambda - \mu)\mathsf{R}(\lambda, \mu)$ does also. The monodromy matrix T generated by this L-operator possesses the property

$$\mathsf{T}^*(\lambda^*) = \sigma_y\mathsf{T}(\lambda)\sigma_y \tag{5.13}$$

where the asterisk denotes Hermitian conjugation of quantum operators (without matrix transposition) and complex conjugation of c-numbers. The spin at the k-th site is described by the matrices σ^k (the quantum operators). The trace identities are

$$H = \pm 2i\frac{d}{d\mu}\ln\tau(\mu)\Big|_{\mu=-i/2} \pm 2M - 2hS_z. \tag{5.14}$$

Thus, the Hamiltonian of the isotropic XXX model considered in Chapter II is obtained:

$$H = \mp\sum_{k=1}^{M}\left(\sigma_x^k\sigma_x^{k+1} + \sigma_y^k\sigma_y^{k+1} + \sigma_z^k\sigma_z^{k+1} - 1\right) - 2hS_z. \tag{5.15}$$

The upper sign corresponds to the ferromagnetic case and the lower one to the antiferromagnetic case. S_z is the operator of the third component of the total spin:

$$S_z = \frac{1}{2}\sum_{n=1}^{M}\sigma_z^n \tag{5.16}$$

which commutes with the Hamiltonian and the transfer matrix:

$$[H, S_z] = [\tau(\lambda), S_z] = 0. \tag{5.17}$$

Let us now consider the anisotropic XXZ magnet. The Hamiltonian (see Chapter II)

$$H = -\sum_{k=1}^{M}\left[\sigma_x^k\sigma_x^{k+1} + \sigma_y^k\sigma_y^{k+1} + \Delta\left(\sigma_z^k\sigma_z^{k+1} - 1\right) + h\sigma_z^k\right] \tag{5.18}$$

is generated by the monodromy matrix of the fundamental spin model, which is constructed from the XXZ R-matrix (4.3). The trace identities are

$$H = 2i(\sin 2\eta)\frac{\partial}{\partial\mu}\ln\tau(\mu)\Big|_{\mu=i\pi/2-i\eta} +2M\cos 2\eta - 2hS_z. \qquad (5.19)$$

Here the coupling constant η, related to γ in the R-matrix (4.3), (4.4), is defined as

$$\gamma = -2\eta \qquad (5.20)$$

where the anisotropy parameter is $\Delta = \cos 2\eta$. The operator S_z commutes with the Hamiltonian and the transfer matrix as before. The L-operator is constructed by means of the R-matrix and is given by

$$\mathsf{L}(n|\lambda) = -i\begin{pmatrix} \cosh(\lambda - i\eta\sigma_z^n) & \sigma_-^n\sin 2\eta \\ \sigma_+^n\sin 2\eta & \cosh(\lambda + i\eta\sigma_z^n) \end{pmatrix}. \qquad (5.21)$$

The corresponding monodromy matrix satisfies the involution

$$\sigma_y\mathsf{T}(\lambda)\sigma_y = \mathsf{T}^*(\lambda^* + i\pi). \qquad (5.22)$$

The isotropic XXX magnet may be considered as the limiting case of the XXZ magnet. In the ferromagnetic case, the Hamiltonian (5.15), L-operator (5.11) and R-matrix (3.17), (3.18) are obtained from the corresponding quantities in the XXZ case (5.18), (5.21), (4.3), (4.4) in the limit (for even M)

$$2\eta = \kappa \to 0; \qquad \lambda_{XXZ} = i\frac{\pi}{2} + \kappa\lambda_{XXX} \qquad (5.23)$$

(λ_{XXX} finite, $c = 1$). In this case, the trace identities (5.19) reproduce formula (5.14).

The following comment must be made in conclusion. We can shift the spectral parameter in the L-operator (5.11) by a complex number ν_n which may depend on the number n of the site, passing to the operator $\lambda(n|\lambda - \nu_n)$ (see (2.5)). This shift corresponds to the inhomogeneous XXX model since the L's differ at each site of the lattice. This L-operator also satisfies the bilinear relation (5.2). The corresponding monodromy matrix is

$$\mathsf{T}(\lambda) = \mathsf{L}(M|\lambda - \nu_M)\cdots\mathsf{L}(1|\lambda - \nu_1). \qquad (5.24)$$

VI.6 Fundamental models of classical statistical physics

This section is an introduction to classical statistical physics on a lattice. The coverage will be brief and will only touch on the construction of the partition function. For the interested reader, we recommend the excellent book by Baxter [5]. Here we shall introduce the extremely important partition function of the six-vertex model with domain wall boundary conditions. In section VII.10 it will be represented as a determinant. Finally this will lead to the solution of the problem of correlation functions.

Any R-matrix satisfying the Yang-Baxter equation generates an exactly solvable model of classical statistical physics on a two-dimensional lattice. The partition function of this model can be calculated by means of QISM. Let us consider the $N \times M$ square lattice (see Figure VI.1). Vertical lines will be enumerated by Latin indices ($k = 1, \ldots, M$) while horizontal lines will be labeled by Greek indices ($\alpha = 1, \ldots, N$). Each spin variable is situated on a lattice edge (link) and takes k different values (this k should not be confused with the index of the vertical lines). The spin variables will be denoted a and b, with subscripts indicating the number of the line carrying them: a_k, b_k, a_α, b_α.

The statistical weight $\mathsf{L}_{\alpha k}$ is associated with the intersection of the α-th and k-th lines and depends on the values of the spin variables situated on the adjoining edges of the vertex:

$$(\mathsf{L}_{\alpha k}(\lambda_\alpha, \nu_k))^{a_k b_k}_{a_\alpha b_\alpha} \tag{6.1}$$

Here λ_α is the spectral parameter (or rapidity) associated with the horizontal line α, and ν_k is the other spectral parameter, associated with the vertical line k (see Figure VI.2). Let us associate the weight with the spin L-operator (6.1) (see also (5.1)):

$$(\mathsf{L}_{\alpha k}(\lambda_\alpha, \nu_k))^{a_k b_k}_{a_\alpha b_\alpha} = \mathsf{R}^{a_k b_\alpha}_{a_\alpha b_k}(\lambda_\alpha, \nu_k)\varphi(\lambda_\alpha, \nu_k) = \widetilde{\mathsf{R}}^{a_\alpha b_\alpha}_{a_k b_k}(\lambda_\alpha, \nu_k)\varphi(\lambda_\alpha, \nu_k) \tag{6.2}$$

where α and k denote the spaces where $\widetilde{\mathsf{R}}$ acts nontrivially. The function $\varphi(\lambda, \nu)$ controls the normalization of the R-matrix. The partition function is equal to

$$Z = \sum_{\text{spins}} \prod_{\text{vertices}} (\mathsf{L}_{\alpha k}(\lambda_\alpha, \nu_k))^{a_k b_k}_{a_\alpha b_\alpha} \tag{6.3}$$

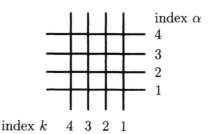

index α

index k 4 3 2 1

Figure VI.1

where the product is taken over all vertices of the lattice and the summation is over all values of the spins, taken independently. For a finite lattice, the partition function depends on the values of the spins on the boundary of the lattice. Usually periodic boundary conditions are imposed. We shall also consider some different types of boundary conditions (domain wall boundary conditions).

The R-matrix (4.3), (4.4) generates the well known ice model. The model associated with the R-matrix (3.17), (3.18) is a special case of the ice model (isotropic in spin space). Solutions for many similar models have been found. We shall not discuss them in detail, but only mention that Baxter's model on the honeycomb lattice [6] is a special case of a model generated by the R-matrix (4.8).

Let us construct a special partition function which plays an important role in the calculation of correlation functions. The statistical weight is given by the L-operator, (5.11), (5.12) of the *XXX* model (5.9):

$$L_{\alpha k}(\lambda_\alpha - \nu_k) = \lambda_\alpha - \nu_k - i\frac{c}{2}\sigma_z^\alpha \sigma_z^k - ic\left(\sigma_-^\alpha \sigma_+^k + \sigma_+^\alpha \sigma_-^k\right) \qquad (6.4)$$

where α and k denote the corresponding spaces. In this case, each spin variable takes only two values: ± 1. Following tradition, we shall denote them by up and down arrows $\uparrow$ $(+1)$ or $\downarrow$ (-1). Consider the square $N \times N$ lattice. Each spectral parameter λ_α corresponds to a horizontal line, and each ν_k to a vertical one (see Figure VI.3).

The statistical weight $L_{\alpha k}(\lambda_\alpha - \nu_k)$ (6.4) corresponds to the vertex at the intersection of these lines. The partition function is given by (6.3) with special boundary conditions, which we shall call domain wall bound-

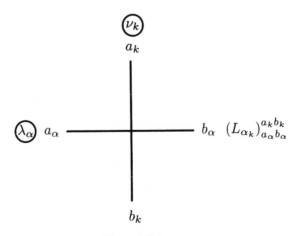

Figure VI.2

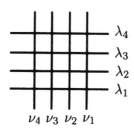

Figure VI.3

ary conditions (see Figure VI.4). On the upper and right boundaries, the spins are down ($\downarrow_k$ and $\downarrow_\alpha$), while on the lower and left boundaries the spins are up ($\uparrow_k$ and $\uparrow_\alpha$) (see Figure VI.4). Thus, the partition function $Z_N(\{\lambda_\alpha\}, \{\nu_k\})$ depends on $2N$ arguments:

$$
Z_N(\{\lambda_\alpha\}, \{\nu_k\}) = \left\{ \prod_{\beta=1}^{N} \uparrow_\beta \right\} \left\{ \prod_{j=1}^{N} \downarrow_j \right\}
$$

$$
\times \left\{ \prod_{\alpha=1}^{N} \prod_{k=1}^{N} \mathsf{L}_{\alpha k}(\lambda_\alpha - \nu_k) \right\} \left\{ \prod_{\beta=1}^{N} \downarrow_\beta \right\} \left\{ \prod_{j=1}^{N} \uparrow_j \right\}. \quad (6.5)
$$

It must be mentioned that the L-operator (6.2) is a matrix in the "vertical" space (a_k, b_k) as well as in the "horizontal" space (a_α, b_α); hence, the summation in (6.3) can be interpreted as matrix multiplication. Here the double product indicates space-ordering:

$$
\prod_{\alpha=1}^{N} \prod_{k=1}^{N} \mathsf{L}_{\alpha k} \equiv (\cdots \mathsf{L}_{23}\mathsf{L}_{22}\mathsf{L}_{21})(\cdots \mathsf{L}_{13}\mathsf{L}_{12}\mathsf{L}_{11}).
$$

Thus defined, Z_N is called the partition function on the inhomogeneous lattice [4]. (This quantity is also called the partition function of the isotropic 6-vertex model.)

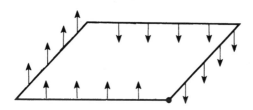

Figure VI.4

The product of L-operators in (6.5) may be rearranged as

$$
Z_N = \left\{ \prod_{j=1}^{N} \downarrow_j \right\} \left\{ \prod_{\beta=1}^{N} \uparrow_\beta \right\}
$$

$$
\times \left\{ \prod_{k=1}^{N} \prod_{\alpha=1}^{N} \mathsf{L}_{\alpha k}(\lambda_\alpha - \nu_k) \right\} \left\{ \prod_{j=1}^{N} \uparrow_j \right\} \left\{ \prod_{\beta=1}^{N} \downarrow_\beta \right\}. \quad (6.6)
$$

Let us express the partition function using the monodromy matrix (5.26):

$$
\mathsf{T}_\alpha(\lambda) = \prod_{k=1}^{N} \mathsf{L}_{\alpha k}(\lambda - \nu_k) = \begin{pmatrix} A(\lambda) & B(\lambda) \\ C(\lambda) & D(\lambda) \end{pmatrix} \quad (6.7)
$$

where the product is taken as space-ordered as above. Here $\mathsf{T}_\alpha(\lambda)$ is a 2×2 matrix in the α-th space; $\mathsf{T}_\alpha(\lambda)$ is also the quantum operator in the spaces $k = 1, \ldots, N$. The operator $\mathsf{L}_{\alpha k}$ is intertwined by $\tilde{\mathsf{R}}$ (5.8), (5.9) in each space (α or k):

$$
\tilde{\mathsf{R}}_{\alpha\beta}(\lambda_\alpha, \lambda_\beta) \mathsf{L}_{\alpha k}(\lambda_\alpha - \nu_k) \mathsf{L}_{\beta k}(\lambda_\beta - \nu_k)
$$
$$
= \mathsf{L}_{\beta k}(\lambda_\beta - \nu_k) \mathsf{L}_{\alpha k}(\lambda_\alpha - \nu_k) \tilde{\mathsf{R}}_{\alpha\beta}(\lambda_\alpha, \lambda_\beta). \quad (6.8)
$$
$$
\tilde{\mathsf{R}}_{kl}(\nu_l, \nu_k) \mathsf{L}_{\alpha k}(\lambda_\alpha - \nu_k) \mathsf{L}_{\alpha l}(\lambda_\alpha - \nu_l)
$$
$$
= \mathsf{L}_{\alpha l}(\lambda_\alpha - \nu_l) \mathsf{L}_{\alpha k}(\lambda_\alpha - \nu_k) \tilde{\mathsf{R}}_{kl}(\nu_l, \nu_k). \quad (6.9)
$$

Hence, for $\mathsf{T}_\alpha(\lambda)$ we have

$$
\tilde{\mathsf{R}}_{\alpha\beta}(\lambda, \mu) \mathsf{T}_\alpha(\lambda) \mathsf{T}_\beta(\mu) = \mathsf{T}_\beta(\mu) \mathsf{T}_\alpha(\lambda) \tilde{\mathsf{R}}_{\alpha\beta}(\lambda, \mu) \quad (6.10)
$$

with the same R-matrix. Here α and β denote vector spaces. Now we can rewrite Z_N in the form

$$
Z_N = \left\{ \prod_{j=1}^{N} \downarrow_j \right\} \left\{ \prod_{\alpha=1}^{N} (\uparrow_\alpha) \mathsf{T}_\alpha(\lambda_\alpha) (\downarrow_\alpha) \right\} \left\{ \prod_{j=1}^{N} \uparrow_j \right\}. \quad (6.11)
$$

Using (6.7) we see that

$$
(\uparrow_\alpha) \mathsf{T}_\alpha(\lambda) (\downarrow_\alpha) = B(\lambda)
$$

and, therefore,

$$
Z_N = \left\{ \prod_{j=1}^{N} \downarrow_j \right\} \left\{ \prod_{\alpha=1}^{N} B(\lambda_\alpha) \right\} \left\{ \prod_{j=1}^{N} \uparrow_j \right\}. \quad (6.12)
$$

Let us rewrite Z_N with the help of the monodromy matrix in the "orthogonal" direction. Introducing the "vertical" monodromy matrix t_j

one has (as a space ordered product):

$$t_j(\nu_j) = \prod_{\alpha=1}^{N} L_{\alpha j}(\lambda_\alpha - \nu_j) = \begin{pmatrix} A(\nu_j) & B(\nu_j) \\ C(\nu_j) & D(\nu_j) \end{pmatrix} \qquad (6.13)$$

which is a 2×2 matrix in the j-th space and quantum operator in the spaces $\alpha = 1, \ldots, N$. We can express Z_N, starting from (6.6), as

$$Z_N = \left\{ \prod_{\beta=1}^{N} \uparrow_\beta \right\} \left\{ \prod_{j=1}^{N} (\downarrow_j) \, t_j(\nu_j) \, (\uparrow_j) \right\} \left\{ \prod_{\beta=1}^{N} \downarrow_\beta \right\}. \qquad (6.14)$$

Due to $(\downarrow_j) \, t_j(\nu) \, (\uparrow_j) = C(\nu)$ we have

$$Z_N = \left\{ \prod_{\beta=1}^{N} \uparrow_\beta \right\} \left\{ \prod_{j=1}^{N} C(\nu_j) \right\} \left\{ \prod_{\beta=1}^{N} \downarrow_\beta \right\}. \qquad (6.15)$$

For the monodromy matrix, we also have

$$\widetilde{R}_{kl}(\nu_l, \nu_k) t_k(\nu_k) t_l(\nu_l) = t_l(\nu_l) t_k(\nu_k) \widetilde{R}_{kl}(\nu_l, \nu_k) \qquad (6.16)$$

as a consequence of (6.9). Thus, different representations are obtained for Z_N. Later we shall analyze Z_N as a function of $\{\lambda_\alpha\}$ and $\{\nu_k\}$. This partition function will play an important role in the investigation of correlation functions.

We would like to emphasize that a similar partition function Z_N can be constructed for the XXZ model (5.21), see for example [38].

Conclusion

In this chapter we have seen that the main concepts of the quantum inverse scattering method (QISM) can be obtained by direct quantization of the classical case (see Chapter V). The quantum R-matrix is extremely important. It is easy to evaluate it because one needs only to know the explicit expression for the L-operator. Nevertheless, it is very powerful since it gives the commutation relations with the monodromy matrix (which is highly nontrivial). In the next chapter, we shall use it to construct the algebraic Bethe Ansatz and to reproduce the results of Part I.

We have also explained how to apply the ideas of QISM to different types of models—quantum field theory, spin models, and lattice models of classical statistical physics. In the next chapter, we shall explicitly evaluate the partition function with domain wall boundary conditions (introduced in section 6 of this chapter). It will play an important role later when we study correlation functions, norms and scalar products.

We have stated the main ideas of QISM ([23], [24], [54], [55] and [63]). We also recommend the following reviews and papers: [15], [26], [31], [43], [46], [58], [60] and [61].

The Yang-Baxter relation plays a central role in modern mathematical physics. It is important in the theory of factorized S-matrices ([14], [65]–[70]). The study of the Bethe wave function in [48] (see also [16], [12] and [13]) led to the discovery of the factorization of a many-body scattering matrix into the product of two-body ones. In papers [65], [66] the one-dimensional quantum mechanics of particles with arbitrary statistics interacting via a δ potential was solved and the "nested Bethe Ansatz" was discovered. The Yang-Baxter relation appears there in its present form as the consistency condition for factorization. The Yang-Baxter relation is important also in statistical physics (see [1]–[9], [27], [35], [49], [50], [51] and [64], especially the book by Baxter [5]). The role of the Yang-Baxter relation in QISM can be understood from [10], [18], [19], [23]–[31], [33]–[35], [37]–[47], [50]–[59] and [63]. One can find a splendid collection of pioneering works and references on the subject in the book [36]. Extensive literature is devoted to the search for solutions of the Yang-Baxter equation ([9]–[11], [18], [19], [34], [35], [43]–[47], [49], [52], [68] and [70]). In the paper [53] it was shown that the Yang-Baxter relation for the scattering matrix follows from an infinite set of conservation laws.

QISM was first applied to the nonlinear Schrödinger (NS) model ([23], [55] and [56]). The comparison of classical and quantum results was performed in [47]. For the trace identities, see references [15], [17], [20], [21], [22], [32], [57], [62] and [63].

The sine-Gordon model was treated by QISM in [54]. The quantum L-operator and R-matrix of the Zhiber-Shabat-Mikhailov model were constructed in [29].

Let us mention that the models considered in this chapter by no means exhaust the number of quantum field models solvable by QISM. Due to lack of space, we refer the reader to [5], [7]–[9], [19], [25]–[27], [34], [35], [37], [43], [44], [49]–[52], [59] and [64].

Fundamental spin models are examined in [5] and [58]. An interesting spin Hamiltonian is generated by the R-matrix (4.8) of the Zhiber-Shabat-Mikhailov model. This Hamiltonian describes the interaction of spins in the presence of an external magnetic field. Its explicit form is given in [29]. Later, this model was solved in [51] and [64]. It is connected with a classical statistical physics model on a honeycomb lattice [6].

VII

The Algebraic Bethe Ansatz

Introduction

The algebraic Bethe Ansatz is presented in this chapter. This is an important generalization of the coordinate Bethe Ansatz presented in Part I, and is one of the essential achievements of QISM. The algebraic Bethe Ansatz is based on the idea of constructing eigenfunctions of the Hamiltonian via creation and annihilation operators acting on a pseudovacuum. The matrix elements of the monodromy matrix play the role of these operators. The transfer matrix (the sum of the diagonal elements of the monodromy matrix) commutes with the Hamiltonian; thus constructing eigenfunctions of $\tau(\mu)$ determines the eigenfunctions of the Hamiltonian.

The basis of the algebraic Bethe Ansatz is stated in section 1. The commutation relations between matrix elements of the monodromy matrix are specified by the R-matrix. The explicit form of the commutation relations allows the construction of eigenfunctions of the transfer matrix (the trace of the monodromy matrix). (Recall that the Hamiltonian may also be obtained from the transfer matrix via the trace identities.) Further developments of the algebraic Bethe Ansatz necessary for the computation of correlation functions are given in section 2. The general scheme is illustrated with some examples in section 3. The NS model, the sine-Gordon model and spin models are considered in detail. The Pauli principle for interacting one-dimensional bosons plays an important role in constructing the ground state of the system and is discussed in section 4. The eigenvalues of the shift operator acting on the monodromy matrix are calculated in section 5. The classification of monodromy matrices possessing a given R-matrix is given in sections 6 and 7. It is shown that they are parametrized by two arbitrary functions. The important concept of the monodromy matrix determinant in the quantum case is introduced in section 8. The properties of the partition function Z_N of the XXX model with domain wall boundary conditions (introduced in section 6 of Chapter VI), are examined in section 9. This partition function which can be explicitly calculated is presented in section 10 where the solution is given as a determinant. We want to emphasize the importance of this determinant representation, because this is the beginning of evaluation of correlation functions, which will be continued in Part III.

Let us mention that only models having the R-matrices of the XXX and XXZ models (VI.3.17), (VI.3.18) and (VI.4.3), (VI.4.4) are considered in this chapter. However, models with different R-matrices may be treated in a similar fashion.

VII.1 The algebraic Bethe Ansatz

Models connected with the simplest R-matrices, those for the XXX and XXZ models (VI.3.17), (VI.4.3) will be considered in this section. In these cases, the monodromy matrix is a 2×2-matrix:

$$\mathsf{T}(\lambda) = \begin{pmatrix} A(\lambda) & B(\lambda) \\ C(\lambda) & D(\lambda) \end{pmatrix}. \tag{1.1}$$

The commutation relations between its matrix elements are given by the bilinear relation

$$\mathsf{R}(\lambda,\mu)\left(\mathsf{T}(\lambda)\bigotimes\mathsf{T}(\mu)\right) = \left(\mathsf{T}(\mu)\bigotimes\mathsf{T}(\lambda)\right)\mathsf{R}(\lambda,\mu) \tag{1.2}$$

with

$$\mathsf{R}(\lambda,\mu) = \begin{pmatrix} f(\mu,\lambda) & 0 & 0 & 0 \\ 0 & g(\mu,\lambda) & 1 & 0 \\ 0 & 1 & g(\mu,\lambda) & 0 \\ 0 & 0 & 0 & f(\mu,\lambda) \end{pmatrix} \tag{1.3}$$

where, for the XXX case,

$$f(\mu,\lambda) = 1 + \frac{ic}{\mu-\lambda}, \qquad g(\mu,\lambda) = \frac{ic}{\mu-\lambda}, \tag{1.4}$$

and for the XXZ case

$$f(\mu,\lambda) = \frac{\sinh(\mu-\lambda+2i\eta)}{\sinh(\mu-\lambda)}, \qquad g(\mu,\lambda) = i\frac{\sin 2\eta}{\sinh(\mu-\lambda)} \tag{1.5}$$

(we put $\eta = -\gamma/2$ in (VI.4.3), (VI.4.4)). The Hamiltonian and the transfer matrix have the same eigenfunctions; this follows from the trace identities (see sections 3 and 5 of Chapter VI) and the commutativity of the transfer matrix at different values of the spectral parameter. To construct eigenfunctions of $\tau(\lambda)$, we should have a generating vector (pseudovacuum $|0\rangle$), which must satisfy the following requirements:

$$A(\lambda)|0\rangle = a(\lambda)|0\rangle; \qquad D(\lambda)|0\rangle = d(\lambda)|0\rangle;$$
$$C(\lambda)|0\rangle = 0. \tag{1.6}$$

Here $a(\lambda)$ and $d(\lambda)$ are complex-valued functions called the "vacuum eigenvalues." The pseudovacuum $|0\rangle$ is similar to the the highest-weight vector in the theory of representations of Lie algebras. The action of A, B, C and D on the vacuum can be understood by looking at the L-operator for the NS model (VI.3.14). For A and D the vacuum $|0\rangle$ is an eigenvector and C annihilates it. B acts like a creation operator and is treated as such below.

It should be noted that the existence of an R-matrix does not automatically guarantee the existence of a pseudovacuum. The latter property should be established for each given model independently. However, the pseudovacuum vector does indeed exist for a large class of integrable models (see examples in section 3). When constructing the pseudovacuum for a model, the following remark is useful. Let the monodromy matrix be the matrix product of two factors:

$$T(\lambda) = T(2|\lambda)T(1|\lambda) \tag{1.7}$$

where the commutation relations between matrix elements of each matrix $T(1|\lambda)$, $T(2|\lambda)$ are given by (1.2) and matrix elements of different matrices commute: $[T^{ab}(1|\lambda), T^{cd}(2|\lambda)] = 0$, for any a, b, c, d. (Thus, $T(\lambda)$ consists of two "commuting" matrix factors.) If $|0\rangle_1$ is the pseudovacuum for $T(1|\lambda)$, and $|0\rangle_2$ is that for $T(2|\lambda)$, with eigenvalues $a_1(\lambda)$, $d_1(\lambda)$, and $a_2(\lambda)$, $d_2(\lambda)$, respectively, then the pseudovacuum $|0\rangle$ exists and is given by

$$|0\rangle = |0\rangle_2 \bigotimes |0\rangle_1$$

and

$$a(\lambda) = a_1(\lambda)a_2(\lambda), \qquad d(\lambda) = d_1(\lambda)d_2(\lambda). \tag{1.8}$$

To prove this statement, it is sufficient to note that the product of triangular matrices is a triangular matrix. This statement is easily generalized to the case when there are more than two commuting factors in the above product. The vacuum eigenvalues $a(\lambda)$ and $d(\lambda)$ of the monodromy matrix $T(\lambda) = L(M|\lambda) \cdots L(1|\lambda)$ (VI.1.1) are

$$a(\lambda) = \prod_{j=1}^{M} a_j(\lambda), \qquad d(\lambda) = \prod_{j=1}^{M} d_j(\lambda) \tag{1.9}$$

where $a_j(\lambda)$, $d_j(\lambda)$ are the vacuum eigenvalues of the L-operator $L(j|\lambda)$.

Later we will see that the eigenfunctions of $\tau(\lambda) = A(\lambda) + D(\lambda)$ are of the form

$$|\Psi_N(\{\lambda_j\})\rangle = \prod_{j=1}^{N} B(\lambda_j)|0\rangle \tag{1.10}$$

where the λ_j satisfy a system of Bethe equations.

We begin by explicitly writing down the commutation relations (1.2) of the matrix elements of the monodromy matrix:

$$[B(\lambda), B(\mu)] = 0; \qquad [C(\lambda), C(\mu)] = 0; \tag{1.11}$$

$$A(\mu)B(\lambda) = f(\mu, \lambda)B(\lambda)A(\mu) + g(\lambda, \mu)B(\mu)A(\lambda); \tag{1.12}$$

$$D(\mu)B(\lambda) = f(\lambda, \mu)B(\lambda)D(\mu) + g(\mu, \lambda)B(\mu)D(\lambda); \tag{1.13}$$

$$C(\lambda)A(\mu) = f(\mu, \lambda)A(\mu)C(\lambda) + g(\lambda, \mu)A(\lambda)C(\mu); \tag{1.14}$$

$$C(\lambda)D(\mu) = f(\lambda, \mu)D(\mu)C(\lambda) + g(\mu, \lambda)D(\lambda)C(\mu); \tag{1.15}$$

$$[C(\lambda), B(\mu)] = g(\lambda, \mu) \left\{ A(\lambda)D(\mu) - A(\mu)D(\lambda) \right\}; \tag{1.16}$$

$$[A(\lambda), A(\mu)] = 0; \qquad [D(\lambda), D(\mu)] = 0; \tag{1.17}$$

$$B(\mu)A(\lambda) = f(\mu, \lambda)A(\lambda)B(\mu) + g(\lambda, \mu)A(\mu)B(\lambda); \tag{1.18}$$

$$D(\mu)C(\lambda) = f(\mu, \lambda)C(\lambda)D(\mu) + g(\lambda, \mu)C(\mu)D(\lambda); \tag{1.19}$$

$$A(\lambda)C(\mu) = f(\mu, \lambda)C(\mu)A(\lambda) + g(\lambda, \mu)C(\lambda)A(\mu); \tag{1.20}$$

$$B(\lambda)D(\mu) = f(\mu, \lambda)D(\mu)B(\lambda) + g(\lambda, \mu)D(\lambda)B(\mu); \tag{1.21}$$

$$[D(\lambda), A(\mu)] = g(\lambda, \mu) \left\{ B(\lambda)C(\mu) - B(\mu)C(\lambda) \right\}; \tag{1.22}$$

$$[A(\lambda), D(\mu)] = g(\lambda, \mu) \left\{ C(\lambda)B(\mu) - C(\mu)B(\lambda) \right\}; \tag{1.23}$$

$$[B(\lambda), C(\mu)] = g(\lambda, \mu) \left\{ D(\lambda)A(\mu) - D(\mu)A(\lambda) \right\}. \tag{1.24}$$

Starting with equations (1.11)–(1.13) we can calculate the result of the action of operators $A(\mu)$ and $D(\mu)$ on the state $\prod B(\lambda_j)|0\rangle$:

$$A(\mu) \prod_{j=1}^{N} B(\lambda_j)|0\rangle = \Lambda \prod_{j=1}^{N} B(\lambda_j)|0\rangle$$

$$+ \sum_{n=1}^{N} \Lambda_n B(\mu) \prod_{\substack{j=1 \\ j \neq n}}^{N} B(\lambda_j)|0\rangle; \tag{1.25}$$

$$D(\mu) \prod_{j=1}^{N} B(\lambda_j)|0\rangle = \widetilde{\Lambda} \prod_{j=1}^{N} B(\lambda_j)|0\rangle$$

$$+ \sum_{n=1}^{N} \widetilde{\Lambda}_n B(\mu) \prod_{\substack{j=1 \\ j \neq n}}^{N} B(\lambda_j)|0\rangle. \tag{1.26}$$

The coefficients Λ and $\widetilde{\Lambda}$ are

$$\Lambda = a(\mu) \prod_{j=1}^{N} f(\mu, \lambda_j); \quad \Lambda_n = a(\lambda_n)g(\lambda_n, \mu) \prod_{\substack{j=1 \\ j \neq n}}^{N} f(\lambda_n, \lambda_j); \tag{1.27}$$

$$\widetilde{\Lambda} = d(\mu) \prod_{j=1}^{N} f(\lambda_j, \mu); \quad \widetilde{\Lambda}_n = d(\lambda_n)g(\mu, \lambda_n) \prod_{\substack{j=1 \\ j \neq n}}^{N} f(\lambda_j, \lambda_n). \tag{1.28}$$

We shall prove these formulæ following the arguments of paper [2].

The commutation relation (1.12) allows us to "move" the operator A from the left to the right through the operators B in (1.10). The first term on the right hand side of (1.12) corresponds to preserving the arguments of the operators, and the second term corresponds to an exchange of the arguments. To compute Λ in (1.25), one always has to use the first term on the right hand side of (1.12) when moving $A(\mu)$. If one applies the second term on the right hand side of (1.12), then $B(\mu)$ will appear (giving a contribution to the next term of (1.25)) and never vanish. So the second term in (1.12) does not contribute to the coefficient Λ. Repeating this procedure, one can move A to the right of all the B (1.10) and then use the relation (1.6) to obtain (1.27) for Λ.

Next we compute Λ_n. Using (1.11) we can rewrite (1.10) as

$$|\Psi_N(\{\lambda_j\})\rangle = B(\lambda_n) \prod_{\substack{j=1 \\ j\neq n}}^{N} B(\lambda_j)|0\rangle. \tag{1.29}$$

It is clear that for the first step, moving $A(\mu)$ past $B(\lambda_n)$, one has only to take into account the second term in (1.12), since this term does not contain $B(\lambda_n)$. After this first step, we obtain:

$$-g(\mu,\lambda_n)B(\mu)A(\lambda_n) \prod_{j\neq n} B(\lambda_j)|0\rangle. \tag{1.30}$$

The next step is the commuting of $A(\lambda_n)$ with $B(\lambda_j)$. Only the first term in (1.12) must be used, as the second one contains $B(\lambda_n)$, which should not be present. Therefore, it is clear that the final result will be given by (1.27). The formulæ (1.26), (1.28) can be obtained similarly. Due to the commutativity of the operators $B(\lambda)$ (1.11), the function $|\Psi_N\rangle$ is symmetric in the set $\{\lambda_j\}$. The linearly independent vectors $|\Psi_N\rangle$ correspond to different sets $\{\lambda_j\}$. Hence, $|\Psi_N\rangle$ is an eigenfunction of $\tau(\mu)$ if and only if $\Lambda_n + \widetilde{\Lambda}_n = 0$. This requirement leads to the Bethe equations

$$r(\lambda_n) \prod_{\substack{j=1 \\ j\neq n}}^{N} \frac{f(\lambda_n,\lambda_j)}{f(\lambda_j,\lambda_n)} = 1, \qquad n=1,\ldots,N \tag{1.31}$$

where

$$r(\lambda) \equiv \frac{a(\lambda)}{d(\lambda)}. \tag{1.32}$$

The logarithmic form of the Bethe equations is often used:

$$\varphi_k = 2\pi n_k, \qquad k=1,\ldots,N. \tag{1.33}$$

Here n_k is an arbitrary set of N integers and

$$\varphi_k = i \ln r(\lambda_k) + i \sum_{\substack{j=1 \\ j \neq n}}^{N} \ln \left[\frac{f(\lambda_k, \lambda_j)}{f(\lambda_j, \lambda_k)} \right]. \tag{1.34}$$

The eigenvalues, $\theta(\mu)$, of the transfer matrix $\tau(\mu)$ are

$$\theta(\mu, \{\lambda_j\}) = a(\mu) \prod_{j=1}^{N} f(\mu, \lambda_j) + d(\mu) \prod_{j=1}^{N} f(\lambda_j, \mu), \tag{1.35}$$

$$\tau(\mu) |\Psi_N(\{\lambda_j\})\rangle = \theta(\mu, \{\lambda_j\}) |\Psi_N(\{\lambda_j\})\rangle.$$

The eigenvalues of the Hamiltonian are computed by means of trace identities.

We have supposed that all $\{\lambda_j\}$ $(j = 1, \ldots, N)$ are different. The case where some of them are equal will be considered in section 4.

VII.2 Comments on the algebraic Bethe Ansatz

In the previous section, the result of the action of the operators $A(\mu)$, $D(\mu)$ and $B(\mu)$ on the state (Bethe vector)

$$\prod_{j=1}^{N} B(\lambda_j) |0\rangle \tag{2.1}$$

was calculated.

To investigate correlation functions, we must also compute the action of $C(\mu)$ on (2.1):

$$C(\mu) \prod_{j=1}^{N} B(\lambda_j) |0\rangle = \sum_{n=1}^{N} M_n \prod_{\substack{j=1 \\ j \neq n}}^{N} B(\lambda_j) |0\rangle$$

$$+ \sum_{k>n} M_{kn} B(\mu) \prod_{\substack{j=1 \\ j \neq k,n}}^{N} B(\lambda_j) |0\rangle. \tag{2.2}$$

The coefficients M_n, M_{kn} are given by

$$M_n = g(\mu, \lambda_n) a(\mu) d(\lambda_n) \prod_{j \neq n} f(\lambda_j, \lambda_n) f(\mu, \lambda_j)$$

$$+ g(\lambda_n, \mu) a(\lambda_n) d(\mu) \prod_{j \neq n} f(\lambda_j, \mu) f(\lambda_n, \lambda_j) \tag{2.3}$$

$$M_{kn} = d(\lambda_k) a(\lambda_n) g(\mu, \lambda_k) g(\lambda_n, \mu) f(\lambda_n, \lambda_k) \prod_{j \neq k,n} f(\lambda_j, \lambda_k) f(\lambda_n, \lambda_j)$$

$$+ d(\lambda_n) a(\lambda_k) g(\mu, \lambda_n) g(\lambda_k, \mu) f(\lambda_k, \lambda_n) \prod_{j \neq k,n} f(\lambda_j, \lambda_n) f(\lambda_k, \lambda_j). \tag{2.4}$$

These formulæ can be obtained analogously to (1.25)–(1.28). In addition to the relations (1.11)–(1.13), one uses (1.16) and $C(\lambda)|0\rangle = 0$.

Let us now consider the dual pseudovacuum $\langle 0|$: $\langle 0| = |0\rangle^\dagger$, $\langle 0|0\rangle = 1$. It is easy to verify that

$$\langle 0|B(\mu) = 0, \quad \langle 0|A(\mu) = a(\mu)\langle 0|, \quad \langle 0|D(\mu) = d(\mu)\langle 0|. \tag{2.5}$$

In complete analogy with the results of section 1, the action of the operators $A(\mu)$, $D(\mu)$ and $B(\mu)$ on the state conjugate to (1.10),

$$\langle \widetilde{\Psi}_N(\{\lambda_j\})| = \langle 0| \prod_{j=1}^{N} C(\lambda_j), \tag{2.6}$$

can be calculated:

$$\langle 0| \prod_{j=1}^{N} C(\lambda_j) A(\mu) = \Lambda \langle 0| \prod_{j=1}^{N} C(\lambda_j)$$
$$+ \sum_{n=1}^{N} \Lambda_n \langle 0|C(\mu) \prod_{\substack{j=1 \\ j \neq n}}^{N} C(\lambda_j); \tag{2.7}$$

$$\langle 0| \prod_{j=1}^{N} C(\lambda_j) D(\mu) = \widetilde{\Lambda} \langle 0| \prod_{j=1}^{N} C(\lambda_j)$$
$$+ \sum_{n=1}^{N} \widetilde{\Lambda}_n \langle 0|C(\mu) \prod_{\substack{j=1 \\ j \neq n}}^{N} C(\lambda_j). \tag{2.8}$$

$$\langle 0| \prod_{j=1}^{N} C(\lambda_j) B(\mu) = \sum_{n=1}^{N} M_n \langle 0| \prod_{\substack{j=1 \\ j \neq n}}^{N} C(\lambda_j)$$
$$+ \sum_{k>n} M_{kn} \langle 0|C(\mu) \prod_{\substack{j=1 \\ j \neq k,n}}^{N} C(\lambda_j). \tag{2.9}$$

The coefficients here are the same as in (1.25)–(1.28), (2.2)–(2.4). If the Bethe equations are valid, then $\langle \widetilde{\Psi}_N|$ is also an eigenfunction of $\tau(\mu)$:

$$\langle \widetilde{\Psi}_N|\tau(\mu) = \theta(\mu)\langle \widetilde{\Psi}_N| \tag{2.10}$$

with the previous eigenvalue (1.35).

The eigenfunctions (2.1), (2.6) are orthogonal:

$$\langle \Psi_N(\{\lambda_j^C\})|\Psi_N(\{\lambda_j^B\})\rangle = \langle 0| \prod_{j=1}^{N} C(\lambda_j^C) \prod_{k=1}^{N} B(\lambda_k^B)|0\rangle = 0, \tag{2.11}$$

$$(\{\lambda_j^C\} \neq \{\lambda_j^B\}),$$

if $\{\lambda_j^C\}$ and $\{\lambda_j^B\}$ are different sets of solutions of the Bethe equations. To prove this statement, one considers the matrix elements

$$\langle 0| \prod_{j=1}^{N} C(\lambda_j^C)\tau(\mu) \prod_{k=1}^{N} B(\lambda_k^B)|0\rangle. \tag{2.12}$$

Taking into account that

$$\theta(\mu, \{\lambda_j^B\}) \neq \theta(\mu, \{\lambda_j^C\}), \qquad (\{\lambda_j^C\} \neq \{\lambda_j^B\}) \tag{2.13}$$

and then comparing (1.35) and (2.10) one obtains (2.11).

As a concluding remark, let us mention that the R-matrix commutes with the matrix $\hat{\varepsilon} \otimes \hat{\varepsilon}$, where

$$\hat{\varepsilon} = \begin{pmatrix} \varepsilon & 0 \\ 0 & \varepsilon^{-1} \end{pmatrix} \tag{2.14}$$

and ε is a c-number. Hence, if the commutation relations of the matrix elements of the monodromy matrix $T(\lambda)$ are given by the R-matrix (1.3), then the commutation relations of the elements of the matrix

$$T_\varepsilon(\lambda) = \begin{pmatrix} \varepsilon A(\lambda) & \varepsilon B(\lambda) \\ \varepsilon^{-1}C(\lambda) & \varepsilon^{-1}D(\lambda) \end{pmatrix} \tag{2.15}$$

are given by the same R-matrix. The matrix T_ε possesses the same pseudovacuum $|0\rangle$ and its vacuum eigenfunctions are

$$a_\varepsilon(\lambda) = \varepsilon a(\lambda), \qquad d_\varepsilon(\lambda) = \varepsilon^{-1}d(\lambda). \tag{2.16}$$

VII.3 Examples

Now we shall calculate eigenfunctions for some important models of two-dimensional quantum field theory and statistical mechanics.

(1) The nonlinear Schrödinger (NS) model. The pseudovacuum is just the Fock vacuum $|0\rangle$ ($\Psi(x)|0\rangle = 0$, $\forall x$). The vacuum eigenvalues (1.6) of the L-operator (VI.3.14) are

$$a_j(\lambda) = 1 - i\frac{\lambda\Delta}{2}; \qquad d_j(\lambda) = 1 + i\frac{\lambda\Delta}{2}. \tag{3.1}$$

Using

$$\lim_{M \to \infty} \left(1 + i\frac{\lambda\Delta}{2}\right)^M = \exp\{i\lambda L/2\}, \qquad L = M\Delta, \tag{3.2}$$

and (1.9), one obtains the vacuum eigenvalues of the monodromy matrix for the continuum model:

$$a(\lambda) = \exp\left\{-i\frac{\lambda L}{2}\right\}; \qquad d(\lambda) = \exp\left\{i\frac{\lambda L}{2}\right\}. \tag{3.3}$$

The Bethe equations (1.31)

$$\exp\{i\lambda_n L\} = \prod_{\substack{j=1\\j\neq n}}^{N} \left(\frac{\lambda_n - \lambda_j + ic}{\lambda_n - \lambda_j - ic}\right); \qquad n = 1,\ldots,N \qquad (3.4)$$

coincide with (I.2.2). The eigenvalues of the transfer matrix (1.35) are

$$\theta(\mu, \{\lambda_j\}) = \exp\left\{-i\frac{\mu L}{2}\right\} \prod_{j=1}^{N} f(\mu, \lambda_j)$$

$$+ \exp\left\{i\frac{\mu L}{2}\right\} \prod_{j=1}^{N} f(\lambda_j, \mu). \qquad (3.5)$$

To calculate the eigenvalues of H, P and Q one has to use the trace identities (VI.3.6) which give:

$$\ln\left[\exp\left\{i\frac{\mu L}{2}\right\}\theta_N(\mu, \{\lambda_j\})\right] \underset{\mu\to i\infty}{=} \frac{ic}{\mu}\left\{Q_N + \mu^{-1}\left[P_N - (ic/2)Q_N\right]\right.$$

$$\left. +\mu^{-2}\left[E_N - icP_N - (c^2/3)Q_N\right] + O\left(\mu^{-3}\right)\right\}. \qquad (3.6)$$

The second term in (3.5) is negligibly small as $\mu \to i\infty$, and one obtains

$$Q_N = N; \qquad P_N = \sum_{j=1}^{N} \lambda_j; \qquad E_N = \sum_{j=1}^{N} \lambda_j^2 \qquad (3.7)$$

which coincide with (I.1.29).

We have thus reproduced all the corresponding results of Chapter I within the context of QISM. The coefficients of the expansion (3.6) in inverse powers of μ can be expressed in terms of

$$I_n = \sum_{j=1}^{N} \lambda_j^n. \qquad (3.8)$$

Conservation of these quantities in scattering processes prevents multiparticle production.

(2) *XXX* Heisenberg model. The monodromy matrix $\mathsf{T}(\lambda)$ of the *XXX* model (VI.5.15) is constructed in a standard way by means of the L-operator (VI.5.11). The R-matrix (1.3), (1.4) coincides with that of the NS model. Now we have two pseudovacua $|0\rangle$, $|0'\rangle$:

$$|0\rangle = \prod_{j=1}^{M} (\uparrow)_j, \qquad C(\lambda)|0\rangle = 0; \qquad (3.9)$$

$$|0'\rangle = \prod_{j=1}^{M} (\downarrow)_j, \qquad B(\lambda)|0'\rangle = 0. \tag{3.10}$$

The vacuum eigenvalues for the first pseudovacuum are

$$a(\lambda) = \left(\lambda - i\frac{c}{2}\right)^M, \qquad d(\lambda) = \left(\lambda + i\frac{c}{2}\right)^M. \tag{3.11}$$

Below we shall consider the lattice of even length M.

The eigenfunctions and eigenvalues of the transfer matrix are constructed as for the NS model. It follows from (VI.5.13) that

$$a^*(\lambda^*) = d(\lambda), \qquad B^\dagger(\lambda^*) = -C(\lambda). \tag{3.12}$$

The Bethe equations are:

$$\left(\frac{\lambda_j + ic/2}{\lambda_j - ic/2}\right)^M = \prod_{\substack{k=1 \\ k \neq j}}^{N} \left(\frac{\lambda_j - \lambda_k + ic}{\lambda_j - \lambda_k - ic}\right), \qquad j = 1, \ldots, N. \tag{3.13}$$

Without loss of generality we can put $c = 1$ everywhere, since this corresponds to a rescaling of the spectral parameter $\lambda \to c\lambda$.

(3) Inhomogeneous XXX Heisenberg model. The inhomogeneous generalization of the XXX monodromy matrix (VI.5.26) is a 2×2 matrix similar to (1.1). The vacuum is (3.9) (as is (3.10)), with eigenvalues

$$a(\lambda) = \prod_{j=1}^{M} \left(\lambda - \nu_j - i\frac{c}{2}\right); \qquad d(\lambda) = \prod_{j=1}^{M} \left(\lambda - \nu_j + i\frac{c}{2}\right). \tag{3.14}$$

The operator for the number of particles is essentially the operator for the third component of total spin,

$$S_z = \frac{1}{2} \sum_{k=1}^{N} \sigma_z^{(k)}, \tag{3.15}$$

with the two pseudovacua (3.9), (3.10) as eigenfunctions:

$$S_z|0\rangle = \frac{M}{2}|0\rangle, \qquad S_z|0'\rangle = -\frac{M}{2}|0'\rangle. \tag{3.16}$$

The function (1.10) with arbitrary $\{\lambda_j\}$ is also an eigenfunction of S_z:

$$S_z \prod_{j=1}^{N} B(\lambda_j)|0\rangle = \left(\frac{M}{2} - N\right) \prod_{j=1}^{N} B(\lambda_j)|0\rangle \tag{3.17}$$

where the largest and smallest eigenvalues are $+M/2$ and $-M/2$, respectively.

Considering $M = N$, we obtain

$$\prod_{j=1}^{N} B(\lambda_j)|0\rangle = Z_N|0'\rangle. \tag{3.18}$$

The numerical coefficient

$$Z_N = \langle 0'| \prod_{j=1}^{N} B(\lambda_j)|0\rangle \tag{3.19}$$

coincides with the partition function with domain wall boundary conditions introduced in section 6, Chapter VI. To see this it is sufficient to compare the representations (VI.6.12) and (3.19).

(4) The XXZ Heisenberg model (VI.5.18), (VI.5.21). The analysis is similar to that of the XXX model in sections 1 and 2. We need only replace the XXX R-matrix by the XXZ R-matrix.

The vacuum eigenvalues are

$$a(\lambda) = (-i)^M \cosh^M(\lambda - i\eta), \qquad d(\lambda) = (-i)^M \cosh^M(\lambda + i\eta) \tag{3.20}$$

with pseudovacuum (3.9).

The Bethe equations are

$$\left(\frac{\cosh(\lambda_n - i\eta)}{\cosh(\lambda_n + i\eta)}\right)^M \prod_{\substack{j=1 \\ j\neq n}}^{N} \frac{\sinh(\lambda_n - \lambda_j + 2i\eta)}{\sinh(\lambda_n - \lambda_j - 2i\eta)} = 1. \tag{3.21}$$

With the help of the trace identities (VI.5.19), the one-particle energy and momentum are calculated:

$$\varepsilon_0(\lambda) = -2\sin^2 2\eta \left[\cosh(\lambda + i\eta)\cosh(\lambda - i\eta)\right]^{-1} + 2h;$$
$$p_0(\lambda) = i\ln\left[\frac{\cosh(\lambda - i\eta)}{\cosh(\lambda + i\eta)}\right]. \tag{3.22}$$

Results (3.21) and (3.22) are the same as (II.1.26) and (II.1.13). In this way, all the results of Chapter II are reproduced within QISM.

(5) The sine-Gordon model (VI.4.2). We shall consider the product of two L-operators on neighboring sites in order to construct the pseudovaccum

$$\mathsf{L}_{(2)}(n|\lambda) = \mathsf{L}(2n|\lambda)\mathsf{L}(2n - 1|\lambda). \tag{3.23}$$

For this L-operator, the pseudovacuum was found in [22]:

$$|0\rangle_n = \delta\left(u_{2n-1} - u_{2n} + (\beta/4) - (2\pi/\beta)\right). \tag{3.24}$$

The vacuum for $\mathsf{T}(\lambda)$ is the product of the $|0\rangle_n$:

$$|0\rangle = \bigotimes_{n=1}^{M/2} |0\rangle_n.$$

For rational values of γ/π, the quantum operators $\exp\{i\beta u_n/2\}$ and $\exp\{i\beta p_n/2\}$ can be represented as finite matrices (see Appendix 1). The normalized vacuum can be constructed and the vacuum eigenvalues of the monodromy matrix found:

$$a(\lambda) = \exp\left\{Ms\cosh(2\lambda - i\gamma)\right\};$$

$$d(\lambda) = \exp\left\{Ms\cosh(2\lambda + i\gamma)\right\}; \qquad s = \left(\frac{m\Delta}{4}\right)^2. \tag{3.25}$$

The Bethe equations are of the form

$$\exp\left\{-2iMs\sin\gamma\,\sinh 2\lambda_n\right\} \prod_{\substack{j=1 \\ j\neq n}}^{N} \frac{\sinh(\lambda_n - \lambda_j - i\gamma)}{\sinh(\lambda_n - \lambda_j + i\gamma)} = 1, \tag{3.26}$$

$$n = 1, \ldots, N.$$

The eigenfunctions are calculated by means of (1.10). The energy and momentum are given by

$$\varepsilon_0(\lambda) = m_0 \cosh 2\lambda,$$
$$p_0(\lambda) = m_0 \sinh 2\lambda. \tag{3.27}$$

Now one can compare the Bethe equations for the massive Thirring (III.1.19) and sine-Gordon models. Examination of (3.26) shows the two models to be equivalent. For more details on their similarity, see [22].

VII.4 The Pauli principle for
one-dimensional interacting bosons

The distinctive feature of one-dimensional quantum systems is that the Pauli principle is fulfilled not only for fermions, but also for interacting bosons (they may not possess equal momenta). Let us consider, for example, the NS model. The eigenfunctions of the transfer matrix (VI.1.3) are constructed by (1.10). The momenta $\{\lambda_j\}$ satisfy the Bethe equations (1.31). When deriving these equations, it was essential to assume that all λ_j are different. Let us now relax this condition and consider the simplest example: λ_1 occurs twice, with $\lambda_1 \equiv \lambda_2$, and the remaining momenta, λ_j $(j = 3, \ldots, N)$ are all different.

The corresponding eigenfunction is

$$|\Psi_N\rangle = B^2(\lambda_1) \prod_{j=3}^{N} B(\lambda_j)|0\rangle. \tag{4.1}$$

Theorem 1. *Eigenfunctions of the form (4.1) do not exist for the NS model.*

Proof: Let $|\Psi_N\rangle$ (4.1) be an eigenfunction of the transfer matrix $\tau(\mu) = A(\mu) + D(\mu)$. This demand results in restrictions on the $\{\lambda_j\}$. One can show that the Bethe equations for (4.1) are

$$\frac{a(\lambda_n)}{d(\lambda_n)} \left[\frac{f(\lambda_n, \lambda_1)}{f(\lambda_1, \lambda_n)}\right]^2 \prod_{\substack{j=3 \\ j \neq n}}^{N} \frac{f(\lambda_n, \lambda_j)}{f(\lambda_j, \lambda_n)} = 1, \qquad n = 3, \ldots, N; \quad (4.2)$$

$$\frac{a(\lambda_1)}{d(\lambda_1)} \prod_{j=3}^{N} \frac{f(\lambda_1, \lambda_j)}{f(\lambda_j, \lambda_1)} = -1; \qquad (4.3)$$

$$\frac{d}{d\lambda_1}\varphi_1 = i\frac{d}{d\lambda_1} \ln \frac{a(\lambda_1)}{d(\lambda_1)} + i\sum_{j=3}^{N} \frac{d}{d\lambda_1} \ln \frac{f(\lambda_1, \lambda_j)}{f(\lambda_j, \lambda_1)} + \frac{4}{c} = 0. \quad (4.4)$$

To prove this, let us take $\lambda_2 \to \lambda_1$ in (1.25)–(1.28); then we find:

$$A(\mu)B^2(\lambda_1) \prod_{j=3}^{N} B(\lambda_j)|0\rangle = \Lambda B^2(\lambda_1) \prod_{j=3}^{N} B(\lambda_j)|0\rangle$$

$$+ B(\mu)B^2(\lambda_1) \sum_{l=3}^{N} \Lambda_l^{(1)} \prod_{\substack{j=3 \\ j \neq l}}^{N} B(\lambda_j)|0\rangle$$

$$+ \Lambda^{(2)} B(\mu)B(\lambda_1) \prod_{j=3}^{N} B(\lambda_j)|0\rangle$$

$$+ \Lambda^{(3)} B(\mu)B'(\lambda_1) \prod_{j=3}^{N} B(\lambda_j)|0\rangle; \qquad (4.5)$$

$$D(\mu)B^2(\lambda_1) \prod_{j=3}^{N} B(\lambda_j)|0\rangle = \tilde{\Lambda} B^2(\lambda_1) \prod_{j=3}^{N} B(\lambda_j)|0\rangle$$

$$+ B(\mu)B^2(\lambda_1) \sum_{l=3}^{N} \tilde{\Lambda}_l^{(1)} \prod_{\substack{j=3 \\ j \neq l}}^{N} B(\lambda_j)|0\rangle$$

$$+ \tilde{\Lambda}^{(2)} B(\mu)B(\lambda_1) \prod_{j=3}^{N} B(\lambda_j)|0\rangle$$

$$+ \tilde{\Lambda}^{(3)} B(\mu)B'(\lambda_1) \prod_{j=3}^{N} B(\lambda_j)|0\rangle. \qquad (4.6)$$

The coefficients are given by

$$\Lambda = a(\mu)f^2(\mu,\lambda_1) \prod_{j=3}^{N} f(\mu,\lambda_j);$$

$$\widetilde{\Lambda} = d(\mu)f^2(\lambda_1,\mu) \prod_{j=3}^{N} f(\lambda_j,\mu); \tag{4.7}$$

$$\Lambda_l^{(1)} = -\, a(\lambda_l)g(\mu,\lambda_l)f^2(\lambda_l,\lambda_1) \prod_{\substack{j=3 \\ j\neq l}}^{N} f(\lambda_l,\lambda_j); \tag{4.8}$$

$$\widetilde{\Lambda}_l^{(1)} = d(\lambda_l)g(\mu,\lambda_l)f^2(\lambda_1,\lambda_l) \prod_{\substack{j=3 \\ j\neq l}}^{N} f(\lambda_j,\lambda_l); \tag{4.9}$$

$$\Lambda^{(2)} = -\, a(\lambda_1)g(\mu,\lambda_1)\left[1+f(\mu,\lambda_1)\right] \prod_{j=3}^{N} f(\lambda_1,\lambda_j)$$

$$-\, icg(\mu,\lambda_1)\frac{\partial}{\partial\lambda_1}\left[a(\lambda_1)\prod_{j=3}^{N} f(\lambda_1,\lambda_j)\right]; \tag{4.10}$$

$$\widetilde{\Lambda}^{(2)} = d(\lambda_1)g(\mu,\lambda_1)\left[1+f(\lambda_1,\mu)\right] \prod_{j=3}^{N} f(\lambda_j,\lambda_1)$$

$$-\, icg(\mu,\lambda_1)\frac{\partial}{\partial\lambda_1}\left[d(\lambda_1)\prod_{j=3}^{N} f(\lambda_j,\lambda_1)\right]; \tag{4.11}$$

$$\Lambda^{(3)} = ica(\lambda_1)g(\mu,\lambda_1) \prod_{j=3}^{N} f(\lambda_1,\lambda_j);$$

$$\widetilde{\Lambda}^{(3)} = icd(\lambda_1)g(\mu,\lambda_1) \prod_{j=3}^{N} f(\lambda_j,\lambda_1). \tag{4.12}$$

Taking into account that on the right hand sides of (4.5) and (4.6) all the vectors are linearly independent, we obtain that $|\Psi_N\rangle$ is an eigenvector of $\tau(\mu)$ with

$$\tau(\mu)|\Psi_N\rangle = \left(\Lambda + \widetilde{\Lambda}\right)|\Psi_N\rangle \tag{4.13}$$

if the following system of equations is satisfied:

$$\Lambda_l^{(1)} + \widetilde{\Lambda}_l^{(1)} = 0; \qquad \Lambda^{(2)} + \widetilde{\Lambda}^{(2)} = 0; \qquad \Lambda^{(3)} + \widetilde{\Lambda}^{(3)} = 0. \tag{4.14}$$

This system coincides with (4.2)–(4.4). So we have proved that the Bethe equations for $|\Psi_N\rangle$ can be written in the form (4.2)–(4.4).

We shall now prove that these equations for the NS model are unsolvable. Equations (4.2) and (4.3) for this model are

$$e^{-i\lambda_n L} \left(\frac{\lambda_n - \lambda_1 + ic}{\lambda_n - \lambda_1 - ic}\right)^2 \prod_{\substack{j=3 \\ j \neq n}}^{N} \left(\frac{\lambda_n - \lambda_j + ic}{\lambda_n - \lambda_j - ic}\right) = 1,$$

$$n = 3, \ldots, N \qquad (4.15)$$

$$e^{-i\lambda_1 L} \prod_{j=3}^{N} \left(\frac{\lambda_1 - \lambda_j + ic}{\lambda_1 - \lambda_j - ic}\right) = -1, \qquad (4.16)$$

and possess solutions for real $\{\lambda_j\}$ only (the proof is similar to that of Theorem 1 in section 2, Chapter I). On the other hand, equation (4.4) looks like:

$$\frac{4}{c} + L + \sum_{l=3}^{N} \frac{2c}{c^2 + (\lambda_1 - \lambda_l)^2} = 0 \qquad (4.17)$$

and has no solutions for real $\{\lambda_j\}$. The theorem is proved.

We have thus proved that two momenta cannot coincide. The case when several momenta coincide is treated in a similar way [5]. Thus we see that in the set $\{\lambda_j\}$ (1.10), there are no repeated momenta if $|\Psi_N\rangle$ is an eigenfunction of $\tau(\mu)$. This theorem plays an important role since, as was shown in Chapter I, the state of the NS Hamiltonian with minimal energy could be constructed as a Dirac sea. The Pauli principle ensures the stability of the Dirac sea. One may show that the unsolvability of equations (4.17) is connected with the convexity of Yang's action, which was discussed in Chapter I. This leads to the conjecture that in all integrable models possessing a convex Yang-Yang action, the Pauli principle holds. If the Yang-Yang action is not convex, the Pauli principle may not be valid.

VII.5 The shift operator

Let us consider the monodromy matrix consisting of two "commuting" factors (1.7)

$$\mathsf{T}(\lambda) = \mathsf{T}(2|\lambda)\mathsf{T}(1|\lambda) = \begin{pmatrix} A(\lambda) & B(\lambda) \\ C(\lambda) & D(\lambda) \end{pmatrix}. \qquad (5.1)$$

Each factor $\mathsf{T}(j|\lambda)$ $(j = 1, 2)$,

$$\mathsf{T}(j|\lambda) = \begin{pmatrix} A_j(\lambda) & B_j(\lambda) \\ C_j(\lambda) & D_j(\lambda) \end{pmatrix}, \qquad (5.2)$$

is intertwined by the R-matrix (1.3) and has vacuum $|0\rangle_j$:

$$A_j(\lambda)|0\rangle_j = a_j(\lambda)|0\rangle_j; \qquad D_j(\lambda)|0\rangle_j = d_j(\lambda)|0\rangle_j;$$
$$C_j(\lambda)|0\rangle_j = 0. \qquad\qquad\qquad\qquad\qquad (5.3)$$

(The case of more than two factors is considered in Appendix 2.) Let us define the shift operator O by the following:

$$OT(\lambda)O^{-1} = \tilde{T}(\lambda) = T(1|\lambda)T(2|\lambda) = \begin{pmatrix} \tilde{A}(\lambda) & \tilde{B}(\lambda) \\ \tilde{C}(\lambda) & \tilde{D}(\lambda) \end{pmatrix}. \qquad (5.4)$$

(Note that tr $T(\lambda) = $ tr $\tilde{T}(\lambda)$.) The commutation relations for the matrix elements of $\tilde{T}(\lambda)$ are given by the same R-matrix (1.3). The vacuum is $|0\rangle = |0\rangle_1 \otimes |0\rangle_2$. The vacuum eigenvalues of $\tilde{T}$ and T coincide (1.8): $a(\lambda) = \tilde{a}(\lambda) = a_1(\lambda)a_2(\lambda)$, $d(\lambda) = \tilde{d}(\lambda) = d_1(\lambda)d_2(\lambda)$.

Consider now the wave functions

$$|\Psi_N\rangle = \prod_{j=1}^{N} B(\lambda_j)|0\rangle, \qquad\qquad\qquad (5.5)$$

$$|\tilde{\Psi}_N\rangle = \prod_{j=1}^{N} \tilde{B}(\lambda_j)|0\rangle. \qquad\qquad\qquad (5.6)$$

Theorem 2. *The wave functions (5.5) and (5.6) are proportional to each other if and only if the Bethe equations are satisfied:*

$$\frac{a(\lambda_n)}{d(\lambda_n)} \prod_{\substack{j=1 \\ j\neq n}}^{N} \frac{f(\lambda_n, \lambda_j)}{f(\lambda_j, \lambda_n)} = 1, \qquad n = 1, \ldots, N. \qquad (5.7)$$

Proof: Expressing the wave function (5.5) in terms of the wave functions (5.2) of the individual sites $\prod B_1(\lambda_j)|0\rangle_1$ and $\prod B_2(\lambda_j)|0\rangle_2$, we have for $B(\lambda)$ from (5.1) and (5.2):

$$B(\lambda) = A_2(\lambda)B_1(\lambda) + D_1(\lambda)B_2(\lambda). \qquad (5.8)$$

Applying the logic of the algebraic Bethe Ansatz we get

$$\prod_{j=1}^{N} B(\lambda_j)|0\rangle = \sum_{\{\lambda\}=\{\lambda^I\}\bigcup\{\lambda^{II}\}} \prod_{j\in I}^{n_1} \prod_{k\in II}^{n_2} a_1(\lambda_j^I)d_2(\lambda_k^{II})$$
$$\times f(\lambda_j^I, \lambda_k^{II}) \left(B_2(\lambda_k^{II})|0\rangle_2 \right) \left(B_1(\lambda_j^I)|0\rangle_1 \right) \quad (5.9)$$

where summation is over all decompositions of the set $\{\lambda_j\}$ into two disjoint subsets $\{\lambda^I\}$ and $\{\lambda^{II}\}$, for which

$$\{\lambda^I\}\bigcap\{\lambda^{II}\} = \emptyset; \qquad \{\lambda^I\}\bigcup\{\lambda^{II}\} = \{\lambda\};$$
$$\mathrm{card}\{\lambda^I\} = n_1, \qquad \mathrm{card}\{\lambda^{II}\} = n_2. \tag{5.10}$$

Each term in (5.9) is a double product; the index j belongs to the first set and the index k to the second. The generalization of the formula (5.9) for many commuting factors instead of just two (5.1) is given in Appendix 2. The wave function (5.6) is expressed as

$$\prod_{j=1}^{N} \widetilde{B}(\lambda_j)|0\rangle = \sum_{\{\lambda\}=\{\lambda^I\}\bigcup\{\lambda^{II}\}} \prod_{j\in I}^{n_1} \prod_{k\in II}^{n_2} a_2(\lambda_k^{II}) d_1(\lambda_j^I)$$
$$\times f(\lambda_k^{II}, \lambda_j^I) \left(B_1(\lambda_j^I)|0\rangle_1\right)\left(B_2(\lambda_k^{II})|0\rangle_2\right). \tag{5.11}$$

All the terms on the right hand side of (5.9) are linearly independent. Calculation shows that the right hand sides of (5.9) and (5.11) are proportional if and only if the system (5.7) is fulfilled. In this context,

$$O|\Psi_N\rangle = |\widetilde{\Psi}_N\rangle = \nu|\Psi_N\rangle;$$
$$\nu = \prod_{j=1}^{N} \frac{a_1(\lambda_j)}{d_1(\lambda_j)} = \prod_{j=1}^{N} \left(\frac{a_2(\lambda_j)}{d_2(\lambda_j)}\right)^{-1} \tag{5.12}$$

where ν is the eigenvalue of the shift operator. The theorem is proved.

Theorem 2 means exactly that the Bethe Ansatz equations (5.7) guarantee the periodicity.

The dual wavefunctions are also proportional to each other if and only if the system (5.7) is fulfilled. The proof is based on the representation

$$\langle 0|\prod_{j=1}^{N} C(\lambda_j) = \sum_{\{\lambda\}=\{\lambda^I\}\bigcup\{\lambda^{II}\}} \prod_{j\in I}^{n_1} \prod_{k\in II}^{n_2} \left({}_1\langle 0|C_1(\lambda_j^I)\right) \bigotimes \left({}_2\langle 0|C_2(\lambda_k^{II})\right)$$
$$\times f(\lambda_k^{II}, \lambda_j^I) a_1(\lambda_k^{II}) d_2(\lambda_j^I). \tag{5.13}$$

In addition,

$$\langle 0|\prod_{j=1}^{N} \widetilde{C}(\lambda_j) = \langle 0|\prod_{j=1}^{N} C(\lambda_j) O^{-1} = \nu^{-1} \langle 0|\prod_{j=1}^{N} C(\lambda_j). \tag{5.14}$$

Thus, if the wave function (5.5) is an eigenfunction of the shift operator, it is necessarily an eigenfunction of the transfer matrix.

Theorem 2 may be applied to the calculation of momentum eigenvalues. Let us consider the NS model. One represents the NS monodromy matrix as the product of two "commuting terms": $\mathsf{T}(\lambda) = \mathsf{T}(2|\lambda)\mathsf{T}(1|\lambda)$, where $\mathsf{T}(1|\lambda)$ is the monodromy matrix on the interval $[0, x]$ with vacuum eigenvalues equal to

$$a_1(\lambda) = \exp\left\{-i\frac{\lambda x}{2}\right\}; \qquad d_1(\lambda) = \exp\left\{i\frac{\lambda x}{2}\right\}, \qquad (5.15)$$

and matrix $\mathsf{T}(2|\lambda)$ is the monodromy matrix on the interval $[x, L]$, with vacuum eigenvalues equal to

$$a_2(\lambda) = \exp\left\{-i\frac{\lambda(L - x)}{2}\right\}; \qquad d_2(\lambda) = \exp\left\{i\frac{\lambda(L - x)}{2}\right\}. \qquad (5.16)$$

The operator $O = \exp\{i(L - x)\widehat{P}\}$ is the shift operator on $L - x$ and $\widehat{P}$ is the momentum operator. The eigenvalue of the shift operator (5.12) is in this case

$$\nu = \exp\left\{i(L - x)\sum_{j=1}^{N}\lambda_j\right\}.$$

This means that the eigenvalue of the momentum operator is $P_N = \sum_{j=1}^{N}\lambda_j$, which coincides with (3.7).

Let us consider the homogeneous lattice model where the vacuum eigenfunctions of each L-operator are similar and equal to $a_\mathsf{L}(\lambda)$ and $d_\mathsf{L}(\lambda)$. In such a model, the eigenvalue of the momentum operator is

$$P_N = i\sum_{j=1}^{N}\ln\left(\frac{a_\mathsf{L}(\lambda_j)}{d_\mathsf{L}(\lambda_j)}\right). \qquad (5.17)$$

For $N = 1$, we have reproduced formula (3.22) for the XXZ magnet.

The theorem proved in this section plays an important role when calculating correlation functions. It is also useful when constructing quantum local Hamiltonians on the lattice ([29] and [30]).

VII.6 Classification of monodromy matrices

Let us consider the monodromy matrix

$$\mathsf{T}(\lambda) = \begin{pmatrix} A(\lambda) & B(\lambda) \\ C(\lambda) & D(\lambda) \end{pmatrix} \qquad (6.1)$$

intertwined by the R-matrix (1.3), (1.4):

$$\mathsf{R}(\lambda, \mu)\left(\mathsf{T}(\lambda)\bigotimes\mathsf{T}(\mu)\right) = \left(\mathsf{T}(\mu)\bigotimes\mathsf{T}(\lambda)\right)\mathsf{R}(\lambda, \mu) \qquad (6.2)$$

and assume the existence of a pseudovacuum $|0\rangle$:

$$A(\lambda)|0\rangle = a(\lambda)|0\rangle; \quad D(\lambda)|0\rangle = d(\lambda)|0\rangle; \quad C(\lambda)|0\rangle = 0. \qquad (6.3)$$

Theorem 3. *The monodromy matrix* $\mathsf{T}(\lambda)$ *(6.1) exists for any given functions* $a(\lambda)$ *and* $d(\lambda)$ *if conditions (6.2) and (6.3) are fulfilled.*

Proof: To prove this statement, one has to represent four linear operators $A(\lambda)$, $B(\lambda)$, $C(\lambda)$ and $D(\lambda)$ and the space they act in. The basis vectors of this space are N particle states with "momenta" $\lambda_1, \ldots, \lambda_N$ (the momenta are arbitrary complex numbers):

$$|\lambda_1, \ldots, \lambda_N\rangle_N, \quad \lambda_j \in \mathbb{C}, \quad N = 0, 1, 2, \ldots \tag{6.4}$$

These states are symmetric functions of all λ_j. The $N = 0$ state is identified with the pseudovacuum. The action of the operators A, B, C and D on the basis vectors is defined by the formulæ for the algebraic Bethe Ansatz. The operator $B(\mu)$ adds a particle with momentum μ to the basis vector:

$$B(\mu)|\lambda_1, \ldots, \lambda_N\rangle_N = |\mu, \lambda_1, \ldots, \lambda_N\rangle_{N+1}. \tag{6.5}$$

The action of operators $A(\mu)$ and $D(\mu)$ preserves the number of particles:

$$A(\mu)|\{\lambda_j\}\rangle_N = \Lambda|\{\lambda_j\}\rangle_N + \sum_{n=1}^{N} \Lambda_n|\mu, \{\lambda_{j\neq n}\}\rangle_N; \tag{6.6}$$

$$D(\mu)|\{\lambda_j\}\rangle_N = \widetilde{\Lambda}|\{\lambda_j\}\rangle_N + \sum_{n=1}^{N} \widetilde{\Lambda}_n|\mu, \{\lambda_{j\neq n}\}\rangle_N. \tag{6.7}$$

The coefficients Λ, $\widetilde{\Lambda}$ are given here by (1.27), (1.28). The operator $C(\mu)$ removes a momentum from the basis vector:

$$C(\mu)|\{\lambda_j\}\rangle_N = \sum_{n=1}^{N} M_n|\{\lambda_{j\neq n}\}\rangle_{N-1}$$
$$+ \sum_{k>n} M_{kn}|\mu, \{\lambda_{j\neq k,n}\}\rangle_{N-1}. \tag{6.8}$$

The coefficients M are given by (2.3), (2.4). Thus, the monodromy matrix is built up. It should be emphasized that equations (6.6) and (6.7) are now definitions.

Let us now prove that the monodromy matrix thus obtained satisfies (6.2). Writing down (6.2) as the 16 scalar relations (1.11)–(1.24), we can act with the right and left hand sides of these equalities on some basis vector (6.4).

The fact that these operators standing on both sides of each equality act equally on arbitrary basis vectors is obvious from direct calculation. When calculating identities, the functions $f(\mu, \lambda)$ and $g(\mu, \lambda)$ should be used. For illustration we shall prove that

$$A(\mu_1)A(\mu_2) = A(\mu_2)A(\mu_1). \tag{6.9}$$

Acting with $A(\mu_2)A(\mu_1)$ on $|\{\lambda_j\}\rangle_N$, we obtain

$$
\begin{aligned}
A(\mu_2)A(\mu_1)|\{\lambda_j\}\rangle_N =\; & a(\mu_1)a(\mu_2) \prod_{j=1}^{N} f(\mu_1,\lambda_j)f(\mu_2,\lambda_j)|\{\lambda_j\}\rangle_N \\
& + \sum_{n=1}^{N} a(\mu_1) \prod_{j=1}^{N} f(\mu_1,\lambda_j)a(\lambda_n)g(\lambda_n,\mu_2) \\
& \qquad \times \prod_{j\neq n} f(\lambda_n,\lambda_j)|\mu_2,\{\lambda_{j\neq n}\}\rangle_N \\
& + \sum_{n=1}^{N} a(\mu_2) \prod_{j\neq n} f(\mu_2,\lambda_j)f(\mu_2,\mu_1)a(\lambda_n)g(\lambda_n,\mu_1) \\
& \qquad \times \prod_{j\neq n} f(\lambda_n,\lambda_j)|\mu_1,\{\lambda_{j\neq n}\}\rangle_N \\
& + \sum_{n=1}^{N} a(\lambda_n)g(\lambda_n,\mu_1) \prod_{j\neq n} f(\lambda_n,\lambda_j)a(\mu_1)g(\mu_1,\mu_2) \\
& \qquad \times \prod_{j\neq n} f(\mu_1,\lambda_j)|\mu_2,\{\lambda_{j\neq n}\}\rangle_N \\
& + \sum_{n=1}^{N}\sum_{k\neq n} a(\lambda_n)g(\lambda_n,\mu_1) \prod_{j\neq n} f(\lambda_n,\lambda_j)a(\lambda_k)g(\lambda_k,\mu_2) \\
& \qquad \times \prod_{j\neq n,k} f(\lambda_k,\lambda_j)f(\lambda_k,\mu_1)|\mu_1,\mu_2,\{\lambda_{j\neq n,k}\}\rangle_N
\end{aligned}
\tag{6.10}
$$

assuming that the momenta μ_1, μ_2 and $\{\lambda_j\}$ are different. If (6.9) is correct, then the right hand side of (6.10) must not change when $\mu_1 \leftrightarrow \mu_2$. Let us check all coefficients separately. The symmetry of $a(\mu_1)a(\mu_2)|\{\lambda_j\}\rangle_N$ is obvious. Next, the coefficient of $a(\mu_1)a(\lambda_n)|\mu_2, \{\lambda_{j\neq n}\}\rangle_N$ transfers into the coefficient of $a(\mu_2)a(\lambda_n)|\mu_1, \{\lambda_{j\neq n}\}\rangle_N$. Canceling the common factor $(\prod_{j\neq n} f(\mu_2,\lambda_j)f(\lambda_n,\lambda_j))$, we get the equality

$$g(\lambda_n,\mu_1)f(\mu_2,\lambda_n) + g(\lambda_n,\mu_2)g(\mu_2,\mu_1) = f(\mu_2,\mu_1)g(\lambda_n,\mu_1) \tag{6.11}$$

which is identically true for both the R-matrices (1.4) and (1.5). The coefficient of $a(\lambda_n)a(\lambda_k)|\mu_1,\mu_2,\{\lambda_{j\neq n,k}\}\rangle_N$ should not change when

$\mu_1 \leftrightarrow \mu_2$. Testing this statement leads to the identity:

$$g(\lambda_n, \mu_1)g(\lambda_k, \mu_2)f(\lambda_k, \mu_1)f(\lambda_n, \lambda_k)$$
$$+ g(\lambda_k, \mu_1)g(\lambda_n, \mu_2)f(\lambda_n, \mu_1)f(\lambda_k, \lambda_n)$$
$$= g(\lambda_n, \mu_2)g(\lambda_k, \mu_1)f(\lambda_k, \mu_2)f(\lambda_n, \lambda_k)$$
$$+ g(\lambda_k, \mu_2)g(\lambda_n, \mu_1)f(\lambda_n, \mu_2)f(\lambda_k, \lambda_n) \quad (6.12)$$

which is indeed valid. Thus, both sides of (6.9) act equally on the basis vector (6.4). We assumed at the start that in the set $\{\mu_1, \mu_2, \lambda_j\}$ all momenta were different. One may continue the equality

$$\left(\mathsf{R}(\lambda, \mu) \, \mathsf{T}(\lambda) \bigotimes \mathsf{T}(\mu) - \mathsf{T}(\mu) \bigotimes \mathsf{T}(\lambda) \, \mathsf{R}(\lambda, \mu) \right) |\lambda_1, \ldots, \lambda_N\rangle_N = 0$$
$$(6.13)$$

for coinciding momenta and verify that it is always valid.

Thus, the commutation relations are proved. The relations (6.3) are also fulfilled. The theorem is proved. Similar statements are also true for the XXZ R-matrix (1.5) (see [8]).

VII.7 Comments on the classification of monodromy matrices

(1) The following equality can be proved by applying the techniques of the previous section:

$$A\left(\lambda - i\frac{c}{2}\right) D\left(\lambda + i\frac{c}{2}\right) - B\left(\lambda - i\frac{c}{2}\right) C\left(\lambda + i\frac{c}{2}\right)$$
$$= a\left(\lambda - i\frac{c}{2}\right) d\left(\lambda + i\frac{c}{2}\right). \quad (7.1)$$

The left hand side is a quadratic combination of operators, while the right hand side is a complex-valued function (a c-number). Acting with both parts of this equality on an arbitrary basis vector, one obtains the same result. The essential notion of the determinant of the monodromy matrix in the quantum case will be introduced in the next section with the help of relation (7.1).

(2) The following identification may be performed:

$$|\lambda_1, \ldots, \lambda_N\rangle_N = \prod_{j=1}^{N} B(\lambda_j)|0\rangle. \quad (7.2)$$

Due to the linearity of the intertwining relations, there is a trivial arbitrariness connected with the multiplication of $\mathsf{T}(\mu)$ by an arbitrary complex-valued function. Thus, essentially different monodromy matrices, $\mathsf{T}(\mu)$, can be parametrized by one arbitrary function $r(\lambda) = a(\lambda)/d(\lambda)$. The action of the operators A, B, C and D on the basis

(7.2) is defined by the formulæ (1.25)–(1.28), (2.2)–(2.4) for the algebraic Bethe Ansatz and depends only on the vacuum eigenvalues $a(\lambda)$ and $d(\lambda)$. This leads to the uniqueness theorem which states that for a given R-matrix, the functions $a(\lambda)$ and $d(\lambda)$ uniquely determine the model.

(3) Linear functionals in the space considered are of interest. Let us define $\langle 0|$ (the dual pseudovacuum) by the relations

$$\langle 0|0 \rangle = 1, \qquad \langle 0|\{\lambda_j\}_N \rangle = 0, \quad N = 1, 2, \dots . \tag{7.3}$$

It is easy to evaluate the result of the action of certain operators on a dual pseudovacuum:

$$\langle 0|A(\mu) = a(\mu)\langle 0|; \qquad \langle 0|D(\mu) = d(\mu)\langle 0|; $$
$$\langle 0|B(\mu) = 0. \tag{7.4}$$

Let us consider the linear functionals

$$\langle 0|C(\lambda_1) \cdots C(\lambda_N). \tag{7.5}$$

We shall call them *basis* functionals. These linear functionals (7.5) are symmetric functions of the $\{\lambda_j\}$ (see (1.11)). The action of the operators $A(\mu)$, $B(\mu)$, $C(\mu)$ and $D(\mu)$ on the basis functionals (7.5) is easily calculated by means of (1.11)–(1.24); the answers coincide with (2.7)–(2.9).

(4) Let us introduce the particle number operator, Q, in the constructed space. On the basis vectors, it acts as follows:

$$Q|0\rangle = 0; \qquad Q|\{\lambda_j\}\rangle_N = N|\{\lambda_j\}\rangle_N;$$
$$\langle 0|\left(\prod_{j=1}^{N} C(\lambda_j)\right) Q = N\langle 0|\left(\prod_{j=1}^{N} C(\lambda_j)\right); \tag{7.6}$$
$$\langle 0|Q = 0.$$

The momenta $\{\lambda_j\}$ are arbitrary (and not subject to the Bethe equations). The commutation relations are easily calculated:

$$[Q, B(\lambda)] = B(\lambda); \qquad [Q, C(\lambda)] = -C(\lambda);$$
$$[Q, A(\lambda)] = [Q, D(\lambda)] = 0, \tag{7.7}$$

or, in matrix form, $2[Q, \mathsf{T}(\lambda)] = [\sigma_z, \mathsf{T}(\lambda)]$. The action of the operator $\exp\{\alpha Q\}$ (where α is an arbitrary constant) on the basis vector is

$$\exp\{\alpha Q\}\,|\{\lambda_j\}\rangle_N = \exp\{\alpha N\}\,|\{\lambda_j\}\rangle_N. \tag{7.8}$$

(5) The properties of the basis (7.2) are similar to those of the basis (7.5). This allows the introduction of Hermitian conjugation into the space as follows. Let us demand that

$$B^\dagger(\lambda^*) = \pm C(\lambda); \qquad \langle 0| = |0\rangle^\dagger \tag{7.9}$$

which is possible if

$$a^*(\lambda^*) = d(\lambda) \tag{7.10}$$

and the bases (7.2) and (7.5) are connected by Hermitian conjugation, and

$$A^\dagger(\lambda^*) = D(\lambda). \tag{7.11}$$

In this case, the value of the linear functional on the basis vector with a similar set of $\{\lambda_j\}$,

$$\langle 0| \prod_{j=1}^{N} C(\lambda_j) \prod_{k=1}^{N} B(\lambda_k)|0\rangle, \tag{7.12}$$

is real for real $\{\lambda_j\}$.

(6) It is shown in [8] that all the results of the previous section are also true for the XXZ R-matrix (1.3), (1.5). In this case, it is natural to consider only the arbitrary periodic function $a(\lambda)/d(\lambda)$ [29]:

$$\frac{a(\lambda)}{d(\lambda)} = \frac{a(\lambda + i\pi)}{d(\lambda + i\pi)}. \tag{7.13}$$

This is connected with the problem of representing $\mathsf{T}(\lambda)$ as a product of L-operators (see section 4, Chapter VIII).

Thus, we have classified all the monodromy matrices for a given R-matrix by the function $r(\lambda)$. Later (section 4, Chapter VIII) we shall classify all the L-operators for the same R-matrix. The connection between $\mathsf{T}(\lambda)$ and $\mathsf{L}(\lambda)$ is given by (VI.1.1). We shall be able to construct an L-operator depending only on four complex parameters (the matrix $\mathsf{T}(\lambda)$ is a functional of $r(\lambda)$). Matrix elements of the L-operator will act in a rather narrow subspace of the Fock space constructed above; this L-operator generates the most general monodromy matrix of the form (VI.1.1). In regard to this relation, the parallel between the theory of Lie group representations and QISM mentioned in (VI.2) should be emphasized.

(7) Let us discuss the consequences of the theorem proved in the previous section. Consider the Bethe equations (1.31):

$$r(\lambda_j) \prod_{\substack{k=1 \\ k \neq j}}^{N} \frac{f(\lambda_j, \lambda_k)}{f(\lambda_k, \lambda_j)} = 1, \qquad j = 1, \ldots, N \tag{7.14}$$

and fix an arbitrary set of N different numbers $\{\lambda_j\}$.

Are there any integrable models for which this set is the solution of the Bethe equations? The answer is positive; many such models are available. They are parametrized by a function $r(\lambda)$ with a (rather loose) restriction: for the points λ_j, $r(\lambda)$ must take the fixed values

$$r(\lambda_j) = \prod_{\substack{k=1 \\ k \neq j}}^{N} \frac{f(\lambda_k, \lambda_j)}{f(\lambda_j, \lambda_k)}, \qquad j = 1, \ldots, N. \tag{7.15}$$

This makes it possible to consider the solutions $\{\lambda_j\}$ of the Bethe equations as free independent variables; this is used frequently in Part III.

(8) Finally, let us discuss the scattering matrix for the case of arbitrary functions $a(\lambda)$ and $d(\lambda)$. Consider the Bethe equations (1.31), (7.14). The factors $f(\lambda_j, \lambda_k)/f(\lambda_k, \lambda_j)$ on the left hand side have the meaning of the bare scattering matrices for a particle with spectral parameter λ_j and a particle with spectral parameter λ_k. Thus, the bare scattering matrix does not depend on the arbitrary functions $a(\lambda)$ and $d(\lambda)$, but is defined entirely by the R-matrix. The dressed scattering phase is obtained from the bare one by means of the dressing equations considered in section 4 of Chapter I. The kernel of these integral equations is defined by the R-matrix:

$$K(\lambda, \mu) = i \frac{\partial}{\partial \lambda} \ln \frac{f(\lambda, \mu)}{f(\mu, \lambda)};$$

the same is true for the right hand side of the dressing equations (I.4.39) for the phase. Thus, the dressed scattering matrix is also entirely defined by the R-matrix. To be more precise, let us consider two models with the same R-matrix but different functions $a(\lambda)$ and $d(\lambda)$. If the values of the spectral parameters on the Fermi boundary in these models coincide, the dressed scattering matrices also coincide.

It should be emphasized that in the above we have considered the S-matrices as functions of the spectral parameter λ. As functions of the dressed (physical) momenta they are, in general, different.

VII.8 The quantum determinant

The generalization of the determinant of the monodromy matrix in the quantum case is introduced here. Let us consider the monodromy matrix $\mathsf{T}(\lambda)$ (1.1) intertwined by the XXX R-matrix (1.3), (1.4) and possessing the vacuum (1.6). When $\lambda = \mu + ic$, the commutation relations (1.11)–(1.24) between the matrix elements of $\mathsf{T}(\lambda)$ and $\mathsf{T}(\mu)$ are essentially simplified due to the fact that $f(\mu, \lambda) = 0$, $g(\mu, \lambda) = -1$, and $\mathsf{R}(\lambda, \mu)$ is proportional to the one-dimensional projector

$$\mathsf{R}(\mu + ic, \mu) = \Pi - I. \tag{8.1}$$

Examples of these commutation relations are:

$$C(\mu)D(\mu + ic) = D(\mu)C(\mu + ic);$$
$$A(\mu)B(\mu + ic) = B(\mu)A(\mu + ic). \tag{8.2}$$

The determinant of the monodromy matrix in the quantum case is defined as

$$\det{}_q T(\lambda) = A\left(\lambda - i\frac{c}{2}\right) D\left(\lambda + i\frac{c}{2}\right) - B\left(\lambda - i\frac{c}{2}\right) C\left(\lambda + i\frac{c}{2}\right). \tag{8.3}$$

Using the commutation relations between $T(\lambda)$ and $T(\lambda + ic)$, this can also be rewritten in the form

$$\det{}_q T(\lambda) = D\left(\lambda - i\frac{c}{2}\right) A\left(\lambda + i\frac{c}{2}\right) - C\left(\lambda - i\frac{c}{2}\right) B\left(\lambda + i\frac{c}{2}\right). \tag{8.4}$$

The quantum determinant commutes with the operators $A(\mu)$, $B(\mu)$, $C(\mu)$ and $D(\mu)$, i.e., with any matrix element of the monodromy matrix:

$$[\det{}_q T(\lambda), T(\mu)] = 0 \tag{8.5}$$

(the commutator here is in the quantum space).

It follows from (7.1) that the quantum determinant is a complex-valued function, but not a quantum operator (more exactly, it is proportional to the unit operator in the quantum space, the coefficient being a c-number)

$$\det{}_q T(\lambda) = a\left(\lambda - i\frac{c}{2}\right) d\left(\lambda + i\frac{c}{2}\right). \tag{8.6}$$

In matrix form, the commutation relations (8.2)–(8.5) are

$$T\left(\lambda - i\frac{c}{2}\right) \sigma_y T^T\left(\lambda + i\frac{c}{2}\right) \sigma_y$$
$$= \det{}_q T(\lambda) \begin{pmatrix} 1 & 0 \\ 0 & 1 \end{pmatrix} = a\left(\lambda - i\frac{c}{2}\right) d\left(\lambda + i\frac{c}{2}\right) \begin{pmatrix} 1 & 0 \\ 0 & 1 \end{pmatrix} \tag{8.7}$$

or,

$$T^{-1}(\lambda) = \sigma_y T^T(\lambda + ic)\sigma_y \left[\det{}_q T\left(\lambda + i\frac{c}{2}\right)\right]^{-1}. \tag{8.8}$$

This formula is the natural generalization of the classical Cramer's formula for the inverse of a 2×2 matrix. Using this, we can prove that the determinant of the "commuting" matrix product $T(\lambda) = T(2|\lambda)T(1|\lambda)$ (1.7) is the product of determinants:

$$\det{}_q T(\lambda) = \det{}_q T(1|\lambda) \det{}_q T(2|\lambda) \tag{8.9}$$

which means that the determinant of the monodromy matrix is equal to the product of the determinants of the L-operators:

$$\det{}_q T(\lambda) = \prod_{j=1}^{M} \det{}_q L(j|\lambda). \tag{8.10}$$

It is also possible to define the quantum determinant for the XXZ R-matrix:

$$\det{}_q T(\lambda) = A(\lambda - i\eta)D(\lambda + i\eta) - B(\lambda - i\eta)C(\lambda + i\eta)$$
$$= a(\lambda - i\eta)d(\lambda + i\eta) \tag{8.11}$$

which possesses all the properties mentioned above.

Let us list the quantum determinants $\det{}_q T(\lambda)$ for the models considered above:

(1) NS model (see (3.3)):
$$\det{}_q T(\lambda) = \exp\{-cL/2\} \tag{8.12}$$

(2) XXX chain (see (3.11)):
$$\det{}_q L(\lambda) = (c^2 + \lambda^2);$$
$$\det{}_q T(\lambda) = (c^2 + \lambda^2)^M \tag{8.13}$$

(3) XXZ chain (see (3.20)):
$$\det{}_q L(\lambda) = -\cosh(\lambda + 2i\eta)\cosh(\lambda - 2i\eta);$$
$$\det{}_q T(\lambda) = [\det{}_q L(\lambda)]^M \tag{8.14}$$

(4) sine-Gordon model (see (3.25)):
$$\det{}_q T(\lambda) = \exp\{2Ms\cosh 2\lambda\}, \quad s = (m\Delta/4)^2 \tag{8.15}$$

VII.9 Recursion properties of the partition function

The partition function Z_N (VI.6.6) for the XXX model with domain wall boundary conditions is examined in more detail in this section. An interesting quantity in itself, Z_N is also of importance when studying scalar products. It is a function of $2N$ variables, $\{\lambda_\alpha\}$, $\{\nu_j\}$.

Lemma 1. The partition function $Z_N(\{\lambda_\alpha\}, \{\nu_j\})$ is symmetric in all variables $\{\lambda_\alpha\}$ and $\{\nu_j\}$ separately, and is a polynomial of degree $(N-1)$ in both λ_α and ν_j, with all other variables fixed:

$$\frac{\partial^N Z_N}{\partial \lambda_\alpha^N} = \frac{\partial^N Z_N}{\partial \nu_j^N} = 0. \tag{9.1}$$

The proof follows from the representation (VI.6.12). Since (1.11)

$$[B(\lambda), B(\mu)] = 0,$$

(this follows because $B(\lambda)$ is an element of the monodromy matrix (VI.6.7)), Z_N is a symmetric matrix of all λ_α. The relations (VI.6.4) and (VI.6.7) imply that:

$$\frac{\partial^N T_\alpha(\lambda)}{\partial \lambda^N} = N! \, I \equiv N! \begin{pmatrix} 1 & 0 \\ 0 & 1 \end{pmatrix}; \tag{9.2}$$

then $\partial^N B(\lambda)/\partial \lambda^N = 0$ and hence $B(\lambda)$ is also a polynomial of degree $(N-1)$ in each λ_α. The representation (VI.6.15) results in similar properties for each ν_j. Lemma 1 is thus proved.

For further investigation of Z_N, we list some properties of the L-operator (VI.6.4):

(1) $\mathsf{L}_{\alpha k}(\nu_k - \lambda_\alpha = ic/2) = -ic\Pi_{\alpha k}$. $\qquad$ (9.3)

Here $\Pi_{\alpha k}$ is the permutation matrix

$$\Pi_{\alpha k} v_k w_\alpha = w_k v_\alpha \qquad (9.4)$$

with subscripts α and k denoting the spaces to which vectors v and w belong.

(2) The L-operator possesses two simple eigenvectors $(\uparrow_\alpha \uparrow_k)$ and $(\downarrow_\alpha \downarrow_k)$:

$$\mathsf{L}_{\alpha k}(\lambda_\alpha - \nu_k)\,(\uparrow_\alpha \uparrow_k) = \left(\lambda_\alpha - \nu_k - i\frac{c}{2}\right)(\uparrow_\alpha \uparrow_k)\,; \qquad (9.5)$$

$$\mathsf{L}_{\alpha k}(\lambda_\alpha - \nu_k)\,(\downarrow_\alpha \downarrow_k) = \left(\lambda_\alpha - \nu_k - i\frac{c}{2}\right)(\downarrow_\alpha \downarrow_k)\,. \qquad (9.6)$$

(3) $[\mathsf{L}_{\alpha k}(\lambda_\alpha - \nu_k),\, \sigma_x^{(k)}\sigma_x^{(\alpha)}] = 0.$ $\qquad$ (9.7)

Let us introduce the operator W defined by

$$W = \prod_{\alpha=1}^{N} \sigma_x^{(\alpha)} \prod_{k=1}^{N} \sigma_x^{(k)}.$$

This operator has the following properties:

$$W^2 = 1;$$

and

$$\left[W,\, \prod_{\alpha=1}^{N} \prod_{k=1}^{N} \mathsf{L}_{\alpha k}(\lambda_\alpha - \nu_k)\right] = 0, \qquad (9.8)$$

which follow from property (3) of the L-operator. Here the double product indicates space ordering as in section VI.6:

$$\prod_{\alpha=1}^{N} \prod_{k=1}^{N} \mathsf{L}_{\alpha k} \equiv (\cdots \mathsf{L}_{23}\mathsf{L}_{22}\mathsf{L}_{21})\,(\cdots \mathsf{L}_{13}\mathsf{L}_{12}\mathsf{L}_{11}).$$

Lemma 2. The function Z_N can be represented in the following form:

$$Z_N(\{\lambda_\alpha\}, \{\nu_j\}) = \langle 0| \prod_{j=1}^{N} C(\lambda_j)|0'\rangle. \qquad (9.9)$$

The proof is as follows. From (VI.6.5) we have

$$
Z_N = \left\{ \prod_{\beta=1}^{N} \uparrow_\beta \right\} \left\{ \prod_{j=1}^{N} \downarrow_j \right\} W
$$

$$
\times \left\{ \prod_{\alpha=1}^{N} \prod_{k=1}^{N} \mathsf{L}_{\alpha k}(\lambda_\alpha - \nu_k) \right\} W \left\{ \prod_{\beta=1}^{N} \downarrow_\beta \right\} \left\{ \prod_{j=1}^{N} \uparrow_j \right\}
$$

$$
= \left\{ \prod_{\beta=1}^{N} \downarrow_\beta \right\} \left\{ \prod_{j=1}^{N} \uparrow_j \right\}
$$

$$
\times \left\{ \prod_{\alpha=1}^{N} \prod_{k=1}^{N} \mathsf{L}_{\alpha k}(\lambda_\alpha - \nu_k) \right\} \left\{ \prod_{\beta=1}^{N} \uparrow_\beta \right\} \left\{ \prod_{j=1}^{N} \downarrow_j \right\}.
$$

$$(9.10)$$

Repeating the arguments of section VI.6 ((VI.6.11)–(VI.6.12)), we obtain (9.9). The lemma is proved.

Lemma 3. The function Z_N satisfies the following recursion relation:

$$
Z_N \Big|_{\lambda_\beta = \nu_l - ic/2} = -ic \prod_{\substack{k=1 \\ k \neq l}}^{N} \left(\lambda_\beta - \nu_k - i\frac{c}{2} \right) \prod_{\substack{\alpha=1 \\ \alpha \neq \beta}}^{N} \left(\lambda_\alpha - \nu_l - i\frac{c}{2} \right)
$$

$$
\times Z_{N-1}\left(\{\lambda_{\alpha \neq \beta}\}, \{\nu_{j \neq l}\} \right).
$$

$$(9.11)$$

It is sufficient to prove (9.11) for $\beta = l = 1$ since Z_N is symmetric in all the λ_α and in all the ν_j. Notice that the operator L_{11} is the furthest to the right in the product chain of L-operators in the representation (VI.6.5), and at $\lambda_1 = \nu_1 - ic/2$, it is equal to the permutation matrix:

$$
\mathsf{L}_{11}(\nu_1 - \lambda_1 = ic/2) = -ic\Pi_{11}. \tag{9.12}
$$

Hence we can calculate the following equality

$$
\mathsf{L}_{11} \left\{ \prod_{j=1}^{N} \uparrow_j \right\} \left\{ \prod_{\alpha=1}^{N} \downarrow_\alpha \right\}
$$

$$
= -ic \left(\downarrow_1 \left\{ \prod_{j=2}^{N} \uparrow_j \right\} \right) \left(\uparrow_1 \left\{ \prod_{\alpha=2}^{N} \downarrow_\alpha \right\} \right). \tag{9.13}
$$

The vector on the right hand side here is the eigenstate of all the $\mathsf{L}_{1j}(\lambda_1 - \nu_j)$, $(j = 2, \ldots, N)$ (9.5) and of all the $\mathsf{L}_{\alpha 1}(\lambda_\alpha - \nu_1)$ $(\alpha = 2, \ldots, N)$ (see (9.6)). After application of these L-operators to the vector (9.13), we have

$$Z_N \bigg|_{\lambda_1 = \nu_1 - ic/2} = -ic \prod_{k=2}^{N} \left(\lambda_1 - \nu_k - i\frac{c}{2} \right) \prod_{\alpha=2}^{N} \left(\lambda_\alpha - \nu_1 - i\frac{c}{2} \right)$$
$$\times Z_{N-1} \left(\{\lambda_{\alpha \neq 1}\}, \{\nu_{j \neq 1}\} \right). \tag{9.14}$$

The lemma is proved.

Lemma 4. The properties which uniquely determine the partition function are:

(a) $Z_1 = -ic$.

(b) Z_N is a symmetric function of $\{\lambda_\alpha\}$ and $\{\nu_j\}$ separately.

(c) Z_N is a polynomial of degree $(N-1)$ in each variable λ_α or ν_j when all others are assumed fixed.

(d) Z_N satisfies the recursion relation (9.11).

This statement is proved by induction. For

$$Z_1 = -ic \tag{9.15}$$

it is obvious. If the function Z_{N-1} is known, then the function Z_N is fixed uniquely. Z_N is an $(N-1)$th-degree polynomial in λ_N. The values of this polynomial at $\lambda_N = \nu_j$ $(j = 1, \ldots, N)$ are known, (9.11). Thus, it is fixed uniquely. The lemma is proved.

It will be suitable to use functions $G_N(\{\lambda_\alpha^B\}, \{\lambda_j^C\})$:

$$G_N \left(\{\lambda_\alpha^B\}, \{\lambda_j^C\} \right) \equiv Z_N \left(\{\lambda_\alpha^B\}, \{\lambda_j^C + i\frac{c}{2}\} \right). \tag{9.16}$$

These functions are easily calculated for small N starting from the recursive properties of Z_N:

$$G_1 = -ic;$$
$$G_2(\lambda_1^B, \lambda_2^B | \lambda_1^C, \lambda_2^C) = c^4 + ic^3(\lambda_1^B + \lambda_2^B - \lambda_1^C - \lambda_2^C)$$
$$- c^2(\lambda_1^B - \lambda_2^C)(\lambda_2^B - \lambda_1^C) - c^2(\lambda_1^B - \lambda_1^C)(\lambda_2^B - \lambda_2^C).$$
$$\tag{9.17}$$

VII.10 Z_N as a determinant

The determinant representation obtained in this section initiates the calculation of correlation functions which will be continued in Part III. It is of interest that the recursion relations of the previous section can be solved explicitly. The answer is given by the determinant of an $N \times N$

matrix:

$$Z_N \left(\{\lambda_\alpha\}, \{\nu_j\} \right) = (-1)^N \frac{\prod\limits_{j}^{N} \prod\limits_{\alpha}^{N} \left(\nu_j - \lambda_\alpha - i\frac{c}{2} \right) \left(\nu_j - \lambda_\alpha + i\frac{c}{2} \right)}{\prod\limits_{N \geq j > k \geq 1} (\nu_k - \nu_j) \prod\limits_{N \geq \alpha > \beta \geq 1} (\lambda_\alpha - \lambda_\beta)} \det \mathcal{M}$$

(10.1)

with

$$\mathcal{M}_{j\alpha} = \frac{ic}{\left(\nu_j - \lambda_\alpha + ic/2 \right) \left(\nu_j - \lambda_\alpha - ic/2 \right)}.$$

(10.2)

One can prove a similar formula for the six-vertex (ice) model [6] which provides a trigonometric version of formulae (10.1) and (10.2). We shall use this expression when constructing the main coefficient of the scalar product in Part III.

We will now show that this expression for Z_N satisfies the properties stated in Lemma 4. This will prove that (10.1) is a unique expression for the partition function.

(a) This result follows trivially from (10.1) by setting $N = 1$. The double product in the numerator contains only one term. $\mathcal{M}$ consists of only one element which cancels the double product leaving a factor of ic. A factor of -1 comes from $(-1)^N$. Hence $Z_1 = -ic$.

(b) Exchanging two λ's ($\lambda_\gamma \leftrightarrow \lambda_\delta$) in (10.2) is the same as interchanging two columns in $\mathcal{M}$ which gives rise to a factor of -1 in $\det \mathcal{M}$. This factor is compensated by a factor of -1 from the exchange of two λ's in

$$\prod_{N \geq \alpha > \beta \geq 1} (\lambda_\alpha - \lambda_\beta),$$

which leaves Z_N unchanged. A similar argument holds for the exchange of two ν's. Thus, Z_N is a symmetric function of $\{l_\alpha\}$ and $\{\nu_j\}$ separately.

(c) This item of Lemma 4 involves proving two statements. First we will show that Z_N is a polynomial rather than a rational function. Then, we will show that Z_N is a polynomial of degree $(N - 1)$.

To prove Z_N is a polynomial rather than a rational function, we need to demonstrate that Z_N has zero residue at its poles. An analysis of (10.1) shows that there are two sources of poles. The first source is in $\mathcal{M}$. The poles here are given by

$$\nu_k - \lambda_\gamma = \pm i\frac{c}{2}.$$

However, these poles are also zeros of the double product in the numerator of (10.1). Thus the residue of these poles is zero. The other source of poles is when two λ's coincide in the denominator. Again this pole has zero residue since $\mathcal{M}$ is degenerate when two λ's coincide (making the determinant of $\mathcal{M}$ zero). A similar argument holds for coinciding ν's. Thus Z_N is a polynomial.

To find the degree of Z_N we should use the symmetry proved in (b) which allows us to check the degree of λ_1 only. The double product in the numerator of (10.1)

$$\prod_j^N \prod_\alpha^N \left(\nu_j - \lambda_\alpha - i\frac{c}{2}\right)\left(\nu_j - \lambda_\alpha + i\frac{c}{2}\right) \tag{10.3}$$

is of order λ_1^{2N} while the product in the denominator of (10.1)

$$\prod_{N \geq \alpha > \beta \geq 1} (\lambda_\alpha - \lambda_\beta) \tag{10.4}$$

is of order λ_1^{N-1}. In $\mathcal{M}$, λ_1 only occurs in the first column which shows that $\det \mathcal{M}$ is of order λ_1^{-2}. Combining these results we see that Z_N is of degree $N-1$ in λ_1. The above can also be shown to be true for ν_1 since Z_N is also symmetric in $\{\nu_j\}$. Hence, Z_N is a polynomial of degree $N-1$ in each variable λ_α or ν_j when all others are assumed fixed.

(d) To show that Z_N satisfies the recursion relation (9.11), we shall use formula (10.1) and the symmetry property proved above. Thus it is sufficient to prove (9.11) for $\beta = l = 1$ which is given by

$$Z_N\Big|_{\lambda_1 = \nu_1 - ic/2} = -ic \prod_{k=2}^N \left(\lambda_1 - \nu_k - i\frac{c}{2}\right) \prod_{\alpha=2}^N \left(\lambda_\alpha - \nu_1 - i\frac{c}{2}\right)$$
$$\times Z_{N-1}\left(\{\lambda_{\alpha \neq 1}\}, \{\nu_{j \neq 1}\}\right). \tag{10.5}$$

Since we are interested in the parameters λ_1 and ν_1, we can isolate their contribution to Z_N as follows:

$$Z_N = (-1)^N \left[(\nu_1 - \lambda_1 - i\frac{c}{2})(\nu_1 - \lambda_1 + i\frac{c}{2})\right]$$

$$\times \frac{\prod_{j=2}^N (\nu_j - \lambda_1 - i\frac{c}{2})(\nu_j - \lambda_1 + i\frac{c}{2}) \prod_{\alpha=2}^N (\nu_1 - \lambda_a - i\frac{c}{2})(\nu_1 - \lambda_a + i\frac{c}{2})}{\prod_{j=2}^N (\nu_1 - \nu_j) \prod_{\alpha=2}^N (\lambda_\alpha - \lambda_1)}$$

$$\times \frac{\prod_{j=2}^N \prod_{\alpha=2}^N \left(\nu_j - \lambda_\alpha - i\frac{c}{2}\right)\left(\nu_j - \lambda_\alpha + i\frac{c}{2}\right)}{\prod_{N \geq j > k \geq 2} (\nu_k - \nu_j) \prod_{N \geq \alpha > \beta \geq 2} (\lambda_\alpha - \lambda_\beta)} \det \mathcal{M}. \tag{10.6}$$

To further isolate the contribution from the parameters λ_1 and ν_1, we need to understand how $\det \mathcal{M}$ behaves when $\lambda_1 \to \nu_1 - ic/2$. By examining (10.2), it is easily seen that $\mathcal{M}$ has a pole in this limit. Thus the determinant will be dominated by this contribution and we can write

$$
\det \mathcal{M} = \det \mathcal{M}_{N-1} \left(\frac{1}{\nu_1 - \lambda_1 - i\frac{c}{2}} \right) \Bigg|_{\lambda_1 = \nu_1 - ic/2}. \tag{10.7}
$$

where $\mathcal{M}_{N-1}$, the $(N-1) \times (N-1)$ minor of $\mathcal{M}$, is independent of λ_1 and ν_1. This pole is cancelled by a zero in (10.6). Using (10.1) for Z_{N-1}, it easily follows that (10.6) reduces to (10.5) which implies that Z_N satisfies the recursion relation (9.11).

We have shown that the determinant formula for Z_N satisfies the requirements of Lemma 4 from section 9 which uniquely specify the partition function. Hence (10.1) is the unique expression of the partition function Z_N.

Conclusion

In this chapter we have shown how to reproduce the results of the Bethe Ansatz starting from the quantum inverse scattering method. The algebraic Bethe Ansatz provides new opportunities for investigation of the various models. It allows for the classification of exactly solvable models and construction of the quantum determinant. The partition function with domain wall boundary conditions Z_N (evaluated in section 10) will be used later for the evaluation of correlation functions, norms of eigenfunctions and scalar products. The partition function for the six-vertex model with periodic boundary conditions was found in [15], [16], [17], [26].

The algebraic Bethe Ansatz was constructed for the first time in [2] and [27] where the Bethe equations and the spectrum of eigenvalues of the energy and momentum operators over the Fock vacuum were derived for the Heisenberg magnet, sine-Gordon model and Bose gas. The coincidence of the formulæ for the eigenfunctions of the Hamiltonian obtained within QISM with the coordinate wave functions of Part I may be proved by generalizing formula (5.9) for the case where the monodromy matrix consists of an arbitrary number of "commuting" factors (see [7] and Appendix 2). It is shown in a similar way that all observable values in the quantum sine-Gordon [22] and massive Thirring models are the same as in Part I. Let us mention that the normalized vacuum of the sine-Gordon model was constructed in [5].

The action of the operator $C(\mu)$ on the Bethe vector (2.1) was investigated in [9] and [10]. The Pauli principle for one-dimensional interacting bosons was formulated in [5].

The algebraic Bethe Ansatz allows us to go further in the investigation of integrable models. It is of particular interest in that it reveals the structure of the monodromy matrix to be of the general form connected with a given R-matrix. The theorem on the classification of monodromy matrices (section 6) in the XXX and XXZ cases was proved in [8]. The concept of the quantum determinant was introduced in [3], [8]. It plays an important role in QISM and was used by many authors. Formula (8.8) was used in [25] when deriving the quantum Gel'fand-Levitan equation. The concept of the quantum determinant was extended for matrices of greater dimensions in [13]. It was used for the generation of new R-matrices in [11] and for the construction of the Casimir operator in quadratic algebras in [23] and [24]. It is related to the antipode in the theory of quantum groups [1].

The recursive properties of the partition function Z_N were investigated in [9]. Its representation in the determinant form was given in [6].

We should also mention important issues (which are not in the book but related) such as the analytic Bethe Ansatz ([18]) and the Bethe Ansatz hierarchy ([12] and [27]). The ideas of the algebraic Bethe Ansatz were applied to the Hubbard model by S. Shastry in [19], [20] and [21].

Appendix VII.1: Matrix representation of quantum operators

Let us consider the rational coupling constant in the quantum sine-Gordon model

$$\frac{\gamma}{\pi} = \frac{Q}{P} \tag{A.1.1}$$

with coprime integers Q and P, $Q < P$.

The relations

$$\pi\chi = \chi\pi \exp(i\gamma) \tag{A.1.2}$$

$$\chi = \exp(i\beta u/2); \qquad \pi = \exp(i\beta p/4) \tag{A.1.3}$$

give

$$\chi^{2P} = \pi^{2P} = 1. \tag{A.1.4}$$

Thus χ and π may be represented as $2P \times 2P$ matrices

$$\chi_{ab} = \delta_{ab} \exp\{i\pi(a-1)/P\},$$
$$\pi_{ab} = \delta_{a+Q,b} \qquad (a, b = 1, \ldots, 2P; \ a + 2P \equiv a). \tag{A.1.5}$$

This representation may be used in the sine-Gordon model since χ and π enter the monodromy matrix via integer powers. For the pseudovacuum (3.24), we then have the Kronecker δ-symbol:

$$\delta_{a_{2n}+P, a_{2n-1}+Q}. \tag{A.1.6}$$

Appendix VII.2: Multisite model

Let us consider a multisite model. In other words, the monodromy matrix $\mathsf{T}(\lambda)$ in this case can be represented as a product of L factors:

$$\mathsf{T}(\lambda) = \mathsf{T}(\mathsf{L}|\lambda) \cdots \mathsf{T}(2|\lambda)\mathsf{T}(1|\lambda) = \begin{pmatrix} A(\lambda) & B(\lambda) \\ C(\lambda) & D(\lambda) \end{pmatrix}. \tag{A.2.1}$$

This is a generalization of section 5. Each factor has the same R-matrix and matrix elements of different factors commute as quantum operators. Each factor can be represented in the form (5.2)

$$\mathsf{T}(\mathsf{j}|\lambda) = \begin{pmatrix} A_j(\lambda) & B_j(\lambda) \\ C_j(\lambda) & D_j(\lambda) \end{pmatrix}. \tag{A.2.2}$$

and has its own pseudovacuum (5.3)

$$\begin{aligned} A_j(\lambda)|0\rangle_j &= a_j(\lambda)|0\rangle_j; \qquad D_j(\lambda)|0\rangle_j = d_j(\lambda)|0\rangle_j; \\ C_j(\lambda)|0\rangle_j &= 0. \end{aligned} \tag{A.2.3}$$

The multisite generalization of formula (5.9) is

$$\prod_{\alpha=1}^{N} B(\lambda_\alpha)|0\rangle = \sum_{\{\lambda\}=\bigcup_{j=1}^{L}\{\lambda_j\}} \prod_{j=1}^{L} B_j(\lambda_j)|0\rangle_j$$
$$\times \prod_{1 \leq j < k \leq L} a_k(\lambda_j)d_j(\lambda_k)f(\lambda_j, \lambda_k). \tag{A.2.4}$$

Here the summation is with respect to the partition of the set of all λ into L subsets; j and k label the subsets. On the right hand side of (A.2.4) we have a product with respect to numbers of the subsets. Each factor (corresponding to one subset) means the product with respect to all the λ's entering this subset. To illustrate this one should compare (A.2.4) for $L = 2$ with (5.9). These two formulæ coincide.

The development of this idea helped to identify the coordinate and algebraic Bethe Ansatzes for the Heisenberg model.

VIII

Lattice Integrable Models of Quantum Field Theory

Introduction

Lattice variants of integrable models, both classical and quantum, are formulated in the present chapter. The nonlinear Schrödinger (NS) equation and the sine-Gordon (SG) model are considered. QISM makes it possible to put continuous models of field theory on a lattice while preserving the property of integrability. In addition, the explicit form of the R-matrix is kept unchanged; this means that the structure of the action-angle variables for the classical models is unchanged. For quantum models, the analogue is the preservation of the S-matrix (see the end of section VII.7). The critical exponents, which characterize the power-law decay of correlators for large distances, are also preserved. For relativistic models of quantum field theory, lattice models may be used to rigorously solve the problem of ultraviolet divergences. The construction of local Hamiltonians for lattice models in quantum field theory is given much attention in this chapter. It is of interest to note that the L-operators of lattice models depend on some additional parameter Δ (which is absent in the R-matrix). This is the lattice spacing. However, the L-operator can be continued in Δ to the whole complex plane. Based on this fact, the most general L-operator may be constructed which is intertwined by a given R-matrix. This solves the problem of enumerating all the integrable models connected with a given R-matrix.

The preservation of the R-matrix when going to the lattice is the most important feature of the constructed lattice models. Let us also note that the lattice models constructed in this chapter are formulated in terms of initial Bose fields. This makes our approach different from other discretizations ([1], [8], [9] and [21]). Our approach is from the point of view of group theory. Going from the continuum to the lattice, we keep the group (R-matrix) but change the representation (L-operator).

172

The classical lattice models are constructed in sections 1 and 2. The explicit form of the L-operator on the lattice is obtained. The local Hamiltonians on the lattice are constructed by means of trace identities. The construction is based on the fact that the L-operator becomes a one-dimensional projector at some value of the spectral parameter.

The classical lattice NS equation is constructed in section 1, while the sine-Gordon equation is constructed in section 2.

The quantum discrete NS model is constructed in section 3. The local Hamiltonian on the lattice can be constructed using (as in the classical case) the degeneracy of the L-operator at values of the spectral parameter. The Hamiltonian is an elementary function of local Bose fields.

The L-operator of the quantum discrete NS model is considered in section 4. It depends on the continuous parameter Δ. At special values of this parameter, it is converted into the L-operator of the XXX Heisenberg model of arbitrary spin and is the most general form of the XXX R-matrix. This L-operator exhausts all possible L-operators related to the XXX R-matrix. The connection of the L-operator with representations of the $SU(2)$ algebra is also discussed. The discrete quantum sine-Gordon model is constructed in section 5.

VIII.1 Classical lattice nonlinear Schrödinger equation

Let us start with the NS model. The continuous variant of this model was discussed in section V.1. The lattice model describes the interaction of fields on a lattice Ψ_n, Ψ_n^* (n being the site number), with Poisson brackets $\{\Psi_n, \Psi_m^*\} = i\delta_{nm}/\Delta$, $\{\Psi_n, \Psi_m\} = \{\Psi_n^*, \Psi_m^*\} = 0$. This model is defined by the L-operator

$$
\mathsf{L}(n|\lambda) = \begin{pmatrix} 1 - \frac{i\lambda\Delta}{2} + \frac{c\Delta^2}{2}\Psi_n^*\Psi_n & -i\sqrt{c}\Delta\Psi_n^*\rho_n^+ \\ i\sqrt{c}\Delta\rho_n^-\Psi_n & 1 + \frac{i\lambda\Delta}{2} + \frac{c\Delta^2}{2}\Psi_n^*\Psi_n \end{pmatrix}.
\tag{1.1}
$$

The quantities $\rho^\pm$ are functions of the products of fields Ψ_n^*, Ψ_n only and

$$
\rho_n^+\rho_n^- = 1 + \frac{c\Delta^2}{4}\Psi_n^*\Psi_n.
\tag{1.2}
$$

The Poisson brackets of this L-operator are given by the classical r-matrix (V.2.11), which coincides with that for the continuous case (V.2.16), and the bilinear relation (V.2.11) is now exact. This can be checked by simple calculations.

Though $\rho^\pm$ are not uniquely defined by (1.2), it will prove convenient to use the following form:

$$
\rho_n^+ = \rho_n^- = \rho_n = \sqrt{1 + \frac{c\Delta^2}{4}\Psi_n^*\Psi_n}.
\tag{1.3}
$$

The L-operator (1.1) differs from the continuous one (V.1.19) only at second order in Δ as $\Delta \to 0$. However, the symmetry properties are the same:

$$\sigma_x \mathsf{L}^*(\lambda^*)\sigma_x = \mathsf{L}(\lambda); \qquad \mathsf{L}^T(-\lambda) = \sigma_x \mathsf{L}(\lambda)\sigma_x; \tag{1.4}$$

$$\det \mathsf{L}(n|\lambda) = \frac{\Delta^2}{4}(\lambda - \nu)(\lambda - \nu^*), \quad \nu = -\frac{2i}{\Delta}. \tag{1.5}$$

The monodromy matrix is constructed as usual (V.1.8) and possesses the same symmetry properties: $\sigma_x \mathsf{T}^*(\lambda^*)\sigma_x = \mathsf{T}(\lambda)$, $\det \mathsf{T}(\lambda) = (\det \mathsf{L})^M$ (V.1.18). In section 4 we shall use another representation of $\rho^{\pm}$:

$$\rho_n^+ = 1; \qquad \rho_n^- = 1 + \frac{c\Delta^2}{4}\Psi_n^*\Psi_n. \tag{1.6}$$

Let us now construct the local Hamiltonian using the trace identities. We shall follow the papers [7] and [15]–[17]. Consider the Hamiltonian of the lattice model

$$H = \sum_{n=1}^{M} H_n \tag{1.7}$$

with the Hamiltonian density H_n depending only on the dynamical variables Ψ_j, Ψ_j^* in a certain neighborhood of the n-th site ($n-m \le j \le n+l$). If, in the limit $M \to \infty$ (interesting for physical applications), $m + l$ remains finite, $m + l \le i$, the Hamiltonian is local and describes the interaction of $i + 1$ neighbors (for i the minimal possible value should be taken).

The Hamiltonian can be obtained as a linear combination of logarithmic derivatives of the transfer matrix $\tau(\lambda) = \mathrm{tr}\, \mathsf{T}(\lambda) = A(\lambda) + D(\lambda)$. Locality is achieved by the following method. At points $\lambda = \nu, \nu^*$, $\det \mathsf{L}(n|\lambda) = \det \mathsf{T}(\lambda) = 0$, i.e., the matrices $\mathsf{L}(n|\lambda)$ and $\mathsf{T}(\lambda)$ are proportional to one-dimensional projectors. Namely,

$$\mathsf{L}_{ik}(n|\nu) = \alpha_i(n)\alpha_k^*(n) \tag{1.8}$$

where α is a two-component vector, with components

$$\alpha_1(n) = -i\Delta\sqrt{\frac{c}{2}}\Psi_n^*,$$

$$\alpha_2(n) = \sqrt{2\left(1 + \frac{c\Delta^2}{4}\Psi_n^*\Psi_n\right)} = \sqrt{2}\rho_n. \tag{1.9}$$

At $\lambda = \nu^*$, the L-operator is also a one-dimensional projector:

$$\mathsf{L}_{ik}(n|\nu^*) = (\sigma_x \alpha^*(n))_i (\sigma_x \alpha(n))_k. \tag{1.10}$$

Using this representation, one can easily put the transfer matrix (V.1.9) into factorized form at $\lambda = \nu$, ν^*:

$$\tau(\lambda) = \prod_{n=1}^{M} \tau_n,$$

$$\tau_n = (\alpha^*(n+1)\,\alpha(n)), \qquad \alpha(M+1) = \alpha(1),$$

(1.11)

where the parentheses denote the scalar product of two one-dimensional vectors:

$$(\alpha^* \, \alpha) = \alpha_1^* \alpha_1 + \alpha_2^* \alpha_2.$$

(1.12)

Thus, the conservation law is local,

$$\ln \tau(\lambda) = \sum_{n=1}^{M} \ln \tau_n,$$

(1.13)

and describes the interactions of nearest neighbors. The logarithmic derivatives $d^m \ln \tau(\lambda)/d\lambda^m$ at $\lambda = \nu$, ν^* are also local [15]. The first derivative is

$$\frac{\partial}{\partial \lambda} \ln \tau(\lambda) \bigg|_{\lambda=\nu} = -i \frac{\Delta}{2} \sum_{n=1}^{M} \frac{(\alpha^*(n+1)\sigma_z \alpha(n-1))}{(\alpha^*(n+1)\alpha(n))(\alpha^*(n)\alpha(n-1))}.$$

(1.14)

We define the Hamiltonian of the model as follows:

$$H = -\frac{8i}{3c\Delta^4} \frac{\partial}{\partial \lambda} \ln \left[\tau(\lambda)\tau_0^{-1}(\lambda) \right] \bigg|_{\lambda=\nu}$$

$$+ \frac{8i}{3c\Delta^4} \frac{\partial}{\partial \lambda} \ln \left[\tau(\lambda)\tau_0^{-1}(\lambda) \right] \bigg|_{\lambda=\nu^*} + \frac{4}{3\Delta^2} Q. \quad (1.15)$$

Here Q is the operator of the number of particles

$$Q = \Delta \sum_{n=1}^{M} \Psi_n^* \Psi_n$$

(1.16)

which commutes with $\tau(\lambda)$. The value of $\tau(\lambda)$ at zero field ($\Psi = 0$) is $\tau_0(\lambda)$:

$$\tau_0(\lambda) = \left(1 - i\frac{\lambda\Delta}{2} \right)^M + \left(1 + i\frac{\lambda\Delta}{2} \right)^M.$$

(1.17)

However, one can take as the Hamiltonian $\ln \tau(\lambda)$ itself, or some other combination of logarithmic derivatives at ν, ν^*. The demand for the proper continuous limit is a rather weak restriction.

Direct calculation gives the following explicit expression for the Hamiltonian (1.15):

$$H = -\frac{4}{3c\Delta^3} \sum_{n=1}^{M} \left[\frac{(\alpha^*(n+1)\sigma_z\alpha(n-1))}{(\alpha^*(n+1)\alpha(n))(\alpha^*(n)\alpha(n-1))} \right.$$
$$\left. + \frac{(\alpha(n+1)\sigma_z\alpha^*(n-1))}{(\alpha(n+1)\alpha^*(n))(\alpha(n)\alpha^*(n-1))} + 1 - c\Delta^2 \Psi_n^* \Psi_n \right]. \quad (1.18)$$

This Hamiltonian is real and invariant under spatial reflection. It has the same structure of action-angle variables as the corresponding continuous model considered in section V.1. One can easily prove by direct calculation that, as $\Delta \to 0$, (1.18) turns into the continuous Hamiltonian (V.1.11), while the charge (1.16) becomes (V.1.13). This transition is smooth. It is also remarkable that the Hamiltonian is an elementary function of local Bose fields.

Having the Hamiltonian and Poisson brackets, we can easily obtain the nonlinear completely integrable equations of motion (the nonlinearity is of rational nature). The r-matrix and the trace identities are obtained above, the time evolution operator $U(n|\lambda)$ is constructed by the generating functional (V.2.20) and is local. Thus the classical equations of motion can be represented in the Lax form.

VIII.2 Classical lattice sine-Gordon model

The lattice sine-Gordon model is defined by local lattice fields p_n and u_n with Poisson brackets $\{p_n, u_n\} = \delta_{nm}$ and by the L-operator

$$\mathsf{L}(n|\lambda) = \begin{pmatrix} e^{-i\beta p_n/8}\rho_n e^{-i\beta p_n/8} & \frac{m\Delta}{2}\sinh(\lambda - i\beta u_n/2) \\ \frac{-m\Delta}{2}\sinh(\lambda + i\beta u_n/2) & e^{i\beta p_n/8}\rho_n e^{i\beta p_n/8} \end{pmatrix} \quad (2.1)$$

where $\rho_n = \sqrt{1 + 2s\cos\beta u_n}$, $s = (m\Delta/4)^2$, with $0 < s < 1/2$. In the continuum limit ($p_n \to \Delta\pi(x_n)$, $u_n \to u(x_n)$), the L-operator turns into the infinitesimal L-operator (V.3.5). The Poisson brackets of matrix elements of the L-operator (2.1) are given by exactly the same r-matrix (V.3.7) as in the continuous case.

The properties of the monodromy matrix $\mathsf{T}(\lambda)$ are the same as in the continuous case (V.3.9). Considering an even number of sites, we have:

$$\sigma_y \mathsf{T}^*(\lambda^*)\sigma_y = \mathsf{T}(\lambda), \quad (2.2)$$

$$\det\mathsf{T}(\lambda) = \det{}^M\mathsf{L}(\lambda); \quad \det\mathsf{L}(\lambda) = 1 + 2s\cosh 2\lambda. \quad (2.3)$$

The Hamiltonian is defined by the trace identities

$$
H - P = \frac{m^2\Delta}{4\gamma} \frac{\partial}{\partial e^{2\lambda}} \ln\left[\tau(\lambda)\tau_0^{-1}(\lambda)\right]\Big|_{e^{2\lambda}=-b};
$$

$$
H + P = \frac{m^2\Delta}{4\gamma} \frac{\partial}{\partial e^{-2\lambda}} \ln\left[\tau(\lambda)\tau_0^{-1}(\lambda)\right]\Big|_{e^{2\lambda}=-b^{-1}};
$$

(2.4)

Here $\tau_0(\lambda)$ is the trace of $\mathsf{T}(\lambda)$ at $p_n = u_n = 0$. The reality of H and P is due to the involution (2.2); b is given by

$$
b = 2s\left(1 + \sqrt{1 - 4s^2}\right)^{-1};
$$

(2.5)

$\tau(\lambda)$ is a meromorphic function of $\exp(2\lambda)$. At $\exp(2\lambda) = -b^{\pm 1}$, the determinant of $\mathsf{T}(\lambda)$ (2.3) is equal to zero; hence H and P are local. The trace identities (2.4) have the correct continuous limit ($\Delta \to 0$, $s \to 0$, $b \to 0$) (see, for example, reference [2] from the Introduction to Part II). The Hamiltonian (2.4) describes the interaction of three nearest neighbors. It is convenient to write out the explicit form of H and P using the variables

$$
\chi_n^{\pm} = \sqrt{\frac{s}{b}}\left(\frac{e^{\pm i\beta u_n/2} + be^{\mp i\beta u_n/2}}{\rho_n}\right)e^{-i\beta p_n/4},
$$

$$
|\chi_n^{\pm}|^2 = 1.
$$

(2.6)

Then,

$$
H \mp P = \frac{m^2\Delta}{4\gamma b(1 - b^2)} \sum_{k=1}^{M}\left[\frac{(\chi_{k\mp1}^+ - b\chi_{k\pm1}^-)(\chi_k^- - b\chi_k^+)}{(1 + \chi_{k\pm1}^-\chi_k^+)(1 + \chi_{k\mp1}^+\chi_k^-)} - \frac{(1-b)^2}{4}\right].
$$

(2.7)

The phase space of the constructed model is the direct product of toruses. The Hamiltonian and momentum are periodic functions of p_n and u_n with periods $2\pi/\beta$ and $4\pi/\beta$, respectively. Let us define the charge Q by

$$
Q = \frac{\beta}{\pi} \sum_{k=1}^{M/2}(u_{2k} - u_{2k-1}).
$$

(2.8)

The Poisson brackets of Q and $\mathsf{T}(\lambda)$ are easily calculated:

$$
\{Q, \mathsf{T}(\lambda)\} = i\frac{\gamma}{\pi}[\sigma_z, \mathsf{T}(\lambda)].
$$

(2.9)

Hence, Q is an integral of motion: $\{Q, \tau(\lambda)\} = \{Q, H\} = 0$. The continuum limit of the L-operator (2.1) and the Hamiltonian (2.7) of the lattice sine-Gordon model is correct. When $\Delta \to 0$, we obtain (V.3.3). The lattice and continuous models have the same structure of action-angle variables.

VIII.3 Quantum lattice nonlinear Schrödinger equation

The quantum NS model on a lattice is defined by the same L-operator as the classical NS model on a lattice (1.1). However, Ψ_n and $\Psi_n^\dagger$ are now quantum operators with commutation relations $[\Psi_n, \Psi_m^\dagger] = \delta_{nm}/\Delta$. The commutation relations of the lattice L-operator of the NS model (1.1) are given by the same matrix

$$R(\lambda, \mu) = \Pi - \frac{ic}{\lambda - \mu} I$$

as in the continuous case. Thus, the relation

$$R(\lambda, \mu) \left(L(\lambda) \bigotimes L(\mu) \right) = \left(L(\mu) \bigotimes L(\lambda) \right) R(\lambda, \mu)$$

remains exactly true on the lattice.

Using the arbitrariness in $\rho_n^\pm$ (see (1.2)), we shall fix $\rho_n = \rho_n^+ = \rho_n^-$ as in (1.3). The quantum L-operator possesses symmetry properties which are similar to the classical ones:

$$\sigma_x L^*(\lambda^*)\sigma_x = L(\lambda). \tag{3.1}$$

The asterisk here denotes Hermitian conjugation of the matrix elements, without matrix transposition, and complex conjugation of c-numbers. The quantum determinant is

$$\det{}_q L(\lambda) = \frac{\Delta^2}{4} \left(\lambda - \nu - i\frac{c}{2} \right) \left(\lambda + \nu + i\frac{c}{2} \right),$$

$$\nu = -\frac{2i}{\Delta}. \tag{3.2}$$

The operator of the number of particles is given by

$$Q = \sum_{n=1}^{M} \Delta\, \Psi_n^\dagger \Psi_n, \tag{3.3}$$

and is the same as for the classical model.

Now one has to define the Hamiltonian by means of trace identities. To do this, let us modify the L-operator (1.1) by making it different at even and odd sites of the lattice:

$$L(n|\lambda) = \begin{pmatrix} 1 + (-1)^n \left(\frac{c\Delta}{4} \right) - \frac{i\lambda\Delta}{2} \\ + \frac{c\Delta^2}{2} \Psi_n^\dagger \Psi_n & -i\sqrt{c}\,\Delta\Psi_n^\dagger\rho_n \\ \\ i\sqrt{c}\,\Delta\rho_n\Psi_n & 1 + (-1)^n \left(\frac{c\Delta}{4} \right) + \frac{i\lambda\Delta}{2} \\ & + \frac{c\Delta^2}{2} \Psi_n^\dagger \Psi_n \end{pmatrix},$$

$$\tag{3.4}$$

$$\rho_n = \sqrt{1 + (-1)^n \frac{c\Delta}{4} + \frac{c\Delta^2}{4} \Psi_n^\dagger \Psi_n}. \tag{3.5}$$

This L-operator is intertwined by the XXX R-matrix (VI.3.17). The transfer matrix constructed from these L-operators has the correct continuous limit. By multiplying the L-operators from two adjacent sites, we see that when $\Delta \to 0$, the resulting L-operator is a product of two of the usual infinitesimal L-operators (VI.3.14) up to terms of order Δ^2. The following properties of $L(n|\lambda)$ can be easily established:

$$\sigma_x L^*(n|\lambda^*)\sigma_x = L(n|\lambda), \tag{3.6}$$

$$L^T(n| -\lambda) = \sigma_x L(n|\lambda)\sigma_x. \tag{3.7}$$

We can invert the quantum L-operator using the quantum determinant (see (VII.8.8)):

$$L(n|\lambda)\sigma_y L^T(n|\lambda + ic)\sigma_y = \frac{\Delta^2}{4}(\lambda - \nu_1^{(n)})(\lambda - \nu_2^{(n)}), \tag{3.8}$$

$$\nu_1^{(n)} = -\frac{2i}{\Delta}\left[1 + (-1)^n \frac{c\Delta}{4}\right]; \quad \nu_2^{(n)} = -ic + \frac{2i}{\Delta}\left[1 + (-1)^n \frac{c\Delta}{4}\right] \tag{3.9}$$

The L-operator can be represented (as in the classical case) as a projector (1.8) at the points $\nu_{1,2}^{(n)}$. However, the components of α in this case are noncommuting quantum operators; hence, we must distinguish the "direct" projector

$$L_{ik} = \alpha_i \beta_k \tag{3.10}$$

from the "inverse" one

$$L_{ik} = \delta_k \gamma_i. \tag{3.11}$$

To be more precise, let us introduce the two-component vector α (its components being quantum operators). In the quantum case, α looks like the classical version (1.9), but now a dependence on the parity of the site operator appears. At even sites ($n = 0 \pmod 2$)

$$\alpha_1(n) = -i\sqrt{\frac{c}{2}}\Delta\Psi_n^\dagger,$$

$$\alpha_2(n) = \sqrt{2\left(1 + \frac{c\Delta^2}{4}\Psi_n^\dagger \Psi_n\right)}, \tag{3.12}$$

while at odd sites ($n = 1 \pmod 2$)

$$\alpha_1(n) = -i\sqrt{\frac{c}{2}}\Delta\Psi_n^\dagger,$$

$$\alpha_2(n) = \sqrt{2\left(1 - \frac{c\Delta}{4} + \frac{c\Delta^2}{4}\Psi_n^\dagger\Psi_n\right)}.$$

(3.13)

Taking λ at the point where the quantum determinant (3.8) vanishes,

$$\nu = -\frac{2i}{\Delta} + \frac{ic}{2},$$

(3.14)

we see that at odd sites, the L-operator is the "direct" projector

$$\mathsf{L}_{ik}(n|\nu) = \alpha_i(n)\alpha_k^\dagger(n) \qquad (n \text{ odd}),$$

(3.15)

while at even sites, it is the "inverse" one:

$$\mathsf{L}_{ik}(n|\nu) = \alpha_k^\dagger(n)\alpha_i(n) \qquad (n \text{ even}).$$

(3.16)

Taking the conjugated zero of the quantum determinant, $\lambda = \nu^*$, where

$$\nu^* = -\nu = \frac{2i}{\Delta} - \frac{ic}{2},$$

(3.17)

and using (3.6), we have the L-operator as the "inverse" projector at odd sites:

$$\mathsf{L}_{ik}(n|\nu^*) = (\sigma_x\alpha(n))_k(\sigma_x\alpha^\dagger(n))_i \qquad (n \text{ odd}),$$

(3.18)

and as the "direct" projector at even sites:

$$\mathsf{L}_{ik}(n|\nu^*) = (\sigma_x\alpha^\dagger(n))_i(\sigma_x\alpha(n))_k \qquad (n \text{ even}).$$

(3.19)

We shall consider a lattice with an even number of sites, M. The operator Q (3.3) commutes with the transfer matrix:

$$[\tau(\lambda),\, Q] = 0.$$

(3.20)

The first logarithmic derivative at $\lambda = \nu$ is local. Using the representation

$$\tau(\nu) = (\alpha^\dagger(M)\,\alpha(M-1))$$
$$\times \left\{\prod_{n=1}^{M/2-1}(\alpha^\dagger(2n)\,\alpha(2n-1))(\alpha^\dagger(2n+1)\,\alpha(2n))\right\}(\alpha^\dagger(1)\,\alpha(M)).$$

(3.21)

Here the product is ordered from left to right. We can calculate the explicit form of the first logarithmic derivative:

$$\frac{\partial}{\partial\lambda}\ln\tau(\lambda)\Big|_{\lambda=\nu} = -i\frac{\Delta}{2}\sum_{n=1}^{M}t_n. \tag{3.22}$$

The local density, t_n, depends on the parity of the site; for odd sites

$$t_n = (\alpha^\dagger(n+2)\,\alpha(n+1))^{-1}$$
$$\times \left\{(\alpha^\dagger(n)\,\alpha(n-1))^{-1}(\alpha^\dagger(n+1)\,\alpha(n))^{-1}(\alpha^\dagger(n+1)\sigma_z\alpha(n-1))\right\}$$
$$\times (\alpha^\dagger(n+2)\,\alpha(n+1)), \tag{3.23}$$

while for even sites

$$t_n = (\alpha^\dagger(n-1)\,\alpha(n-2))^{-1}$$
$$\times \left\{(\alpha^\dagger(n+1)\,\alpha(n))^{-1}(\alpha^\dagger(n)\,\alpha(n-1))^{-1}(\alpha^\dagger(n+1)\sigma_z\alpha(n-1))\right\}$$
$$\times (\alpha^\dagger(n-1)\,\alpha(n-2)). \tag{3.24}$$

(We note that the quantum t_n differ from the classical ones (1.14) only by a similarity transformation.) Similar expressions for $\partial\ln\tau(\lambda)/\partial\lambda$ are valid for $\lambda=\nu^*$. It can be proved that

$$\left[\frac{\partial\ln\tau(\lambda)}{\partial\lambda}\Big|_{\lambda=\nu}\right]^\dagger = \frac{\partial\ln\tau(\lambda)}{\partial\lambda}\Big|_{\lambda=\nu^*}, \tag{3.25}$$

$$\tilde{P}\frac{\partial\ln\tau(\lambda)}{\partial\lambda}\Big|_{\lambda=\nu}\tilde{P} = -\frac{\partial\ln\tau(\lambda)}{\partial\lambda}\Big|_{\lambda=\nu^*}. \tag{3.26}$$

(Here $\tilde{P}$ is the operator of spatial reflection.) The Hamiltonian of the lattice NS model is defined similarly to the classical one:

$$H = -\frac{8i}{3c\Delta^4}\frac{\partial}{\partial\lambda}\ln\left[\tau(\lambda)\tau_0^{-1}(\lambda)\right]\Big|_{\lambda=\nu} + \frac{8i}{3c\Delta^4}\frac{\partial}{\partial\lambda}\ln\left[\tau(\lambda)\tau_0^{-1}(\lambda)\right]\Big|_{\lambda=\nu^*}$$
$$+ \left(\frac{4}{3\Delta^2(1-\frac{\Delta^2c^2}{16})}\right)Q. \tag{3.27}$$

Here Q is the operator of the number of particles (3.3), and $\tau_0(\lambda)$ is the transfer matrix at zero field:

$$\tau_0(\lambda) = \left(1+\frac{c\Delta}{4}-\frac{i\lambda\Delta}{2}\right)^{M/2}\left(1-\frac{c\Delta}{4}-\frac{i\lambda\Delta}{2}\right)^{M/2}$$
$$+ \left(1+\frac{c\Delta}{4}+\frac{i\lambda\Delta}{2}\right)^{M/2}\left(1-\frac{c\Delta}{4}+\frac{i\lambda\Delta}{2}\right)^{M/2}. \tag{3.28}$$

It is interesting to mention that in the continuous limit, the contribution of $c\Delta/4$ can be dropped.

We have thus constructed the local Hamiltonian in terms of initial Bose fields:

$$H = -\frac{4}{3c\Delta^3} \sum_{n=1}^{M} \left(t_n + t_n^\dagger + \frac{8 - c\Delta}{8 - 2c\Delta} \right)$$
$$+ \left(\frac{4}{3\Delta^2(1 - \frac{\Delta^2 c^2}{16})} \right) \sum_{n=1}^{M} \Delta \Psi_n^\dagger \Psi_n, \quad (3.29)$$

where t_n is given by (3.23), (3.24). The Hamiltonian is Hermitian and possesses the proper continuous limit.

It is clear that the corresponding model is integrable. The Fock vacuum is the pseudovacuum $|0\rangle$ ($\Psi_n|0\rangle = 0$):

$$A(\lambda)|0\rangle = a(\lambda)|0\rangle, \quad D(\lambda)|0\rangle = d(\lambda)|0\rangle,$$
$$C(\lambda)|0\rangle = 0; \quad (3.30)$$
$$a(\lambda) = \left(1 + \frac{c\Delta}{4} - i\frac{\lambda\Delta}{2} \right)^{M/2} \left(1 - \frac{c\Delta}{4} - i\frac{\lambda\Delta}{2} \right)^{M/2};$$
$$d(\lambda) = \left(1 + \frac{c\Delta}{4} + i\frac{\lambda\Delta}{2} \right)^{M/2} \left(1 - \frac{c\Delta}{4} + i\frac{\lambda\Delta}{2} \right)^{M/2}. \quad (3.31)$$

The Bethe equations are

$$\left[\frac{\left(1 + \frac{c\Delta}{4} - i\frac{\lambda_j\Delta}{2} \right) \left(1 - \frac{c\Delta}{4} - i\frac{\lambda_j\Delta}{2} \right)}{\left(1 + \frac{c\Delta}{4} + i\frac{\lambda_j\Delta}{2} \right) \left(1 - \frac{c\Delta}{4} + i\frac{\lambda_j\Delta}{2} \right)} \right]^{M/2}$$
$$= \prod_{\substack{k \neq j \\ k=1}}^{N} \left(\frac{\lambda_j - \lambda_k - ic}{\lambda_j - \lambda_k + ic} \right). \quad (3.32)$$

The eigenvalues of the Hamiltonian can be interpreted in terms of particles:

$$E_N = \sum_{k=1}^{N} \varepsilon_0(\lambda_k),$$
$$\varepsilon_0(\lambda) = \left(\frac{4}{3\Delta^2(1 - \frac{\Delta^2 c^2}{16})} \right) - \frac{4}{3\Delta^2} \left[h(\lambda) + \frac{\lambda^2\Delta^2}{h(\lambda)} \right]^{-1}, \quad (3.33)$$
$$h(\lambda) = 1 - \frac{\Delta^2}{16} \left(4\lambda^2 + c^2 \right).$$

The eigenvalue of momentum is

$$
p_0(\lambda) = \frac{i}{2\Delta} \ln \left[\frac{\left(1 + \frac{c\Delta}{4} - i\frac{\lambda\Delta}{2}\right)\left(1 - \frac{c\Delta}{4} - i\frac{\lambda\Delta}{2}\right)}{\left(1 + \frac{c\Delta}{4} + i\frac{\lambda\Delta}{2}\right)\left(1 - \frac{c\Delta}{4} + i\frac{\lambda\Delta}{2}\right)} \right]. \tag{3.34}
$$

Thus, we have constructed the quantum lattice NS model with a local Hamiltonian. Its distinctive feature is that it is formulated in terms of initial local bose fields. We see that the main idea behind the construction of local quantum Hamiltonians is the same as in the classical case (see section 1).

It must be mentioned that with the help of trace identities (generalizing those for the fundamental spin models), local Hamiltonians of different types may be constructed. This was done for the lattice NS model and for the sine-Gordon model in [27].

VIII.4 Classification of quantum L-operators

It appears that the most general L-operator for the XXX R-matrix can be obtained from the L-operator of the lattice NS model. Let us fix $\rho_n^{\pm}$ in (1.1) as follows: $\rho_n^+ = 1$, $\rho_n^- = 1 + (c\Delta^2/4)\Psi_n^\dagger\Psi_n$ (see (1.6)). This L-operator depends on an arbitrary complex variable Δ. It is easy to introduce three more complex variables. Shifting the spectral parameter in (1.1), we shall pass to the operator $\mathsf{L}(\lambda - \nu)$ (VI.2.5). Multiplying the L-operator by an arbitrary complex number and also multiplying on the left by the matrix $\exp\{\varepsilon\sigma_z\}$ (VII.2.14), (VII.2.15) (ε is an arbitrary parameter), we obtain the following L-operator:

$$
\mathsf{L}(n|\lambda) = \begin{pmatrix} a_n^{(1)}(\lambda + ic\Psi_n^\dagger\Psi_n) + a_n^{(0)} & \Psi_n^\dagger \\ c\rho_n\Psi_n & d_n^{(1)}(\lambda - ic\Psi_n^\dagger\Psi_n) + d_n^{(0)} \end{pmatrix} \tag{4.1}
$$

where

$$
\rho_n = i\left[a_n^{(1)}d_n^{(0)} - a_n^{(0)}d_n^{(1)}\right] + cd_n^{(1)}a_n^{(1)}\Psi_n^\dagger\Psi_n,
$$
$$
[\Psi_n, \Psi_m^\dagger] = \delta_{nm}, \quad [\Psi_n, \Psi_m] = [\Psi_n^\dagger, \Psi_m^\dagger] = 0.
$$

This L-operator possesses the vacuum ($\Psi_n|0\rangle = 0$) and is intertwined by the XXX R-matrix. The vacuum eigenvalues are

$$
a_n(\lambda) = a_n^{(1)}\lambda + a_n^{(0)},
$$
$$
d_n(\lambda) = d_n^{(1)}\lambda + d_n^{(0)}, \tag{4.2}
$$

where $a_n^{(0)}, d_n^{(0)}, a_n^{(1)}, d_n^{(1)}$ are four arbitrary complex parameters.

Theorem 1. *A monodromy matrix with arbitrary rational function* $a(\lambda)/d(\lambda)$ *is generated by the* L*-operator (4.1).*

Proof: Let us construct the monodromy matrix

$$\mathsf{T}(\mathsf{L}) = \mathsf{L}(M|\lambda)\cdots\mathsf{L}(1|\lambda). \tag{4.3}$$

The numbers a_n and d_n are in general different at each lattice site. Hence, $\mathsf{T}(\lambda)$ is a function of $4M$ arbitrary complex parameters. For $\mathsf{T}(\lambda)$ the ratio $a(\lambda)/d(\lambda)$ (VII.1.9) is equal to

$$\frac{a(\lambda)}{d(\lambda)} = \prod_{n=1}^{M} \frac{a_n^{(1)}\lambda + a_n^{(0)}}{d_n^{(1)}\lambda + d_n^{(0)}}. \tag{4.4}$$

The right hand side can be made equal to any arbitrary rational function since M is arbitrary. The theorem is proved.

It should be noted that monodromy matrices are parametrized by the function $a(\lambda)/d(\lambda)$ (see section VII.6). Using the L-operator (4.1), the monodromy matrix $\mathsf{T}(\lambda)$ with arbitrary analytic function $a(\lambda)/d(\lambda)$ may be constructed. Doing this, one must approximate the analytic functions by rational ones. A function with cuts may also be obtained; this can be dealt with by approximating the cuts by convergent sequences of zeros and poles.

The L-operator of the lattice NS model is the most general one and contains the L-operator of the XXX chain (VI.5.11) as a special case (as well as its generalization to higher spins [20]). This is quite natural since the classical continuous Heisenberg magnet and the NS equation are gauge-invariant (see reference [2] in the Introduction to Part II). To illustrate this, let us use the L-operator (1.1), with

$$\rho_n^+ = \rho_n^- = \sqrt{1 + \frac{c\Delta^2}{4}\Psi_n^\dagger\Psi_n}. \tag{4.5}$$

Or, after a simple transformation:

$$\mathsf{L}(n|\lambda) = -\frac{2}{\Delta}\sigma_z\mathsf{L}(n|\lambda) = i\lambda + c\sum_{l=x,y,z}\sigma_l t_l^{(n)}. \tag{4.6}$$

This L-operator is intertwined by the same XXX R-matrix (since the XXX R-matrix commutes with the matrix $\sigma_z \otimes \sigma_z$). We shall call this L-operator the L-operator of the generalized XXX model. In (4.6) σ_l are the Pauli matrices, and the t_l are quantum operators:

$$t_1^{(n)} = \frac{i}{\sqrt{c}}\left(\Psi_n^\dagger \rho_n + \rho_n \Psi_n\right),$$

$$t_2^{(n)} = \frac{1}{\sqrt{c}}\left(\rho_n \Psi_n - \Psi_n^\dagger \rho_n\right), \tag{4.7}$$

$$t_3^{(n)} = -\frac{2}{c}\left(\frac{1}{\Delta} + \frac{\Delta c}{2}\Psi_n^\dagger \Psi_n\right).$$

These operators generate a representation of the $SU(2)$ algebra:

$$[t_i, t_k] = i\epsilon_{ikl}t_l \tag{4.8}$$

with spin $t^2 = s(s+1)$, $s = -2/c\Delta$. This is the well known Holstein-Primakoff representation [11]. In general, this representation is infinite-dimensional, but for special values of Δ it is finite-dimensional. Namely, s integer or half-integer corresponds to XXX models; higher spins were introduced in [20]. For $s = 1/2$, the operator (4.6) becomes the usual XXX L-operator (VI.5.11).

VIII.5 Quantum lattice sine-Gordon model

The quantum sine-Gordon (SG) model on a lattice allows us to solve the problem of ultraviolet divergences in a rigorous way. The L-operator is given by (2.1):

$$\mathsf{L}(n|\lambda) = \begin{pmatrix} e^{-i\beta p_n/8}\rho_n e^{-i\beta p_n/8} & \frac{m\Delta}{2}\sinh(\lambda - i\beta u_n/2) \\ \frac{-m\Delta}{2}\sinh(\lambda + i\beta u_n/2) & e^{i\beta p_n/8}\rho_n e^{i\beta p_n/8} \end{pmatrix} \tag{5.1}$$

$$\rho_n = \sqrt{1 + 2s\cos\beta u_n}, \quad \beta = \sqrt{8\gamma}, \quad s = \left(\frac{m\Delta}{4}\right)^2,$$

with quantum operators u_n and p_n: $[u_n, p_m] = i\delta_{nm}$. This L-operator is intertwined by the XXZ R-matrix exactly; its symmetry properties are analogous to the classical ones (2.2):

$$\sigma_y \mathsf{L}^*(\lambda^*)\sigma_y = \mathsf{L}(\lambda). \tag{5.2}$$

The asterisk here means Hermitian conjugation of the quantum operators. The quantum determinant is

$$\det{}_q\mathsf{L}(\lambda) = 1 + 2s\cosh 2\lambda, \tag{5.3}$$

and is equal to the classical determinant (2.3) (this should be compared with the quantum continuous case (VII.8.15)). The continuous limit of the L-operator is proper (VI.4.2). The construction of different Hamiltonians for the lattice SG model ([2], [3], [5], [6], [17] and [27]) with a proper continuous ($\Delta \to 0$) quasiclassical limit is based on this L-operator.

The model considered above is solved by means of the algebraic Bethe Ansatz. To construct the pseudovacuum, let us use the two-site monodromy matrix $L_2(n|\lambda) = L(2n|\lambda)L(2n - 1|\lambda)$. The pseudovacuum is constructed in a similar manner to the continuous case (VII.3.24):

$$|0\rangle_n = \left\{1 - 2s\cos\left[\frac{\beta}{2}(u_{2n} + u_{2n-1})\right]\right\}^{-1/2}$$

$$\times \delta\left(u_{2n} - u_{2n-1} - \frac{\beta}{4} + \frac{2\pi}{\beta}\right) \quad (5.4)$$

(in addition to the pseudovacuum $|0\rangle_n$, the operator L_2 also possesses another pseudovacuum). For rational γ/π, the vacuum may be normalized with quantum operators represented as finite matrices (see Appendix 1 in the previous chapter). The vacuum eigenvalues of L_2 are:

$$a_n(\lambda) = 1 + 2s\cosh(2\lambda - i\gamma), \quad d_n(\lambda) = 1 + 2s\cosh(2\lambda + i\gamma). \quad (5.5)$$

The Bethe equations are of the form

$$\left(\frac{1 + 2s\cosh(2\lambda_l - i\gamma)}{1 + 2s\cosh(2\lambda_l + i\gamma)}\right)^{M/2} = -\prod_{k=1}^{N}\frac{\sinh(\lambda_l - \lambda_k + i\gamma)}{\sinh(\lambda_l - \lambda_k - i\gamma)}. \quad (5.6)$$

The operator for the number of particles may be introduced for the lattice model as

$$Q = \frac{4}{\beta}\sum_{n=1}^{M/2}(u_{2n} - u_{2n-1}) + M\frac{(\pi - \gamma)}{2\gamma}. \quad (5.7)$$

It is easy to check that Q commutes with the transfer matrix and hence with the Hamiltonian. The Bethe wave function Ψ_N (VII.1.10) is an eigenfunction of the operator Q with eigenvalue N. The continuous limit of the physical quantities that have been introduced is proper.

The problem of ultraviolet divergences was solved in [2], [3], [5], [6], [12], [14], [16] and [17] by means of the lattice SG model. The bound states in the lattice model are described by the same inequalities as in the continuous case (see (III.1.15)). The ground state of the lattice model is a rather complicated combination of bound states, the latter being dependent on the arithmetic structure of the coupling constant. In the continuous limit, the picture is simplified, and is in agreement with that of the massive Thirring model obtained there with the help of standard perturbation theory (see section III.2).

The dependence of the L-operator of the lattice SG model on a continuous parameter allows one to construct the most general L-operator intertwined by the XXZ R-matrix as was done in section 4 for the lattice NS model.

Conclusion

Continuous models of quantum field theory have been put on a lattice without spoiling their dynamical structure. Our approach (based on the preservation of the R-matrix) guarantees that the continuous limit of the lattice model will be smooth (no phase transitions as $\Delta \to 0$). The most important dynamical characteristics of lattice models will coincide with those of the continuous model.

The construction of local Hamiltonians on a lattice made use of the trace identities, which are similar to the continuous models. The Hamiltonian was expressed at points where the L-operator was degenerate. This guaranteed that interactions were limited to nearest neighbors even in the limit of an infinite lattice.

The lattice models considered in this chapter, playing an important role in the development of QISM, were introduced in [12], [15], [16] and [17], and studied in [2]–[7], [13], [14], [18], [22], [25]–[28]. There exist different approaches to the construction of lattice models starting from continuous models ([1], [8]–[10] and [21]). The distinctive feature of our approach is the coincidence of the Hamiltonian structures (the R-matrices for continuous and lattice cases are the same.)

The general method of constructing local Hamiltonians stated here was worked out in papers [14], [15] and [17]. The Hamiltonian (1.15) is suggested in [7].

The explicit form of the time evolution operator, the local equations of motion, the Gel'fand-Levitan-Marchenko equations and soliton solutions for the lattice SG model are given in [25].

The solution of the quantum lattice NS model is given in [7]. Let us note that other local lattice Hamiltonians may be constructed by means of trace identities from the quantum L-operator (1.1) (generalizing trace identities for the fundamental spin models). This was done in [27]. On the basis of the same L-operator, the quantum Gel'fand-Levitan-Marchenko equation for the NS model was constructed in [24], and [20] is devoted to the generalization of the multicomponent NS model. The thermodynamics of the lattice model was investigated in [4].

Theorem 1 in section 4 was proved in [18]; in [26] it was generalized to the case of the XXZ R-matrix. The quadratic algebras [22], [23] closely connected with the L-operators we have considered should be mentioned here.

The problem of ultraviolet divergences for the SG model was solved in [2], [3], [5], [6], [12], [14], [16] and [17] with the help of the lattice SG model. The ground state of the lattice SG model was constructed in the same papers and the equations describing its thermodynamics were derived.

In the limiting case, the L-operator of the lattice SG model generates the lattice L-operator of the Liouville equation, which plays an important role when investigating conformal quantum field theory [28].

The lattice L-operator of the sine-Gordon equation is the most general L-operator related to R-matrices of the XXZ model ([25] and [26]).

Part III
The Determinant Representation
for Quantum Correlation Functions

Introduction to Part III

The algebraic Bethe Ansatz is a powerful method for the calculation of quantum correlation functions. In Part III we shall start this calculation. We shall arrive at an extremely important conclusion: quantum correlation functions can be represented as determinants of certain matrices. The dimension of these matrices is equal to the number of particles in the ground state. In order to better understand the nature of this matrix, one should look once more through section 6 of Chapter VI and section 10 of Chapter VII, where the determinant representation of the partition function for the six-vertex model with a special type of boundary conditions (domain wall) was obtained. Starting from determinant formulæ (VII.10.1) and (VII.10.2) and using also the formula for the determinant of the sum of two matrices (in the Appendix to Chapter IX), one can reproduce all the determinant formulæ in Part III. In the thermodynamic limit, the correlation functions will be represented as determinants of a Fredholm integral operator. In Part IV, we shall explain how to use this determinant representation to write down the differential equation for a quantum correlation function. In Part IV we shall also discover that the differential equation for a quantum correlation function is closely related to the initial classical nonlinear differential equation, which was quantized (in our example this is the nonlinear Schrödinger equation). The quantum correlation functions play the role of the τ functions of a classical differential equation. The differential equations for the quantum correlation functions are classical completely integrable differential equations. They have a Lax representation (see section 1 of Chapter V) and admit formulation in terms of the Riemann-Hilbert problem. This makes it possible to evaluate explicitly the asymptotics of the correlation functions, which is the most important problem from the physical point of view. Part III provides the algebraic background for this outlined program.

In order to obtain all of the above mentioned results in Part IV, we need to go through combinatorics in Part III to get the determinant representation of quantum correlation functions. Our main example in this Part and the next Part will be the nonlinear Schrödinger model. The algebraic Bethe Ansatz for this model was described in Chapter VII of Part II. Let us remind the reader that the nonlinear Schrödinger model can be described by the 2-dimensional monodromy matrix

$$\mathsf{T}(\lambda) = \begin{pmatrix} A(\lambda) & B(\lambda) \\ C(\lambda) & D(\lambda) \end{pmatrix}. \tag{0.1}$$

Here A, B, C and D are quantum operators acting in the Fock space with commutation relations given by the R-matrix

$$\mathsf{R}(\lambda, \mu) \left(\mathsf{T}(\lambda) \bigotimes \mathsf{T}(\mu) \right) = \left(\mathsf{T}(\mu) \bigotimes \mathsf{T}(\lambda) \right) \mathsf{R}(\lambda, \mu). \tag{0.2}$$

The R-matrix is given by

$$\mathsf{R}(\lambda, \mu) = \begin{pmatrix} f(\mu, \lambda) & 0 & 0 & 0 \\ 0 & g(\mu, \lambda) & 1 & 0 \\ 0 & 1 & g(\mu, \lambda) & 0 \\ 0 & 0 & 0 & f(\mu, \lambda) \end{pmatrix} \tag{0.3}$$

with

$$f(\mu, \lambda) = 1 + \frac{ic}{\mu - \lambda}, \quad g(\mu, \lambda) = \frac{ic}{\mu - \lambda}. \tag{0.4}$$

An important role is played by the pseudovacuum, $|0\rangle$. This is the vector in the Fock space which has the following properties:

$$A(\lambda)|0\rangle = a(\lambda)|0\rangle, \quad D(\lambda)|0\rangle = d(\lambda)|0\rangle,$$
$$C(\lambda)|0\rangle = 0. \tag{0.5}$$

For the nonlinear Schrödinger model

$$a(\lambda) = \exp\{-i\lambda L/2\}, \quad d(\lambda) = \exp\{i\lambda L/2\}. \tag{0.6}$$

But this is not the way we are going to proceed. In section 6 of Chapter VII, we classified all the monodromy matrices (0.1) which have commutation relations (0.2) with the R-matrix fixed by (0.3) and (0.4) (we also demanded the existence of a pseudovacuum $|0\rangle$). We proved that there exist infinitely many such monodromy matrices—they are parametrized by two arbitrary c-number functions $a(\lambda)$ and $d(\lambda)$ (see (0.5)). So if $a(\lambda)$ and $d(\lambda)$ are some functions different from (0.6) the model which we consider (the generalized model) does not coincide with the nonlinear Schrödinger model, but is closely related. The relation of these models is similar to the relation between different representations of the same group (fixing the R-matrix is similar to fixing the structure constants of the group).

In Part III we shall consider a generalized model with arbitrary functions $a(\lambda)$ and $d(\lambda)$. This is very convenient for recursion relations. When we decide to come back to the nonlinear Schrödinger model, it will be sufficient to put $a(\lambda)$ and $d(\lambda)$ equal to (0.6). This will be the point of view of Part III.

In Chapter IX we shall derive the determinant representation for scalar products

$$\langle 0| \prod_{j=1}^{N} C(\lambda_j^C) \prod_{k=1}^{N} B(\lambda_k^B)|0\rangle. \tag{0.7}$$

In order to do this we shall introduce an auxiliary Fock space (of the Bose type). We shall use corresponding Bose fields (auxiliary quantum fields) in order to solve some combinatoric problems.

In Chapter X we shall consider norms of the eigenfunctions of the Hamiltonian. We shall prove that the square of the norm is equal to the determinant of some matrix (see formula (2.19), from section 2 of Chapter I). The physical meaning of this matrix is very clear. In order to obtain the matrix, one should linearize the periodic boundary conditions in the logarithmic form (Bethe equations) near the solution. See for example formulæ (1.33), (1.34) of section 1, Chapter VII. In other words, the matrix is equal to the matrix of second derivatives of the Yang-Yang action (see formulæ (2.15), (2.16) of section 2, Chapter I).

In Chapter XI, the correlation function of local densities

$$j(x) = \Psi^\dagger(x)\Psi(x)$$

is considered (it is called the current correlator). Here we explain that the study of this correlation function can be replaced by the study of the ground state mean value of the operator

$$\exp\left\{ \alpha \int_0^x \Psi^\dagger(y)\Psi(y)\,dy \right\}. \tag{0.8}$$

Here α is a free parameter. The mean value of the operator (0.8) is represented as a determinant.

This is a quite remarkable operator. A lot of interesting correlation functions can be extracted from

$$\left\langle \exp\left\{ \alpha \int_0^x \Psi^\dagger(y)\Psi(y)\,dy \right\} \right\rangle.$$

For small α it gives current correlations, at $\alpha = i\pi$ it is related to field correlations, at $\alpha = -\infty$ it gives the probability that, due to fluctuations, there will be no particles in the space interval $[0, x]$.

In Chapter XII, the determinant representation is obtained for the field correlator $\langle \Psi^\dagger(x)\Psi(0)\rangle$. It should be emphasized that the methods of Part III are very general and can be applied to other correlators and other integrable models. An evident generalization is the XXZ model and other models with the same R-matrix. The corresponding R-matrix has the form (0.3) with

$$f(\mu, \lambda) = \frac{\sinh(\mu - \lambda + 2i\eta)}{\sinh(\mu - \lambda)}$$

and

$$g(\mu, \lambda) = i\frac{\sin 2\eta}{\sinh(\mu - \lambda)}.$$

(0.9)

IX

Scalar Products

Introduction

Those properties of scalar products that are basic for the calculation of the norms of Bethe wave functions, as well as of correlators, are investigated here. Scalar products are considered as functionals of "vacuum eigenvalues" $a(\lambda)$, $d(\lambda)$ (see section VII.6). These functionals are rather special, since they depend on values of the functions $a(\lambda)$, $d(\lambda)$ at a finite number of points. The case of the R-matrix of the XXX model is considered in this section, the generalization to the case of the XXZ R-matrix being quite obvious. The reader should study this chapter carefully as the mathematical apparatus given here will be used further for correlation functions.

The contents of the chapter are as follows. The definition of the scalar product is given in section 1. In section 2, properties of coefficients in the expansion for scalar products are investigated. In section 3, we prove the formula for the residue of scalar products which is essential for calculating norms in the next chapter. In section 4 the recursion formula for scalar products

$$\langle 0| \prod_{j=1}^{N} C(\lambda_j^C) \ \prod_{k=1}^{N} B(\lambda_k^B)|0\rangle$$

is obtained; this relates the scalar product in the N-particle case with the scalar product in the $(N-1)$-particle case. The introduction of auxiliary quantum fields in section 5 permits one to solve this recursion explicitly. The formula for the scalar product is obtained in a determinant form (the dimension of the corresponding matrix is equal to the number of particles). This is a central point in our consideration of scalar products and correlation functions. In section 6, the same determinant representation for scalar products is obtained in another way. It is based on the representation of the determinant of the sum of two matrices in terms of minors of individual matrices. This formula is explicitly explained in the Appendix. In section 7, some remarks concerning the norms are made.

IX.1 The scalar product

Consider (in the scheme of the algebraic Bethe Ansatz, see Chapter VII) the scalar product S_N:

$$\langle 0| \prod_{j=1}^{N} C(\lambda_j^C) \ \prod_{k=1}^{N} B(\lambda_k^B)|0\rangle = S_N. \tag{1.1}$$

It is supposed here that the $2N$ complex numbers $\{\lambda_j^C\}$ and $\{\lambda_k^B\}$ are independent variables and do not satisfy any equations. In particular, the Bethe equations are not assumed to be satisfied.

The scalar product is a symmetric function of N variables λ_j^C and also a symmetric function of the N variables λ_k^B. This follows from the commutation relations (VII.1.11)

$$[C(\lambda),\, C(\mu)] = [B(\lambda),\, B(\mu)] = 0.$$

It is easy to verify that the number of "annihilation" operators $C(\lambda)$ should be equal to the number of "creation" operators $B(\lambda)$ in the scalar product, otherwise it is equal to zero.

The scalar product can be calculated by means of commutation relations (VII.1.16), (VII.1.12–15). In the simplest case ($N = 1$), the scalar product is equal to

$$\begin{aligned} S_1 &= \langle 0|C(\lambda^C)\, B(\lambda^B)|0\rangle \\ &= g(\lambda^C, \lambda^B) \left\{ a(\lambda^C)d(\lambda^B) - a(\lambda^B)d(\lambda^C) \right\}. \end{aligned} \tag{1.2}$$

Let us investigate the properties of the scalar product for any N (see section VII.2). (The integer value N has the physical meaning of the number of particles.) Repeatedly using formula (VII.2.2), one establishes that the scalar product can be represented in the following form:

$$\begin{aligned} S_N &= \langle 0| \prod_{j=1}^{N} C(\lambda_j^C) \ \prod_{k=1}^{N} B(\lambda_k^B)|0\rangle \\ &= \sum \left\{ \prod_{j=1}^{N} a(\lambda_j^A) \ \prod_{k=1}^{N} d(\lambda_k^D) \right\} K_N \left(\begin{matrix} \{\lambda^C\}, \{\lambda^B\} \\ \{\lambda^A\}, \{\lambda^D\} \end{matrix} \right). \end{aligned} \tag{1.3}$$

The sum here is with respect to all partitions of the set $\{\lambda^C\}\cup\{\lambda^B\}$ (consisting of $2N$ elements) into two disjoint sets $\{\lambda^A\}$ and $\{\lambda^D\}$ possessing equal numbers of elements:

$$\{\lambda^C\} \cup \{\lambda^B\} = \{\lambda^A\} \cup \{\lambda^D\}; \quad \{\lambda^A\} \cap \{\lambda^D\} = \emptyset. \tag{1.4}$$

The number of elements in each of the sets $\{\lambda^C\}$, $\{\lambda^B\}$, $\{\lambda^A\}$ and $\{\lambda^D\}$ is equal to N:

$$\text{card}\{\lambda^C\} = \text{card}\{\lambda^B\} = \text{card}\{\lambda^A\} = \text{card}\{\lambda^D\} = N. \tag{1.5}$$

Here "card" denotes the number of elements in the set. The ordering of elements inside the set is not important. The number of terms in the sum on the right hand side of (1.3) is equal to

$$\binom{2N}{N} = \frac{(2N)!}{(N!)^2}.$$

The quantities K_N are rational functions of the variables λ_j and do not depend on the vacuum eigenvalues $a(\lambda)$ and $d(\lambda)$. To investigate the explicit form of K_N, it is convenient to introduce the following sets of spectral parameters λ_j:

$$\{\lambda^{AC}\} = \{\lambda^A\} \cap \{\lambda^C\}; \quad \{\lambda^{DC}\} = \{\lambda^D\} \cap \{\lambda^C\};$$
$$\{\lambda^{AB}\} = \{\lambda^A\} \cap \{\lambda^B\}; \quad \{\lambda^{DB}\} = \{\lambda^D\} \cap \{\lambda^B\}. \tag{1.6}$$

It should be noted that the following two identities are valid:

$$\{\lambda^{AC}\} \cup \{\lambda^{AB}\} = \{\lambda^A\}; \quad \{\lambda^{AC}\} \cup \{\lambda^{DC}\} = \{\lambda^C\};$$
$$\{\lambda^{DC}\} \cup \{\lambda^{DB}\} = \{\lambda^D\}; \quad \{\lambda^{AB}\} \cup \{\lambda^{DB}\} = \{\lambda^B\}. \tag{1.7}$$

The number of elements in each set will be denoted as

$$n_0 = \text{card}\{\lambda^{DC}\} = \text{card}\{\lambda^{AB}\},$$
$$N - n_0 = \text{card}\{\lambda^{AC}\} = \text{card}\{\lambda^{DB}\}. \tag{1.8}$$

Sometimes the number of elements of the set will be written as a subscript, e.g., $\{\lambda^{DC}\}_{n_0}$ means that $\text{card}\{\lambda^{DC}\} = n_0$.

Let us now give the following definition. Consider the partition with $n_0 = 0$:

$$\{\lambda^A\} = \{\lambda^C\}, \quad \{\lambda^D\} = \{\lambda^B\} \tag{1.9}$$

such that

$$\{\lambda^{AC}\} = \{\lambda^C\}, \quad \{\lambda^{AB}\} = \emptyset, \quad \{\lambda^{DC}\} = \emptyset, \quad \{\lambda^{DB}\} = \{\lambda^B\}. \tag{1.10}$$

The coefficient K_N in (1.3) for this partition will be called the highest coefficient,

$$K_N \left(\begin{matrix} \{\lambda^C\}, \{\lambda^B\} \\ \{\lambda^C\}, \{\lambda^B\} \end{matrix} \right). \tag{1.11}$$

The scalar product S_N is given by formula (1.3) for arbitrary N. Below we give the explicit form of S in the $(N-1)$-particle case:

$$S_{N-1} = \langle 0| \prod_{j=1}^{N-1} C(\lambda_j^C) \prod_{k=1}^{N-1} B(\lambda_k^B)|0\rangle$$

$$= \sum \left\{ \prod_{j=1}^{N-1} a(\lambda_j^A) \prod_{k=1}^{N-1} d(\lambda_k^D) \right\} K_{N-1} \left(\begin{matrix} \{\lambda^C\}, \{\lambda^B\} \\ \{\lambda^A\}, \{\lambda^D\} \end{matrix} \right).$$

$$(1.12)$$

It should be also mentioned that the definition and properties of scalar products for the XXZ case are similar (see paper [2]).

IX.2 Properties of the coefficients K_N

As it was shown in section VII.6, the functions $a(\lambda)$ and $d(\lambda)$ in (1.3) can be considered as arbitrary functions; these functions will be our parameters. Now the idea is to choose these functions $a(\lambda)$, $d(\lambda)$ in such a way that all the terms in the sum in (1.3), except one, become equal to zero. This allows one to express the coefficients K_N in terms of the partition function Z_N which was investigated in detail in sections VI.6, VII.9 and VII.10.

Theorem 1. *The highest coefficient K_N (1.11) in (1.3) is related to the partition function Z_N as follows*

$$K_N \left(\begin{matrix} \{\lambda^C\}, \{\lambda^B\} \\ \{\lambda^C\}, \{\lambda^B\} \end{matrix} \right) = \left(\prod_{j=1}^{N} \prod_{k=1}^{N} \left(\lambda_k^B - \lambda_j^C \right) \right)^{-1}$$

$$\times Z_N \left(\{\lambda_j^B\}, \{\lambda_j^C + ic/2\} \right). \qquad (2.1)$$

(The index C of λ_j is related to the operator $C(\lambda)$ and the term $ic/2$ contains the coupling constant c of the nonlinear Schrödinger model. They should not be confused.)

Sometimes, instead of the partition function Z_N it will be more convenient to use the function G_N which differs from Z_N by a shift of the second argument:

$$G_N \left(\{\lambda_j^B\}, \{\lambda_j^C\} \right) = Z_N \left(\{\lambda_j^B\}, \{\lambda_j^C + ic/2\} \right). \qquad (2.2)$$

Proof: Consider the scalar product (1.3) for the particular case of the "inhomogeneous" monodromy matrix of the XXX model (VI.5.24). The vacuum eigenvalues $a(\lambda)$, $d(\lambda)$ are given by formula (VII.3.14). Consider the case $M = N$, i.e., the number of sites is equal to the number of operators B (or C) in (1.3) (and also the number of particles). Let us calculate in this model the scalar product S_N (1.3) for some sets $\{\lambda^B\}_N$, $\{\lambda^C\}_N$. The parameters of non-homogeneity ν_j are chosen as follows:

$$\nu_j = \lambda_j^C + i\frac{c}{2} \tag{2.3}$$

Here the label C is not related to the coupling constant c.

The vacuum eigenvalues are then equal to

$$a(\lambda) = \prod_{j=1}^{N} (\lambda - \lambda_j^C - ic), \quad d(\lambda) = \prod_{j=1}^{N} (\lambda - \lambda_j^C). \tag{2.4}$$

In this particular model we have $d(\lambda_j^C) = 0$, and only one term (containing the highest coefficient (1.11)) survives in the sum (1.3):

$$\langle 0| \prod_{j=1}^{N} C(\lambda_j^C) \prod_{k=1}^{N} B(\lambda_k^B)|0\rangle$$

$$= \left(\prod_{j,k=1}^{N} (\lambda_k^B - \lambda_j^C)(\lambda_k^C - \lambda_j^C - ic) \right) K_N \left(\begin{matrix} \{\lambda^C\}, \{\lambda^B\} \\ \{\lambda^C\}, \{\lambda^B\} \end{matrix} \right). \tag{2.5}$$

This is true only for our particular example: the inhomogeneous XXX model with the special choice (2.3) of ν_j.

Using (VII.3.18), (VI.6.12) and (VI.6.15) one has

$$\langle 0| \prod_{j=1}^{N} C(\lambda_j^C) \prod_{k=1}^{N} B(\lambda_k^B)|0\rangle = \langle 0| \prod_{j=1}^{N} C(\lambda_j^C)|0'\rangle\langle 0'| \prod_{k=1}^{N} B(\lambda_k^B)|0\rangle$$

$$= Z_N \left(\{\lambda_j^C\}, \{\nu_j\} \right) Z_N \left(\{\lambda_j^B\}, \{\nu_j\} \right). \tag{2.6}$$

It should be remembered that in this model the vector $|0\rangle$ is pure ferromagnetic with all spins up. The product of B's overturns all the spins and the result is another ferromagnetic state, $|0'\rangle$, with all the spins down.

Taking into account condition (2.3), this relation can be rewritten as

$$\langle 0| \prod_{j=1}^{N} C(\lambda_j^C) \prod_{k=1}^{N} B(\lambda_k^B)|0\rangle = Z_N \left(\{\lambda_j^C\}, \{\lambda_j^C + ic/2\} \right)$$

$$\times Z_N \left(\{\lambda_j^B\}, \{\lambda_j^C + ic/2\} \right). \tag{2.7}$$

The first factor on the right hand side can be calculated explicitly by using the recursion relation (VII.9.11) for Z_N:

$$Z_N\left(\{\lambda_j^C\}, \{\lambda_j^C + ic/2\}\right) = \prod_{j,k=1}^{N}\left(\lambda_k^C - \lambda_j^C - ic\right). \qquad (2.8)$$

Now putting formulæ (2.7), (2.8) into the left hand side of (2.5) one comes to formulæ (2.1), which finishes the proof of the theorem.

Let us consider the analytical properties of the highest coefficient K_N. It has poles. If one of the λ_j^C coincides with one of the λ_k^B, the highest coefficient possesses a first order pole. The highest coefficient K_N is a symmetric function of all the λ_j^C as well as of all the λ_k^B (separately). Thus, it is sufficient to consider the case $\lambda_N^C \to \lambda_N^B \to \lambda_N$. Using the recursion relations (VII.9.11), one obtains for the singular part:

$$K_N\left(\begin{matrix}\{\lambda^C\}_N, \{\lambda^B\}_N\\ \{\lambda^C\}_N, \{\lambda^B\}_N\end{matrix}\right) \longrightarrow$$

$$-i\frac{c}{\lambda_N^B - \lambda_N^C}\left(\prod_{j=1}^{N-1} f(\lambda_j^C, \lambda_N)f(\lambda_N, \lambda_j^B)\right)$$

$$\times K_{N-1}\left(\begin{matrix}\{\lambda^C\}_{N-1}, \{\lambda^B\}_{N-1}\\ \{\lambda^C\}_{N-1}, \{\lambda^B\}_{N-1}\end{matrix}\right) + O\left(1\right) \qquad (2.9)$$

where we have used the representation (2.1) and recursion relations (VII.9.11) for the partition function Z_N. Formula (2.9) shows that the residues in the poles of K_N can be simply related to K_{N-1}. This is an important recursion property.

In section VII.10, the representation of the partition function Z_N in the form of a determinant was given. Using this representation as well as formula (2.1) one arrives at the following representation for the highest coefficient:

$$K_N\left(\begin{matrix}\{\lambda^C\}, \{\lambda^B\}\\ \{\lambda^C\}, \{\lambda^B\}\end{matrix}\right) = \left\{\prod_{j>k} g(\lambda_j^B, \lambda_k^B)g(\lambda_k^C, \lambda_j^C)\right\}$$

$$\times \left\{\prod_{j,k} h(\lambda_j^C, \lambda_k^B)\right\} \det{}_N M. \qquad (2.10)$$

The matrix M is an $N \times N$-dimensional matrix with elements

$$M_{jk} = \frac{g(\lambda_j^C, \lambda_k^B)}{h(\lambda_j^C, \lambda_k^B)} \equiv t(\lambda_j^C, \lambda_k^B). \qquad (2.11)$$

The functions $h(\lambda, \mu)$ and $t(\lambda, \mu)$ are given by

$$t(\lambda, \mu) = \frac{(ic)^2}{(\lambda - \mu)(\lambda - \mu + ic)}$$

$$h(\lambda, \mu) = \frac{f(\lambda, \mu)}{g(\lambda, \mu)} = \frac{\lambda - \mu + ic}{ic}. \tag{2.12}$$

Here f and g are equal to

$$f(\lambda, \mu) = 1 + \frac{ic}{\lambda - \mu}, \qquad g(\lambda, \mu) = \frac{ic}{\lambda - \mu}.$$

It should be mentioned that formula (2.10) for K_N is also valid in the XXZ case (with the corresponding functions f and g, see the Introduction to Part, formula (0.9)). Formula (2.10) will be very important below.

Theorem 2. *Arbitrary coefficients K_N in (1.3) are expressed in terms of the highest coefficient as follows:*

$$K_N \left(\begin{matrix} \{\lambda^C\}, \{\lambda^B\} \\ \{\lambda^A\}, \{\lambda^D\} \end{matrix} \right) = \left\{ \prod_{j \in (AC)} \prod_{k \in (DC)} f(\lambda_j^{AC}, \lambda_k^{DC}) \right\}$$

$$\times \left\{ \prod_{l \in (AB)} \prod_{m \in (DB)} f(\lambda_l^{AB}, \lambda_m^{DB}) \right\}$$

$$\times K_{n_0} \left(\begin{matrix} \{\lambda^{AB}\}, \{\lambda^{DC}\} \\ \{\lambda^{AB}\}, \{\lambda^{DC}\} \end{matrix} \right)$$

$$\times K_{N-n_0} \left(\begin{matrix} \{\lambda^{AC}\}, \{\lambda^{DB}\} \\ \{\lambda^{AC}\}, \{\lambda^{DB}\} \end{matrix} \right). \tag{2.13}$$

In the double products here, each of the indices j, k, l, m run independently over the corresponding subset. The proof of this theorem is similar to that of Theorem 1.

It should be mentioned that the following brief notation will often be used below:

$$f_{jk} \equiv f(\lambda_j, \lambda_k), \qquad K_{al} = K(\lambda_a, \lambda_l) = \frac{2c}{c^2 + (\lambda_a - \lambda_l)^2}. \tag{2.14}$$

Remark 1. If all the λ_j entering the set $\{\lambda^C\} \cup \{\lambda^B\}$ are fixed except for one, denoted further as $\tilde{\lambda}$, and $\tilde{\lambda} \to \infty$, then the arbitrary coefficients K_N in (1.3) decrease as $1/\tilde{\lambda}$. This follows from formula (VII.9.1).

Remark 2. If

$$d(\lambda_1^C) = a(\lambda_2^C) = 0, \quad f(\lambda_1^C, \lambda_2^C) = 0 \qquad (2.15)$$

then the scalar product is equal to zero:

$$\langle 0| \prod_{j=1}^{N} C(\lambda_j^C) \prod_{k=1}^{N} B(\lambda_k^B)|0\rangle = 0. \qquad (2.16)$$

This fact is a direct consequence of formulæ (1.3) and (2.13). In other words, one can write that under the conditions (2.15)

$$C(\lambda_1)C(\lambda_2) = 0. \qquad (2.17)$$

Analogously, if

$$a(\lambda_2^B) = d(\lambda_1^B) = 0, \quad f(\lambda_1^B, \lambda_2^B) = 0 \qquad (2.18)$$

then

$$\langle 0| \prod_{j=1}^{N} C(\lambda_j^C) \prod_{k=1}^{N} B(\lambda_k^B)|0\rangle = 0 \qquad (2.19)$$

which is equivalent to the relation

$$B(\lambda_1^B)B(\lambda_2^B) = 0. \qquad (2.20)$$

Remark 3. The arbitrary coefficients K_N in (1.3) are rational functions of all the variables $\lambda_j \in \{\lambda^C\}\cup\{\lambda^B\}$ and can be represented in the following form:

$$K_N\left(\begin{array}{cc}\{\lambda^C\}, \{\lambda^B\}\\ \{\lambda^A\}, \{\lambda^D\}\end{array}\right) = \left(\prod_{j,k=1}^{N}(\lambda_j^D - \lambda_k^A)^{-1}\right) P_N\left(\begin{array}{cc}\{\lambda^C\}, \{\lambda^B\}\\ \{\lambda^A\}, \{\lambda^D\}\end{array}\right) \qquad (2.21)$$

where P_N is a polynomial. Consider the case for which $\lambda_N^C \in \{\lambda^A\}$, $\lambda_N^B \in \{\lambda^D\}$, and $\lambda_N^C \to \lambda_N^B \to \lambda_N$. Then the arbitrary coefficients possess a first order pole with the residue reduced to the corresponding coefficient in the scalar product S_{N-1} (1.12) for one less particle:

$$K_N\left(\begin{array}{cc}\{\lambda^C\}, \{\lambda^B\}\\ \{\lambda^A\}, \{\lambda^D\}\end{array}\right) \longrightarrow$$

$$\frac{-ic}{\lambda_N^B - \lambda_N^C}\left(\prod_{j=1}^{N-1} f(\lambda_j^A, \lambda_N)f(\lambda_N, \lambda_j^D)\right) K_{N-1}\left(\begin{array}{cc}\{\lambda^C\}, \{\lambda^B\}\\ \{\lambda^A\}, \{\lambda^D\}\end{array}\right). \qquad (2.22)$$

This formula can be derived directly from (2.9) and (2.13).

IX.3 The residue formula

Formula (1.3) for the scalar product contains only the values of the arbitrary functions $a(\lambda)$ and $d(\lambda)$ at $2N$ points λ_j^C, λ_k^B (entering the set $\{\lambda^C\} \cup \{\lambda^B\}$). These values can be considered as $4N$ independent complex variables

$$a_j^C = a(\lambda_j^C), \quad a_j^B = a(\lambda_j^B), \quad d_j^C = d(\lambda_j^C), \quad d_j^B = d(\lambda_j^B). \tag{3.1}$$

So the scalar product S is a function of $6N$ independent complex variables, namely:

$$\langle 0| \prod_{j=1}^{N} C(\lambda_j^C) \prod_{k=1}^{N} B(\lambda_k^B)|0\rangle \tag{3.2}$$
$$= S_N \left(\{a_j^C\}, \{a_j^B\}, \{d_j^C\}, \{d_j^B\}, \{\lambda_j^B\}, \{\lambda_j^C\} \right).$$

Now consider the scalar product in the special case when one of the variables λ_j^B coincides with one of the variables λ_k^C. Due to the symmetry of S_N in the variables λ_j^B and λ_k^C, it is sufficient to consider the case

$$\lambda_N^C \to \lambda_N^B \to \lambda_N. \tag{3.3}$$

All other variables λ_j^C, λ_k^B are taken to be different and to have fixed values of the variables $\{a_j\}$, $\{d_j\}$ (generally speaking, $a_N^C \neq a_N^B$, $d_N^C \neq d_N^B$). Formulæ (1.3), (2.22) lead to the fact that the scalar product possesses a first order pole; e.g., one has for (1.2) (the scalar product in the one-particle case $N = 1$):

$$\langle 0|C(\lambda^C)\, B(\lambda^B)|0\rangle \xrightarrow[\lambda^C \to \lambda^B]{} \frac{ic}{\lambda^C - \lambda^B} \left(a^C d^B - a^B d^C \right). \tag{3.4}$$

It should be remembered that we are now considering the discontinuous functions $a(\lambda)$ and $d(\lambda)$ such that the expression in brackets is not in general equal to zero, as $\lambda^C \to \lambda^B$.

Theorem 3. *At arbitrary N, the scalar product S_N in the limit (3.3) possesses a first order pole. The residue at the pole can be expressed in terms of the scalar product with one less particle, S_{N-1}:*

$$\langle 0| \prod_{j=1}^{N} C(\lambda_j^C) \prod_{k=1}^{N} B(\lambda_k^B)|0\rangle \xrightarrow[\lambda_N^C \to \lambda_N^B]{}$$
$$\frac{ic}{\lambda_N^C - \lambda_N^B} \left(a_N^C d_N^B - a_N^B d_N^C \right) \langle 0| \prod_{j=1}^{N-1} C(\lambda_j^C) \prod_{k=1}^{N-1} B(\lambda_k^B)|0\rangle^{\mathrm{mod}}.$$
$$\tag{3.5}$$

The scalar product on the right hand side is given by the expression (1.12) with the "modified" vacuum eigenvalues $\tilde{a}(\lambda)$, $\tilde{d}(\lambda)$:

$$\tilde{a}(\lambda) = a(\lambda)f(\lambda, \lambda_N), \quad \tilde{d}(\lambda) = d(\lambda)f(\lambda_N, \lambda) \qquad (3.6)$$

i.e.,

$$\langle 0| \prod_{j=1}^{N-1} C(\lambda_j^C) \prod_{k=1}^{N-1} B(\lambda_k^B)|0\rangle^{\text{mod}}$$
$$= S_{N-1}(\tilde{a}_j^C, \tilde{a}_j^B, \tilde{d}_j^C, \tilde{d}_j^B, \lambda_j^B, \lambda_j^C) \qquad (j=1,\ldots,N-1) \qquad (3.7)$$

with

$$\tilde{a}_j^C = a_j^C f(\lambda_j^C, \lambda_N) \quad \tilde{a}_j^B = a_j^B f(\lambda_j^B, \lambda_N) \quad (j=1,\ldots,N-1)$$
$$\tilde{d}_j^C = d_j^C f(\lambda_N, \lambda_j^C), \quad \tilde{d}_j^B = d_j^B f(\lambda_N, \lambda_j^B). \qquad (3.8)$$

Proof: It is obvious from equation (2.22) that the only singular terms on the right hand side of (1.3) are those corresponding to the following two possibilities

(a) $\lambda_N^C \in \{\lambda^A\}, \quad \lambda_N^B \in \{\lambda^D\}$; $\qquad\qquad\qquad\qquad\qquad\qquad$ (3.9)

(b) $\lambda_N^C \in \{\lambda^D\}, \quad \lambda_N^B \in \{\lambda^A\}$. $\qquad\qquad\qquad\qquad\qquad\qquad$ (3.10)

Considering the sum of all terms of type (a) in (1.3) one can write this sum in the form

$$a_N^C d_N^B \sum K_N \left(\begin{matrix} \{\lambda^C\}, \{\lambda^B\} \\ \{\lambda^A\}, \{\lambda^D\} \end{matrix} \right) \prod_{j=1}^{N-1} a(\lambda_j^A)d(\lambda_j^D). \qquad (3.11)$$

The sum here is taken over all partitions of the set $\{\lambda_j^C\} \cup \{\lambda_j^B\}$ into two sets $\{\lambda_j^A\}$, $\{\lambda_j^D\}$. All the coefficients K_N in formula (3.11) possess first order poles. The explicit form (2.22) of the residue allows one to put the singular part of the sum (3.11) in the form

$$-ic\frac{a_N^C d_N^B}{\lambda_N^B - \lambda_N^C}\langle 0| \prod_{j=1}^{N-1} C(\lambda_j^C) \prod_{k=1}^{N-1} B(\lambda_k^B)|0\rangle^{\text{mod}}. \qquad (3.12)$$

The modified scalar product here is the same as that entering the right hand side of (3.5). Similar considerations for the sum of terms of type (b) lead to the following expression for the corresponding singular part:

$$ic\frac{a_N^B d_N^C}{\lambda_N^B - \lambda_N^C}\langle 0| \prod_{j=1}^{N-1} C(\lambda_j^C) \prod_{k=1}^{N-1} B(\lambda_k^B)|0\rangle^{\text{mod}}. \qquad (3.13)$$

Now summing expressions (3.12) and (3.13), one obtains the right hand side of expression (3.5) which completes the proof of the theorem.

It is to be emphasized that formula (3.5) plays a central role in calculating norms of Bethe wave functions and correlation functions.

The dependence of the scalar product on some of the variables in (3.1) is trivial. Indeed, it is obvious from equation (1.3) that the quantity

$$\langle 0| \prod_{j=1}^{N} \frac{C(\lambda_j^C)}{d(\lambda_j^C)} \prod_{k=1}^{N} \frac{B(\lambda_k^B)}{d(\lambda_k^B)} |0\rangle \tag{3.14}$$

depends on $4N$ independent variables only, namely

$$r_j^C = r(\lambda_j^C), \quad r_j^B = r(\lambda_j^B), \quad \lambda_j^C, \quad \lambda_j^B \quad (j = 1, \dots, N) \tag{3.15}$$

(recall that $r(\lambda) = a(\lambda)/d(\lambda)$). Thus, it is useful to introduce "renormalized" creation and annihilation operators

$$\mathbb{B}(\lambda) = \frac{B(\lambda)}{d(\lambda)}, \qquad \mathbb{C}(\lambda) = \frac{C(\lambda)}{d(\lambda)}. \tag{3.16}$$

Formula (3.5) describing a first order pole in S_N can be rewritten using the "renormalized" operators in (3.14) to give the following:

$$\langle 0| \prod_{j=1}^{N} \mathbb{C}(\lambda_j^C) \prod_{k=1}^{N} \mathbb{B}(\lambda_k^B) |0\rangle \xrightarrow[\lambda_N^C \to \lambda_N^B]{}$$

$$ic\frac{r_N^C - r_N^B}{\lambda_N^C - \lambda_N^B} \left(\prod_{j=1}^{N-1} f_{Nj}^B f_{Nj}^C \right) \langle 0| \prod_{j=1}^{N-1} \mathbb{C}(\lambda_j^C) \prod_{k=1}^{N-1} \mathbb{B}(\lambda_k^B) |0\rangle^{\text{mod}}$$

$$\tag{3.17}$$

where the following are also used

$$f_{Nj}^B = f(\lambda_N, \lambda_j^B), \quad f_{Nj}^C = f(\lambda_N, \lambda_j^C). \tag{3.18}$$

The modification on the right hand side of (3.17) means the change

$$r(\lambda) \longrightarrow \tilde{r}(\lambda) = r(\lambda)\frac{f(\lambda, \lambda_N)}{f(\lambda_N, \lambda)}. \tag{3.19}$$

It should be noted that for concrete physical models (such as the Bose gas and the XXX magnet), the complex variables r_j (3.15) are the values of the smooth function $r(\lambda)$ at different points. In this case, the residue in formula (3.17) becomes equal to zero (the corresponding limit is thus finite). In the limit $\lambda_N^C \to \lambda_N^B \to \lambda_N$ (all other λ_j^C, λ_k^B are taken to be different), the scalar product depends on the following variables:

$$\lambda_j^C, \lambda_j^B, r_j^C, r_j^B \quad (j = 1, \dots, N-1), \quad \lambda_N, r_N, \bar{\bar{z}}_N,$$

$$\bar{\bar{z}}_N = i\frac{\partial \ln r(\lambda)}{\partial \lambda}\Big|_{\lambda=\lambda_N}. \tag{3.20}$$

The dependence on the variable $\bar{\bar{z}}_N$ is linear. The coefficient at $\bar{\bar{z}}_N$ is defined by the residue (3.17):

$$\frac{\partial}{\partial \bar{\bar{z}}_N} \langle 0| \left(\prod_{j=1}^{N-1} \mathbb{C}(\lambda_j^C) \right) \mathbb{C}(\lambda_N)\mathbb{B}(\lambda_N) \left(\prod_{j=1}^{N-1} \mathbb{B}(\lambda_j^B) \right) |0\rangle$$

$$= c r_N \left(\prod_{k=1}^{N-1} f_{Nk}^B f_{Nk}^C \right) \langle 0| \prod_{j=1}^{N-1} \mathbb{C}(\lambda_j^C) \prod_{k=1}^{N-1} \mathbb{B}(\lambda_k^B)|0\rangle^{\mathrm{mod}}. \quad (3.21)$$

We obtained this formula by using l'Hôpital's rule in (3.17).

Consider now the scalar product in the limit $\lambda_j^C \to \lambda_j^B \to \lambda_j$ ($j = 1,\dots,N$). All the λ_j are taken to be different. In this case, the scalar product depends on $3N$ independent variables $\lambda_j, r_j, \bar{\bar{z}}_j$:

$$r_j = r(\lambda_j), \quad \bar{\bar{z}}_j = i \frac{\partial \ln r(\lambda)}{\partial \lambda}\bigg|_{\lambda=\lambda_j}. \quad (3.22)$$

The scalar product is now a linear function of each $\bar{\bar{z}}_j$. The coefficient at $\bar{\bar{z}}_N$ is equal to (see (3.21))

$$\frac{\partial}{\partial \bar{\bar{z}}_N} \langle 0| \prod_{j=1}^{N} \mathbb{C}(\lambda_j) \prod_{j=1}^{N} \mathbb{B}(\lambda_j)|0\rangle$$

$$= c r_N \left(\prod_{k=1}^{N-1} f(\lambda_N, \lambda_k)f(\lambda_N, \lambda_k) \right) \langle 0| \prod_{j=1}^{N-1} \mathbb{C}(\lambda_j) \prod_{j=1}^{N-1} \mathbb{B}(\lambda_j)|0\rangle^{\mathrm{mod}}.$$

$$\quad (3.23)$$

The modification in this case refers to the following:

$$\tilde{r}_j = \tilde{r}(\lambda_j) = r(\lambda_j)\frac{f(\lambda_j, \lambda_N)}{f(\lambda_N, \lambda_j)};$$

$$\bar{\bar{z}}_j = \bar{\bar{z}} + K_{jN} \quad (3.24)$$

$$K_{jN} = K(\lambda_j, \lambda_N) = i\frac{\partial}{\partial \lambda_j} \ln\left(\frac{f_{jN}}{f_{Nj}}\right) = \frac{2c}{c^2 + (\lambda_j - \lambda_N)^2} \quad (3.25)$$

$$(j = 1,\dots,N-1, \quad f_{jN} \equiv f(\lambda_j, \lambda_N))$$

where the modification rule for the variable z_j is obtained from (3.19). (3.22). The coefficients at other z_j are readily recoverd from the symmetry (it is easily seen that the scalar product in this case is invariant under the transposition of triples $(z_j, \lambda_j, r_j) \leftrightarrow (z_k, \lambda_k, r_k)$).

Next consider the case where some λ_j^C coincide with some λ_j^B; however, all the $\{\lambda_j^C\}$ are different and all the $\{\lambda_j^B\}$ are different. In this case, the scalar product depends on the variables r_j and z_j which correspond to coincident λ_j and on the variables r_j^B and r_j^C which correspond to non-coincident λ_j^B and λ_j^C. Formula (3.21) is still valid.

Finally, consider the scalar product in the case where $\lambda_j^C = \lambda_j^B = \lambda_j$ $(j = 1, \ldots, N)$. All the λ_j's are different and satisfy the system of Bethe equations (VII.1.31). In this case (if involutions (VII.7.9) are fulfilled) the scalar product is reduced to the norm of the Bethe wavefunction (the eigenfunction of the trace of the monodromy matrix and of the Hamiltonian). The scalar product now depends on $2N$ variables λ_j, z_j only, because all the variables r_j can be expressed as functions of the variables λ_j using the Bethe equations:

$$r_j = \prod_{\substack{k \neq j \\ k=1}}^{N} \frac{f(\lambda_k, \lambda_j)}{f(\lambda_j, \lambda_k)}. \tag{3.26}$$

The scalar product is, as before, a linear function of each variable z_j. Formula (3.23) is also still valid, with the variables z_j modified according to the rule (3.24). It is essential that the squared norm also enters into the right hand side of (3.23), i.e., the spectral parameters λ_j $(j = 1, \ldots, N-1)$ satisfy the modified system of Bethe equations

$$\tilde{r}_j \prod_{\substack{k=1 \\ k \neq j}}^{N-1} \frac{f(\lambda_j, \lambda_k)}{f(\lambda_k, \lambda_j)} = 1, \qquad \tilde{r}_j = r_j \frac{f(\lambda_j, \lambda_N)}{f(\lambda_N, \lambda_j)} \quad (j = 1, \ldots, N-1). \tag{3.27}$$

It is worth mentioning that the squared norm of the Bethe eigenfunction is invariant under the transposition $(z_j, \lambda_j) \leftrightarrow (z_k, \lambda_k)$.

The formulæ obtained in this chapter will allow us to calculate the norm of the Bethe eigenfunction in the next chapter. It should be noted that the monodromy matrix of the XXZ model possesses quite similar properties; the corresponding theorems are proved in paper [2].

IX.4 The recursion formula for scalar products

In the previous sections, the properties of the scalar product were investigated. The main result obtained was formula (3.5) connecting the residue of the scalar product S_N (3.2) at coinciding λ_N^C, λ_N^B with the value of the "modified" scalar product S_{N-1}. In the present section, a new formula expressing S_N in terms of scalar products S_{N-1} is established. To make the notation more compact, we shall denote S_N (defined by (3.2)) as

$$S_N = S_N \left([a(\lambda), d(\lambda)] \right). \tag{4.1}$$

One should emphasize that $[a(\lambda), d(\lambda)]$ is the set of two c-number functions $a(\lambda)$ and $d(\lambda)$ (the bracket has nothing to do with commutators).

It should be kept in mind, however, that the scalar product S_N depends on the values of the functions $a(\lambda)$, $d(\lambda)$ at the points λ_j^C, λ_j^B ($j = 1, \ldots, N$) (and also on the variables λ_j^C, λ_j^B themselves). The following theorem is valid:

Theorem 4. *The scalar products S_N and S_{N-1} are related by the following recursion formula:*

$$S_N([a(\lambda), d(\lambda)])$$

$$= a(\lambda_1^C) \sum_{n=1}^{N} d(\lambda_n^B) g(\lambda_1^C, \lambda_n^B) \prod_{j \neq 1}^{N} g(\lambda_1^C, \lambda_j^C)$$

$$\times \prod_{k \neq n} g(\lambda_k^B, \lambda_n^B) S_{N-1}([a(\lambda)h(\lambda, \lambda_n^B), d(\lambda)h(\lambda_1^C, \lambda)])$$

$$+ d(\lambda_1^C) \sum_{n=1}^{N} a(\lambda_n^B) g(\lambda_n^B, \lambda_1^C) \prod_{j \neq n} g(\lambda_n^B, \lambda_j^B)$$

$$\times \prod_{k \neq 1} g(\lambda_k^C, \lambda_1^C) S_{N-1}([a(\lambda)h(\lambda, \lambda_1^C), d(\lambda)h(\lambda_n^B, \lambda)]). \qquad (4.2)$$

The set $\{\lambda^C\}_{N-1}$ entering the scalar product S_{N-1} on the right hand side is obtained from the set $\{\lambda^C\}_N$ by removing λ_1^C. Analogously, the set $\{\lambda^B\}_{N-1}$ is obtained by removing λ_n^B from the set $\{\lambda^B\}_N$. Functional arguments $a(\lambda)$, $d(\lambda)$ in the scalar product S_{N-1} are modified according to the rules written in (4.2) explicitly. Functions g and h are defined as previously,

$$g(\lambda, \mu) = \frac{ic}{\lambda - \mu}, \quad f(\lambda, \mu) = \frac{\lambda - \mu + ic}{\lambda - \mu}, \quad t(\lambda, \mu) = \frac{g(\lambda, \mu)}{h(\lambda, \mu)},$$

$$t(\lambda, \mu) = \frac{(ic)^2}{(\lambda - \mu)(\lambda - \mu + ic)}, \qquad (4.3)$$

$$h(\lambda, \mu) = \frac{f(\lambda, \mu)}{g(\lambda, \mu)} = \frac{1}{ic}(\lambda - \mu + ic).$$

It should be emphasized that the relation (4.2) can be regarded as a recursion relation for calculating S_N and that this relation together with the condition

$$S_0 = \langle 0|0 \rangle = 1 \qquad (4.4)$$

fix S_N in a unique way.

Proof: Combining representation (1.3) for the scalar product with the explicit expressions (2.13), (2.11) and (2.10) for the coefficients K_N, one arrives at the following formula for S_N:

$$S_N([a(\lambda), d(\lambda)]) = \sum \left\{ \prod_{j=1}^{N} a(\lambda_j^A) d(\lambda_j^D) \right\} K_N \qquad (4.5)$$

(summation here is as in (1.12)) where the coefficients K_N are given explicitly as a product of two determinants and some factors f, g, h

$$K_N \sim \det{}_n \left(M_{DC}^{AB} \right) \det{}_{N-n} \left(M_{DB}^{AC} \right) \qquad (4.6)$$

and the $n \times n$ matrix M_{DC}^{AB} and the $(N-n) \times (N-n)$ matrix M_{DB}^{AC} are given by

$$\begin{aligned}
\left(M_{DC}^{AB} \right)_{jk} &= t(\lambda_j^{AB}, \lambda_k^{DC}) & (j,k = 1, \ldots, n) \\
\left(M_{DB}^{AC} \right)_{jk} &= t(\lambda_j^{AC}, \lambda_k^{DB}) & (j,k = 1, \ldots, N-n)
\end{aligned} \qquad (4.7)$$

($\mathrm{card}\{\lambda^{AB}\} = \mathrm{card}\{\lambda^{DC}\} = n$, $\mathrm{card}\{\lambda^{AC}\} = \mathrm{card}\{\lambda^{DB}\} = N-n$). The product of the vacuum eigenvalues $a(\lambda)$, $d(\lambda)$ in (4.5) can be written as

$$\prod_{(AC)} a(\lambda^{AC}) \prod_{(AB)} a(\lambda^{AB}) \prod_{(DC)} d(\lambda^{DC}) \prod_{(DB)} d(\lambda^{DB})$$

($\prod\limits_{(AC)}$ stands for the product over all $\lambda \in \{\lambda^{AC}\}$, etc.) It is obvious that one can put all the vacuum values "inside" the determinants by introducing matrices $\widetilde{M}$ given by

$$\begin{aligned}
\left(\widetilde{M}_{DC}^{AB} \right)_{jk} &= t(\lambda_j^{AB}, \lambda_k^{DC}) d(\lambda_k^{DC}) a(\lambda_j^{AB}), \\
\left(\widetilde{M}_{DB}^{AC} \right)_{jk} &= t(\lambda_j^{AC}, \lambda_k^{DB}) a(\lambda_j^{AC}) d(\lambda_k^{DB})
\end{aligned} \qquad (4.8)$$

so that the explicit representation for S_N can be put in the following form:

$$
\begin{aligned}
S_N&([a(\lambda), d(\lambda)]) \\
&= \sum \left\{ \prod_{(AC)} \prod_{(DC)} f(\lambda^{AC}, \lambda^{DC}) \right\} \left\{ \prod_{(AB)} \prod_{(DB)} f(\lambda^{AB}, \lambda^{DB}) \right\} \\
&\quad \times \left\{ \prod_{p>q} g(\lambda_p^{DC}, \lambda_q^{DC}) \right\} \left\{ \prod_{p>q} g(\lambda_q^{AB}, \lambda_p^{AB}) \right\} \\
&\quad \times \left\{ \prod_{t>s} g(\lambda_t^{DB}, \lambda_s^{DB}) \right\} \left\{ \prod_{t>s} g(\lambda_s^{AC}, \lambda_t^{AC}) \right\} \\
&\quad \times \left\{ \prod_{(AB)} \prod_{(DC)} h(\lambda^{AB}, \lambda^{DC}) \right\} \left\{ \prod_{(AC)} \prod_{(DB)} h(\lambda^{AC}, \lambda^{DB}) \right\} \\
&\quad \times \det{}_n \widetilde{M}_{DC}^{AB} \det{}_{N-n} \widetilde{M}_{DB}^{AC}.
\end{aligned} \qquad (4.9)
$$

The sum here is taken over all the partitions of

$$\{\lambda^C\}_N \cup \{\lambda^B\}_N = \{\lambda^A\}_N \cup \{\lambda^D\}_N,$$

as explained in (1.4)–(1.7). A similar formula can be written for S_{N-1}.

It is obvious that, due to the arbitrariness of the functions $a(\lambda)$, $d(\lambda)$, one can represent the scalar product in the following form by extracting the dependence on the vacuum eigenvalues at the point λ_1^C:

$$
\begin{aligned}
S_N([a(\lambda), d(\lambda)]) \longrightarrow\ & a(\lambda_1^C) \sum_{n=1}^{N} d(\lambda_n^B)\sigma^{(1n)} \\
& + d(\lambda_1^C) \sum_{n=1}^{N} a(\lambda_n^B)\sigma^{(2n)}.
\end{aligned}
\tag{4.10}
$$

The quantities $\sigma^{(1n)}$ do not depend on $a(\lambda_1^C)$, $d(\lambda_1^C)$, $d(\lambda_n^B)$ and the quantities $\sigma^{(2n)}$ do not depend on $a(\lambda_1^C)$, $d(\lambda_1^C)$, $a(\lambda_n^B)$; they are uniquely defined. Explicit expressions for them are easily obtained from formula (4.9). For example, to get $\sigma^{(11)}$, one notices that $a(\lambda_1^C)$ $(\lambda_1^C \in \{\lambda^{AC}\})$ and $d(\lambda_1^B)$ $(\lambda_1^B \in \{\lambda^{DB}\})$ should enter $\det_{N-n} \widetilde{M}_{DB}^{AC}$, the coefficient being essentially a factor of the matrix element $\left(\widetilde{M}_{DB}^{AC}\right)_{11}$ so that

$$
\begin{aligned}
& a(\lambda_1^C)d(\lambda_1^B)\sigma^{(11)} \\
&= \sum{}' \left\{ \prod_{(D'C')} f(\lambda_1^C, \lambda^{D'C'}) \right\} \left\{ \prod_{(A'C')} \prod_{(D'C')} f(\lambda^{A'C'}, \lambda^{D'C'}) \right\} \\
&\quad \times \left\{ \prod_{(A'B')} \prod_{(D'B')} f(\lambda^{A'B'}, \lambda^{D'B'}) \right\} \left\{ \prod_{(A'B')} f(\lambda^{A'B'}, \lambda_1^B) \right\} \\
&\quad \times \left\{ \prod_{p>q} g(\lambda_p^{D'C'}, \lambda_q^{D'C'}) \right\} \left\{ \prod_{p>q} g(\lambda_q^{A'B'}, \lambda_p^{A'B'}) \right\} \\
&\quad \times \left\{ \prod_{(D'B')} g(\lambda^{D'B'}, \lambda_1^B) \right\} \left\{ \prod_{t>s} g(\lambda_t^{D'B'}, \lambda_s^{D'B'}) \right\} \\
&\quad \times \left\{ \prod_{(A'C')} g(\lambda_1^C, \lambda^{A'C'}) \right\} \left\{ \prod_{t>s} g(\lambda_s^{A'C'}, \lambda_t^{A'C'}) \right\} \\
&\quad \times \left\{ \prod_{(A'B')} \prod_{(D'C')} h(\lambda^{A'B'}, \lambda^{D'C'}) \right\} \\
&\quad \times \left\{ h(\lambda_1^C, \lambda_1^B) \right\} \left\{ \prod_{(D'B')} h(\lambda_1^C, \lambda^{D'B'}) \right\} \\
&\quad \times \left\{ \prod_{(A'C')} h(\lambda^{A'C'}, \lambda_1^B) \right\} \left\{ \prod_{(A'C')} \prod_{(D'B')} h(\lambda^{A'C'}, \lambda^{D'B'}) \right\}
\end{aligned}
$$

$$\times \det{}_n(\widetilde{M}_{D'C'}^{A'B'}) \det{}_{N-n-1}(\widetilde{M}_{D'B'}^{A'C'})$$
$$\times \frac{g(\lambda_1^C, \lambda_1^B)}{h(\lambda_1^C, \lambda_1^B)} \times a(\lambda_1^C)d(\lambda_1^B). \tag{4.11}$$

The sum $\sum'$ here is taken over all the partitions

$$\{\lambda^{C'}\}_{N-1} \cup \{\lambda^{B'}\}_{N-1} = \{\lambda^{A'}\}_{N-1} \cup \{\lambda^{D'}\}_{N-1}, \tag{4.12}$$

where λ_1^C, λ_1^B do not enter the sets $\{\cdots\}_{N-1}$. Comparing (4.11) with the formula analogous to (4.9) for the scalar product S_{N-1}, one obtains that

$$a(\lambda_1^C)d(\lambda_1^B)\sigma^{(11)} = a(\lambda_1^C)d(\lambda_1^B)g(\lambda_1^C, \lambda_1^B)$$
$$\times \prod_{j=2}^{N} g(\lambda_1^C, \lambda_j^C) \prod_{k=2}^{N} g(\lambda_k^B, \lambda_1^B)$$
$$\times S_{N-1}([a(\lambda)h(\lambda, \lambda_1^B), d(\lambda)h(\lambda_1^C, \lambda)]) \tag{4.13}$$

such that the corresponding contributions in (4.10) and (4.2) are the same. The equality of the contributions containing $a(\lambda_1^C)d(\lambda_n^B)$ for $n \neq 1$ as well as the contributions of the kind $d(\lambda_1^C)a(\lambda_k^B)$ is also easily established. Thus Theorem 4 is proved.

IX.5 The determinant representation of scalar products in terms of auxiliary quantum fields

In this section recursion relations (4.2) for the scalar products are solved explicitly in terms of auxiliary quantum fields acting in an auxiliary Fock space. Let us introduce two "coordinates" $q_A(\lambda)$, $q_D(\lambda)$ and two momenta $\pi_A(\lambda)$, $\pi_D(\lambda)$ satisfying the following commutation relations:

$$[\pi_D(\lambda), q_D(\mu)] = \ln h(\lambda, \mu),$$
$$[\pi_A(\lambda), q_A(\mu)] = \ln h(\mu, \lambda) \tag{5.1}$$

with all the other commutators between the fields being equal to zero:

$$[\pi_a(\lambda), \pi_b(\mu)] = [q_a(\lambda), q_b(\mu)] = 0, \quad \forall a, b$$
$$[\pi_a(\lambda), q_b(\mu)] = 0, \quad a \neq b \quad (a = A, D, \ b = A, D).$$

The vacuum in this auxiliary Fock space where the newly introduced operators act is denoted $|0\rangle$. It is annihilated by the momenta, i.e,

$$\pi_D(\lambda)|0\rangle = 0, \quad \pi_A(\lambda)|0\rangle = 0. \tag{5.2}$$

The dual vacuum $(0|$ is an eigenfunction of the coordinates:

$$(0|q_A(\lambda) = \ln a(\lambda)(0|, \quad (0|q_D(\lambda) = \ln d(\lambda)(0|,$$
$$(0|0) = 1. \tag{5.3}$$

It is important to mention that our auxiliary quantum fields can be represented in terms of canonical Bose fields $\psi_k(\lambda)$, $\psi_k^\dagger(\mu)$ in the canonical Fock space. The index of all ψ runs through two values, $k = A, D$. The commutation relations are the standard ones

$$[\psi_k(\lambda), \psi_l^\dagger(\mu)] = \delta_{kl}\delta(\lambda - \mu)$$
$$[\psi_k(\lambda), \psi_l(\mu)] = [\psi_k^\dagger(\lambda), \psi_l^\dagger(\mu)] = 0.$$

The Fock vacuum $|0)$ is also the standard one

$$\psi_k(\lambda)|0) = 0, \quad (0|\psi_k^\dagger(\mu) = 0.$$

Our auxiliary quantum fields can be expressed in terms of the canonical fields:

$$\pi_A(\lambda) = \psi_A(\lambda), \quad \pi_D(\lambda) = \psi_D(\lambda);$$

$$q_A(\mu) = \ln a(\mu) + \int\limits_{-\infty}^{+\infty} \ln h(\mu, \nu)\psi_A^\dagger(\nu)\, d\nu$$

$$q_D(\mu) = \ln d(\mu) + \int\limits_{-\infty}^{+\infty} \ln h(\nu, \mu)\psi_D^\dagger(\nu)\, d\nu.$$

It should be remembered that $a(\lambda)$ and $d(\lambda)$ are complex-valued functions and $h(\lambda, \mu) = (\lambda - \mu + ic)/ic$. A similar representation is also valid for all the auxiliary quantum fields which appear in this Part.

The quantum fields $\Phi_A(\lambda)$, $\Phi_D(\lambda)$ are defined as follows:

$$\Phi_A(\lambda) = q_A(\lambda) + \pi_D(\lambda); \quad \Phi_D(\lambda) = q_D(\lambda) + \pi_A(\lambda). \tag{5.4}$$

It is emphasized that in these definitions the coordinate is added to the momentum conjugate to the other coordinate. The remarkable property of the fields thus introduced is that they commute with one another (this is readily checked by means of (5.1)):

$$[\Phi_A(\lambda), \Phi_A(\mu)] = [\Phi_D(\lambda), \Phi_D(\mu)] = [\Phi_A(\lambda), \Phi_D(\mu)] = 0. \tag{5.5}$$

Let us now define the $N \times N$ matrix S with matrix elements given by

$$S_{jk} = t(\lambda_j^C, \lambda_k^B) \exp\{\Phi_A(\lambda_j^C) + \Phi_D(\lambda_k^B)\}$$
$$+ t(\lambda_k^B, \lambda_j^C) \exp\{\Phi_A(\lambda_k^B) + \Phi_D(\lambda_j^C)\}, \tag{5.6}$$

$$t(\lambda, \mu) \equiv \frac{g(\lambda, \mu)}{h(\lambda, \mu)} = \frac{(ic)^2}{(\lambda - \mu)(\lambda - \mu + ic)}.$$

The matrix elements commute with each other, so that the determinant $\det_N S$ of this matrix is well defined.

The vacuum mean value of this determinant taken in the auxiliary Fock space,

$$(0|\det{}_N S|0), \tag{5.7}$$

will be of primary importance. To calculate it, one first uses the commutation relations (5.1) to put $\det_N S$ into the normal order form (all the momenta π should be put to the right and all the coordinates to the left). After this, the action of the momenta (5.2) and of the coordinates (5.3) on the vacuum states $|0)$ and $(0|$, correspondingly, are easily calculated.

The main result of the investigation of the scalar product given below is that it is proportional to the mean value (5.7):

Theorem 5. *The scalar product S_N is equal to*

$$S_N = \langle 0| \prod_{j=1}^{N} C(\lambda_j^C) \prod_{k=1}^{N} B(\lambda_k^B)|0\rangle$$

$$= \left\{ \prod_{j>k=1}^{N} g(\lambda_j^C, \lambda_k^C) g(\lambda_k^B, \lambda_j^B) \right\} (0|\det{}_N S|0). \tag{5.8}$$

Proof: Let us develop the determinant entering (5.8) in the first line (associated with λ_1^C, see (5.6)):

$$\det_N S = \sum_{n=1}^{N} (-1)^{n-1} S_{1n} \det_{N-1} \widetilde{S}_{(1n)} \tag{5.9}$$

where the $(N-1) \times (N-1)$ matrix $\widetilde{S}_{(1n)}$ is obtained from the $N \times N$ matrix S by removing the first column and the n-th row. Now we shall develop the determinant in (5.8) with respect to its first line. We shall use the explicit expression (5.6) of S_{1n}, the commutation relations (5.1) and the formulæ

$$\exp x \exp y = \exp y \exp x \exp\{[x, y]\},$$
$$(\exp x)y = (y + [x, y]) \exp x \tag{5.10}$$

(valid if the commutator $[x, y]$ is a c-number). We finally obtain for

$(0|\det_N S|0)$ in (5.8):

$$(0|\det_N S|0) = \sum_{n=1}^{N}(-1)^{n-1}a(\lambda_1^C)d(\lambda_n^B)g(\lambda_1^C,\lambda_n^B)$$
$$\times (0|\exp\{\pi_D(\lambda_1^C) + \pi_A(\lambda_n^B)\}\det_{N-1}\widetilde{S}_{(1n)}|0)$$
$$+ \sum_{n=1}^{N}(-1)^{n-1}d(\lambda_1^C)a(\lambda_n^B)g(\lambda_n^B,\lambda_1^C)$$
$$\times (0|\exp\{\pi_D(\lambda_n^B) + \pi_A(\lambda_1^C)\}\det_{N-1}\widetilde{S}_{(1n)}|0). \quad (5.11)$$

Here we have also used (5.3).

Consider now the state $(\widetilde{0}|$ defined by

$$(\widetilde{0}| = (0|\exp\{\pi_D(\lambda^D) + \pi_A(\lambda^A)\}. \quad (5.12)$$

Due to (5.2), $(\widetilde{0}|0) = 1$. Using (5.1), (5.3) and (5.10), one computes that

$$(\widetilde{0}|q_A(\lambda) = \ln \widetilde{a}(\lambda)(\widetilde{0}|,$$
$$(\widetilde{0}|q_D(\lambda) = \ln \widetilde{d}(\lambda)(\widetilde{0}| \quad (5.13)$$

with

$$\widetilde{a}(\lambda) = a(\lambda)h(\lambda,\lambda_A), \quad \widetilde{d}(\lambda) = d(\lambda)h(\lambda_D,\lambda). \quad (5.14)$$

Thus, state $(\widetilde{0}|$ can be viewed as a new dual vacuum corresponding to the modified vacuum eigenvalues of the coordinates. Taking this into account, one concludes that the matrix elements of $\det_{N-1}\widetilde{S}_{(1n)}$ in (5.11) can be calculated with respect to the old vacuum but with vacuum eigenvalues modified in the way needed by the recursion formula (4.2). Now it is quite easy to see that putting (5.11) into (5.8), one exactly reproduces the recursion relation (4.2). As $(0|0) = 1$, the functions S_N defined by (5.8) and (4.2) are the same for any N. The theorem is thus proved.

IX.6 Another proof of the determinant representation for scalar products

Here an alternative proof of the determinant representation (5.8) for scalar products is given. It is based on the representation of the determinant of a sum of two matrices in terms of minors of the individual matrices. The explicit formula is given in the Appendix to this chapter.

We shall start from the representation (4.9) for scalar products, which follows from (1.3), (2.13) and (2.10). The following expression for the scalar product S_N,

$$
S_N = \left\{ \prod_{j>k} g(\lambda_j^C, \lambda_k^C) g(\lambda_k^B, \lambda_j^B) \right\} \sum (-1)^{[P_C]+[P_B]}
$$

$$
\times \left[\prod_{j,k} h(\lambda_j^{AC}, \lambda_k^{DB}) \prod_{j,m} h(\lambda_j^{AC}, \lambda_m^{DC}) \right.
$$

$$
\left. \times \prod_{l,k} h(\lambda_l^{AB}, \lambda_k^{DB}) \prod_{l,m} h(\lambda_l^{AB}, \lambda_m^{DC}) \right]
$$

$$
\times \det_{N-n} \left(\widetilde{M}_{DC}^{AB} \right) \det_n \left(\widetilde{M}_{DB}^{AC} \right), \tag{6.1}
$$

is equivalent to representation (4.9). We have used here that

$$
f(\lambda, \mu) = g(\lambda, \mu) h(\lambda, \mu) \text{ and } g(\lambda, \mu) = -g(\mu, \lambda).
$$

The sum here is taken as in (1.3), i.e., over partitions

$$
\{\lambda^C\}_N \cup \{\lambda^B\}_N = \{\lambda^A\}_N \cup \{\lambda^D\}_N,
$$
$$
\{\lambda^A\}_N = \{\lambda^{AC}\}_n \cup \{\lambda^{AB}\}_{N-n}, \tag{6.2}
$$
$$
\{\lambda^D\}_N = \{\lambda^{DC}\}_{N-n} \cup \{\lambda^{DB}\}_n.
$$

To fix the order of rows and columns in the determinant, we assume that the enumeration of λ's in the sets is made in correspondence with the original enumeration in $\{\lambda^C\}$, $\{\lambda^B\}$, e.g.,

$$
\{\lambda^{AC}\}_n = \{\lambda_1^{AC} \equiv \lambda_{j_1}^C, \lambda_2^{AC} \equiv \lambda_{j_2}^C, \dots, \lambda_n^{AC} \equiv \lambda_{j_n}^C \};
$$

it is supposed that $j_1 < j_2 < \cdots < j_n$. The quantity $(-1)^{[P_C]}$ is the parity of the permutation P_C,

$$
P_C \colon \{\lambda_1^{AC} \equiv \lambda_{j_1}^C, \dots, \lambda_n^{AC} \equiv \lambda_{j_n}^C, \lambda_1^{DC} \equiv \lambda_{j_{n+1}}^C, \dots, \lambda_{N-n}^{DC} \equiv \lambda_{j_N}^C \}
$$
$$
\to \{\lambda_1^C, \dots, \lambda_N^C\}, \tag{6.3}
$$

and $(-1)^{[P_B]}$ is the parity of the permutation P_B,

$$
P_B \colon \{\lambda_1^{DB} \equiv \lambda_{k_1}^B, \dots, \lambda_n^{DB} \equiv \lambda_{k_n}^B, \lambda_1^{AB} \equiv \lambda_{k_{n+1}}^B, \dots, \lambda_{N-n}^{AB} \equiv \lambda_{k_N}^B \}
$$
$$
\to \{\lambda_1^B, \dots, \lambda_N^B\}. \tag{6.4}
$$

The matrices $\widetilde{M}_{DB}^{AC}$ (dimension $n \times n$) and $\widetilde{M}_{DC}^{AB}$ (dimension $(N-n) \times (N-n)$) are defined as (see (2.11))

$$
\left(\widetilde{M}_{DB}^{AC} \right)_{jk} = t(\lambda_j^{AC}, \lambda_k^{DB}) a(\lambda_j^{AC}) d(\lambda_k^{DB})
$$
$$
(j, k = 1, \dots, n) \tag{6.5}
$$

$$\left(\widetilde{M}_{DC}^{AB}\right)_{ml} = t(\lambda_l^{AB}, \lambda_m^{DC})a(\lambda_l^{AB})d(\lambda_m^{DC}) \tag{6.6}$$

$$(l, m = 1, \ldots, N - n).$$

Let us introduce quantum fields $\mathring{\Phi}_A(\lambda)$, $\mathring{\Phi}_D(\lambda)$ which are represented as the sum of "coordinates" Q and "momenta" P:

$$\mathring{\Phi}_A(\lambda) = \mathring{Q}_A(\lambda) + \mathring{P}_D(\lambda), \quad \mathring{\Phi}_D(\lambda) = \mathring{Q}_D(\lambda) + \mathring{P}_A(\lambda). \tag{6.7}$$

The coordinates and momenta satisfy the following commutation relations:

$$[\mathring{P}_D(\lambda), \mathring{Q}_D(\mu)] = \ln h(\lambda, \mu);$$
$$[\mathring{P}_A(\lambda), \mathring{Q}_A(\mu)] = \ln h(\mu, \lambda) \tag{6.8}$$

with all other commutators of these operators being zero. The fields (6.7) also commute with one another:

$$[\mathring{\Phi}_a(\lambda), \mathring{\Phi}_b(\mu)] = 0, \quad a = A, D; \; b = A, D. \tag{6.9}$$

The vacuum $|0)$ and dual vacuum $(0|$ in the auxiliary Fock space where the operators $\mathring{P}$, $\mathring{Q}$ act are annihilated by $\mathring{P}$ and $\mathring{Q}$, respectively:

$$\mathring{P}_a|0) = 0, \quad (0|\mathring{Q}_a = 0. \tag{6.10}$$

Of course, the fields we have introduced here are closely related to those introduced earlier, (5.1)–(5.5), which satisfy the same commutation relations: we have simply extracted the vacuum mean values

$$\mathring{P}(\lambda) = \pi(\lambda), \quad \mathring{Q}(\lambda) = q(\lambda) - (0|q(\lambda)|0).$$

Define now the $N \times N$ matrix $\widetilde{S}$

$$\widetilde{S}_{jk} = t(\lambda_j^C, \lambda_k^B)a(\lambda_j^C)d(\lambda_k^B) \exp\{\mathring{\Phi}_A(\lambda_j^C) + \mathring{\Phi}_D(\lambda_k^B)\}$$
$$+ t(\lambda_k^B, \lambda_j^C)a(\lambda_k^B)d(\lambda_j^C) \exp\{\mathring{\Phi}_A(\lambda_k^B) + \mathring{\Phi}_D(\lambda_j^C)\}. \tag{6.11}$$

The evident reformulation of theorem (5.8) is:

$$S_N = \langle 0| \prod_{j=1}^{N} C(\lambda_j^C) \prod_{k=1}^{N} B(\lambda_k^B)|0\rangle$$

$$= \left\{\prod_{j>k}^{N} g(\lambda_j^C, \lambda_k^C)g(\lambda_k^B, \lambda_j^B)\right\}(0|\det {}_N\widetilde{S}|0). \tag{6.12}$$

We will give, however, a different proof based on representation (6.11).

Proof: Let us make use of the well known Laplace theorem for the determinant of the sum of two matrices. One readily rewrites formula (6.12) for S_N (with the matrix $\widetilde{S}$, (6.11)) in a form similar to (6.1):

$$S_N = \left\{ \prod_{j>k} g(\lambda_j^C, \lambda_k^C) g(\lambda_k^B, \lambda_j^B) \right\} \sum (-1)^{[P_C]+[P_B]}$$

$$\times (0| \exp\{ \sum_{j=1}^{n} \mathring{\Phi}_A(\lambda_j^{AC}) + \sum_{k=1}^{n} \mathring{\Phi}_D(\lambda_k^{DB})$$

$$+ \sum_{l=1}^{N-n} \mathring{\Phi}_A(\lambda_l^{AB}) + \sum_{m=1}^{N-n} \mathring{\Phi}_D(\lambda_m^{DC}) \} |0)$$

$$\times \det_{N-n} \left(M_{DC}^{AB} \right) \det_{n} \left(M_{DB}^{AC} \right) \qquad (6.13)$$

where the sum is taken as in (6.1). The permutations P_C and P_B are also the same. (It can be shown that in our enumeration of elements of the sets, we have

$$(-1)^{[P_C]+[P_B]} = (-1)^{j_1+\cdots+j_n+k_1+\cdots+k_n},$$

(see (6.3), (6.4)). However, this is unimportant at present.)

The only difference between the exact formula (6.1) and formula (6.13) is that the vacuum expectation value $(0|\cdots|0)$ stands in (6.13) instead of the square bracket in (6.1). This vacuum expectation value is, however, easy to calculate. One represents the fields Φ as sums of "momenta" and "coordinates" (6.7) and then uses the commutation relations (6.8) (see also (5.10)) to move all the momenta $\mathring{P}$ to the right and all the coordinates $\mathring{Q}$ to the left. As the action of $\mathring{P}$ on the vacuum $|0)$ and $\mathring{Q}$ on the dual vacuum $(0|$ is known (6.10), one readily obtains that the vacuum expectation value $(0|\cdots|0)$ entering (6.13) is indeed equal to the expression in the square bracket on the right hand side of (6.1). The theorem (6.12), (5.8) is again proved.

Formula (6.12) can be further simplified. Let us rewrite the expression for the vacuum mean value of $\det_N \widetilde{S}$ as

$$(0|\det_N \widetilde{S}|0) = \left\{ \prod_{j=1}^{N} a(\lambda_j^C) d(\lambda_j^B) \right\}$$

$$\times (0| \prod_{j=1}^{N} \exp\{ \mathring{\Phi}_A(\lambda_j^C) + \mathring{\Phi}_D(\lambda_j^B) \} \det_N \overline{\overline{s}} |0) \quad (6.14)$$

where the $N \times N$ matrix $\overline{\overline{s}}$ is

$$\overline{\overline{s}}_{jk} = t(\lambda_j^C, \lambda_k^B) + r(\lambda_k^B) r^{-1}(\lambda_j^C) t(\lambda_k^B, \lambda_j^C) \exp\{ \varphi(\lambda_k^B) - \varphi(\lambda_j^C) \}.$$
$$(6.15)$$

The matrix $\bar{\bar{s}}$ depends only on the quantum field $\varphi(\lambda)$,

$$\varphi(\lambda) = \overset{\circ}{\Phi}_A(\lambda) - \overset{\circ}{\Phi}_D(\lambda) = q(\lambda) + p(\lambda),$$
$$q(\lambda) = \overset{\circ}{Q}_A(\lambda) - \overset{\circ}{Q}_D(\lambda); \quad p(\lambda) = -\overset{\circ}{P}_A(\lambda) + \overset{\circ}{P}_D(\lambda), \tag{6.16}$$

where the coordinate q and momentum p satisfy the commutation relations

$$[q(\lambda),\, q(\mu)] = [p(\lambda),\, p(\mu)] = 0,$$
$$[p(\lambda),\, q(\mu)] = H(\lambda, \mu) \tag{6.17}$$

with the function H given by

$$H(\lambda, \mu) = -\ln(h(\lambda, \mu)h(\mu, \lambda)) = \ln \frac{c^2}{(\lambda - \mu)^2 + c^2}. \tag{6.18}$$

Using the commutation relations (6.8) (and (5.10)) one obtains

$$(0| \prod_{j=1}^{N} \exp\{\overset{\circ}{\Phi}_A(\lambda_j^C) + \overset{\circ}{\Phi}_D(\lambda_j^B)\} = (\tilde{0}| \prod_{j=1}^{N} \prod_{k=1}^{N} h(\lambda_j^C, \lambda_k^B) \tag{6.19}$$

where the state $(\tilde{0}|$,

$$(\tilde{0}| \equiv (0| \prod_{j=1}^{N} \exp\{\overset{\circ}{P}_D(\lambda_j^C) + \overset{\circ}{P}_A(\lambda_j^B)\}, \tag{6.20}$$

can be considered as a new dual vacuum for the field $\varphi(\lambda)$. Indeed, it is easily verified that $(\tilde{0}|$ is the dual eigenvector of the coordinate $q(\lambda)$ with a modified (nonzero) vacuum value:

$$(\tilde{0}|q(\lambda) = \alpha(\lambda)(\tilde{0}|,$$
$$\alpha(\lambda) = \sum_{n=1}^{N} \ln \frac{h(\lambda, \lambda_n^B)}{h(\lambda_n^C, \lambda)}, \quad (\tilde{0}|0) = 1. \tag{6.21}$$

Thus it is convenient to introduce the field $\overset{\circ}{\varphi}(\lambda)$ with zero vacuum mean value:

$$\overset{\circ}{\varphi}(\lambda) = \varphi(\lambda) - (\tilde{0}|\varphi(\lambda)|0) = \varphi(\lambda) - \alpha(\lambda) \tag{6.22}$$

which is represented as

$$\overset{\circ}{\varphi}(\lambda) = \overset{\circ}{p}(\lambda) + \overset{\circ}{q}(\lambda),$$
$$\overset{\circ}{p}(\lambda) = p(\lambda); \quad \overset{\circ}{q}(\lambda) = q(\lambda) - (\tilde{0}|q(\lambda)|0) = q(\lambda) - \alpha(\lambda). \tag{6.23}$$

The new coordinate and momentum satisfy the same commutation relations (6.17)

$$[\overset{\circ}{q}(\lambda),\, \overset{\circ}{q}(\mu)] = [\overset{\circ}{p}(\lambda),\, \overset{\circ}{p}(\mu)] = 0,$$
$$[\overset{\circ}{p}(\lambda),\, \overset{\circ}{q}(\mu)] = H(\lambda, \mu) = \ln \frac{c^2}{(\lambda - \mu)^2 + c^2} \tag{6.24}$$

and also

$$\mathring{p}(\lambda)|0\rangle = 0, \quad (\widetilde{0}|\mathring{q}(\lambda) = 0. \tag{6.25}$$

As a result, the representation (6.12) can be put into the following form:

$$S_N = \langle 0| \prod_{j=1}^{N} C(\lambda_j^C) \prod_{k=1}^{N} B(\lambda_k^B)|0\rangle$$

$$= \left\{ \prod_{j>k} g(\lambda_j^C, \lambda_k^C) g(\lambda_k^B, \lambda_j^B) \right\}$$

$$\times \left\{ \prod_{j=1}^{N} a(\lambda_j^C) d(\lambda_j^B) \right\} \left\{ \prod_{j=1}^{N} \prod_{k=1}^{N} h(\lambda_j^C, \lambda_k^B) \right\}$$

$$\times (\widetilde{0}|\det {}_N \overline{\overline{s}}|0) \tag{6.26}$$

where the expression for the matrix $\overline{\overline{s}}$ in terms of the field $\mathring{\varphi}$ is obtained from (6.15), (6.22):

$$\overline{\overline{s}}_{jk} = t(\lambda_j^C, \lambda_k^B) + r(\lambda_k^B) r^{-1}(\lambda_j^C) t(\lambda_k^B, \lambda_j^C)$$

$$\times \exp\{\mathring{\varphi}(\lambda_k^B) - \mathring{\varphi}(\lambda_j^C)\} \prod_{m=1}^{N} \frac{h(\lambda_k^B, \lambda_m^B) h(\lambda_m^C, \lambda_j^C)}{h(\lambda_m^C, \lambda_k^B) h(\lambda_j^C, \lambda_m^B)}. \tag{6.27}$$

This expression for the scalar product contains only one auxiliary quantum field.

IX.7 Remarks about norms

In this section, formula (6.26) is applied to calculate "norms" of the states. Let us put $\{\lambda^C\}_N = \{\lambda^B\}_N = \{\lambda\}_N$, i.e., consider the diagonal matrix element. This matrix element is still not, generally speaking, the norm of a Bethe eigenfunction since the parameters λ_j are not yet taken to satisfy the Bethe equations. We suppose, however, that all the λ_j's are different. Taking the corresponding case in (6.26), (6.27) one obtains

$$\langle 0| \prod_{j=1}^{N} C(\lambda_j) \prod_{k=1}^{N} B(\lambda_k)|0\rangle = \left\{ \prod_{j=1}^{N} a(\lambda_j) d(\lambda_j) \right\}$$

$$\times \prod_{j \neq k} f(\lambda_j, \lambda_k) (\widetilde{0}|\det {}_N \overline{\overline{s}}|0) \tag{7.1}$$

with

$$\overline{\overline{s}}_{jk} = t_{jk} + \exp\{\mathring{\varphi}(\lambda_k) - \mathring{\varphi}(\lambda_j)\} t_{kj} r(\lambda_k) r^{-1}(\lambda_j) \prod_{n=1}^{N} \frac{h_{kn} h_{nj}}{h_{nk} h_{jn}}. \tag{7.2}$$

For $j = k$ one should understand (7.2) in the sense of l'Hôpital's rule:

$$\bar{\bar{s}}_{jj} = ic\frac{\partial}{\partial\lambda}\left[\ln r(\lambda) + \sum_{n=1}^{N}\frac{h(\lambda, \lambda_n)}{h(\lambda_n, \lambda)}\right]\Bigg|_{\lambda=\lambda_j}$$
$$- 2 + ic\frac{\partial\mathring{\varphi}(\lambda)}{\partial\lambda}\Bigg|_{\lambda=\lambda_j}. \tag{7.3}$$

Now we discuss briefly the norms of the Bethe eigenfunctions $|\varphi_N(\{\lambda\})\rangle$ of the transfer matrix $\tau(\mu)$, i.e., the scalar products of states $|\varphi_N\rangle$:

$$|\varphi_N(\{\lambda\})\rangle = \prod_{j=1}^{N} B(\lambda_j)|0\rangle,$$
$$\langle\varphi_N(\{\lambda\})| = \langle 0|\prod_{j=1}^{N} C(\lambda_j) \tag{7.4}$$

where the spectral parameters λ_j are taken to be the solutions of the system of Bethe equations

$$r(\lambda_k)\prod_{k=1}^{N}\frac{h_{kn}}{h_{nk}} = (-1)^{N-1} \quad (k = 1, \ldots, N). \tag{7.5}$$

(The eigenfunctions $|\varphi_N(\{\lambda\})\rangle$ should not be confused with the auxiliary quantum fields.) The system of Bethe equations permits one to simplify expression (7.2):

$$\langle\varphi_N(\{\lambda\})|\varphi_N(\{\lambda\})\rangle = \left(\prod_{j=1}^{N} a(\lambda_j)d(\lambda_j)\right)\left(\prod_{j\neq k} f_{jk}\right)(\tilde{0}|\det n|0) \tag{7.6}$$

where the matrix n is given by

$$n_{jk} = ic\delta_{jk}\frac{\partial}{\partial\lambda_j}\left[\ln r(\lambda_j) + \sum_{n=1}^{N}\ln\frac{h_{jn}}{h_{nj}}\right]$$
$$+ t_{jk} + t_{kj}\exp\{\mathring{\varphi}(\lambda_k) - \mathring{\varphi}(\lambda_j)\}. \tag{7.7}$$

The last two terms on the right hand side should be understood in the sense of l'Hôpitale's rule as

$$\lim_{\lambda\to\lambda_j}\left[t(\lambda_j, \lambda) + t(\lambda, \lambda_j)\exp\{\mathring{\varphi}(\lambda) - \mathring{\varphi}(\lambda_j)\}\right]$$
$$= -2 + ic\frac{\partial\mathring{\varphi}(\lambda)}{\partial\lambda}\Bigg|_{\lambda=\lambda_j}. \tag{7.8}$$

The auxiliary quantum field $\mathring{\varphi}(\lambda)$ is still present in (7.6). A detailed analysis of the vacuum mean value of the determinant in (7.6) is given in the next chapter. This shows that one can put the field $\mathring{\varphi}(\lambda)$ equal to zero, so that one can replace $(\widetilde{0}|\det_N n|0)$ by $\det_N n_0$. The matrix n_0 is obtained from n at $\overset{\circ}{\varphi}(\lambda) \equiv 0$. After obvious transformations, one then obtains

$$\langle \varphi_N(\{\lambda\})|\varphi_N(\{\lambda\})\rangle = c^N \prod_{j=1}^{N} a(\lambda_j)d(\lambda_j) \prod_{j<k} f(\lambda_j, \lambda_k)f(\lambda_k, \lambda_j) \det_N(\varphi')$$

(7.9)

where the $N \times N$ matrix φ' is given as

$$\varphi'_{jk} = \delta_{jk}\left(i\frac{\partial}{\partial\lambda_j}\ln r(\lambda_j) + \sum_{l=1}^{N} K(\lambda_j, \lambda_l)\right) - K(\lambda_j, \lambda_k)$$

(7.10)

with

$$K(\lambda, \mu) = \frac{2c}{(\lambda - \mu)^2 + c^2}$$

(7.11)

which is exactly the generalized Gaudin hypothesis in the form (X.1.10). Let us emphasize once more that the proof of the Gaudin hypothesis is given in the next chapter.

Remark. The problem of evaluating

$$(\widetilde{0}|\det_N \overline{\overline{s}}|0)$$

was explained earlier as a problem of normal ordering. One can use standard methods of quantum field theory to evaluate this mean value. Namely, we replace the quantum field $\overset{\circ}{\varphi}(\lambda)$ in the expression for $\det_N \overline{\overline{s}}$ by a complex-valued function $\overset{\circ}{\varphi}(\lambda)$. Then the mean value in (7.1) can be represented as a functional integral

$$\int \prod d\overset{\circ}{\overline{\varphi}}(\lambda) \det_N \overline{\overline{s}}[\overset{\circ}{\overline{\varphi}}(\lambda)] \exp\left\{-\frac{1}{2}\int d\lambda d\mu\, \overset{\circ}{\overline{\varphi}}(\lambda)\left(H^{-1}\right)(\lambda, \mu)\overset{\circ}{\overline{\varphi}}(\mu)\right\}$$

(7.12)

where $(H^{-1})(\lambda, \mu)$ is the kernel of the linear operator inverse to the operator $\widehat{H}$ with kernel $H(\lambda, \mu)$ (6.18).

Conclusion

In this chapter, we have studied in detail the properties of the scalar product

$$\langle 0|\prod_{j=1}^{N} C(\lambda_j^C) \prod_{k=1}^{N} B(\lambda_k^B)|0\rangle.$$

In the next chapter, this will be used to prove the Gaudin hypothesis about the norms of the Bethe wave functions. In this chapter we have also obtained a determinant representation for the scalar product (using auxiliary quantum Bose fields). This approach will be extended to correlation functions in Chapters XI and XII.

Scalar products (generated by the algebraic Bethe Ansatz) were first studied in detail in paper [2]. The determinant representation for the partition function Z_N and the highest coefficient K_N were first obtained in paper [1]. Auxiliary Bose fields were first introduced in paper [3], where the determinant representation for scalar products was obtained.

Appendix IX.1: Determinant of the sum of two matrices

In this Appendix, we consider the Laplace formula for the determinant of the sum of two matrices.

Consider two matrices A and B. Their matrix elements are complex numbers and the dimension of the matrices is $N \times N$. The object of interest is the determinant of their sum

$$\det(A + B). \tag{A.1.1}$$

We would like to know if it is possible to express this as a sum of a few terms, each term being the product of some minor of the matrix A and another minor of the matrix B, thus

$$\det(A + B) = \sum \text{minor}(A) \times \text{minor}(B). \tag{A.1.2}$$

The answer is yes. To describe the generic term on the right hand side of (A.1.2), we need to consider partitions. Consider rows $j = 1, \ldots, N$ and columns $k = 1, \ldots, N$ (the corresponding matrix element being denoted by $(A + B)_{jk}$). Consider a partition P_L of the rows j into two subsets: rows of type A and rows of type B. Consider a similar partition P_C of the columns k into columns of type A and columns of type B. Now each row (and column) is of type A or B. There exists one constraint between partitions P_L and P_C: the number of rows of type A is equal to the number of columns of type A (the same is true for B). Otherwise the partitions P_L and P_C are independent. (In (A.1.2) we shall sum with respect to these partitions.)

Now consider the following deformation of the matrix $A + B$ generated by these two partitions. Consider a matrix element $(A + B)_{jk}$. If row j and column k are of type A then replace $(A + B)_{jk}$ by A_{jk}. If j and k are both of type B then replace $(A + B)_{jk}$ by B_{jk}. If j is of type A and k is of type B replace $(A + B)_{jk}$ by 0, and if j is of type B and k is of type A also replace $(A + B)_{jk}$ by 0. Let us denote the matrix obtained

221

after such a deformation from $A + B$ by

$$(A + B)_{P_L P_C}.$$

(A.1.3)

The determinant of this matrix, $\det(A + B)_{P_L P_C}$, is important.

Theorem.

$$\det(A + B) = \sum_{P_L P_C} \det(A + B)_{P_L P_C}.$$

(A.1.4)

Here, on the right hand side, the summation is with respect to all partitions P_L and P_C, which are independent, except for the condition that the number of A rows should be equal to the number of A columns. We shall denote this number by n_A (n_B will be the number of B rows and B columns, and $n_A + n_B = N$).

Proof: We shall prove the theorem by induction with respect to the dimension of the matrix, N. The basis of induction can be explicitly checked for 2×2 matrices:

$$\det(A + B) = \det A + A_{11}B_{22} + B_{11}A_{22}$$
$$- B_{21}A_{12} - A_{21}B_{12} + \det B.$$

(A.1.5)

Now comes the induction step. Suppose the formula (A.1.4) is valid for matrices of dimension $N = 1, 2, 3, \ldots, M$. Let us prove that it is also valid for $N = M + 1$. In order to do this, develop $\det(A + B)$ (for $N = M + 1$) with respect to the first row. This reduces the problem to that of matrices of dimension M. The coefficient of A_{11} will be $\det(\check{A} + \check{B})$. Here the matrix $\check{A} + \check{B}$ can be obtained from $A + B$ by removing the first row and the first column. (The dimension of $\check{A} + \check{B}$ is $M \times M$, so (A.1.4) is valid.) Now let us evaluate the coefficient of A_{11} in the sum

$$\sum_{P_L, P_C} \det(A + B)_{P_L P_C}$$

(A.1.6)

for $N = M + 1$. The coefficient of A_{11} is given by a similar sum (similar to (A.1.6)) but for $N = M + 1$. In this case, due to the inductive assumption, we know that this is equal to $\det(\check{A} + \check{B})$. So we have proved that for $N = M + 1$ the coefficient of A_{11} is the same both for the left hand side and right hand side of (A.1.4). In a similar way one can prove that the coefficients of A_{1k} ($k = 1, \ldots, N$) coincide. The same is true for B_{1k}. Each time we have used the inductive assumption. It is also clear that if $A_{1k} = B_{1k} = 0$ for $k = 1, \ldots, N$ both sides of (A.1.4) are equal to zero. This finishes the induction step (equation (A.1.4) is valid for $N = M + 1$), and the proof of the theorem.

Now let us consider the consequences of the theorem. Consider an individual term on the right hand side of (A.1.4),

$$\det(A + B)_{P_L P_C}. \tag{A.1.7}$$

We can exchange the positions of the rows in such a way that first all the A rows will be moved to the beginning and then all the B rows will be moved to the end. During this permutation, we shall not change the internal ordering of the A rows or of the B rows. The parity of this permutation we shall denote by $(-1)^{[P_L]}$. In a similar way we can change the order of the columns. We shall define the permutation which puts all the A columns at the beginning (without changing their internal ordering) and all B columns at the end (keeping their internal order). The parity of this permutation we shall denote by $(-1)^{[P_C]}$. After this transposition of lines and columns, the determinant in (A.7) will acquire a factor

$$(-1)^{[P_L]+[P_C]} \tag{A.1.8}$$

and the matrix $(A + B)_{P_L P_C}$ will be block diagonal:

$$\begin{pmatrix} A_{P_L P_C} & 0 \\ 0 & B_{P_L P_C} \end{pmatrix}. \tag{A.1.9}$$

The matrix $A_{P_L P_C}$ has dimension $n_A \times n_A$ and can be obtained from A by removing all B lines and B columns (according to the partitions P_L and P_C). The matrix $B_{P_L P_C}$ has dimension $n_B \times n_B$ and can be obtained by removing all the A lines and A columns (according to the partitions P_L and P_C). The determinant of the matrix (A.1.9) is equal to the product of $\det A_{P_L P_C}$ and $\det A_{P_L P_C}$.

Finally, $\det(A + B)$ can be rewritten in the form

$$\det(A + B) = \sum_{P_L, P_C} (-1)^{[P_L]+[P_C]} \det A_{P_L P_C} \det B_{P_L P_C}. \tag{A.1.10}$$

Here P_L is the partition of the rows into subsets A and B and P_C is the partition of the columns into subsets A and B. Formula (A.10) follows from (A.4) and the above considerations.

X

Norms of Bethe Wave Functions

Introduction

The norms of Bethe wave functions, i.e., the eigenfunctions of the transfer matrix and the Hamiltonian, are calculated in this chapter. The norms are considered as functionals of the function $\ln r(\lambda)$ which is an arbitrary function (see section VII.6, Theorem 3, $r(\lambda) = a(\lambda)/d(\lambda)$). Norms are functionals of a very special kind: they depend only on the values of the function $\partial \ln r(\lambda)/\partial\lambda$ at a finite number of points.

The question of calculating the norms of Bethe wave functions was first considered in the case of the NS model. In 1972, M. Gaudin [1] (treating the model by means of the coordinate Bethe Ansatz) formulated the hypothesis that the squared norm of the wave function is equal to some Jacobian. The squared norm of the Bethe wave function in the Bose gas case can be written in terms of the wave function χ (see Chapter I, formulæ (I.1.25), (I.2.19)):

$$\langle \psi_N | \psi_N \rangle = \int\limits_0^L d^N z \, |\chi_N(z_1, \ldots, z_N)|^2, \qquad (0.1)$$

and Gaudin's hypothesis was

$$\langle \psi_N | \psi_N \rangle = \det_N \left(\frac{\partial^2 S}{\partial\lambda_j \partial\lambda_k} \right) \qquad (0.2)$$

or, in our notation,

$$\langle 0| \prod_{j=1}^N C(\lambda_j) \, \prod_{k=1}^N B(\lambda_k) |0\rangle$$

$$= c^N \left\{ \prod_{j\neq k}^N f(\lambda_j, \lambda_k) \right\} \det_N \left(\frac{\partial^2 S}{\partial\lambda_j \partial\lambda_k} \right) \left\{ \prod_{j=1}^N a(\lambda_j) d(\lambda_j) \right\} \qquad (0.3)$$

where $\partial^2 S / \partial \lambda_j \partial \lambda_k$ is an $N \times N$-dimensional matrix of the second derivatives of the Yang-Yang action (see section I.2, formula (I.2.16)) evaluated at the solutions of the Bethe equations (I.2.13)

$$\frac{\partial^2 S}{\partial \lambda_j \partial \lambda_k} = \delta_{jk} \left[L + \sum_{l=1}^{N} K(\lambda_k, \lambda_l) \right] - K(\lambda_j, \lambda_k)$$

$$K(\lambda_j, \lambda_k) = \frac{2c}{c^2 + (\lambda_j - \lambda_k)^2}.$$

(0.4)

Its explicit form is given by formula (I.2.16). Let us introduce the matrix φ' as follows:

$$\varphi'_{jk} = \frac{\partial^2 S}{\partial \lambda_j \partial \lambda_k}.$$

(0.5)

This matrix can also be written in terms of the variables φ_j introduced in Chapter VII, formula (VII.1.34):

$$\varphi'_{jk} = \frac{\partial \varphi_j}{\partial \lambda_k}.$$

In terms of these variables, the Bethe equations are written as

$$\varphi_j = 0 \pmod{2\pi}.$$

Thus, one can say that the matrix φ' is obtained by linearization of the system of Bethe equations in the vicinity of its solutions.

It should be remembered that for positive coupling constant $c > 0$, the matrix φ' is positive definite and $\det_N(\varphi') > 0$ (see Chapter I, section 2, formula (I.2.18)). The system of Bethe equations is non-singular, and all its solutions are real for the NS model with $\lambda_j \neq \lambda_k$ for $j \neq k$. The scalar product of eigenfunctions corresponding to different sets of solutions λ_j of the Bethe system is equal to zero due to their orthogonality property (see section VII.2, formula (VII.2.11)). It should also be mentioned that the formula for the "squared norm" of the Bethe wave function has already been obtained for the generalized model (IX.7.6). This formula is readily rewritten for the NS model. However, the auxiliary quantum field φ enters the right hand side of (IX.7.6). In this chapter, it is shown that the quantum field φ can be put equal to zero, and formulæ (IX.7.9), (IX.7.10) proved.

In the proof of Gaudin's hypothesis, the results of the first three sections of the previous chapter are used. The proof of Gaudin's hypothesis was given originally in paper [3].

The contents of this chapter are as follows. The generalized Gaudin hypothesis is formulated in section 1 (the squared norm of the eigenfunction is proportional to some Jacobian). In section 2, the properties of this Jacobian are studied and the properties which the norms should satisfy for the hypothesis to be valid are obtained. In section 3, Gaudin's hypothesis is proved. The thermodynamic limits of the norms of the eigenfunctions are calculated in section 4. In section 5, a special case of the scalar product is considered: one of the states is an eigenstate of the Hamiltonian, the other is not. The determinant formula (without the auxiliary quantum fields) is then presented.

X.1 Generalized Gaudin hypothesis

To prove Gaudin's hypothesis, let us formulate the hypothesis for integrable models with arbitrary functions $a(\lambda)$ and $d(\lambda)$ (see section VII.6). The Bethe eigenfunctions of the transfer matrix in this model are of the form (section VII.1)

$$|\Psi_N(\{\lambda\})\rangle = |\Psi_N(\lambda_1,\ldots,\lambda_N)\rangle = \prod_{j=1}^{N} \mathbb{B}(\lambda_j)|0\rangle \qquad (1.1)$$

$$\langle\Psi_N(\{\lambda\})| = \langle 0| \prod_{j=1}^{N} \mathbb{C}(\lambda_j). \qquad (1.2)$$

The "renormalized" (IX.3.16) creation and annihilation operators are used here:

$$\mathbb{B}(\lambda) = \frac{B(\lambda)}{d(\lambda)}; \quad \mathbb{C}(\lambda) = \frac{C(\lambda)}{d(\lambda)}.$$

The set $\{\lambda\}$ in (1.1), (1.2) should satisfy the system of Bethe equations

$$r(\lambda_j) \prod_{\substack{k=1 \\ k\neq j}}^{N} \frac{f(\lambda_j,\lambda_k)}{f(\lambda_k,\lambda_j)} = 1. \qquad (1.3)$$

The arbitary function $r(\lambda)$ is given as the ratio of the vacuum eigenvalues

$$r(\lambda) = \frac{a(\lambda)}{d(\lambda)}.$$

The generalized Gaudin hypothesis can be written in the following way:

$$\langle\Psi_N(\{\lambda\})|\Psi_N(\{\lambda\})\rangle = c^N \left(\prod_{\substack{j,k=1 \\ j\neq k}}^{N} f(\lambda_j,\lambda_k) \right) \det{}_N \varphi' \qquad (1.4)$$

where φ' is the $N \times N$ matrix

$$\varphi'_{jk} = \frac{\partial \varphi_j}{\partial \lambda_k}$$

$$= \delta_{jk}\left(z_k + \sum_{l=1}^{N} K(\lambda_k, \lambda_l)\right) - K(\lambda_j, \lambda_k) \tag{1.5}$$

and

$$z_k = i\frac{\partial \ln r(\lambda)}{\partial \lambda}\bigg|_{\lambda=\lambda_k}$$

(IX.3.22) and the function $K(\lambda, \mu)$ is given by (IX.3.25). Using the representation (I.2.16), it is not difficult to show that the matrix φ' is positive definite for z_j positive and $c > 0$:

$$\varphi' > 0, \quad \det{}_N\varphi' > 0, \quad \text{if } z_j > 0. \tag{1.6}$$

An inequality similar to (I.2.18) shows that all eigenvalues of the matrix φ' are positive if $c > 0$ and $z_j > 0$. We shall prove formula (1.4) for a generalized model. This means that $a(\lambda)$ and $d(\lambda)$ are arbitrary functions, but the R-matrix is fixed (see the Introduction to Part III, (0.3), (0.4), (0.9)). The nonlinear Schrödinger model is a special case as well as the sine-Gordon and Heisenberg models.

In the special case of the NS model, the involution $B(\lambda) = C^{\dagger}(\lambda^*)$ is present and the vacuum eigenvalues are defined by

$$a(\lambda) = d^*(\lambda^*) = \exp\{-i\lambda L/2\}.$$

Due to these properties, the left hand side of formula (1.4) is indeed the squared norm of the wave function. It should also be mentioned that it follows from the Bethe system (1.3) in the case of the NS, model that

$$\prod_{j=1}^{N} d^2(\lambda_j) = 1, \quad \lambda_j = \lambda_j^*.$$

This is true only for the Bose gas (nonlinear Schrödinger model). Thus the generalized formula (1.4) in this case is reduced to the Gaudin hypothesis formulated in the beginning of this chapter. In the general case, the left hand side of equation (1.4) will also be called the "squared norm" of the wave function.

We now turn again to the generalized model. Due to the arbitrariness of the function $r(\lambda)$, one can consider the λ_j's as free variables, even though the Bethe system (1.5) is fulfilled (see also the discussion in section VII.7 and formula (VII.7.15)). The quantities $\{z_j\}$ are also independent variables. This fact, too, is a consequence of the arbitrariness of the function $r(\lambda)$. Hence the Jacobian on the right hand side of (1.4) is a function of $2N$ independent variables $\{\lambda_j\}$ and $\{z_j\}$. It should be noted that the subscript N of the determinant is related to the dimension of the matrix φ'. For $N = 0$ we put

$$\det_0 \varphi' \equiv 1 \tag{1.7}$$

by definition.

If one now uses Bethe eigenvectors $|\varphi_N\rangle$ normalized as

$$|\varphi_N(\{\lambda\})\rangle = \prod_{j=1}^{N} B(\lambda_j)|0\rangle = \prod_{j=1}^{N} d(\lambda_j)|\Psi_N(\{\lambda\})\rangle, \tag{1.8}$$

$$\langle\varphi_N(\{\lambda\})| = \langle 0|\prod_{j=1}^{N} C(\lambda_j) = \prod_{j=1}^{N} d(\lambda_j)\langle\Psi_N(\{\lambda\})| \tag{1.9}$$

then the generalized Gaudin hypothesis (1.4) is rewritten as

$$\langle\varphi_N(\{\lambda\})|\varphi_N(\{\lambda\})\rangle = c^N \left(\prod_{j=1}^{N} a(\lambda_j)d(\lambda_j)\right) \left(\prod_{\substack{j,k=1 \\ j\neq k}}^{N} f(\lambda_j, \lambda_k)\right) \det{}_N \varphi' \tag{1.10}$$

which is equivalent to (1.4) due to the following identity obtained from the Bethe system (1.3):

$$\prod_{j=1}^{N} r(\lambda_j) \equiv \prod_{j=1}^{N} \frac{a(\lambda_j)}{d(\lambda_j)} = 1. \tag{1.11}$$

Once more, we consider all the models described by the XXX or XXZ R-matrices (see the Introduction to Part III, formulæ (0.3), (0.4), (0.9)). These models are parametrized by two arbitrary functions, $a(\lambda)$ and $d(\lambda)$. For each of these models, formulæ (1.4) and (1.10) are valid. We shall prove this below. The interesting models (the nonlinear Schrödinger, sine-Gordon and Heisenberg antiferromagnet models) can all be obtained by specification of the functions $a(\lambda)$ and $d(\lambda)$.

X.2 Properties of the Jacobian

Let us change the normalization of the eigenfunctions and introduce the "renormalized" square of the norm of the wave function

$$\| \lambda_1, \ldots, \lambda_N \|_N = \| \{\lambda\} \|_N = \frac{\langle\Psi_N(\{\lambda\})|\Psi_N(\{\lambda\})\rangle}{c^N \prod_{j\neq k}^{N} f(\lambda_j, \lambda_k)}. \tag{2.1}$$

This is now the definition. Here the Bethe vectors $|\Psi_N\rangle$ and $\langle\Psi_N|$ are given by formulæ (1.1), (1.2). The hypothesis (1.4) in the new notation is rewritten as

$$\| \lambda_1, \ldots, \lambda_N \|_N = \det{}_N \varphi' \tag{2.2}$$

which we shall prove.

Consider now the following five characteristic properties of the Jacobian $\det_N(\varphi')$:

(1) The Jacobian is not changed under the following interchange of variable pairs

$$(z_j, \lambda_j) \longleftrightarrow (z_k, \lambda_k) \quad (j, k = 1, \ldots, N). \tag{2.3}$$

(2) The Jacobian is a linear function of each z_j.
(3) The corresponding coefficient of z_N is equal to

$$\frac{\partial \det_N(\varphi')}{\partial z_N} = \det{}_{N-1} \varphi'_{\text{mod}}. \tag{2.4}$$

Here

$$
\begin{aligned}
(\varphi'_{\text{mod}})_{jk} &= \frac{\partial \tilde{\varphi}_j}{\partial \lambda_k} \\
&= \delta_{jk} \left(\tilde{z}_k + \sum_{l=1}^{N-1} K_{kl} \right) - K_{jk}
\end{aligned}
\tag{2.5}
$$

where, for the NS model,

$$K_{jk} = K(\lambda_j, \lambda_k) = \frac{2c}{c^2 + (\lambda_j - \lambda_k)^2}$$

and

$$\tilde{\varphi}_j = i \ln \tilde{r}(\lambda_j) + i \sum_{\substack{k=1 \\ k \neq j}}^{N-1} \ln \frac{f(\lambda_j, \lambda_k)}{f(\lambda_k, \lambda_j)}, \tag{2.6}$$

$$\tilde{r}(\lambda) = r(\lambda) \frac{f(\lambda, \lambda_N)}{f(\lambda_N, \lambda)}, \quad \tilde{z}_k = z_k + K_{kN}.$$

(4) If all $z_j = 0$ $(j = 1, \ldots, N)$ then the Jacobian is equal to zero for $N > 0$:

$$\det{}_N \varphi' \Big|_{z_j = 0, (j=1,\ldots,N)} = \delta_{0N}. \tag{2.7}$$

Indeed, the matrix

$$\delta_{jk} \left(\sum_{l=1}^{N} K_{kl} \right) - K_{jk} \tag{2.8}$$

possesses an eigenvector with zero eigenvalue. This is the vector with all components equal to one.

(5) For $N = 1$, it is easy to calculate that

$$\det{}_1 \varphi' = z_1. \tag{2.9}$$

This exhausts the list of characteristic properties. To prove that $\| \lambda_1, \ldots, \lambda_N \|_N$ (2.1) possesses the same properties is equivalent to proving the generalized Gaudin hypothesis.

Theorem 1. *To prove equality (2.2), it is sufficient to show that the quantity $\| \lambda_1, \ldots, \lambda_N \|_N$ (2.1) possesses the following five properties:*

(1) *It is invariant under the permutation of pairs*

$$(z_j, \lambda_j) \longleftrightarrow (z_k, \lambda_k). \tag{2.10}$$

(2) *It is linear in each z_k.*

(3) *The corresponding coefficient of z_N is equal to*

$$\frac{\partial \| \lambda_1, \ldots, \lambda_N \|_N}{\partial z_N} = \| \lambda_1, \ldots, \lambda_{N-1} \|_{N-1}^{\text{mod}}. \tag{2.11}$$

The modification here consists of the replacement $z \to \tilde{z}$ (2.6).

(4) *For λ_j fixed,*

$$\| \lambda_1, \ldots, \lambda_N \|_N = \delta_{0N}, \ \text{if } z_j = 0. \tag{2.12}$$

(5) *For $N = 0$ and $N = 1$*

$$\| \lambda \|_0 = 1, \quad \| \lambda \|_1 = z_1. \tag{2.13}$$

Proof: The proof of this theorem can be obtained using induction in N. The basis of induction holds (see properties (2.13), (1.7), (2.7)). The next step in the induction is constructed as follows. Suppose that formula (2.2) is true for $N = 1, \ldots, q - 1$. Let us prove then that it is also true for $N = q$. Consider the difference

$$\Delta_q = \| \lambda_1, \ldots, \lambda_q \|_q - \det{}_q \varphi'. \tag{2.14}$$

This is a linear function of z_q. The corresponding derivative is equal to

$$\frac{\partial \Delta_q}{\partial z_q} = \| \lambda_1, \ldots, \lambda_{q-1} \|_{q-1}^{\text{mod}} - \det{}_{q-1} \varphi'_{\text{mod}} = 0 \tag{2.15}$$

where the right hand side is equal to zero due to the induction assumption. Hence, Δ_q does not depend on z_q. Because of the symmetry of Δ_q

(property (2)) it does not depend on z_j ($j = 1, \ldots, q$) at all. On the other hand, property (4) ensures that

$$\Delta_q = 0 \text{ if } z_j = 0, \ j = 1, \ldots, N. \tag{2.16}$$

This means that Δ_q is equal to zero identically at any z_j. The induction step is thus constructed. The theorem is proved.

Hence, to prove hypothesis (2.2), (1.4), (1.10) it is sufficient to establish properties (1)–(5) of Theorem 1 for the quantity $\| \lambda_1, \ldots, \lambda_N \|_N$, which differs from the square of the norm of the Bethe wave function (1.1) by the factor $(1/c^N) \prod_{j \neq k} f(\lambda_j, \lambda_k)$ (see formula (2.1)).

X.3 Proof of Gaudin's hypothesis

It is proved in this section that the quantity $\| \lambda_1, \ldots, \lambda_N \|_N$ (2.1) indeed possesses the five properties of Theorem 1 of the previous section, formulæ (2.10)–(2.13), which completes the proof of the generalized Gaudin hypothesis.

The equality $\| \lambda \|_0 = 1$ is equivalent to the normalization $\langle 0 | 0 \rangle = 1$ (2.13), which is valid. Taking the limit $\lambda^C \to \lambda^B$ in formula (IX.1.2) we obtain

$$\langle 0 | C(\lambda) B(\lambda) | 0 \rangle = ica(\lambda) d(\lambda) \frac{\partial}{\partial \lambda} \ln \frac{a(\lambda)}{d(\lambda)}$$

and

$$r(\lambda) = \frac{a(\lambda)}{d(\lambda)} = 1,$$

$$\langle \Psi_1(\lambda) | \Psi_1(\lambda) \rangle = ic \frac{\partial}{\partial \lambda} \ln r(\lambda) = cz_1,$$

$$\| \lambda_1 \| = z_1.$$

So property (5) is established.

Due to the commutation relations (see (VII.1.11)),

$$[\mathbb{C}(\lambda), \mathbb{C}(\mu)] = [\mathbb{B}(\lambda), \mathbb{B}(\mu)] = 0$$

the symmetry required in property (1) is obvious.

In the end of section IX.3 the properties of the squared norm

$$\langle \Psi_N | \Psi_N \rangle = \langle 0 | \prod_j \mathbb{C}(\lambda_j) \prod_k \mathbb{B}(\lambda_k) | 0 \rangle$$

were investigated. The quantity $\langle \Psi_N | \Psi_N \rangle$ differs from $\| \lambda_1, \ldots, \lambda_N \|_N$ by a trivial factor only (see (2.1), where property (2) was established, in particular the sentence after formula (IX.3.22)). Property (3) follows from formula (IX.3.23) and the Bethe equations (IX.3.26).

Let us prove that property (4) is valid. This property is equivalent to the following statement:

$$\langle \Psi_N(\{\lambda\}) | \Psi_N(\{\lambda\}) \rangle = 0 \text{ if } z_j = 0 \tag{3.1}$$

for $N > 0$ with λ_j fixed by the system of Bethe equations. Consider the set $\{\lambda_j\}$ consisting of different numbers λ_j. Introduce the number

$$s = \min_{j \neq k} \left(|\lambda_j - \lambda_k|/4 \right). \tag{3.2}$$

The derivatives of the function $r(\lambda)$ are equal to zero at the points λ_j, the values of this function at points λ_j being equal to r_j. Define a new function $\widetilde{r}(\lambda)$ which is equal to r_j not only at points λ_j but also in the vicinity of radius s, supposing that in the rest of the complex plane it is defined in some smooth way (see Figure X.1).

The squared norms $\langle \Psi_N(\{\lambda\}) | \Psi_N(\{\lambda\}) \rangle$ for $r(\lambda)$ and $\widetilde{r}(\lambda)$ are equal. Indeed, each of the norms depends only on $2N$ variables, λ_j, z_j, which are the same in both cases. The spectral parameters λ_j are also solutions of the Bethe equations involving the new function $\widetilde{r}(\lambda)$:

$$\widetilde{r}(\lambda_j) \prod_{\substack{k \neq j \\ k=1}}^{N} \frac{f(\lambda_j, \lambda_k)}{f(\lambda_k, \lambda_j)} = 1. \tag{3.3}$$

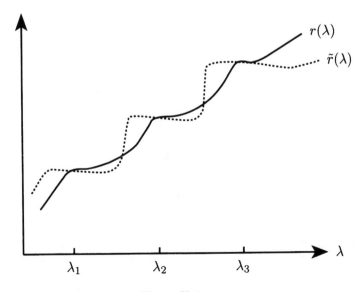

Figure X.1

By definition, $\widetilde{\widetilde{r}}(\lambda_j) = r(\lambda_j) = r_j$. It should, however, be emphasized that for the function $\widetilde{\widetilde{r}}(\lambda)$ not only the numbers λ_j $(j = 1, \ldots, N)$ solve the Bethe equations but also the numbers $\lambda_j + y$, where y is any complex number with $|y| < s$. This is obvious because the function $f(\lambda_j, \lambda_k)$ depends on the difference $\lambda_j - \lambda_k$ only and $\widetilde{\widetilde{r}}(\lambda + y) = \widetilde{\widetilde{r}}(\lambda)$. We now prove that the scalar product

$$\langle 0| \prod_{j=1}^{N} \mathbb{C}(\lambda_j) \prod_{k=1}^{N} \mathbb{B}(\lambda_k + y)|0\rangle = S(y) \tag{3.4}$$

is equal to zero. To this end, one calculates the matrix element

$$\langle 0| \prod_{j=1}^{N} \mathbb{C}(\lambda_j) \, \tau(\mu) \prod_{k=1}^{N} \mathbb{B}(\lambda_k + y)|0\rangle \tag{3.5}$$

where the transfer matrix $\tau(\mu) = A(\mu) + D(\mu)$ is the trace of the monodromy matrix (see sections VI.1, formula (VI.1.12), and VII.1, formula (VII.1.35)). This matrix element can be calculated in two different ways: acting with the operator $\tau(\mu)$ to the left or to the right. The states with the vectors standing to the right of $\tau(\mu)$ in (3.5) and to the left are both eigenvectors of the operator $\tau(\mu)$ but the eigenvalues are different:

$$\langle 0| \prod_{j=1}^{N} \mathbb{C}(\lambda_j) \, \tau(\mu) \prod_{k=1}^{N} \mathbb{B}(\lambda_k + y)|0\rangle = \theta(\mu, \{\lambda_j\}) S(y)$$

$$= \theta(\mu, \{\lambda_j + y\}) S(y) \tag{3.6}$$

where $\theta(\mu)$ is the eigenvalue of $\tau(\mu)$ (VII.1.35). As

$$\theta(\mu, \{\lambda_j\}) \neq \theta(\mu, \{\lambda_j + y\})$$

one concludes that $S(y) = 0$. In this way we have proved (3.1). This concludes the proof of property (4).

Therefore it is established that the quantity $\| \lambda_1, \ldots, \lambda_N \|_N$ possesses all five properties (2.10)–(2.13) required by Theorem 1 of section 2. The proof of Gaudin's hypothesis is completed. We have proved also formulæ (2.1), (1.10), (1.4). The nonlinear Schrödinger model is a special case.

It should be mentioned that the formula for the norms of the Bethe eigenfunctions in the XXZ case is proved in a similar way; this proof is given in paper [3]. The generalization of (1.4) for the XXZ case is:

$$\langle \Psi_N(\{\lambda\}) | \Psi_N(\{\lambda\}) \rangle = (\sin 2\eta)^N \prod_{j \neq k} f(\lambda_j, \lambda_k) \, \det{}_N \varphi'$$

where

$$|\Psi_N(\{\lambda\})\rangle = \prod j = 1^N \mathbb{B}(\lambda_j)|0\rangle;$$

$$\langle\Psi_N(\{\lambda\})| = \langle 0| \prod_{j=1}^{N} \mathbb{C}(\lambda_j);$$

$$r(\lambda_j) \prod_{\substack{k\neq j \\ k=1}}^{N} \frac{f(\lambda_j, \lambda_k)}{f(\lambda_k, \lambda_j)} = 1; \tag{3.7}$$

$$f(\lambda, \mu) = \frac{\sinh(\lambda - \mu + 2i\eta)}{\sinh(\lambda - \mu)}$$

and

$$\varphi_k = i \ln \frac{a(\lambda_k)}{d(\lambda_k)} - i \sum_{\substack{j=1 \\ j\neq k}}^{N} \ln \frac{f(\lambda_j, \lambda_k)}{f(\lambda_k, \lambda_j)};$$

$$\varphi'_{jk} = \frac{\partial\varphi_j}{\partial\lambda_k} = \delta_{jk}\left\{i\frac{\partial}{\partial\lambda_k} \ln \frac{a(\lambda_k)}{d(\lambda_k)}\right. \tag{3.8}$$

$$\left. + \sum_{l=1}^{N} \frac{\sin 4\eta}{\sinh(\lambda_k - \lambda_l + 2i\eta)\sinh(\lambda_k - \lambda_l - 2i\eta)}\right\}$$

$$- \frac{\sin 4\eta}{\sinh(\lambda_k - \lambda_j + 2i\eta)\sinh(\lambda_k - \lambda_j - 2i\eta)}.$$

We emphasize that formulæ (1.4) and (3.7) are valid in fact for any integrable model possessing an R-matrix of the XXX or XXZ type. For the XXX Heisenberg magnet, using the involution

$$B^\dagger(\lambda^*) = -C(\lambda)$$

(see (VI.5.13)) and the explicit form of the functions $a(\lambda)$ and $d(\lambda)$ (given in (VII.3.11) and (VII.3.12)), one gets from (1.4) the explicit form for the norm of the wave function. Analogously, formula (3.7) leads to an expression for the norm of the wave function of the XXZ magnet. In the sine-Gordon model at rational values of the coupling constant γ/π, the pseudovacuum is normalizable, i.e., one can put $\langle 0|0\rangle = 1$ (see section VII.3, Appendix 1 to Chapter VII and section VIII.5). In this case, one can also derive an explicit formula for the norm.

X.4 Thermodynamic limit

Here the norm of the wave function in the thermodynamic limit at zero temperature (see section I.3) is discussed. This is interesting from the physical point of view. The following calculations are all done using the

NS model as an example. The norm of the wave function in this model is given by

$$\langle \Psi_N(\{\lambda\})|\Psi_N(\{\lambda\})\rangle = c^N \prod_{j \neq k} f(\lambda_j, \lambda_k) \det{}_N \varphi'. \tag{4.1}$$

Now $r(\lambda) = a(\lambda)/d(\lambda) = \exp\{-i\lambda L\}$ and $z(\lambda) = i\partial \ln r(\lambda)/\partial \lambda = L$. The matrix φ' (see (1.5)) simplifies to

$$\varphi'_{jl} = \delta_{jl}\left[L + \sum_{m=1}^{N} K(\lambda_j, \lambda_m)\right] - K(\lambda_j, \lambda_l),$$

$$K(\lambda_j, \lambda_l) = \frac{2c}{c^2 + (\lambda_j - \lambda_l)^2} \tag{4.2}$$

(see also (I.4)).

We now discuss the behavior of the Jacobian $\det_N \varphi'$ in the thermodynamic limit. To do this, one writes the $N \times N$ matrix φ' as a product of two matrices,

$$\varphi' = \widehat{G}\theta, \quad \det{}_N \varphi' = \det\theta \det\widehat{G}. \tag{4.3}$$

Here θ is a diagonal matrix

$$\theta_{jl} = \delta_{jl}\vartheta_l, \quad \vartheta_l = L + \sum_{m=1}^{N} K_{lm},$$

$$K_{lm} = K(\lambda_l - \lambda_m) \tag{4.4}$$

and the matrix $\widehat{G}$ is given by

$$\widehat{G}_{jl} = \delta_{jl} - \frac{K_{jl}}{\vartheta_l}. \tag{4.5}$$

The thermodynamic limit at zero temperature was described in Chapter I, section 3. The thermodynamic limit of the quantity ϑ_l in (4.4) is easily calculated to be

$$\vartheta_l = 2\pi L\rho(\lambda_l). \tag{4.6}$$

Here $\rho(\lambda)$ is the density of particles in momentum space, defined in Chapter I, see formulæ (3.4)–(3.7) (one changes the sum to an integral as in (I.3.6) and uses equation (I.3.7)). Thus the asymptotic expression for $\det\theta$ is of the form

$$\det\theta = \prod_{j=1}^{N}\Big(2\pi L\rho(\lambda_j)\Big). \tag{4.7}$$

The matrix $\widehat{G}$ in the thermodynamic limit turns into an integral operator

$$\widehat{G} \to 1 - \frac{\widehat{K}}{2\pi}, \quad \left(\widehat{K}\varphi\right)(\lambda) = \int\limits_{-q}^{q} K(\lambda, \mu)\varphi(\mu)\, d\mu \qquad (4.8)$$

(see (I.3.9)). In the thermodynamic limit, the Jacobian has the following representation:

$$\lim \frac{\det{}_N \varphi'}{\prod\limits_{j=1}^{N} (2\pi L \rho(\lambda_j))} = \det\left(1 - \frac{1}{2\pi}\widehat{K}\right). \qquad (4.9)$$

The determinant of the integral operator stands on the right hand side here. When calculating correlation functions, the following asymptotic expression will often be used:

$$\det{}_N \varphi' = \prod\limits_{j=1}^{N}\left(2\pi L \rho(\lambda_j)\right)\det\left(1 - \frac{1}{2\pi}\widehat{K}\right). \qquad (4.10)$$

The density of excited states $\rho(\lambda)$ is slightly changed. Formula (4.10), however, can be used if one takes for ρ the new value of the vacancy density. The thermodynamics of the model at nonzero temperature was constructed in section 5, Chapter I. Let us consider the thermodynamic limit of the norm in this case.

Let us also discuss the situation where the eigenvector $|\Psi_N\rangle$ goes in the thermodynamic limit to the state $|\Omega_T\rangle$ which is one of the eigenfunctions present in the state of thermal equilibrium (see section I.8); it is characterized macroscopically by the "particle density" $\rho_p(\lambda)$ and by the total density of vacancies $\rho_t(\lambda)$ (see (I.5.3)–(I.5.6)). The densities are related by the equation (I.5.24),

$$1 + \int\limits_{-\infty}^{+\infty} K(\lambda, \mu)\rho_p(\mu)\, d\mu = 2\pi\rho_t(\lambda). \qquad (4.11)$$

The thermodynamic limit of $\det\theta$ is now

$$\lim \det\theta = \prod\limits_{j=1}^{N}\left(2\pi L \rho_t(\lambda_j)\right) \qquad (4.12)$$

instead of (4.7). The matrix $\widehat{G}$ in the thermodynamic limit turns into a linear integral operator:

$$
\widehat{G} = 1 - \frac{1}{2\pi}\widehat{K}_\rho, \quad \frac{\rho_p(\mu)}{\rho_t(\mu)} = \frac{1}{1 + e^{\varepsilon(\mu)/T}}
$$

$$
\left(\widehat{K}_\rho \xi\right)(\lambda) = \int\limits_{-\infty}^{+\infty} K(\lambda,\mu)\frac{\rho_p(\mu)}{\rho_t(\mu)}\xi(\mu)\,d\mu. \tag{4.13}
$$

Here $\xi(\lambda)$ is some c-number function.

One obtains for the thermodynamic limit of the determinant in this case

$$
\lim \det{}_N \varphi' = \prod_{j=1}^{N}\left(2\pi L\rho_t(\lambda_j)\right)\det\left(1 - \frac{1}{2\pi}\widehat{K}_\rho\right). \tag{4.14}
$$

So the norms of eigenfunctions $|\Omega_T\rangle$ present in the state of thermal equilibrium are the same (because all of the $|\Omega_T\rangle$ have the same $\rho_t(\lambda)$). The same will happen with mean values of local operators.

X.5 A special case of the scalar product

In Chapter IX we considered the determinant representation for the scalar product

$$
\langle 0|\prod_{j=1}^{N} C(\lambda_j^C)\; \prod_{k=1}^{N} B(\lambda_k^B)|0\rangle \tag{5.1}
$$

in the generic situation when $\{\lambda_j^C\}$ and $\{\lambda_k^B\}$ were free and did not satisfy any equations. The corresponding determinant representation included auxiliary quantum fields (see IX.5.8). In this section we consider the squares of the norms of the eigenfunctions. This means that the sets $\{\lambda_j^C\}$ and $\{\lambda_j^B\}$ each satisfy the periodic boundary conditions (Bethe equations), see (1.3), and coincide with one another.

The determinant representation for the square of the norm of the eigenfunction (1.4) does not include auxiliary quantum fields. This is nice, because it is simpler. A.N. Kirilov, F.A. Smirnov [2], [5] and N.A. Slavnov [4] noted that there exists an intermediate case when one of the vectors (say $\langle 0|\prod_{j=1}^{N} C(\lambda_j^C)$) is an eigenvector of the Hamiltonian (and transfer matrix) and the other one (say $\prod_{j=1}^{N} B(\lambda_j^B)|0\rangle$) is not. In this case, the determinant representation does not include auxiliary quantum fields.

Let us be more precise. Consider the scalar product (5.1) in the case when the λ_j^C satisfy periodic boundary conditions (Bethe equations)

$$\frac{a(\lambda_j^C)}{d(\lambda_j^C)} \prod_{\substack{l=1 \\ l \neq j}}^{N} \frac{f(\lambda_j^C, \lambda_l^C)}{f(\lambda_l^C, \lambda_j^C)} = 1 \tag{5.2}$$

but the λ_k^B do not. The variables λ_k^B are free and do not obey any equation or constraint. The following determinant representation is valid for such scalar products:

$$\langle 0| \prod_{j=1}^{N} C(\lambda_j^C) \prod_{k=1}^{N} B(\lambda_k^B)|0\rangle = \left\{ \prod_{j=1}^{N} d(\lambda_j^B) d(\lambda_j^C) \right\} \tag{5.3}$$

$$\times \left\{ \prod_{N \geq j > k \geq 1} g(\lambda_j^B, \lambda_k^B) g(\lambda_k^C, \lambda_j^C) \right\} \left\{ \prod_{j=1}^{N} \prod_{k=1}^{N} h(\lambda_j^C, \lambda_k^B) \right\} \det M.$$

The matrix elements of M are given by the formula

$$M_{lk} = \frac{g(\lambda_k^C, \lambda_l^B)}{h(\lambda_k^C, \lambda_l^B)} - \frac{a(\lambda_l^B)}{d(\lambda_l^B)} \frac{g(\lambda_l^B, \lambda_k^C)}{h(\lambda_l^B, \lambda_k^C)} \left(\prod_{m=1}^{N} \frac{f(\lambda_l^B, \lambda_m^C)}{f(\lambda_m^C, \lambda_l^B)} \right). \tag{5.4}$$

The proof of this formula can be found in paper [4]. Hence, we see that no auxiliary quantum fields are necessary, and the scalar product is merely proportional to the determinant of a matrix involving the standard functions $a(\lambda)$, $d(\lambda)$, $f(\lambda, \mu)$, $g(\lambda, \mu)$ and $h(\lambda, \mu)$.

Conclusion

In this chapter we have proved Gaudin's hypothesis for the norms (0.3) of the eigenfunctions of the Hamiltonian. This has a very deep physical meaning: if the system of Bethe equations (periodic boundary conditions) becomes degenerate the corresponding wave function vanishes. Remarkably, the determinant formula does not contain auxiliary dual fields. This formula will be used for the evaluation of correlation functions in the next chapter.

M. Gaudin formulated his hypothesis in his thesis in 1972; it was published only in 1983 [1]. Gaudin's hypothesis was first proved in paper [3] for both the NS and XXZ Heisenberg antiferromagnet.

XI
Correlation Functions of Currents

Introduction

In this chapter we consider the operator

$$j(x) = \Psi^\dagger(x)\Psi(x) \tag{0.1}$$

which is defined in the context of the quantum nonlinear Schrödinger equation (Bose gas). This operator $j(x)$ is called the local density operator. Below we prefer the terminology of quantum field theory, so we will refer to (0.1) as the current. The determinant representation for the correlation function of currents $j(x)$ is obtained in this chapter.

We shall consider the equal-time correlation function of two currents $\langle j(x)j(0)\rangle$ at zero and finite temperature. In section 1, we reduce the problem of evaluating this correlation function to the problem of evaluating the mean value of the operator

$$\exp\{\alpha Q_1(x)\} \tag{0.2}$$

with respect to the ground state. Here α is an arbitrary complex parameter and $Q_1(x)$ is the operator of the number of particles on the interval $[0, x]$:

$$Q_1(x) = \int\limits_0^x \Psi^\dagger(y)\Psi(y)\,dy. \tag{0.3}$$

Next we define the operator $\exp\{\alpha Q_1\}$ in the two-site generalized model, which was defined in section 5 of Chapter VII. The operator $\exp\{\alpha Q_1\}$ has already been introduced for the general case; see Chapter VII, formula (7.8). This permits us to derive the representation of matrix elements of the operator $\exp\{\alpha Q_1\}$ in terms of scalar products. This is quite a remarkable operator. A lot of interesting correlation functions can be extracted from

$$\langle \exp\{\alpha Q_1(x)\}\rangle.$$

For small α it gives current correlations $(j(x) = \Psi^\dagger(x)\Psi(x))$, at $\alpha = i\pi$ it is related to field correlations, at $\alpha = -\infty$ it gives the probability that, due to fluctuations, there will be no particles in the space interval $[0, x]$. The asymptotics for $\langle \exp\{\alpha Q_1(x)\}\rangle$ can be found in section XVII.5. The differential equations for it are given in chapters XIV–XVI.

Our approach is very similar to the one-site generalized model, which was used in section 5 of Chapter IX to derive the determinant representation for the scalar product. Our final formula is similar to (IX.5.8).

In section 2, auxiliary quantum fields are introduced and the determinant representation for the matrix elements of the operator $\exp\{\alpha Q_1\}$ in the generalized model is obtained. The derivation is similar to that of section 6 of Chapter IX for the scalar products. It is based essentially on the representation of the determinant of the sum of two matrices in terms of their minors. The corresponding formula is explicitly described in the Appendix to Chapter IX. The diagonal matrix elements of the operator $\exp\{\alpha Q_1\}$ in the generalized model are studied in section 3. In section 4, the generating functional $\langle \exp\{\alpha Q_1(x)\}\rangle$ of the equal-time current correlators at zero temperature is calculated. We study the thermodynamic limit. Finally, the temperature-dependent correlator is obtained in section 5. In this sections results of Chapter I (sections 5–8) will be used. The calculations will be similar to those of section X.4.

XI.1 Generalized two-site model

Consider the current operator $j(x) = \Psi^\dagger(x)\Psi(x)$ in the NS model. Our aim is to calculate the equal-time current correlator at zero temperature

$$\langle j(x_1)j(x_2)\rangle = \frac{\langle \Omega|j(x_1)j(x_2)|\Omega\rangle}{\langle \Omega|\Omega\rangle}. \tag{1.1}$$

The ground state of the Hamiltonian, $|\Omega\rangle$, at zero temperature in the NS model was explicitly described in section 3 of Chapter I. It is convenient to introduce the operator for the number of particles on the interval $[x_1, x_2]$:

$$Q(x_2, x_1) = \int\limits_{x_1}^{x_2} \Psi^\dagger(y)\Psi(y)\, dy.$$

The currents correlation function (1.1) can be reduced to the mean value of the square of the operator $Q(x_2, x_1)$:

$$\langle \Omega|j(x_1)j(x_2)|\Omega\rangle = -\frac{1}{2}\frac{\partial^2}{\partial x_1 \partial x_2}\langle \Omega|Q^2(x_2, x_1)|\Omega\rangle.$$

Operator ordering in this formula is discussed below. The current correlation function (1.1) depends only on the absolute value of the distance $|x_1 - x_2|$ due to translation invariance and invariance under spatial reflection.

In the following, we shall put $x_1 = 0$, $x_2 = x > 0$. The operator $Q_1(x)$ for the number of particles in the interval $[0, x]$ is

$$Q_1(x) = \int_0^x j(y)\,dy, \qquad j(y) = \Psi^\dagger(y)\Psi(y). \tag{1.2}$$

The spectrum of this operator consists of all integer numbers.

The current correlator can be expressed as follows:

$$\langle :j(x)j(0): \rangle = \frac{1}{2}\frac{\partial^2}{\partial x^2}\langle Q_1^2(x)\rangle. \tag{1.3}$$

The normal ordering is described as follows

$$:j(x)j(0): = \Psi^\dagger(x)\Psi^\dagger(0)\Psi(x)\Psi(0).$$

$$\langle :j(x)j(0): \rangle = \langle j(x)j(0)\rangle - \delta(x)\langle j(0)\rangle \tag{1.4}$$

so that $\langle j(x)j(0)\rangle = \langle :j(x)j(0): \rangle$ for $x > 0$.

It will be convenient to consider a "generating operator" $\exp\{\alpha Q_1(x)\}$ where α is an arbitrary complex number. Matrix elements of the operator Q_1^2 are easily expressed in terms of the matrix elements of the generating operator due to the relation

$$Q_1^n = \frac{\partial^n}{\partial \alpha^n}\exp\{\alpha Q_1\}\Big|_{\alpha=0}.$$

Our scheme for calculating matrix elements of the operator $\exp\{\alpha Q_1(x)\}$ is as follows. First we introduce the two-site generalized integrable model possessing the same XXX R-matrix as the NS model. The operator Q_1 is defined in this model in an algebraic way. Then the general matrix element

$$\langle 0| \prod_{j=1}^N C(\lambda_j^C)\,\exp\{\alpha Q_1\} \prod_{k=1}^N B(\lambda_k^B)|0\rangle \tag{1.5}$$

at finite N is considered in the generalized model. The case where the sets $\{\lambda^C\}$ and $\{\lambda^B\}$ do not coincide is considered first. In this case, spectral parameters λ_j^C, λ_k^B $(j, k = 1, \ldots, N)$ are independent complex variables. Then one sets $\{\lambda^C\} = \{\lambda^B\} = \{\lambda\}$ and demands that the set $\{\lambda\}$ is a solution of the system of Bethe equations. Finally, one goes to the thermodynamic limit.

As in the calculation of norms, it is possible to consider not only the concrete NS model but a whole class of models characterized by arbitrary functional parameters (see section 6 of Chapter VII). To this end, one introduces the two-site generalized model and defines matrix elements (1.5) in this model (see section 5 of Chapter VII). The NS model is a particular case of the generalized model at special values of the functional parameters. The monodromy matrix of such a generalized model is given by the matrix product of two monodromy matrices (a product of two "commuting" monodromy matrices, see section VII.5):

$$T(\lambda) = T(2|\lambda)T(1|\lambda) = \begin{pmatrix} A(\lambda) & B(\lambda) \\ C(\lambda) & D(\lambda) \end{pmatrix}, \tag{1.6}$$

$$T(i|\lambda) = \begin{pmatrix} A_i(\lambda) & B_i(\lambda) \\ C_i(\lambda) & D_i(\lambda) \end{pmatrix}, \quad i = 1, 2. \tag{1.7}$$

The matrix $T(1|\lambda)$ is associated with the first site and the matrix $T(2|\lambda)$ with the second site of some two-site lattice. The matrix elements of $T(i|\lambda)$ are quantum operators commuting at different sites of the lattice. Matrices $T(i|\lambda)$ and $T(i|\mu)$ at the same site are intertwined by the R-matrix of the XXX model (see (VII.1.2)–(VII.1.4)). The monodromy matrix $T(i|\lambda)$ possesses the pseudovacuum $|0\rangle_i$ as well as the dual pseudovacuum $_i\langle 0|$ with the standard properties $(i = 1, 2)$

$$C_i(\lambda)|0\rangle_i = 0, \quad A_i(\lambda)|0\rangle_i = a_i(\lambda)|0\rangle_i,$$
$$D_i(\lambda)|0\rangle_i = d_i(\lambda)|0\rangle_i; \tag{1.8}$$

$$_i\langle 0|B_i(\lambda) = 0, \quad _i\langle 0|A_i(\lambda) = a_i(\lambda)_i\langle 0|,$$
$$_i\langle 0|D_i(\lambda) = d_i(\lambda)_i\langle 0|. \tag{1.9}$$

The state $|0\rangle = |0\rangle_2 \otimes |0\rangle_1$ is the pseudovacuum state for the matrix $T(\lambda)$ with the corresponding vacuum values equal to (see section VII.1, formula (VII.1.9))

$$a(\lambda) = a_1(\lambda)a_2(\lambda), \quad d(\lambda) = d_1(\lambda)d_2(\lambda). \tag{1.10}$$

The functions $a_i(\lambda)$, $d_i(\lambda)$ will be the main functional parameters of the two-site model (see sections 5 and 6 of Chapter VII).

The operator for the number of particles Q_i $(i = 1, 2)$ at the i-th site of the lattice is defined in a standard way $(i = 1, 2)$:

$$Q_i \prod_{k=1}^{n} B_j(\lambda_k)|0\rangle_j = n\delta_{ij} \prod_{k=1}^{n} B_j(\lambda_k)|0\rangle_j, \tag{1.11}$$

$$_j\langle 0| \prod_{k=1}^{n} C_j(\lambda_k)Q_i = n\delta_{ij} \, _j\langle 0| \prod_{k=1}^{n} C_j(\lambda_k). \tag{1.12}$$

It should be emphasized here that $\{\lambda_k\}$ is a set of independent parameters. The complete operator for the number of particles is $Q = Q_1 + Q_2$. Similarly, any function of Q_i can be defined.

Now, using formulæ (VII.5.9) and (VII.5.13), one obtains the following expression for the matrix element of the operator $\exp\{\alpha Q_1\}$:

$$
\langle 0| \prod_{j=1}^{N} C(\lambda_j^C) \exp\{\alpha Q_1\} \prod_{k=1}^{N} B(\lambda_k^B)|0\rangle
$$

$$
= \sum{}_1 \langle 0| \prod_I C_1(\lambda_I^C) \prod_I B(\lambda_I^B)|0\rangle_1
$$

$$
\times {}_2\langle 0| \prod_{II} C_2(\lambda_{II}^C) \prod_{II} B(\lambda_{II}^B)|0\rangle_2
$$

$$
\times \prod_I a_2(\lambda_I^B) \prod_I d_2(\lambda_I^C) \prod_{II} a_1(\lambda_{II}^C) \prod_{II} d_1(\lambda_{II}^B)
$$

$$
\times \prod_I \prod_{II} f(\lambda_I^B, \lambda_{II}^B) \prod_I \prod_{II} f(\lambda_{II}^C, \lambda_I^C)
$$

$$
\times \exp\{\alpha n_1\}. \tag{1.13}
$$

The spectral parameters λ_j^C, λ_k^B here are independent complex variables. The sum is taken over all the partitions of the set $\{\lambda_j^B,\ j = 1, \ldots, N\}$ into two disjoint sets $\{\lambda_I^B\}$ and $\{\lambda_{II}^B\}$, and over similar partitions of the set $\{\lambda^C\}$. These partitions are independent except for the following restrictions on the number of elements:

$$
\begin{aligned}
\mathrm{card}\{\lambda_I^B\} &= \mathrm{card}\{\lambda_I^C\} = n_1, \\
\mathrm{card}\{\lambda_{II}^B\} &= \mathrm{card}\{\lambda_{II}^C\} = n_2 = N - n_1.
\end{aligned} \tag{1.14}
$$

The product $\prod_I$ denotes the product over all $\{\lambda_j\} \in \{\lambda_I\}$ containing n_1 factors. Similarly, the product $\prod_{II}$ denotes the product over all $\lambda \in \{\lambda_{II}\}$.

So we have defined matrix element (1.5) in the generalized model. Remember, the functions $a_1(\lambda)$, $a_2(\lambda)$ and $d_1(\lambda)$, $d_2(\lambda)$ are now arbitrary. The mean value (1.13) depends on the values of these functions at the points λ_j^C, λ_k^B $(j, k = 1, \ldots, N)$ only. These values (as in the case of scalar products in section IX.3, formula (IX.3.1)) can be considered as $8N$ independent variables

$$
\begin{aligned}
a_{1,j}^C &\equiv a_1(\lambda_j^C), \quad a_{1,j}^B \equiv a_1(\lambda_j^B); \\
d_{1,j}^C, d_{1,j}^B; \quad & a_{2,j}^C, a_{2,j}^B; \quad d_{2,j}^C, d_{2,j}^B.
\end{aligned} \tag{1.15}
$$

We now discuss in more detail the relation of the generalized model constructed above to the nonlinear Schrödinger (NS) model, which was a special case of the generalized model. To obtain the NS model from the generalized model one should make the following specifications. In the NS model, the matrix $\mathsf{T}(1|\lambda)$ is the monodromy matrix for the interval

$[0, x]$ and $\mathsf{T}(2|\lambda)$ is the monodromy matrix for the interval $[x, L]$ such that

$$
\begin{aligned}
a(\lambda) &= \exp\{-i\lambda L/2\}, \quad d(\lambda) = \exp\{i\lambda L/2\}, \quad L > 0; \\
a_1(\lambda) &= \exp\{-i\lambda x/2\}, \quad d_1(\lambda) = \exp\{i\lambda x/2\}, \quad x > 0; \\
a_2(\lambda) &= \exp\{-i\lambda(L-x)/2\}, \quad d_2(\lambda) = \exp\{i\lambda(L-x)/2\}, \quad L > x.
\end{aligned}
\tag{1.16}
$$

With these identifications, the operator Q_1 of the generalized model becomes the operator $Q_1(x)$ (1.2) of the NS model.

The problem of calculating the current correlator is thus reduced to the problem of calculating the mean value of the operator $\exp\{\alpha Q_1\}$ which is formulated in the generalized model in a natural way. In turn, matrix elements of the operator $\exp\{\alpha Q_1\}$ are expressed in terms of the scalar products studied in Chapter IX.

Especially simple is the limiting case $\alpha \to -\infty$, $\exp\alpha = 0$. The physical meaning of this limit is the following. In the ground state of the Bose gas, particles move randomly. There exists a probability $P(x)$ that on the interval $[0, x]$ there will be no particles. This is related to the $\alpha = -\infty$ case as follows:

$$
P(x) = \left. \frac{\langle 0| \prod\limits_{j=1}^{N} C(\lambda_j) \exp\{\alpha Q_1\} \prod\limits_{j=1}^{N} B(\lambda_j)|0\rangle}{\langle 0| \prod\limits_{j=1}^{N} C(\lambda_j) \prod\limits_{j=1}^{N} B(\lambda_j)|0\rangle} \right|_{\alpha=-\infty}.
\tag{1.17}
$$

Here $\{\lambda_j\}$ satisfies the Bethe equations (with periodic boundary conditions). We shall call $P(x)$ the probability of forming an empty interval. This probability can be reduced to scalar products using (1.13). Let us start with a different bra and ket:

$$
\langle 0| \prod\limits_{j=1}^{N} C(\lambda_j^C) \exp\{\alpha Q_1\} \prod\limits_{j=1}^{N} B(\lambda_j^B)|0\rangle \Big|_{\alpha=-\infty}
$$

$$
= \prod\limits_{j=1}^{N} d_1(\lambda_j^B) a_1(\lambda_j^C) \, {}_2\langle 0| \prod\limits_{j=1}^{N} C_2(\lambda_j^C) \prod\limits_{j=1}^{N} B_2(\lambda_j^B)|0\rangle_2.
\tag{1.18}
$$

Here we have used formula (1.13) and the fact that $\exp\alpha = 0$. Now we can use formula (5.8) from Chapter IX to write down the determinant representation. Modifying this formula a little we get:

$$\langle 0| \prod_{j=1}^{N} C(\lambda_j^C) \exp\{\alpha Q_1\} \prod_{j=1}^{N} B(\lambda_j^B)|0\rangle$$

$$= \left\{ \prod_{N \geq j > k \geq 1} g(\lambda_j^C, \lambda_k^C) g(\lambda_k^B, \lambda_j^B) \right\} \times (0|\det{}_N \widetilde{S}_2|0) \prod_{j=1}^{N} d_1(\lambda_j^B) a_1(\lambda_j^C). \tag{1.19}$$

Here the matrix elements of $\widetilde{S}_2$ depend on the auxiliary quantum fields

$$\varphi_{A_2}(\lambda) = Q_{A_2}(\lambda) + P_{D_2}(\lambda),$$
$$\varphi_{D_2}(\lambda) = Q_{D_2}(\lambda) + P_{A_2}(\lambda). \tag{1.20}$$

The fields $\varphi_{A_2}(\lambda)$, $\varphi_{D_2}(\lambda)$ commute with one another (they form an abelian subgroup). The momenta and coordinates satisfy the following commutation relations:

$$[P_{D_2}(\lambda), Q_{D_2}(\mu)] = \ln h(\lambda, \mu)$$
$$[P_{A_2}(\lambda), Q_{A_2}(\mu)] = \ln h(\mu, \lambda), \tag{1.21}$$

and all other commutators are zero.

The vacuum $|0)$ is different from $|0\rangle$ because it belongs to a different Fock space:

$$P_{A_2}(\lambda)|0) = 0, \quad P_{D_2}(\lambda)|0) = 0,$$
$$(0|Q_{A_2}(\lambda) = 0, \quad (0|Q_{D_2}(\lambda) = 0, \tag{1.22}$$
$$(0|0) = 1.$$

Finally, the matrix elements of the $N \times N$ matrix $\widetilde{S}_2$ are equal to

$$\left(\widetilde{S}_2\right)_{jk} = a_2(\lambda_j^C) d_2(\lambda_k^B) t(\lambda_j^C, \lambda_k^B) \exp\{\varphi_{A_2}(\lambda_j^C) + \varphi_{D_2}(\lambda_k^B)\}$$
$$+ a_2(\lambda_k^B) d_2(\lambda_j^C) \exp\{\varphi_{A_2}(\lambda_k^B) + \varphi_{D_2}(\lambda_j^C)\} t(\lambda_k^B, \lambda_j^C). \tag{1.23}$$

Here

$$t(\lambda, \mu) = \frac{g(\lambda, \mu)}{h(\lambda, \mu)} = \frac{(ic)^2}{(\lambda - \mu)(\lambda - \mu + ic)}.$$

The norms of the eigenfunctions in the denominator of (1.17) were evaluated in Chapter X (see, for example, (X.1.10)). In section 4 of Chapter X it was explained how to evaluate the thermodynamic limit of our determinants. Let us apply these ideas to the probability of an empty interval $P(x)$ (1.17). One should also recall the modification of the auxiliary quantum fields which occurs when $\{\lambda^C\} = \{\lambda^B\}$ (see (IX.6.17)). The most convenient quantum field in this case is

$$\mathring{\varphi}(\lambda) = \mathring{p}(\lambda) + \mathring{q}(\lambda) \tag{1.24}$$

(see Chapter IX, formulæ (6.23)–(6.25)), where

$$[\mathring{q}(\lambda), \mathring{q}(\mu)] = [\mathring{p}(\lambda), \mathring{p}(\mu)] = 0,$$

$$[\mathring{p}(\lambda), \mathring{q}(\mu)] = H(\lambda, \mu) = \ln \frac{c^2}{(\lambda - \mu)^2 + c^2}, \qquad (1.25)$$

$$\mathring{p}(\lambda)|0) = 0, \quad (0|\mathring{q}(\lambda) = 0.$$

We should also use the Bethe equations

$$\frac{a(\lambda_j)}{d(\lambda_j)} \prod_{\substack{k \neq j \\ k=1}}^{N} \frac{f(\lambda_j, \lambda_k)}{f(\lambda_k, \lambda_j)} = 1.$$

In the thermodynamic limit, the probability of an empty interval is equal to the ratio of two Fredholm determinants:

$$P(x) = \frac{(0| \det(1 + \widehat{V})|0)}{\det(1 - \widehat{K}/2\pi)}. \qquad (1.26)$$

Here $\widehat{K}$ and $\widehat{V}$ are integral operators on the interval $[-q, q]$ (q is the value of the spectral parameter on the Fermi bounday, see section 3 of Chapter I, formula (I.3.3)). We consider the zero temperature case. The kernel of $\widehat{K}$ is equal to

$$K(\lambda_1, \lambda_2) = \frac{2c}{c^2 + (\lambda_1 - \lambda_2)^2}, \quad \lambda_1, \lambda_2 \in [-q, q] \qquad (1.27)$$

where c is the coupling constant of the nonlinear Schrödinger model and the kernel of the integral operator $\widehat{V}$ is equal to

$$V(\lambda_1, \lambda_2) = -\frac{c}{2\pi} \left[\frac{\exp\{ix\lambda_1/2 + \mathring{\varphi}(\lambda_1)/2\} \exp\{-ix\lambda_2/2 - \mathring{\varphi}(\lambda_2)/2\}}{(\lambda_1 - \lambda_2)(\lambda_1 - \lambda_2 + ic)} \right.$$

$$\left. + \frac{\exp\{ix\lambda_2/2 + \mathring{\varphi}(\lambda_2)/2\} \exp\{-ix\lambda_1/2 - \mathring{\varphi}(\lambda_1)/2\}}{(\lambda_2 - \lambda_1)(\lambda_2 - \lambda_1 + ic)} \right].$$

$$(1.28)$$

We should mention that the integral operator $\widehat{V}$ is not singular. Here we have skipped many details of the derivation, but these details will be presented in full in the sections below for the more general case of the operator $\exp\{\alpha Q_1\}$ for arbitrary α.

XI.2 The determinant representation for the matrix elements of the operator $\exp\{\alpha Q_1\}$ in the generalized model

In this section, the determinant representation for the matrix elements of the operator $\exp\{\alpha Q_1\}$ (similar to the corresponding representation (IX.5.8), (IX.6.12) for scalar products) is derived. The derivation is in the spirit of section IX.6.

Let us begin by rewriting formula (1.13) for the matrix elements. Using the representation (IX.6.12) for scalar products entering the right hand side, one obtains the following identity:

$$\langle 0| \prod_{j=1}^{N} C(\lambda_j^C) \exp\{\alpha Q_1\} \prod_{k=1}^{N} B(\lambda_k^B)|0\rangle$$

$$= \left\{ \prod_{N \geq j > k \geq 1} g(\lambda_j^C, \lambda_k^C) g(\lambda_k^B, \lambda_j^B) \right\}$$

$$\times \sum (-1)^{[P_C]+[P_B]} \left[\prod_I \prod_{II} h(\lambda_I^B, \lambda_{II}^B) \prod_I \prod_{II} h(\lambda_{II}^C, \lambda_I^C) \right]$$

$$\times \langle 0| \det{}_{n_1} \widetilde{S}_1(\lambda_I^C, \lambda_I^B) \det{}_{n_2} \widetilde{S}_2(\lambda_{II}^C, \lambda_{II}^B)|0\rangle. \qquad (2.1)$$

The sum here is taken exactly as in (1.13), i.e., over all the partitions

$$\{\lambda^C\}_N = \{\lambda_I^C\}_{n_1} \cup \{\lambda_{II}^C\}_{n_2},$$
$$\{\lambda^B\}_N = \{\lambda_I^B\}_{n_1} \cup \{\lambda_{II}^B\}_{n_2}, \qquad (2.2)$$
$$n_1 + n_2 = N.$$

$(-1)^{[P_C]}$ is the parity of permutation P_C,

$$P_C \colon \{\lambda_I^C\}_{n_1} \cup \{\lambda_{II}^C\}_{n_2} \longrightarrow \{\lambda^C\}_N, \qquad (2.3)$$

and $(-1)^{[P_B]}$ is the parity of permutation P_B,

$$P_B \colon \{\lambda_I^B\}_{n_1} \cup \{\lambda_{II}^B\}_{n_2} \longrightarrow \{\lambda^B\}_N \qquad (2.4)$$

(the agreement concerning the ordering of elements in the sets is made similarly to (IX.6.3)). The matrices $\widetilde{S}_1(\lambda^C, \lambda^B)$ and $\widetilde{S}_2(\lambda^C, \lambda^B)$ are defined by

$$\left(\widetilde{S}_1(\lambda^C, \lambda^B) \right)_{jk} = \exp \alpha \, d_2(\lambda_j^C) a_2(\lambda_k^B) \left(S_1(\lambda^C, \lambda^B) \right)_{jk},$$
$$\left(\widetilde{S}_2(\lambda^C, \lambda^B) \right)_{jk} = a_1(\lambda_j^C) d_1(\lambda_k^B) \left(S_2(\lambda^C, \lambda^B) \right)_{jk}; \qquad (2.5)$$

where

$$\left(S_1(\lambda^C, \lambda^B) \right)_{jk} = a_1(\lambda_j^C) d_1(\lambda_k^B) t(\lambda_j^C, \lambda_k^B) \exp \left\{ \Phi_{A_1}(\lambda_j^C) + \Phi_{D_1}(\lambda_k^B) \right\}$$
$$\qquad\qquad + a_1(\lambda_k^B) d_1(\lambda_j^C) t(\lambda_k^B, \lambda_j^C) \exp \left\{ \Phi_{A_1}(\lambda_k^B) + \Phi_{D_1}(\lambda_j^C) \right\}$$
$$\left(S_2(\lambda^C, \lambda^B) \right)_{jk} = a_2(\lambda_j^C) d_2(\lambda_k^B) t(\lambda_j^C, \lambda_k^B) \exp \left\{ \Phi_{A_2}(\lambda_j^C) + \Phi_{D_2}(\lambda_k^B) \right\}$$
$$\qquad\qquad + a_2(\lambda_k^B) d_2(\lambda_j^C) t(\lambda_k^B, \lambda_j^C) \exp \left\{ \Phi_{A_2}(\lambda_k^B) + \Phi_{D_2}(\lambda_j^C) \right\}.$$

The matrix $\widetilde{S}_1(\lambda_I^C, \lambda_I^B)$ is an $(n_1 \times n_1)$-dimensional matrix, and the matrix $\widetilde{S}_2(\lambda_{II}^C, \lambda_{II}^B)$ is an $(n_2 \times n_2)$-dimensional matrix.

The auxiliary quantum fields $\Phi_{A_i}(\lambda)$, $\Phi_{D_i}(\lambda)$ $(i = 1, 2)$ are simply the fields entering the representation (IX.6.12),

$$\Phi_{A_i}(\lambda) = Q_{A_i}(\lambda) + P_{D_i}(\lambda),$$
$$\Phi_{D_i}(\lambda) = Q_{D_i}(\lambda) + P_{A_i}(\lambda) \tag{2.6}$$

with momenta P and coordinates Q satisfying the commutation relations (IX.6.8)

$$[P_{D_i}(\lambda), Q_{D_k}(\mu)] = \delta_{ik} \ln h(\lambda, \mu)$$
$$[P_{A_i}(\lambda), Q_{A_k}(\mu)] = \delta_{ik} \ln h(\mu, \lambda) \tag{2.7}$$

with all other commutators zero.

The vacuum $|0\rangle$ and dual vacuum $\langle 0|$ are annihilated by all momenta and by all coordinates, respectively:

$$P_a(\lambda)|0\rangle = 0, \quad \langle 0|Q_a(\lambda) = 0, \quad a = A_i, D_i,$$
$$\langle 0|0\rangle = 1. \tag{2.8}$$

Obtaining (2.1) from (1.13) is quite straightforward. One extracts the product of g-factors from (1.13) using the relation

$$f(\lambda, \mu) = g(\lambda, \mu)h(\lambda, \mu)$$

and the factor $(-1)^{[P_C]+[P_B]}$ appears due to the antisymmetry property of g, $g(\lambda, \mu) = -g(\mu, \lambda)$. We also put factors a_1, d_1, a_2, d_2 (written in (1.13) explicitly) into the corresponding determinants. The factor $\exp\{n_1\alpha\} = (\exp\{\alpha\})^{n_1}$ now enters the first determinant.

The sum in (2.1) is very similar to the sum for

$$\det_N \left(\widetilde{S}_1(\{\lambda^C\}_N, \{\lambda^B\}_N) + \widetilde{S}_2(\{\lambda^C\}_N, \{\lambda^B\}_N) \right),$$

except for the factors of h in the square brackets. To treat these factors it is natural to introduce new auxiliary quantum fields:

$$\varphi_{D_1}(\lambda) = q_{D_1}(\lambda) + p_{A_2}(\lambda), \quad \varphi_{A_1}(\lambda) = q_{A_1}(\lambda) + p_{D_2}(\lambda),$$
$$\varphi_{A_2}(\lambda) = q_{A_2}(\lambda) + p_{D_1}(\lambda), \quad \varphi_{D_2}(\lambda) = q_{D_2}(\lambda) + p_{A_1}(\lambda). \tag{2.9}$$

The commutation relations of momenta p and coordinates q are given as

$$[p_{D_i}(\lambda), q_{D_k}(\mu)] = \delta_{ik} \ln h(\lambda, \mu),$$
$$[p_{A_i}(\lambda), q_{A_k}(\mu)] = \delta_{ik} \ln h(\mu, \lambda) \tag{2.10}$$

with all other commutators equal to zero.

It should be noted that the operators $p(\lambda)$, $q(\lambda)$ commute with the operators $P(\mu)$, $Q(\mu)$ (2.6), (2.7):

$$[p(\lambda), P(\mu)] = [p(\lambda), Q(\mu)] = [q(\lambda), Q(\mu)] = [q(\lambda), P(\mu)] = 0. \tag{2.11}$$

It is not difficult to verify that all fields φ_a, Φ_b commute with one another:

$$[\varphi_a(\lambda),\ \Phi_b(\mu)] = [\varphi_a(\lambda),\ \varphi_b(\mu)] = [\Phi_a(\lambda),\ \Phi_b(\mu)] = 0. \qquad (2.12)$$

Also, similarly to (2.8),

$$p_a(\lambda)|0\rangle = 0, \quad \langle 0|q_a(\lambda) = 0. \qquad (2.13)$$

Now introduce the $N \times N$ matrix M given by

$$M_{jk} = \left(\tilde{S}_1(\lambda^C, \lambda^B)\right)_{jk} \exp\left\{\varphi_{D_1}(\lambda_j^C) + \varphi_{A_1}(\lambda_k^B)\right\}$$
$$+ \left(\tilde{S}_2(\lambda^C, \lambda^B)\right)_{jk} \exp\left\{\varphi_{A_2}(\lambda_j^C) + \varphi_{D_2}(\lambda_k^B)\right\} \qquad (2.14)$$

where the matrices $\tilde{S}$ were defined in (2.5).

Now everything is ready to formulate the main theorem:

Theorem 1. *The matrix elements of the operator* $\exp\{\alpha Q_1\}$ *are proportional to the vacuum mean value in the auxiliary Fock space of the determinant of the matrix* M,

$$\langle 0| \prod_{j=1}^{N} C(\lambda_j^C) \exp\{\alpha Q_1\} \prod_{k=1}^{N} B(\lambda_k^B)|0\rangle = \left\{ \prod_{j>k=1}^{N} g(\lambda_j^C, \lambda_k^C) g(\lambda_k^B, \lambda_j^B) \right\}$$
$$\times \langle 0| \det{}_N M |0\rangle. \qquad (2.15)$$

The determinant of M *is well defined due to (2.12).*

Proof: The proof is similar to that of the theorem (IX.6.12). Formula (2.14) represents M as a sum of two matrices. The determinant of the sum of two matrices can be represented in terms of minors of individual matrices, see the Appendix to Chapter IX (formula A.1.10). Developing $\det_N M$ as a determinant of the sum of two matrices, one obtains a sum very similar to the sum in (2.1). The only difference is that the expression

$$\left[\prod_{I,II} h(\lambda_I^B, \lambda_{II}^B) \prod_{I,II} h(\lambda_{II}^C, \lambda_I^C)\right] \langle 0| \det{}_{n_1} \tilde{S}_1 \det{}_{n_2} \tilde{S}_2 |0\rangle \qquad (2.16)$$

is now replaced by

$$\langle 0| \prod_I \exp\{\varphi_{D_1}(\lambda_I^C) + \varphi_{A_1}(\lambda_I^B)\} \prod_{II} \exp\{\varphi_{A_2}(\lambda_{II}^C) + \varphi_{D_2}(\lambda_{II}^B)\}$$
$$\times \det{}_{n_1} \tilde{S}_1 \det{}_{n_2} \tilde{S}_2 |0\rangle.$$

Due to the commutativity of p, q with P, Q (2.11), this can be rewritten as

$$(0| \prod_I \exp\{\varphi_{D_1}(\lambda_I^C) + \varphi_{A_1}(\lambda_I^B)\} \prod_{II} \exp\{\varphi_{A_2}(\lambda_{II}^C) + \varphi_{D_2}(\lambda_{II}^B)\}|0)$$

$$\times (0| \det{}_{n_1} \tilde{S}_1 \det{}_{n_2} \tilde{S}_2 |0). \tag{2.17}$$

Using relations (2.10), (2.13) and (IX.5.10), one easily obtains that the first vacuum mean value in (2.17) is exactly equal to the factor in the square brackets in (2.16). The theorem is thus proved.

At $\exp \alpha = 0$ we can reproduce the probability of an empty interval, see (1.19).

XI.3 Mean value of $\exp\{\alpha Q_1\}$

The formula for matrix elements of the operator $\exp\{\alpha Q_1\}$ can be further simplified (similarly to (IX.6.26)). We will do this simplification for its diagonal matrix elements which is not as lengthy as the general case. Our aim is to obtain the mean value with respect to the Bethe eigenfunctions; this is necessary to obtain the correlator. This is done in two steps. First, one puts $\{\lambda^C\}_N = \{\lambda^B\}_N = \{\lambda\}_N$ ($\lambda_j^C = \lambda_j^B = \lambda_j$, $j = 1, \ldots, N$) and then imposes on λ_j the system of Bethe equations.

Let us do the first step. Consider the mean value

$$\langle 0| \prod_{j=1}^N C(\lambda_j) \exp\{\alpha Q_1\} \prod_{k=1}^N B(\lambda_k)|0\rangle \equiv (\exp\{\alpha Q_1\})_N \tag{3.1}$$

where all λ_j are different but otherwise arbitrary. In this case, it is not difficult to transform representation (2.15) as follows:

$$(\exp\{\alpha Q_1\})_N = \{\prod_{j>k}^N g(\lambda_j, \lambda_k) g(\lambda_k, \lambda_j)\}\{\prod_{j=1}^N a(\lambda_j) d(\lambda_j)\}$$

$$\times (0| \prod_{j=1}^N \exp\{\varphi_{A_2}(\lambda_j) + \varphi_{D_2}(\lambda_j)\}$$

$$\exp\{\Phi_{A_2}(\lambda_j) + \Phi_{D_2}(\lambda_j)\} \det_N m|0) \tag{3.2}$$

which is done in complete analogy with how formula (IX.6.14) was obtained. (Recall that $a(\lambda) = a_1(\lambda)a_2(\lambda)$, $d(\lambda) = d_1(\lambda)d_2(\lambda)$.) The matrix m entering (3.2) is given by

$$m_{jk} = t_{jk} + t_{kj} \left(m(\lambda_k)m^{-1}(\lambda_j)\right) \exp\{-\varphi_1(\lambda_j) + \varphi_1(\lambda_k)\}$$

$$+ \exp\{\alpha + \varphi_3(\lambda_j) + \varphi_4(\lambda_k)\} \left(r(\lambda_k)r^{-1}(\lambda_j)\right)$$

$$\times \left(t_{kj} + t_{jk}l(\lambda_j)l^{-1}(\lambda_k) \exp\{\varphi_2(\lambda_j) - \varphi_2(\lambda_k)\}\right) \tag{3.3}$$

where $l(\lambda) = a_1(\lambda)/d_1(\lambda)$, $m(\lambda) = a_2(\lambda)/d_2(\lambda)$, $r(\lambda) = a(\lambda)/d(\lambda)$, $t_{jk} \equiv t(\lambda_j, \lambda_k)$. The new fields φ_a introduced here are defined as

$$\begin{aligned}
\varphi_1(\lambda) &= \Phi_{A_2}(\lambda) - \Phi_{D_2}(\lambda) = q_1(\lambda) + p_1(\lambda) \\
\varphi_2(\lambda) &= \Phi_{A_1}(\lambda) - \Phi_{D_1}(\lambda) = q_2(\lambda) + p_2(\lambda) \\
\varphi_3(\lambda) &= \varphi_{D_1}(\lambda) - \varphi_{A_2}(\lambda) - \Phi_{A_2}(\lambda) + \Phi_{D_1}(\lambda) = q_3(\lambda) + p_3(\lambda) \\
\varphi_4(\lambda) &= \varphi_{A_1}(\lambda) - \varphi_{D_2}(\lambda) - \Phi_{D_2}(\lambda) + \Phi_{A_1}(\lambda) = q_4(\lambda) + p_4(\lambda)
\end{aligned} \tag{3.4}$$

with

$$\begin{aligned}
q_1(\lambda) &= Q_{A_2}(\lambda) - Q_{D_2}(\lambda); \quad p_1(\lambda) = P_{D_2}(\lambda) - P_{A_2}(\lambda); \\
q_2(\lambda) &= Q_{A_1}(\lambda) - Q_{D_1}(\lambda); \quad p_2(\lambda) = P_{D_1}(\lambda) - P_{A_1}(\lambda); \\
q_3(\lambda) &= q_{D_1}(\lambda) - q_{A_2}(\lambda) - Q_{A_2}(\lambda) + Q_{D_1}(\lambda); \\
p_3(\lambda) &= p_{A_2}(\lambda) - p_{D_1}(\lambda) - P_{D_2}(\lambda) + P_{A_1}(\lambda); \\
q_4(\lambda) &= q_{A_1}(\lambda) - q_{D_2}(\lambda) - Q_{D_2}(\lambda) + Q_{A_1}(\lambda); \\
p_4(\lambda) &= p_{D_2}(\lambda) - p_{A_1}(\lambda) - P_{A_2}(\lambda) + P_{D_1}(\lambda).
\end{aligned} \tag{3.5}$$

It is worth mentioning that the fields $\varphi_{1,2}$ are analogous to the field φ (IX.6.16). It is easily understood that φ_a are commuting fields:

$$[\varphi_a(\lambda),\, \varphi_b(\mu)] = 0, \quad a, b = 1, 2, 3, 4. \tag{3.6}$$

Commutation relations of momenta p_i and coordinates q_i are calculated by means of (3.5), (2.7), (2.10). They can be represented in the form

$$\begin{aligned}
[p_a(\lambda),\, p_b(\mu)] &= [q_a(\lambda),\, q_b(\mu)] = 0, \\
[p_a(\lambda),\, q_b(\mu)] &= H_{ab}(\lambda, \mu) \quad (a, b = 1, \ldots, 4).
\end{aligned} \tag{3.7}$$

The 4×4 matrix H with matrix elements H_{ab} can be represented as

$$H(\lambda, \mu) = (\ln h(\lambda, \mu))\, A + (\ln h(\mu, \lambda))\, A^T \tag{3.8}$$

where $\ln h$ is a scalar factor and the matrix A is equal to

$$A = \begin{pmatrix} -1 & 0 & 0 & -1 \\ 0 & -1 & 1 & 0 \\ 1 & 0 & -1 & 1 \\ 0 & -1 & 1 & -1 \end{pmatrix}. \tag{3.9}$$

A^T is the transpose of A, $A^T_{ab} = A_{ba}$. To obtain the commutator of fields $p_1(\lambda)$ and $q_3(\mu)$, for example, one takes the $(1, 3)$ element of the matrix H (intersection of the first row and the third column) with the result

$$[p_1(\lambda),\, q_3(\mu)] = (\ln h(\lambda, \mu)) \times 0 + (\ln h(\mu, \lambda)) \times 1 = \ln h(\mu, \lambda)$$

and so on.

The vacuum $|0)$ is annihilated by all the "momenta" p_a and the dual vacuum $(0|$ is annihilated by all the coordinates q_a,

$$p_a|0) = 0, \quad (0|q_a = 0. \tag{3.10}$$

We have defined the matrix m completely. It is worth mentioning, however, that its diagonal matrix elements m_{jj} should be understood in the sense of taking the corresponding limit, as in (IX.7.3).

We turn now to formula (3.2). The state to the left of $\det_N m$ can be represented as

$$(0| \prod_{j=1}^N \exp\{\varphi_{A_2}(\lambda_j) + \varphi_{D_2}(\lambda_j) + \Phi_{A_2}(\lambda_j) + \Phi_{D_2}(\lambda_j)\} = \left\{ \prod_{j,k=1}^N h_{jk} \right\} (\widetilde{0}| \tag{3.11}$$

where the state $(\widetilde{0}|$ is given by

$$(\widetilde{0}| \equiv (0| \prod_{j=1}^N \exp\{p_{D_1}(\lambda_j) + p_{A_1}(\lambda_j) + P_{D_2}(\lambda_j) + P_{A_2}(\lambda_j)\}. \tag{3.12}$$

To obtain this, one uses (2.6), (2.7), (2.9) and (2.10), applies formula (IX.5.10) and then (2.8), (2.13). State $(\widetilde{0}|$ can be considered as a new dual vacuum for the fields $\varphi_a(\lambda)$ with modified eigenvalues of coordinates $q_a(\lambda)$. Indeed, using the definition of q_a in terms of the old fields (see (3.5)) one obtains

$$(\widetilde{0}|q_1(\lambda) = \left\{ \sum_{m=1}^N \ln \frac{h(\lambda, \lambda_m)}{h(\lambda_m, \lambda)} \right\} (\widetilde{0}| \equiv \alpha_1(\lambda)(\widetilde{0}|;$$

$$(\widetilde{0}|q_2(\lambda) = 0 \equiv \alpha_2(\lambda);$$

$$(\widetilde{0}|q_3(\lambda) = \left\{ \sum_{m=1}^N \ln \frac{h(\lambda_m, \lambda)}{h(\lambda, \lambda_m)} \right\} (\widetilde{0}| \equiv \alpha_3(\lambda)(\widetilde{0}|; \tag{3.13}$$

$$(\widetilde{0}|q_4(\lambda) = \left\{ \sum_{m=1}^N \ln \frac{h(\lambda, \lambda_m)}{h(\lambda_m, \lambda)} \right\} (\widetilde{0}| \equiv \alpha_4(\lambda)(\widetilde{0}|.$$

The functions $\alpha_1(\lambda)$, $\alpha_2(\lambda)$, $\alpha_3(\lambda)$ and $\alpha_4(\lambda)$ are defined by these equalities.

Again, as in section IX.6, it is quite natural to introduce quantum fields $\mathring{\varphi}_a(\lambda)$ with zero vacuum expectation values,

$$\mathring{\varphi}_a(\lambda) = \varphi_a(\lambda) - (\widetilde{0}|\varphi_a(\lambda)|0) = \mathring{p}_a(\lambda) + \mathring{q}_a(\lambda);$$

$$(\widetilde{0}|\varphi_a(\lambda)|0) = (\widetilde{0}|q_a(\lambda)|0) = \alpha_a(\lambda). \tag{3.14}$$

The commutation relations between momenta and coordinates do not change (see (3.7)):

$$[\mathring{p}_a(\lambda), \mathring{p}_b(\mu)] = [\mathring{q}_a(\lambda), \mathring{q}_b(\mu)] = 0,$$
$$[\mathring{p}_a(\lambda), \mathring{q}_b(\mu)] = H_{ab}(\lambda, \mu),$$

(3.15)

and one has instead of (3.10)

$$\mathring{p}_a(\lambda)|0\rangle = 0, \quad (\tilde{0}|\mathring{q}_a(\lambda) = 0, \quad (\tilde{0}|0\rangle = 1.$$

(3.15a)

Thus one can rewrite (3.2) as

$$\langle 0| \prod_{j=1}^{N} C(\lambda_j) \exp\{\alpha Q_1\} \prod_{k=1}^{N} B(\lambda_k)|0\rangle$$

$$= \{\prod_{j \neq k} \varphi_{jk}\}\{\prod_{j=1}^{N} a(\lambda_j)d(\lambda_j)\}(\tilde{0}| \det_N m|0\rangle \qquad (3.16)$$

with matrix elements of the $N \times N$ matrix m given by (3.3); it is assumed that the old fields φ_a should be expressed in terms of the new fields $\mathring{\varphi}_a$.

Now recall the Bethe equations

$$l(\lambda_k)m(\lambda_k) \prod_{j=1}^{N} \frac{h(\lambda_k, \lambda_j)}{h(\lambda_j, \lambda_k)} = (-1)^{N-1}.$$

Let us take (3.3) and use the new fields $\mathring{\varphi}_a$ (3.15) and the Bethe equations to obtain (eigenfunction $|\varphi_N(\{\lambda\})\rangle$ is defined by (X.1.8))

$$\langle \varphi_N(\{\lambda\})| \exp\{\alpha Q_1\}|\varphi_N(\{\lambda\})\rangle$$

$$= \left\{\prod_{j>k}^{N} f_{jk}f_{kj}\right\}\left\{\prod_{j=1}^{N} a(\lambda_j)d(\lambda_j)\right\}(\tilde{0}| \det m|0\rangle \qquad (3.17)$$

with matrix m given by

$$m_{jk} = c\delta_{jk}\left[i\frac{\partial}{\partial\lambda_j}\ln r(\lambda_j) + \sum_{n=1}^{N} K(\lambda_j, \lambda_n)\right]$$
$$+ t_{jk} + t_{kj}l^{-1}(\lambda_k)l(\lambda_j)\exp\{-\mathring{\varphi}_1(\lambda_j) + \mathring{\varphi}_1(\lambda_k)\}$$
$$+ e^\alpha \exp\{\mathring{\varphi}_3(\lambda_j) + \mathring{\varphi}_4(\lambda_k)\}$$
$$\times \left\{t_{kj} + t_{jk}l(\lambda_j)l^{-1}(\lambda_k)\exp\{\mathring{\varphi}_2(\lambda_j) - \mathring{\varphi}_2(\lambda_k)\}\right\}.$$

(3.18)

Here we have also used (3.13) and

$$K(\lambda, \mu) = i\frac{\partial}{\partial\lambda}\ln\frac{h(\lambda, \mu)}{h(\mu, \lambda)} = \frac{2c}{(\lambda - \mu)^2 + c^2}.$$

(3.19)

The diagonal matrix elements m_{jj} should again be understood in the sense of the corresponding limit, as in (IX.7.3), since the functions $t_{jk} = t(\lambda_j, \lambda_k) = g(\lambda_j, \lambda_k)/h(\lambda_j, \lambda_k)$ are singular for $\lambda_j \to \lambda_k$ (but m_{jj} is not singular). Recall that the eigenfunction $|\varphi_N(\{\lambda\})\rangle$ is defined as

$$|\varphi_N(\{\lambda\})\rangle = \prod_{j=1}^{N} B(\lambda_j)|0\rangle.$$

Let us also consider the normalized mean value. Using formula (X.1.10) for the "squared norm" of the Bethe eigenfunctions one has (see (3.18))

$$\frac{\langle \varphi_N(\{\lambda\})| \exp\{\alpha Q_1\}|\varphi_N(\{\lambda\})\rangle}{\langle \varphi_N(\{\lambda\})|\varphi_N(\{\lambda\})\rangle} = \frac{\langle \tilde{0}| \det_N m|0\rangle}{c^N \det_N(\varphi')} \qquad (3.20)$$

with the $N \times N$ matrix φ' given by (see (X.1.5))

$$\varphi'_{jk} = \delta_{jk}\left[i\frac{\partial \ln r(\lambda_j)}{\partial \lambda_j} + \sum_{l=1}^{N} K(\lambda_j, \lambda_l)\right] - K(\lambda_j, \lambda_k). \qquad (3.21)$$

It should be noted that the expressions in the square brackets of (3.21) and the square brackets of (3.18) are the same.

Thus we have obtained the determinant representation for the mean value of the operator $\exp\{\alpha Q_1\}$ in the two-site generalized model, which is the main result of this section.

XI.4 Thermodynamic limit in the one-dimensional Bose gas at zero temperature

In this section we turn to the one-dimensional Bose gas (the NS model) described in detail in Chapter I and consider the thermodynamic limit. Here the case of zero temperature ($T = 0$) is considered.

Consider the correlation function

$$\frac{\langle \Omega| \exp\{\alpha Q_1(x)\}|\Omega\rangle}{\langle \Omega|\Omega\rangle} = \langle \exp\{\alpha Q_1(x)\}\rangle \qquad (4.1)$$

where $|\Omega\rangle$ is the ground state of the gas (the "Dirac sea") constructed in section I.3; the operator $Q_1(x)$,

$$Q_1(x) = \int_0^x j(y)\,dy, \quad j(y) = \Psi^\dagger(y)\Psi(y), \qquad (4.2)$$

is the operator of number of particles in the interval $[0, x]$. As explained in section 1 ((1.1)–(1.4)), the equal-time correlator of currents is easily retrieved from the generating correlator,

$$\langle j(x) \rangle = \frac{\partial}{\partial x} \frac{\partial}{\partial \alpha} \langle \exp\{\alpha Q_1(x)\} \rangle \Big|_{\alpha=0} ; \qquad (4.3)$$

$$\langle j(x)j(0) \rangle = \frac{1}{2} \frac{\partial^2}{\partial x^2} \frac{\partial^2}{\partial \alpha^2} \langle \exp\{\alpha Q_1(x)\} \rangle \Big|_{\alpha=0} . \qquad (4.4)$$

The normalized mean value of the operator $\exp\{\alpha Q_1(x)\}$ at $\alpha = -\infty$ possesses a clear physical meaning, giving the probability that there are no particles in the interval $[0, x]$ in the ground state:

$$P(x) \equiv \langle \exp\{\alpha Q_1(x)\} \rangle \Big|_{\alpha=-\infty} . \qquad (4.5)$$

The function $P(x)$ will be called the "emptiness formation probability," see (1.17)–(1.26).

To obtain the representation for the correlator (4.1), one goes to the thermodynamic limit in formula (3.20), which is valid in the generalized model (the number of particles $N \to \infty$, the length of the box $L \to \infty$, with the gas density $D = N/L$ remaining finite, $|\varphi_N\rangle \to |\Omega\rangle$). It should be taken into account that now, as follows from (1.16), the functions $r(\lambda)$ and $l(\lambda)$ in (3.18) are given by

$$r(\lambda) = \frac{a(\lambda)}{d(\lambda)} = \exp\{-i\lambda L\},$$

$$l(\lambda) = \frac{a_1(\lambda)}{d_1(\lambda)} = \exp\{-ix\lambda\}. \qquad (4.6)$$

The thermodynamic limit of the determinant of matrix φ' in the denominator of (3.20) has already been calculated in section X.4 (see (X.4.10)) with the result

$$\lim \det_N \varphi' = \prod_{j=1}^{N} \left(2\pi L \rho(\lambda_j) \right) \det \left(1 - \frac{1}{2\pi} \widehat{K} \right). \qquad (4.7)$$

The determinant on the right hand side is the Fredholm determinant of the linear integral operator $\widehat{K}$ given by (X.4.8):

$$\left(\widehat{K} \xi \right) (\lambda) = \int_{-q}^{q} K(\lambda, \mu) \xi(\mu) \, d\mu,$$

$$K(\lambda, \mu) = \frac{2c}{(\lambda - \mu)^2 + c^2} \qquad (4.8)$$

(q being the value of the spectral parameter at the Fermi boundary).

The thermodynamic limit of the determinant entering the numerator on the right hand side of (3.20) is obtained in a similar way. The fields

$\overset{\circ}{\varphi}_a$ in the expression (3.18) for the matrix m commute with one another, so the determinant is well defined.

Consider the matrix $\widetilde{m}$:

$$\widetilde{m}_{jk} = \frac{1}{c} m_{jk}, \quad \det{}_N \widetilde{m} = c^{-N} \det{}_N m. \tag{4.9}$$

The term proportional to δ_{jk} in (3.18) is easily calculated for the NS model (1.16), giving a contribution

$$\delta_{jk} \left[L + \sum_{l=1}^{N} K(\lambda_j, \lambda_l) \right] \tag{4.10}$$

to $\widetilde{m}_{jk}$ which appears to be just the same as the corresponding contribution to the matrix φ'_{jk} (3.21), (X.4.2). Representing the matrix $\widetilde{m}$ in a way similar to (X.4.3) with the same diagonal matrix θ, one obtains the obvious answer for the thermodynamic limit of $\widetilde{m}$:

$$\lim \det{}_N \widetilde{m} = \lim \frac{1}{c^N} \det{}_N m$$

$$= \prod_{j=1}^{N} \left(2\pi L \rho(\lambda_j) \right) \det \left(1 + \frac{1}{2\pi} \widehat{V} \right) \tag{4.11}$$

where on the right hand side we have the Fredholm determinant of the linear operator $\widehat{V}$,

$$\left(\widehat{V} \xi \right)(\lambda) = \int_{-q}^{q} V(\lambda, \mu) \xi(\mu) \, d\mu, \tag{4.12}$$

with kernel $V(\lambda, \mu)$ given by

$$V(\lambda, \mu) = \frac{1}{c} \left[t(\lambda, \mu) + t(\mu, \lambda) \exp\{-ix(\lambda - \mu)\} \exp\{\overset{\circ}{\varphi}_1(\mu) - \overset{\circ}{\varphi}_1(\lambda)\} \right.$$
$$+ e^\alpha \exp\{\overset{\circ}{\varphi}_3(\lambda) + \overset{\circ}{\varphi}_4(\mu)\}$$
$$\left. \times \left(t(\mu, \lambda) + t(\lambda, \mu) \exp\{-ix(\lambda - \mu)\} \exp\{\overset{\circ}{\varphi}_2(\lambda) - \overset{\circ}{\varphi}_2(\mu)\} \right) \right] \tag{4.13}$$

where

$$t(\lambda, \mu) = \frac{g(\lambda, \mu)}{h(\lambda, \mu)} = -\frac{c^2}{(\lambda - \mu)(\lambda - \mu + ic)}. \tag{4.14}$$

In (4.12) $\xi(\lambda)$ is some c-number function.

Thus one arrives at the following answer for the normalized mean value (4.1) of the operator $\exp\{\alpha Q_1(x)\}$ in the one-dimensional Bose gas at zero temperature:

$$\langle \exp\{\alpha Q_1(x)\}\rangle = \frac{(\widetilde{0}|\det\left(1 + \frac{1}{2\pi}\widehat{V}\right)|0)}{\det\left(1 - \frac{1}{2\pi}\widehat{K}\right)}. \tag{4.15}$$

The determinants here are Fredholm determinants of the linear integral operators $\widehat{K}$ (4.8) and $\widehat{V}$ (4.12).

Due to the commutativity of the fields $\mathring{\varphi}_a(\lambda)$,

$$[\mathring{\varphi}_a(\lambda), \mathring{\varphi}_b(\lambda)] = 0, \tag{4.16}$$

entering the kernel $V(\lambda, \mu)$ (4.13), (4.14) of the operator $\widehat{V}$, its Fredholm determinant is well defined. Calculating the mean value

$$(\widetilde{0}|\det\left(1 + \frac{1}{2\pi}\widehat{V}\right)|0)$$

in the auxiliary Fock space is the problem of the normal ordering of products of the fields $\mathring{\varphi}_a$. Decomposing (see (3.14))

$$\mathring{\varphi}_a(\lambda) = \mathring{q}_a(\lambda) + \mathring{p}_a(\lambda), \tag{4.17}$$

and using the commutation relations (3.15)

$$[\mathring{q}_a(\lambda), \mathring{q}_b(\mu)] = [\mathring{p}_a(\lambda), \mathring{p}_b(\mu)] = 0,$$
$$[\mathring{p}_a(\lambda), \mathring{q}_b(\mu)] = H_{ab}(\lambda, \mu) \tag{4.18}$$

one should move all the "momenta" $\mathring{p}_a$ to the right and all the "coordinates" $\mathring{q}_a$ to the left. After this is done, they can be put "equal to zero" as their action on the vacuum $|0\rangle$ and on the dual vacuum $(\widetilde{0}|$ is trivial (3.15a):

$$\mathring{p}_a(\lambda)|0\rangle = 0, \quad (\widetilde{0}|\mathring{q}_a(\lambda) = 0, \quad (\widetilde{0}|0) = 1. \tag{4.19}$$

The representation for the generator of current correlators is thus obtained. At $\exp\alpha = 0$ we reproduce the probability of an empty interval (1.26).

XI.5 Temperature-dependent correlation function

The thermodynamics of the Bose gas (NS model) was described in Chapter I, sections 5–8. The correlation function of $\exp\{\alpha Q_1(x)\}$ in the one–dimensional Bose gas at finite temperature is defined as usual:

$$\langle \exp\{\alpha Q_1(x)\}\rangle_T = \frac{\mathrm{tr}\left(\exp\{-H/T\}\exp\{\alpha Q_1(x)\}\right)}{\mathrm{tr}\left(\exp\{-H/T\}\right)}. \tag{5.1}$$

Here $Q_1(x)$ (4.2) is the operator of the number of particles in the interval $[0, x]$. The Hamiltonian H is given by (I.3.14):

$$H = \int dx \left(\partial_x \Psi^\dagger \partial_x \Psi + c \Psi^\dagger \Psi^\dagger \Psi \Psi - h \Psi^\dagger \Psi \right). \tag{5.2}$$

The coupling constant c is taken as positive.

The traces in (5.1) are taken in the quantum space of the gas where the operators Ψ, $\Psi^\dagger$ act (the original Fock space of Bose type). In the thermodynamic limit, the essential contributions to the traces are due to the eigenfunctions of the Hamiltonian, $|\Omega_T\rangle$, each with the same energy E_T, and characterized macroscopically by the same particle density in momentum space, $\rho_p(\lambda)$, and the same density of vacancies $\rho_t(\lambda)$ (I.5.3)–(I.5.6). This follows from evaluation of the functional integrals by, like (I.5.16), the method of steepest descent.

The ratio

$$\vartheta(\lambda) = \frac{\rho_p(\lambda)}{\rho_t(\lambda)} = \frac{1}{1 + e^{\varepsilon(\lambda)/T}} \tag{5.3}$$

is the Fermi weight. The energy $\varepsilon(\lambda)$ is the solution to the Yang-Yang equation (I.5.25)

$$\varepsilon(\lambda) = \lambda^2 - h - \frac{T}{2\pi} \int\limits_{-\infty}^{\infty} K(\lambda, \mu) \ln \left(1 + e^{-\varepsilon(\mu)/T} \right) d\mu \tag{5.4}$$

with kernel $K(\lambda, \mu)$ given by

$$K(\lambda, \mu) = \frac{2c}{(\lambda - \mu)^2 + c^2}. \tag{5.5}$$

It should also be noted that the particle density and the density of vacancies are related by equation (I.5.24):

$$2\pi \rho_t(\lambda) = 1 + \int\limits_{-\infty}^{\infty} K(\lambda, \mu) \rho_p(\mu) \, d\mu. \tag{5.6}$$

The density, D, of the gas in x-space is

$$D = \int\limits_{-\infty}^{\infty} \rho_p(\lambda) \, d\lambda. \tag{5.7}$$

Let us consider the mean value

$$\frac{\langle \Omega_T | \exp\{\alpha Q_1(x)\} | \Omega_T \rangle}{\langle \Omega_T | \Omega_T \rangle}. \tag{5.8}$$

Here $|\Omega_T\rangle$ is one of the eigenvectors which are present in the state of thermodynamic equilibrium (see sections 5–8 of Chapter I). To describe the relation between different eigenvectors $|\Omega_T\rangle$ it is useful to recall the periodic boundary conditions in logarithmic form (see Chapter I, section 2, formula (I.2.13)). Different eigenvectors $|\Omega_T\rangle$ correspond to different sets of integers n_j, but all the $|\Omega_T\rangle$ have the same distribution law $\rho_h(\lambda)$.

The densities (I.5.3)–(I.5.5)

$$\rho_p(\lambda), \rho_t(\lambda), \rho_h(\lambda) \tag{5.9}$$

are the same for all eigenvectors $|\Omega_T\rangle$ in the state of themodynamic equilibrium. According to section I.8 we should consider the mean value (5.8).

To evaluate the mean value (5.8) one takes the corresponding thermodynamic limit in the representation (3.20),

$$\frac{\langle \Omega_T | \exp\{\alpha Q_1(x)\} | \Omega_T \rangle}{\langle \Omega_T | \Omega_T \rangle} = \lim \frac{\langle \tilde{0} | \det_N m | 0 \rangle}{c^N \det_N \varphi'} \tag{5.10}$$

taking into account the relations (4.6), valid for the one-dimensional Bose gas. The limit of the determinant in the denominator has in fact already been calculated. Taking into account that the ratio ρ_p/ρ_t is the Fermi weight $\vartheta(\lambda)$ (5.3), one rewrites (X.4.14) as

$$\lim \det_N \varphi' = \prod_{j=1}^{N} \left(2\pi L \rho_t(\lambda_j) \right) \det \left(1 - \frac{1}{2\pi} \widehat{K}_T \right) \tag{5.11}$$

where $\widehat{K}_T$ is the linear integral operator

$$\left(\widehat{K}_T \xi \right)(\lambda) = \int_{-\infty}^{\infty} K_T(\lambda, \mu) \xi(\mu) \, d\mu \tag{5.12}$$

with kernel $K_T(\lambda, \mu)$ given by

$$K_T(\lambda, \mu) = K(\lambda, \mu) \vartheta(\mu) \tag{5.13}$$

(see (X.4.13); in (5.12) $\xi(\lambda)$ is an arbitrary function).

Calculating the thermodynamic limit of the numerator in (5.10) in a similar way results in the following answer:

$$\lim c^{-N} \langle \tilde{0} | \det_N m | 0 \rangle = \left(\prod_{j=1}^{N} 2\pi L \rho_t(\lambda_j) \right) \langle \tilde{0} | \det \left(1 + \frac{1}{2\pi} \widehat{V}_T \right) | 0 \rangle \tag{5.14}$$

where the linear integral operator $\widehat{V}_T$,

$$\left(\widehat{V}_T \xi \right)(\lambda) = \int_{-\infty}^{\infty} V_T(\lambda, \mu) \xi(\mu) \, d\mu, \tag{5.15}$$

possesses the kernel

$$V_T(\lambda, \mu) = V(\lambda, \mu)\vartheta(\mu) \tag{5.16}$$

with the function $V(\lambda, \mu)$ given by formula (4.13). Finally we get

$$\frac{\langle \Omega_T | \exp\{\alpha Q_1(x)\} | \Omega_T \rangle}{\langle \Omega_T | \Omega_T \rangle} = \frac{(\tilde{0}| \det\left(1 + \frac{1}{2\pi}\widehat{V}_T\right) |0)}{\det\left(1 - \frac{1}{2\pi}\widehat{K}_T\right)}.$$

Now we see that (5.10) and (5.8) depend only on macroscopic characteristics of the thermodynamic equilibrium $(\rho_p(\lambda), \rho_t(\lambda))$:

$$\frac{\rho_p(\lambda)}{\rho_t(\lambda)} = \frac{1}{1 + e^{\varepsilon(\lambda)/T}}.$$

We have proved that (5.8) is the same for all eigenvectors $|\Omega_T\rangle$ of the Hamiltonian that are present in the state of thermodynamic equilibrium.

Now we can apply the results of section I.8 (see formulæ (I.8.19) and (I.8.21)):

$$\langle \exp\{\alpha Q_1(x)\}\rangle_T = \frac{\text{tr}\left(e^{-H/T}\exp\{\alpha Q_1(x)\}\right)}{\text{tr}\, e^{-H/T}} = \frac{\langle \Omega_T | \exp\{\alpha Q_1(x)\} | \Omega_T \rangle}{\langle \Omega_T | \Omega_T \rangle}.$$

Finally, one arrives at the following result for the temperature correlator (5.1):

$$\langle \exp\{\alpha Q_1(x)\}\rangle = \frac{(\tilde{0}| \det\left(1 + \frac{1}{2\pi}\widehat{V}_T\right) |0)}{\det\left(1 - \frac{1}{2\pi}\widehat{K}_T\right)} \tag{5.17}$$

where the Fredholm determinants of the operators $\widehat{K}_T$ (5.12), (5.13) and $\widehat{V}_T$ (5.15), (5.16) enter the denominator and the numerator on the right hand side.

It should be mentioned that formula (5.17) differs from the corresponding formula (4.15) for the zero temperature correlator only in a change of the integration measure (I.8.5)

$$\int\limits_{-q}^{q} d\lambda \longrightarrow \int\limits_{-\infty}^{\infty} d\lambda\, \vartheta(\lambda) \tag{5.18}$$

confirming the discussion in section I.8. This was proved in [1], [2].

It should be noted that instead of operators $\widehat{V}$ and $\widehat{V}_T$ one can use operators $\widehat{V}^{(x)}$, $\widehat{V}^{(x)}$ with kernels

$$V^{(x)}(\lambda, \mu) = \chi(\lambda)V(\lambda, \mu)\chi^{-1}(\mu); \quad V_T^{(x)}(\lambda, \mu) = \chi(\lambda)V_T(\lambda, \mu)\chi^{-1}(\mu) \tag{5.19}$$

where the function $\chi(\lambda)$ is more or less arbitrary (it can also include commuting fields). The Fredholm determinant obviously is not changed under (5.19). In particular, the function χ can be chosen so that the new kernel is symmetric in λ, μ. This is a similarity transformation of the integral operator $\widehat{V}$, which does not change the determinant.

Conclusion

In this chapter we have discovered an extremely important result: quantum correlation functions can be represented as determinants. In Part IV we shall see that this helps us to write down differential equations for quantum correlation functions. The techniques of this chapter are very close to those of Chapter IX (section 6). Similar techniques will be used in the next chapter.

The two-site generalized model and the results of section 1 were obtained in [3] (for the corresponding generalization to the XXZ case see [4]). The main results of this chapter were obtained in papers [5–6]. Temperature dependent correlation functions were first considered in [1] (see also [2]).

XII

Correlation Function of Fields

Introduction

The aim of this chapter is to obtain the determinant representation for the equal-time correlation function $\langle \Psi^\dagger(x)\Psi(0)\rangle$ in the nonrelativistic one-dimensional Bose gas (NS model). To apply the approach of the previous chapters, one has to introduce local fields $\Psi^\dagger(x)$, $\Psi(0)$ into the two-site generalized model. This is done in section 1. In section 2, the formula for the matrix elements of the operator $\Psi^\dagger(x)\Psi(0)$ in the generalized model are derived. In terms of the auxiliary quantum fields introduced in section 3, this formula is transformed to the vacuum mean value of the determinant with respect to the vacuum in the auxiliary Fock space. The determinant representation for the mean value of the operator $\Psi^\dagger(x)\Psi(0)$ with respect to the N-particle Bethe eigenstate is derived in section 4. The derivation is similar to that of the previous chapter. It is also similar to that of section 6 of Chapter IX. The derivation is based on the representation of the determinant of the sum of two matrices in terms of the minors of the individual matrices. The explicit formula is given in the Appendix to Chapter IX. In section 5 expressions for the correlation function in the ground state of the one-dimensional Bose gas (zero and finite temperature) are obtained and the thermodynamic limit is also considered.

XII.1 Local fields in the generalized model

To consider local quantum fields $\Psi^\dagger(x)\Psi(0)$ in the frame of the two-site generalized model, one has to clarify the structure of the monodromy matrices $\mathsf{T}(1|\lambda)$, $\mathsf{T}(2|\lambda)$ (see section XI.1, formulæ (XI.1.6)–(XI.1.12)).

The monodromy matrix of the NS model is the product of the corresponding infinitesimal L-operators (see (VI.1.1), (VI.3.14)) located at the sites of the one-dimensional space lattice (spacing $\Delta \to 0$) with M sites

$$T_{NS}(\lambda) = L_M(\lambda)L_{M-1}(\lambda) \cdots L_1(\lambda) \tag{1.1}$$

where

$$L_n(\lambda) = \begin{pmatrix} 1 - \dfrac{i\lambda\Delta}{2} & -i\sqrt{c}\Psi_n^\dagger\Delta \\ i\sqrt{c}\Psi_n\Delta & 1 + \dfrac{i\lambda\Delta}{2} \end{pmatrix} + O\left(\Delta^2\right). \tag{1.2}$$

The operators Ψ_n, $\Psi_n^\dagger$ here are lattice approximations of the fields $\Psi(x)$, $\Psi^\dagger(x)$

$$\Psi_n \sim \Psi(x = n\Delta), \quad \Psi_n^\dagger \sim \Psi^\dagger(x = n\Delta);$$
$$[\Psi_n, \Psi_m^\dagger] = \frac{1}{\Delta}\delta_{nm}, \quad [\Psi_n, \Psi_m] = [\Psi_n^\dagger, \Psi_m^\dagger] = 0. \tag{1.3}$$

The commutation relations between the matrix elements of the operator L_n are of the same form as those of the monodromy matrix (VI.1.4)–(VI.1.6) and are given by the XXX R-matrix (VII.1.3), (VII.1.4). The matrix elements of L-operators at different sites of the lattice commute. The operator L_n possesses the bare Fock vacuum $|\omega_n\rangle$,

$$\Psi_n|\omega_n\rangle = 0, \tag{1.4}$$

with the corresponding vacuum eigenvalues

$$a_L(\lambda) = 1 - \frac{i\lambda\Delta}{2}, \quad d_L(\lambda) = 1 + \frac{i\lambda\Delta}{2}. \tag{1.5}$$

Now let us turn to the generalized two-site model of section XI.1. Let us explicitly write down the L-operators in the vicinity of points x and 0 where $\Psi^\dagger(x)$ and $\Psi(0)$ are located. (Remember that we are going to calculate the correlation function $\langle\Psi^\dagger(x)\Psi(0)\rangle$.) We shall write the monodromy matrices $T(i|\lambda)$ ($i = 1, 2$) (XI.1.6), (XI.1.7) in the form

$$T(2|\lambda) = \tilde{T}(2|\lambda)L_n(\lambda), \quad n\Delta = x;$$
$$T(1|\lambda) = \tilde{T}(1|\lambda)L_1(\lambda), \quad \Delta \to 0. \tag{1.6}$$

The monodromy matrices $\tilde{T}(i|\lambda)$ are "arbitrary" and possess all the properties of $T(i|\lambda)$, i.e., their matrix elements are intertwined by the same XXX R-matrix and they possess vacua $|\tilde{0}\rangle_i$, $_i\langle\tilde{0}|$ with eigenvalues $\tilde{a}_i(\lambda)$, $\tilde{d}_i(\lambda)$ (see (XI.1.8), (XI.1.9)). It is then evident that the matrices $T(i|\lambda)$ (1.6) possess the following Fock vacua,

$$|0\rangle_2 = |\tilde{0}\rangle_2 \otimes |\omega_n\rangle, \quad _2\langle 0| = {}_2\langle\tilde{0}| \otimes \langle\omega_n|,$$
$$|0\rangle_1 = |\tilde{0}\rangle_1 \otimes |\omega_1\rangle, \quad _1\langle 0| = {}_1\langle\tilde{0}| \otimes \langle\omega_1|, \tag{1.7}$$

with corresponding vacuum eigenvalues

$$a_i(\lambda) = \tilde{a}_i(\lambda) \left(1 - \frac{i\lambda\Delta}{2}\right), \quad d_i(\lambda) = \tilde{d}_i(\lambda) \left(1 + \frac{i\lambda\Delta}{2}\right). \tag{1.8}$$

It should be emphasized that the functions $a_i(\lambda)$, $d_i(\lambda)$ are arbitrary functional parameters of the generalized model.

Further, we are interested in correlation functions for the generalized model in the continuous limit, where $\Delta \to 0$, $n\Delta = x$, $M\Delta = L$ (L is the length of the box), $a_i(\lambda) \to \tilde{a}_i(\lambda)$, $d_i(\lambda) \to \tilde{d}_i(\lambda)$ (see (1.8)). The commutation relations of the lattice fields $\Psi_1 \sim \Psi(0)$, $\Psi_1^\dagger \sim \Psi^\dagger(0)$, $\Psi_n \sim \Psi(x)$, $\Psi_n^\dagger \sim \Psi^\dagger(x)$ with the corresponding monodromy matrices do not, however, vanish in the continuous limit as the commutator in (1.3) contains a factor Δ^{-1}.

As we are interested in the mean value of the operator $\Psi^\dagger(x)\Psi(0)$, it is necessary to have commutation relations of these two operators with the matrix elements of the monodromy matrices. Using (1.6), (1.2), (1.3), one easily obtains in the continuous limit the following relations:

$$[\mathsf{T}(2|\lambda), \Psi^\dagger(x)] = i\sqrt{c}\mathsf{T}(2|\lambda) \begin{pmatrix} 0 & 0 \\ 1 & 0 \end{pmatrix}; \tag{1.9}$$

$$[\mathsf{T}(1|\lambda), \Psi^\dagger(x)] = 0; \tag{1.10}$$

$$[\mathsf{T}(2|\lambda), \Psi(0)] = 0; \tag{1.11}$$

$$[\mathsf{T}(1|\lambda), \Psi(0)] = i\sqrt{c}\mathsf{T}(1|\lambda) \begin{pmatrix} 0 & 1 \\ 0 & 0 \end{pmatrix}.$$

The last two formulæ can be put in the form

$$[\mathsf{T}(\lambda), \Psi(0)] = i\sqrt{c}\mathsf{T}(\lambda) \begin{pmatrix} 0 & 1 \\ 0 & 0 \end{pmatrix}. \tag{1.12}$$

The following relations, which follow from (1.10)–(1.12), will play an important role:

$$[C_2(\lambda), \Psi^\dagger(x)] = i\sqrt{c}D_2(\lambda), \quad [C_1(\lambda), \Psi^\dagger(x)] = 0, \tag{1.13}$$
$$[\Psi(0), B_1(\lambda)] = -i\sqrt{c}A_1(\lambda), \quad [\Psi(0), B_2(\lambda)] = 0, \tag{1.14}$$
$$[\Psi(0), B(\lambda)] = -i\sqrt{c}A(\lambda). \tag{1.15}$$

It is also evident that (see (1.7))

$$\langle 0|\Psi^\dagger(x) = {}_2\langle 0|\Psi^\dagger(x) = 0, \quad \Psi(0)|0\rangle = \Psi(0)|0\rangle_1 = 0. \tag{1.16}$$

Formulæ (1.9)–(1.12) are in fact the definition of local fields $\Psi^\dagger(x)$, $\Psi(0)$ in the two-site generalized model. Thus the problem of calculating the mean value of the operator $\Psi^\dagger(x)\Psi(0)$ is naturally formulated in this model.

XII.2 Matrix elements of the operator
$\Psi^\dagger(x)\Psi(0)$ in the generalized model

Consider the matrix element G_N defined by

$$G_N \equiv \langle 0| \prod_{j=1}^{N} C(\lambda_j^C)\, \Psi^\dagger(x)\Psi(0) \prod_{k=1}^{N} B(\lambda_k^B)|0\rangle \qquad (2.1)$$

in the generalized model. Using the commutator (1.15) and relation (1.16) one easily calculates the action of the operator $\Psi(0)$ to the right:

$$\Psi(0) \prod_{k=1}^{N} B(\lambda_k^B)|0\rangle$$

$$= -i\sqrt{c} \sum_{k=1}^{N} B(\lambda_1^B)\cdots B(\lambda_{k-1}^B)A(\lambda_k^B)B(\lambda_{k+1}^B)\cdots B(\lambda_N^B)|0\rangle. \qquad (2.2)$$

To go further, we use an idea very similar to that used in section VII.1, formulæ (VII.1.25)–(VII.1.30), in the scheme of the algebraic Bethe Ansatz. Let us use the commutation relation (VII.1.12) to move all the operators $A(\lambda_k)$ on the right hand side of (2.2) to the right. As

$$A(\lambda)|0\rangle = a(\lambda)|0\rangle,$$

the result is easily seen to be of the form

$$\Psi(0) \prod_{k=1}^{N} B(\lambda_k^B)|0\rangle = -i\sqrt{c} \sum_{k=1}^{N} \Lambda_k a(\lambda_k^B) \prod_{\substack{m=1\\m\neq k}}^{N} B(\lambda_m^B)|0\rangle \qquad (2.3)$$

where the coefficients Λ_k (expressed in terms of functions $f(\lambda_k^B, \lambda_j^B)$, $g(\lambda_k^B, \lambda_j^B)$) are rational functions of λ^B's. To calculate these coefficients, one first notices that due to the commutativity (VII.1.11) of the operators B, the right hand side is a symmetric function of all λ^B's. Hence, it is sufficient to calculate the coefficient Λ_1 only. The corresponding contribution can be obtained from the first term with $k=1$ in the sum (2.2). Now one has to move the operator $A(\lambda_1^B)$ to the right using only the first term in the commutation relation (VII.1.12). The result is

$$\Lambda_1 = \prod_{m=2}^{N} f(\lambda_1, \lambda_m).$$

Due to the symmetry mentioned above, one has for Λ_k

$$\Lambda_k = \prod_{\substack{m=1\\m\neq k}}^{N} f(\lambda_k, \lambda_m). \qquad (2.4)$$

In what follows we will also need another expression for the right hand side of (2.3). Namely, we will represent it as

$$\Psi(0) \prod_{k=1}^{N} B(\lambda_k^B)|0\rangle$$

$$= -i\sqrt{c} \sum a_1(\lambda_0^B) a_2(\lambda_0^B) \prod_I a_2(\lambda_I^B) \prod_{II} d_1(\lambda_{II}^B) \prod_{I,II} f(\lambda_I^B, \lambda_{II}^B)$$

$$\times \prod_I f(\lambda_0^B, \lambda_I^B) \prod_{II} f(\lambda_0^B, \lambda_{II}^B) \left(\prod_I B_1(\lambda_I^B)|0\rangle_1 \right) \otimes \left(\prod_{II} B_2(\lambda_{II}^B)|0\rangle_2 \right).$$

$$(2.5)$$

The sum here is taken over all partitions

$$\{\lambda_I^B\}_{n_1} \cup \{\lambda_{II}^B\}_{n_2} \cup \{\lambda_0^B\}$$

of the set $\{\lambda^B\}_N$ into three disjoint sets such that there is only one element in the last set $\{\lambda_0^B\}$, and $n_1 + n_2 + 1 = N$. Formula (2.5) is easily obtained by putting expression (VII.5.9) directly into (2.3). Another way of deriving formula (2.5) is to put the state vector $\prod B(\lambda^B)|0\rangle$ on the left hand side of (2.2) into the form (VII.5.9) and then use commutation relations (1.14). In this approach, we only need to use formulæ (2.3), (2.4) for the first site of the lattice:

$$\Psi(0) \prod_I^n B_1(\lambda_I^B)|0\rangle_1 = -i\sqrt{c} \sum_{\{\lambda_0\}} a_1(\lambda_0^B) \prod_{I'} f(\lambda_0^B, \lambda_{I'}^B) \prod_{I'} B_1(\lambda_{I'}^B)|0\rangle_1.$$

$$(2.6)$$

Here the set $\{\lambda_0\}$ consists of only one element and $\{\lambda_I\} = \{\lambda_0\}_1 \cup \{\lambda_{I'}\}_{n-1}$ is a partition into two disjoint sets. This is another way of proving formula (2.5).

Now we shall evaluate the action of the operator $\Psi^\dagger(x)$ on the vector $\langle 0| \prod C(\lambda_j^C)$. We shall use formula (VII.5.13) to represent the vector $\langle 0| \prod C(\lambda^C)$ in terms of $_2\langle 0| \prod C_2(\lambda^C)$ and $_1\langle 0| \prod C_1(\lambda^C)$. Next we shall evaluate the action of $\Psi^\dagger(x)$ on $_2\langle 0| \prod C_2(\lambda^C)$. To do this, we shall use (1.13) and (1.16) in order to obtain

$$_2\langle 0| \prod_{II} C_2(\lambda_{II}^C)\Psi^\dagger(x) = i\sqrt{c} \sum \,_2\langle 0| \prod_{II'} C_2(\lambda_{II'}^C) \prod_{II'} f(\lambda_{II'}^C, \lambda_0^C) d_2(\lambda_0^C)$$

$$(2.7)$$

where the sum is over all partitions

$$\{\lambda_{II}^C\}_n = \{\lambda_{II'}^C\}_{n-1} \cup \{\lambda_0\}_1.$$

One should emphasize that formula (2.7) is very similar to (2.3), (2.4), (2.6). Now combining this formula with (VII.5.13) we obtain

$$\langle 0| \prod_{j=1}^{N} C(\lambda_j^C)\Psi^\dagger(x)$$

$$= i\sqrt{c}\sum a_1(\lambda_0^C)d_2(\lambda_0^C)\left(_1\langle 0| \prod_I C_1(\lambda_I^C)\right) \otimes \left(_2\langle 0| \prod_{II} C_2(\lambda_{II}^C)\right)$$

$$\times \prod_I d_2(\lambda_I^C) \prod_{II} a_1(\lambda_{II}^C) \prod_I f(\lambda_0^C,\lambda_I^C) \prod_{II} f(\lambda_{II}^C,\lambda_0^C) \prod_{I,II} f(\lambda_{II}^C,\lambda_I^C).$$

$$(2.8)$$

The sum here is taken over all partitions of the set $\{\lambda^C\}_N$ into three disjoint sets $\{\lambda_0^C\}_1$, $\{\lambda_I^C\}_{n_1}$, $\{\lambda_{II}^C\}_{n_2}$, $n_1 + n_2 + 1 = N$.

Finally, combining (2.5) and (2.8), one obtains an expression for the matrix element G_N (2.1)

$$G_N = c\sum \left(_1\langle 0\| \prod_I C_1(\lambda_I^C)\prod_I B_1(\lambda_I^B)\rangle 0_1\right)\left(_2\langle 0\| \prod_{II} C_2(\lambda_{II}^C)\prod_{II} B_2(\lambda_{II}^B)\rangle 0_2\right)$$

$$\times a_1(\lambda_0^C)d_2(\lambda_0^C)a_1(\lambda_0^B)a_2(\lambda_0^B) \prod_{I,II} f_{I,II}^{BB} f_{II,I}^{CC}$$

$$\times \prod_I a_2(\lambda_I^B)d_2(\lambda_I^C)f_{0I}^{CC} f_{0I}^{BB}$$

$$\times \prod_{II} d_1(\lambda_{II}^B)a_1(\lambda_{II}^C)f_{II0}^{CC} f_{0II}^{BB}.$$

$$(2.9)$$

The sum here is taken over all the partitions

$$\{\lambda^C\}_N = \{\lambda_I^C\}_{n_1} \cup \{\lambda_{II}^C\}_{n_2} \cup \{\lambda_0^C\}_1,$$
$$\{\lambda^B\}_N = \{\lambda_I^B\}_{n_1} \cup \{\lambda_{II}^B\}_{n_2} \cup \{\lambda_0^B\}_1, \quad n_1 + n_2 + 1 = N.$$

The numbers of elements in the corresponding sets of λ^B and λ^C are the same; if these numbers were different, the corresponding scalar products in (2.9) would be equal to zero. The following definitions

$$f_{I,II}^{BB} = f(\lambda_I^B,\lambda_{II}^B), \quad f_{0I}^{CC} = f(\lambda_0^C,\lambda_I^C), \quad \text{etc.,} \qquad (2.10)$$

are used above.

The representation (2.9), expressing the matrix element G_N in terms of scalar products, will be used further in the derivation of the determinant formula for G_N in terms of auxiliary quantum fields.

XII.3 Auxiliary quantum fields and
the determinant representation

Our aim is again to obtain a determinant representation for the matrix element G_N (2.1). To do this, one introduces auxiliary quantum fields. The considerations below are quite similar to those of Chapters IX and XI. The proofs are rather similar to those given in sections IX.6 and XI.2.

To write down formula (2.9) as a determinant of some $N \times N$ matrix, let us introduce ten auxiliary ("dual") fields acting in the auxiliary Fock space. This is done in a way similar to (XI.2.6):

$$\Phi_{A_k}(\lambda) = Q_{A_k}(\lambda) + P_{D_k}(\lambda), \quad \Phi_{D_k}(\lambda) = Q_{D_k}(\lambda) + P_{A_k}(\lambda)$$
$$\varphi_{A_1}(\lambda) = q_{A_1}(\lambda) + p_{D_2}(\lambda), \quad \varphi_{D_1}(\lambda) = q_{D_1}(\lambda) + p_{A_2}(\lambda),$$
$$\varphi_{A_2}(\lambda) = q_{A_2}(\lambda) + p_{D_1}(\lambda), \quad \varphi_{D_2}(\lambda) = q_{D_2}(\lambda) + p_{A_1}(\lambda), \qquad (3.1)$$
$$\varphi_{A_3}(\lambda) = q_{A_3}(\lambda) + p_{D_3}(\lambda), \quad \varphi_{D_3}(\lambda) = q_{D_3}(\lambda) + p_{A_3}(\lambda);$$
$$(0|0) = 1.$$

The vacuum $|0)$ in the auxiliary Fock space is annihilated by all the "momenta":

$$P_a(\lambda)|0) = p_a(\lambda)|0) = 0. \qquad (3.2)$$

The dual vacuum $(0|$ is an eigenvector of all the "coordinates":

$$(0|Q_{A_k}(\lambda) = (0|q_{A_k}(\lambda) = \ln a_k(\lambda)(0|; \quad k = 1, 2$$
$$(0|Q_{D_k}(\lambda) = (0|q_{D_k}(\lambda) = \ln d_k(\lambda)(0|; \quad k = 1, 2 \qquad (3.3)$$
$$(0|q_{A_3}(\lambda) = (0|q_{D_3}(\lambda) = 0.$$

The inessential difference compared with (XI.2.6), (XI.2.9) is that some of the fields Q, q now possess nonzero vacuum eigenvalues (which could be subtracted). The number k in φ_{A_k}, φ_{D_k} is now related to the eigenvalue of the corresponding coordinate q_{A_k}, q_{D_k} in (3.3) (and not to the number of the corresponding set $\{\lambda\}$ as in Chapter XI).

The nonzero commutators of the coordinates and momenta are

$$[P_{D_i}(\lambda), Q_{D_k}(\mu)] = \delta_{ik} \ln h(\lambda, \mu), \quad i, k = 1, 2$$
$$[P_{A_i}(\lambda), Q_{A_k}(\mu)] = \delta_{ik} \ln h(\mu, \lambda),$$
$$[p_{D_l}(\lambda), q_{D_m}(\mu)] = \delta_{lm} \ln h(\lambda, \mu), \qquad (3.4)$$
$$[p_{A_l}(\lambda), q_{A_m}(\mu)] = \delta_{lm} \ln h(\mu, \lambda); \quad l, m = 1, 2, 3.$$

All the commutators which are not explicitly written here are equal to zero.

We again have the remarkable property that all the fields Φ, φ (3.1) commute with one another:

$$[\varphi_l(\lambda), \varphi_m(\mu)] = [\varphi_l(\lambda), \Phi_i(\mu)] = [\Phi_i(\lambda), \Phi_k(\mu)] = 0$$
$$(i, k = 1, 2; \; l, m = 1, 2, 3).$$

(3.5)

Now turn to the representation (2.9) for the matrix element G_N. The scalar products entering the right hand side can be represented as determinants (see (IX.5.8), (IX.6.12)):

$$_m\langle 0| \prod_{j=1}^{N} C_m(\lambda_j^C) \prod_{k=1}^{N} B_m(\lambda_k^B)|0\rangle_m$$

$$= \left\{ \prod_{1 \le j < k \le N} g_{jk}^{CC} g_{kj}^{BB} \right\} (0|\det {}_N S^{(m)}(\lambda^C, \lambda^B)|0) \qquad (m = 1, 2)$$

(3.6)

where $S^{(m)}(\lambda^C, \lambda^B)$ is an $N \times N$ matrix with the following matrix elements:

$$S_{jk}^{(m)}(\lambda^C, \lambda^B) = t_{jk}^{CB} \exp\{\Phi_{A_m}(\lambda_j^C) + \Phi_{D_m}(\lambda_k^B)\}$$
$$+ t_{kj}^{BC} \exp\{\Phi_{D_m}(\lambda_j^C) + \Phi_{A_m}(\lambda_k^B)\}$$

(3.7)

$(m = 1, 2; \; j, k = 1, \ldots, N)$ (the vacuum mean values a_m, d_m now enter the quantum fields $\Phi_{A,D}$ and are not explicit as in (IX.6.11)). Here

$$t_{jk}^{BC} = t(\lambda_j^B, \lambda_k^C) = \frac{g(\lambda_j^B, \lambda_k^C)}{h(\lambda_j^B, \lambda_k^C)} = -\frac{c^2}{(\lambda_j^B - \lambda_k^C)(\lambda_j^B - \lambda_k^C + ic)}.$$

(3.8)

Using the representation (3.6), one rewrites (2.9) in a form similar to (XI.2.1):

$$G_N = c \left\{ \prod_{N \ge j > k \ge 1} g_{jk}^{CC} g_{kj}^{BB} \right\}$$
$$\times \sum (-1)^{[P_C]+[P_B]+n_1}$$
$$\times (0|\det {}_{n_1} S^{(1)}(\lambda_I^C, \lambda_I^B) \det {}_{n_2} S^{(2)}(\lambda_{II}^C, \lambda_{II}^B)|0)$$
$$\times \Big[a_1(\lambda_0^C) d_2(\lambda_0^C) a_1(\lambda_0^B) a_2(\lambda_0^B)$$
$$\times \prod_{I} a_2(\lambda_I^B) d_2(\lambda_I^C) h_{0I}^{CC} h_{0I}^{BB}$$
$$\times \prod_{II} d_1(\lambda_{II}^B) a_1(\lambda_{II}^C) h_{II0}^{CC} h_{0II}^{BB} \prod_{I,II} h_{II,I}^{CC} h_{I,II}^{BB} \Big].$$

(3.9)

The sum here is taken as in (2.9), $\text{card}\{\lambda_0^{C,B}\} = 1$, $\text{card}\{\lambda_I^{C,B}\} = n_1$, $\text{card}\{\lambda_{II}^{C,B}\} = n_2$, $n_1 + n_2 + 1 = N$. The factors $(-1)^{[P_C]}$, $(-1)^{[P_B]}$ are the parities of the permutations

$$P_C \colon \{\lambda_0^C\} \cup \{\lambda_I^C\} \cup \{\lambda_{II}^C\} \to \{\lambda^C\},$$
$$P_B \colon \{\lambda_0^B\} \cup \{\lambda_I^B\} \cup \{\lambda_{II}^B\} \to \{\lambda^B\}. \tag{3.10}$$

(This should be compared with (IX.6.3)). To obtain (3.9) one uses the relation $f(\lambda, \mu) = g(\lambda, \mu) h(\lambda, \mu)$ taking into account the antisymmetry property of g, $g(\lambda, \mu) = -g(\mu, \lambda)$.

Now everything is ready to prove the following theorem.

Theorem 1. *The matrix element G_N (2.1) can be represented as*

$$G_N = \langle 0| \prod_{j=1}^N C(\lambda_j^C)\, \Psi^\dagger(x) \Psi(0) \prod_{k=1}^N B(\lambda_k^B)|0\rangle$$

$$= \left\{ \prod_{j>k=1}^N g_{jk}^{CC} g_{kj}^{BB} \right\} \frac{\partial}{\partial \alpha} (0|\det{}_N M|0) \bigg|_{\alpha=0} \tag{3.11}$$

where on the right hand side we have the vacuum mean value in the auxiliary Fock space of the determinant of M with matrix elements

$$M_{jk} = \widetilde{S}_{jk}^{(1)}(\lambda^C, \lambda^B) + \widetilde{S}_{jk}^{(2)}(\lambda^C, \lambda^B) + c\alpha A_{jk}(\lambda^C, \lambda^B) \tag{3.12}$$

where

$$\widetilde{S}_{jk}^{(1)}(\lambda^C, \lambda^B) = -S_{jk}^{(1)}(\lambda^C, \lambda^B) \exp\{\varphi_{D_2}(\lambda_j^C) + \varphi_{A_2}(\lambda_k^B) + \varphi_{D_3}(\lambda_k^B)\},$$
$$\widetilde{S}_{jk}^{(2)}(\lambda^C, \lambda^B) = S_{jk}^{(2)}(\lambda^C, \lambda^B) \exp\{\varphi_{A_1}(\lambda_j^C) + \varphi_{D_1}(\lambda_k^B)\},$$
$$A_{jk}(\lambda^C, \lambda^B) = a_1(\lambda_k^B) \exp\{\varphi_{A_3}(\lambda_k^B) + \varphi_{A_2}(\lambda_k^B) + \varphi_{A_1}(\lambda_j^C) + \varphi_{D_2}(\lambda_j^C)\}$$
$$\equiv b(\lambda_j^C, \lambda_k^B). \tag{3.13}$$

The matrices $\widetilde{S}^{(1)}$, $\widetilde{S}^{(2)}$ are defined in (3.7). Again it should be emphasized that the fields φ, Φ (the latter enters $S^{(1)}$, $S^{(2)}$) commute so that the determinant in (3.11) is well defined.

Proof: First, we develop the determinant of M as a determinant of the sum of two matrices:

$$\det{}_N M = \det{}_N(\widetilde{S} + \alpha c A)$$

with $\widetilde{S} = \widetilde{S}^{(1)} + \widetilde{S}^{(2)}$ (see the Appendix to Chapter IX). Taking into account that the rank of A is equal to 1 so that only the first rank minors are not equal to zero, one obtains (using the Laplace theorem)

$$\det{}_N M = \det{}_N \widetilde{S} + \alpha c \sum (-1)^{[Q_C]+[Q_B]}$$
$$\times \left(\det{}_{N-1} \widetilde{S}(\lambda_S^C, \lambda_S^B) \right) b(\lambda_0^C, \lambda_0^B) \tag{3.14}$$

where the sum is taken over all partitions

$$\{\lambda^C\}_N = \{\lambda^C_S\}_{N-1} \cup \{\lambda^C_0\}_1, \quad \{\lambda^B\}_N = \{\lambda^B_S\}_{N-1} \cup \{\lambda^B_0\}_1.$$

The factors $(-1)^{[Q_C]}$ and $(-1)^{[Q_B]}$ are the signs of the permutations

$$Q_C: \{\lambda^C_0\}_1 \cup \{\lambda^C_S\}_{N-1} \rightarrow \{\lambda^C\}_N,$$
$$Q_B: \{\lambda^B_0\}_1 \cup \{\lambda^B_S\}_{N-1} \rightarrow \{\lambda^B\}_N.$$

Thus $\det{}_N M$ is a linear function of α; only the second term contributes to the right hand side of (3.11). Taking the derivative in α and developing the determinant

$$\det{}_{N-1}\left(\tilde{S}(\lambda^C_S, \lambda^B_S)\right) = \det{}_{N-1}\left(\tilde{S}^{(1)}(\lambda^C_S, \lambda^B_S) + \tilde{S}^{(2)}(\lambda^C_S, \lambda^B_S)\right) \quad (3.15)$$

as a determinant of the sum of two matrices, one has

$$\frac{\partial}{\partial\alpha}\det{}_N M = c\sum(-1)^{[P_C]+[P_B]}\det{}_{n_1}\tilde{S}^{(1)}(\lambda^C_I, \lambda^B_I)$$
$$\times \det{}_{n_2}\tilde{S}^{(2)}(\lambda^C_{II}, \lambda^B_{II})\, b(\lambda^C_0, \lambda^B_0), \quad (3.16)$$

where the sum is taken exactly as in (3.9) and P_C, P_B are the same permutations (see Appendix to Chapter IX). Extracting from the determinants the exponents with fields φ and taking into account the minus sign in the definition of $\tilde{S}^{(1)}$ (3.13), one rewrites (3.16) as

$$\frac{\partial}{\partial\alpha}(0|\det{}_N M|0) = c\sum(-1)^{[P_C]+[P_B]+n_1}a_1(\lambda^B_0)$$
$$\times (0|\det{}_{n_1}S^{(1)}(\lambda^C_I, \lambda^B_I)\det{}_{n_2}S^{(2)}(\lambda^C_{II}, \lambda^B_{II})|0)$$
$$\times (0|\exp\{\varphi_{A_3}(\lambda^B_0)+\varphi_{A_2}(\lambda^B_0)+\varphi_{A_1}(\lambda^C_0)+\varphi_{D_2}(\lambda^C_0)\}$$
$$\times \prod_I \exp\{\varphi_{D_2}(\lambda^C_I) + \varphi_{A_2}(\lambda^B_I) + \varphi_{D_3}(\lambda^B_I)\}$$
$$\times \prod_{II} \exp\{\varphi_{A_1}(\lambda^C_{II}) + \varphi_{D_1}(\lambda^B_{II})\}|0) \quad (3.17)$$

where the sum is taken exactly as in (3.16), (3.9). We have also taken into account the fact that the momenta p and coordinates q entering the fields φ commute with the momenta P and coordinates Q entering the fields Φ in the scalar products so that the vacuum mean value can indeed be factorized. It is now possible to use the commutation relations and the known actions of the fields p, q on the auxiliary vacua $|0)$, $(0|$ (see (3.1)–(3.5)) to compute the second mean value in (3.17). The result is that the sums in (3.17) and (3.9) are equal. The second vacuum mean value in (3.17) is equal to the corresponding product of the h's in (3.9). The theorem of (3.11) is thus proved.

It should be noted that we have shown that the mean value

$$(0|\det{}_N M|0)$$

is a linear function of α (see (3.14)). Putting $\alpha = 0$ in (3.11) is not oblig-
atory. Throughout this section, we have considered λ^C, λ^B as arbitary
complex numbers (supposing only that they are all different) and do not
satisfy any set of equations.

XII.4 Mean value of the operator $\Psi^\dagger(x)\Psi(0)$ in the NS model

Our aim here is to use the representation (3.11) to calculate the mean
value of the operator $\Psi^\dagger(x)\Psi(0)$ with respect to the Bethe eigenfunctions
for a finite number of particles N. We shall consider the limit when
$\{\lambda^B\} = \{\lambda^C\}$. The corresponding calculations are essentially very similar
to those in sections IX.6 and XI.3.

We begin by transforming the expression $(0|\det{}_N M|0)$ in (3.11) and
shall give only the general scheme of the calculations. Thus we see

$$(0|\det{}_N M|0) = (0| \prod_{j=1}^{N} \exp\{\Phi_{A_2}(\lambda_j^C) + \Phi_{D_2}(\lambda_j^B)$$

$$+ \varphi_{A_1}(\lambda_j^C) + \varphi_{D_1}(\lambda_j^B)\} \det_N \widetilde{M}|0)$$

$$= \prod_{j=1}^{N} a(\lambda_j^C) d(\lambda_j^B) \prod_{j,k=1}^{N} h_{jk}^{CB} (\widetilde{0}|\det{}_N \widetilde{M}|0) \qquad (4.1)$$

which is quite similar to (IX.6.14), (IX.6.26). Here the new dual vacuum
$(\widetilde{0}|$ is

$$(\widetilde{0}| = (0| \prod_{m=1}^{N} \exp\{P_{D_2}(\lambda_m^C) + P_{A_2}(\lambda_m^B) + p_{D_2}(\lambda_m^C) + p_{A_2}(\lambda_m^B)\} \quad (4.2)$$

and the matrix elements of $\widetilde{M}$ are

$$\widetilde{M}_{jk} = M_{jk} \exp\{-\Phi_{A_2}(\lambda_j^C) - \varphi_{A_1}(\lambda_j^C) - \Phi_{D_2}(\lambda_k^B) - \varphi_{D_1}(\lambda_k^B)\}. \quad (4.3)$$

The state $(\widetilde{0}|$, the new dual vacuum, is an eigenstate of all "coordinates"
with modified eigenvalues (with respect to (3.3)). It satisfies the normalization condition

$$(\widetilde{0} \,|\, 0) = 1. \qquad (4.4)$$

One then introduces new fields $\mathring{\Phi}$, $\mathring{\varphi}$:

$$
\begin{aligned}
\mathring{\Phi}_a(\lambda) &= \Phi_a(\lambda) - (\widetilde{0}|\Phi_a(\lambda)|0) \\
&= \Phi_a(\lambda) - (\widetilde{0}|Q_a(\lambda)|0) \equiv \mathring{P}_{\bar{a}}(\lambda) + \mathring{Q}_a(\lambda); \\
a &= A_i, D_i; \ i = 1, 2; \\
\mathring{\varphi}_a(\lambda) &= \varphi_a(\lambda) - (\widetilde{0}|\varphi_a(\lambda)|0) \\
&= \varphi_a(\lambda) - (\widetilde{0}|q_a(\lambda)|0) \equiv \mathring{p}_a(\lambda) + \mathring{q}_{\bar{a}}(\lambda); \\
\bar{a}, a &= A_l, D_l; \ l = 1, 2, 3
\end{aligned} \tag{4.5}
$$

Indices a and $\bar{a}$ run through the same values as in (3.1), and the new "momenta" and "coordinates"

$$
\begin{aligned}
\mathring{P}_a(\lambda) &= P_a(\lambda), \quad \mathring{Q}_a(\lambda) = Q_a(\lambda) - (\widetilde{0}|Q_a(\lambda)|0), \\
\mathring{p}_a(\lambda) &= p_a(\lambda), \quad \mathring{q}_a(\lambda) = q_a(\lambda) - (\widetilde{0}|q_a(\lambda)|0)
\end{aligned}
$$

satisfy the same commutation relations (3.4) as the old ones. However, the new coordinates annihilate the new dual vacuum:

$$
(\widetilde{0}|\mathring{Q}_a(\lambda) = 0, \quad (\widetilde{0}|\mathring{q}_a(\lambda) = 0. \tag{4.6}
$$

It is rather straightforward to express the matrix elements (4.3) of $\widetilde{M}$ in terms of the new fields, even though the calculations are rather tedious. The result is

$$
\begin{aligned}
\widetilde{M}_{jk} &= t_{jk}^{CB} + t_{kj}^{BC} \exp\{\mathring{\Phi}_{A_2}(\lambda_k^B) - \mathring{\Phi}_{D_2}(\lambda_k^B) \\
&\quad + \mathring{\Phi}_{D_2}(\lambda_j^C) - \mathring{\Phi}_{A_2}(\lambda_j^C)\} Z(\lambda_k^B) Z^{-1}(\lambda_j^C) \\
&\quad - \left[t_{jk}^{CB} \exp\{\mathring{\Phi}_{A_1}(\lambda_j^C) + \mathring{\Phi}_{D_1}(\lambda_k^B)\} \right. \\
&\quad \left. + t_{kj}^{BC} \exp\{\mathring{\Phi}_{A_1}(\lambda_j^B) + \mathring{\Phi}_{D_1}(\lambda_j^C)\} \frac{a_1(\lambda_k^B) d_1(\lambda_j^C)}{d_1(\lambda_k^B) a_1(\lambda_j^C)} \right] \\
&\quad \times Z(\lambda_k^B) Z^{-1}(\lambda_j^C) \exp\{\mathring{\varphi}_{D_2}(\lambda_j^C) - \mathring{\varphi}_{A_1}(\lambda_j^C) \\
&\quad - \mathring{\Phi}_{A_2}(\lambda_j^C) + \mathring{\varphi}_{A_2}(\lambda_k^B) + \mathring{\varphi}_{D_3}(\lambda_k^B) - \mathring{\Phi}_{D_2}(\lambda_k^B) - \mathring{\varphi}_{D_1}(\lambda_k^B)\} \\
&\quad + c\alpha \frac{a_1(\lambda_k^B)}{d_1(\lambda_k^B)} Z(\lambda_k^B) Z^{-1}(\lambda_j^C) \\
&\quad \times \exp\{\mathring{\varphi}_{D_2}(\lambda_j^C) - \mathring{\Phi}_{A_2}(\lambda_j^C) \\
&\quad + \mathring{\varphi}_{A_2}(\lambda_k^B) + \mathring{\varphi}_{A_3}(\lambda_k^B) - \mathring{\Phi}_{D_2}(\lambda_k^B) - \mathring{\varphi}_{D_1}(\lambda_k^B)\}. \tag{4.7}
\end{aligned}
$$

The following quantities are introduced here:

$$Z(\lambda_k^B) \equiv \frac{a_2(\lambda_k^B)}{d_2(\lambda_k^B)} \prod_{m=1}^{N} \frac{h_{km}^{BB}}{h_{mk}^{CB}},$$

$$Z(\lambda_j^C) \equiv \frac{a_2(\lambda_j^C)}{d_2(\lambda_j^C)} \prod_{m=1}^{N} \frac{h_{jm}^{CB}}{h_{mj}^{CC}}.$$

(4.8)

Let us now calculate the mean value of the operator $\Psi^\dagger(x)\Psi(0)$. To this end, one puts $\lambda_j^C = \lambda_j^B = \lambda_j$ $(j = 1, \ldots, N)$ and uses the system of Bethe equations

$$\frac{a(\lambda_n)}{d(\lambda_n)} \prod_{k=1}^{N} \frac{h(\lambda_n, \lambda_k)}{h(\lambda_k, \lambda_n)} = (-1)^{N-1},$$

$$a(\lambda) = a_1(\lambda)a_2(\lambda), \quad d(\lambda) = d_1(\lambda)d_2(\lambda).$$

(4.9)

The Bethe eigenvectors $|\varphi_N\rangle$ are

$$|\varphi_N(\{\lambda\})\rangle = \prod_{j=1}^{N} B(\lambda_j)|0\rangle \langle\varphi_N(\{\lambda\})| = \langle 0| \prod_{j=1}^{N} C(\lambda_j)$$

(4.10)

with λ_j satisfying (4.9). Due to (4.9), the functions Z (4.8) are considerably simplified.

From now on, we restrict ourselves to the NS model, taking functions $a_i(\lambda)$, $d_i(\lambda)$ $(i = 1, 2)$ prescribed by the relations (XI.1.16), so that

$$Z(\lambda) = (-1)^{N-1} \frac{d_1(\lambda)}{a_1(\lambda)} = (-1)^{N-1} \exp\{ix\lambda\}.$$

(4.11)

This results in the following representation for the mean value of the operator $\Psi^\dagger(x)\Psi(0)$ for the one-dimensional Bose gas (compare with (XI.3.17)):

$$\langle\varphi_N(\{\lambda\})|\Psi^\dagger(x)\Psi(0)|\varphi_N(\{\lambda\})\rangle = \left\{ \prod_{\substack{j\neq k \\ j,k=1}}^{N} f_{jk} \right\} \frac{\partial}{\partial\alpha} \langle\tilde{0}|\det{}_N v|0\rangle \Big|_{\alpha=0}$$

(4.12)

(for the NS model, $\prod_{j=1}^{N} a(\lambda_j)d(\lambda_j) = 1$).

The $N \times N$ dimensional matrix v is given by

$$
\begin{aligned}
v_{jk} = c\delta_{jk} & \left[L + \sum_{m=1}^{N} K_{km} \right] \\
& + t_{jk} + t_{kj} \exp\{ ix\lambda_{kj} + \overset{\circ}{\Phi}_{A_2}(\lambda_k) - \overset{\circ}{\Phi}_{D_2}(\lambda_k) + \overset{\circ}{\Phi}_{D_2}(\lambda_j) - \overset{\circ}{\Phi}_{A_2}(\lambda_j) \} \\
& - \left[t_{jk} \exp\{ ix\lambda_{kj} + \overset{\circ}{\Phi}_{A_1}(\lambda_j) + \overset{\circ}{\Phi}_{D_1}(\lambda_k) \} \right. \\
& \quad \left. + t_{kj} \exp\{ \overset{\circ}{\Phi}_{A_1}(\lambda_k) + \overset{\circ}{\Phi}_{D_1}(\lambda_j) \} \right] \\
& \times \exp\{ \overset{\circ}{\varphi}_{D_2}(\lambda_j) - \overset{\circ}{\varphi}_{A_1}(\lambda_j) \\
& \quad - \overset{\circ}{\Phi}_{A_2}(\lambda_j) + \overset{\circ}{\varphi}_{A_2}(\lambda_k) + \overset{\circ}{\varphi}_{D_3}(\lambda_k) - \overset{\circ}{\Phi}_{D_2}(\lambda_k) - \overset{\circ}{\varphi}_{D_1}(\lambda_k) \} \\
& + c\alpha \exp\{ -ix\lambda_j + \overset{\circ}{\varphi}_{A_3}(\lambda_k) + \overset{\circ}{\varphi}_{A_2}(\lambda_k) \\
& \quad - \overset{\circ}{\varphi}_{D_1}(\lambda_k) - \overset{\circ}{\Phi}_{D_2}(\lambda_k) + \overset{\circ}{\varphi}_{D_2}(\lambda_j) - \overset{\circ}{\Phi}_{A_2}(\lambda_j) \} \quad (4.13)
\end{aligned}
$$

where $\lambda_{jk} \equiv \lambda_j - \lambda_k$ and

$$
K_{jk} \equiv K(\lambda_j, \lambda_k), \quad K(\lambda, \mu) = \frac{2c}{(\lambda - \mu)^2 + c^2} \qquad (4.14)
$$

(see (XI.3.19)) are used.

Formula (4.12) gives the final expression for the correlator of two fields in the N-particle eigenstate of the Hamiltonian of the one-dimensional Bose gas.

XII.5 Thermodynamic limit

The correlator of fields in the thermodynamic limit is the most interesting quantity from the physical point of view. This procedure was explained in detail in sections XI.4, XI.5 for the current correlator. Here we shall state the results for the field correlator.

First consider the correlator of fields at zero temperature, $T = 0$. Then the correlator is equal to the normalized mean value of the operator $\Psi^\dagger(x)\Psi(0)$ with respect to the ground state of the gas, $|\Omega\rangle$:

$$
\langle \Psi^\dagger(x)\Psi(0) \rangle = \frac{\langle \Omega | \Psi^\dagger(x)\Psi(0) | \Omega \rangle}{\langle \Omega | \Omega \rangle}. \qquad (5.1)
$$

The expression for this correlator is obtained from (4.12) in the same way as in section XI.4 for the current correlator. Going to the thermodynamic limit, one obtains the following answer:

$$
\langle \Psi^\dagger(x)\Psi(0) \rangle = \left. \frac{\dfrac{\partial}{\partial\alpha}\left(\widetilde{0}\big| \det\left(1 + \dfrac{1}{2\pi}\widehat{W} \right) \big| 0 \right)}{\det\left(1 - \dfrac{1}{2\pi}\widehat{K} \right)} \right|_{\alpha=0}. \qquad (5.2)
$$

The Fredholm determinant of the operator $\widehat{K}$ acting on the interval $[-q, q]$, as given by (XI.4.8), is in the denominator. In the numerator we have the Fredholm determinant of the linear integral operator $\widehat{W}$ which acts on c-number functions $\xi(\lambda)$ as

$$\left(\widehat{W}\xi\right)(\lambda) = \int_{-q}^{q} W(\lambda, \mu)\xi(\mu)\, d\mu \qquad (5.3)$$

with kernel $W(\lambda, \mu)$ given by

$$\begin{aligned}
cW(\lambda, \mu) = {}& t(\lambda, \mu) + t(\mu, \lambda)\exp\{ix(\mu - \lambda) + \overset{\circ}{\Phi}_{A_2}(\mu) \\
& - \overset{\circ}{\Phi}_{D_2}(\mu) + \overset{\circ}{\Phi}_{D_2}(\lambda) - \overset{\circ}{\Phi}_{A_2}(\lambda)\} \\
& - [t(\lambda, \mu)\exp\{ix(\mu - \lambda) + \overset{\circ}{\Phi}_{A_1}(\lambda) + \overset{\circ}{\Phi}_{D_1}(\mu)\} \\
& \quad + t(\mu, \lambda)\exp\{\overset{\circ}{\Phi}_{A_1}(\mu) + \overset{\circ}{\Phi}_{D_1}(\lambda)\}] \\
& \times \exp\{\overset{\circ}{\varphi}_{D_2}(\lambda) - \overset{\circ}{\varphi}_{A_1}(\lambda) - \overset{\circ}{\Phi}_{A_2}(\lambda) + \overset{\circ}{\varphi}_{A_2}(\mu) + \overset{\circ}{\varphi}_{D_3}(\mu) \\
& \quad - \overset{\circ}{\varphi}_{D_1}(\mu) - \overset{\circ}{\Phi}_{D_2}(\mu)\} \\
& + \alpha c \exp\{-ix\lambda + \overset{\circ}{\varphi}_{D_2}(\lambda) - \overset{\circ}{\Phi}_{A_2}(\lambda) + \overset{\circ}{\varphi}_{A_3}(\mu) + \overset{\circ}{\varphi}_{A_2}(\mu) \\
& \quad - \overset{\circ}{\varphi}_{D_1}(\mu) - \overset{\circ}{\Phi}_{D_2}(\mu)\}. \qquad (5.4)
\end{aligned}$$

Formula (5.2) is the result for the zero temperature correlator of two fields in the nonrelativistic Bose gas.

Consider now the correlator in the case of finite temperature T which is defined in a way similar to (XI.5.1):

$$\langle \Psi^{\dagger}(x)\Psi(0)\rangle_T = \frac{\operatorname{tr}\left(\exp\{-H/T\}\Psi^{\dagger}(x)\Psi(0)\right)}{\operatorname{tr}\left(\exp\{-H/T\}\right)}. \qquad (5.5)$$

First let us evaluate the mean value of the operator $\Psi^{\dagger}(x)\Psi(0)$ with respect to one of the eigenfunctions $|\Omega_T\rangle$ present in the state of thermal equilibrium (section I.8). The calculations are similar to those in section XI.5, see formulæ (XI.5.10)–(XI.5.17). The result is

$$\frac{\langle \Omega_T|\Psi^{\dagger}(x)\Psi(0)|\Omega_T\rangle}{\langle \Omega_T|\Omega_T\rangle} = \left.\frac{\frac{\partial}{\partial \alpha}\langle \tilde{0}|\det\left(1 + \frac{1}{2\pi}\widehat{W}_T\right)|0\rangle}{\det\left(1 - \frac{1}{2\pi}\widehat{K}_T\right)}\right|_{\alpha=0} \qquad (5.6)$$

with the operator $\widehat{K}_T$ now acting on the interval $[-\infty, \infty]$ defined in (XI.5.12) and having kernel $K_T(\lambda, \mu)$ given by

$$K_T(\lambda, \mu) = K(\lambda, \mu)\vartheta(\mu). \qquad (5.7)$$

$\vartheta(\mu) = (1 + \exp\{\varepsilon(\mu)/T\})^{-1}$ is the Fermi weight (XI.5.3) and $\varepsilon(\mu)$ is the Yang-Yang energy (XI.5.4).

The operator $\widehat{W}_T$ also acts on the interval $[-\infty, \infty]$. Its kernel is

$$W_T(\lambda, \mu) = W(\lambda, \mu)\vartheta(\mu) \tag{5.8}$$

with the function $W(\lambda, \mu)$ given in (5.4). (Remark: note that both operators $\widehat{W}$ (5.2) and $\widehat{W}_T$ (5.6) could be changed to operators $\widehat{W}^{(x)}$, $\widehat{W}_T^{(x)}$ as in (XI.5.19) without affecting the Fredholm determinant. This transformation can be chosen to symmetrize the kernels. Now we can use the arguments of section (I.8) because we saw that the mean value

$$\frac{\langle \Omega_T | \Psi^\dagger(x)\Psi(0) | \Omega_T \rangle}{\langle \Omega_T | \Omega_T \rangle}$$

depends only on $\varepsilon(\lambda)$ (see (I.5.19)). But it is the same for all $|\Omega_T\rangle$ (eigenstates present in the state of thermal equilibrium), so we can apply formula (I.8.19) and finally get for the temperature correlation function (5.5)

$$\langle \Psi^\dagger(x)\Psi(0)\rangle_T = \frac{\dfrac{\partial}{\partial\alpha}(\widetilde{0}|\det\left(1 + \dfrac{1}{2\pi}\widehat{W}_T\right)|0)}{\det\left(1 - \dfrac{1}{2\pi}\widehat{K}_T\right)}\Bigg|_{\alpha=0}. \tag{5.9}$$

Hence, the equal-time correlator of two fields in the one-dimensional Bose gas is represented as a Fredholm determinant for zero as well as nonzero temperatures. Formulæ (5.2) and (5.6) are the main results of this section.

Conclusion

In this chapter, we have shown that the correlation function of fields can also be represented in terms of a determinant.

In the previous chapter this was shown for the correlation function of local densities. In fact, all correlation functions in the model can be expressed in terms of determinants.

Local fields were introduced into the generalized model in paper [1] where the representation of section 2 was obtained. Sections 3–5 are based on the results of paper [2] .

Conclusion to Part III

In this Part we have reached an extremely important understanding: quantum correlation functions can be represented as determinants. In the thermodynamic limit, correlation functions are equal to determinants of Fredholm integral operators. In the next Part, we shall see that, starting from these determinants, one can write down classical differential equations for quantum correlation functions. These equations will be completely integrable (have a Lax representation). We shall solve these equations and explicitly find the asymptotics of correlation functions. We shall see the close relation between the differential equations for quantum correlation functions and the original classical differential equation which was quantized (in our example this is the nonlinear Schrödinger equation). The first determinant formula appeared as a hypothesis for the square of the norm of the Hamiltonian eigenfunction (see [1], Chapter X). This formula was proved in the paper ([3], Chapter X). The determinant formula for the partition function of the six-vertex model with domain wall boundary conditions is extremely important ([1] of Chapter IX). The special case of the scalar product (one vector being an eigenfunction of the Hamiltonian) is another important example (refs. [2], [4], [5], Chapter X). To move forward, the important concept of auxiliary quantum fields was introduced in ([3], Chapter IX). This permits us to write down determinant formulæ for all form factors and all correlation functions. It is natural to compare Fredholm integral operators, which appear in our approach, with the quantum Gel'fand-Levitan equation approach ([1], [2], [3]). They are similar. Our integral operator depends on quantum dual fields (which are canonical Bose fields) and we shall treat our Fredholm operator later as a Gel'fand-Levitan operator for the differential equations that drive quantum correlation functions. So the two approaches are similar, with one essential difference. In our case all quantum fields are organized into commuting combinations. This permits us to avoid using perturbation theory when we transform our integral operator. This, in turn, helps us to write down the differential equations and explicitly evaluate the asymptotics of quantum correlation functions.

In the next Part, the impenetrable Bose gas ($c = \infty$) will be studied and more determinant formulæ will appear. Let us mention some determinant formulæ that have not appeared in the book. N.A. Slavnov found the determinant representation for the field correlation function, which is space, time and temperature dependent. H. Itoyama, H. Thacker and V.E. Korepin found the determinant representation for the sine-Gordon correlation function. The following should also be emphasized: the determinant representations of Part III are related to explicit formulas for form factors constructed by F.A. Smirnov for the sine-Gordon model ([5], Chapter X).

Part IV
Differential Equations for Quantum Correlation Functions

Introduction to Part IV

The quantum correlation functions for different exactly solvable (completely integrable) models are all constructed in a similar way. In Part IV our main example will be the nonlinear Schrödinger (NS) equation. We shall also comment on the other models: the Heisenberg antiferromagnet, the Hubbard model and so on.

We shall discuss different approaches to quantum correlation functions. Our main approach leads to differential equations for quantum correlation functions. These are classical nonlinear differential equations; they are completely integrable and closely related to the original classical differential equation that was quantized (the NS equation in our case).

In the first stage of calculation of correlation functions we represent them as a determinant of some Fredholm integral operator (here we shall use the results of Part III). The correct language for the description of quantum correlation functions is the language of τ functions, which was developed in [7]. This helps us to relate the differential equations for quantum correlation functions with the hierarchy of the original classical differential equation that was quantized (the NS equation in our example). In this way we solve the most difficult problem; that of the evaluation of time- and temperature-dependent correlation functions.

This approach can be applied not only to correlation functions for the NS equation [3], [4], [5], but also to the sine-Gordon model [2], to nonrelativistic fermions [1], to the Heisenberg antiferromagnetic with special anisotropy and to other models. The simplest example is the impenetrable Bose gas (the NS equation at $c = \infty$). In Chapter XIII we remind the reader of the main properties of the impenetrable Bose gas and derive determinant representations (of Fredholm integral operators) for various different correlation functions. The Fredholm integral operators that appear in these formulæ, have a very special structure (they form an infinite-dimensional group). This permits us to write down differential

281

equations for quantum correlation functions. In Chapter XIV we explain how to do this.

The other remarkable property of these Fredholm integral operators is that each of them is related to some special Riemann problem. This provides an alternative way to derive the differential equations for quantum correlation functions. The Riemann problem arising in this context is explained in detail in Chapter XV. The Riemann problem provides the best way to calculate the asymptotics of quantum correlation functions. To do this one should solve the completely integrable nonlinear differential equations for quantum correlation functions. How to do this is explained in Chapter XVI. The asymptotics of different correlation functions of the impenetrable Bose gas are evaluated explicitly in Chapter XVI. Much attention is paid to temperature correlation functions, so we recommend the reader to look once more through sections I.5–I.8. The most interesting asymptotics are those of time- and temperature-dependent correlation functions of fields.

Chapters XVII and XVIII are devoted to other approaches to quantum correlation functions not related to differential equations. In Chapter XVII we consider a different (non-determinant) representation for quantum correlation functions. The algebraic Bethe Ansatz permits us to write down a series for correlation functions, which becomes very effective for special correlation functions. The main example in Chapter XVII is the correlation function of local densities $j(x) = \Psi^\dagger(x)\Psi(x)$ (currents) for the one-dimensional Bose gas. The asymptotics of the current correlation function $\langle j(x,t)j(0,0)\rangle$ can be extracted directly from our series. This is especially interesting for finite temperature. Temperature correlation functions are considered in detail in this chapter; for completely integrable models they have a lot of very special properties. For example, they can be represented in a form similar to that at zero temperature.

In Chapter XVIII we explain the conformal approach to quantum correlation functions. For gapless models (like the Bose gas) zero temperature is the point of the phase transition. The long distance asymptotics of correlation functions are given by the power of the distance. This enables us to describe the long distance asymptotics of correlation functions by conformal field theory. To calculate the conformal dimensions one should be able to calculate finite size corrections to the energy levels of the models. This has already been done in Part I. This idea helps us to evaluate the long distance asymptotics of correlation functions at zero (and also very small) temperature for the Bose gas, Heisenberg antiferromagnet and Hubbard models.

XIII

Correlation Functions for Impenetrable Bosons. The Determinant Representation

Introduction

Impenetrable bosons in one space dimension are a special case (at coupling constant $c \to \infty$) of the one-dimensional Bose gas (NS model) considered in detail in Chapter I. The results obtained there for the NS model can be specialized to the case of impenetrable bosons. It is to be said that all the formulæ are considerably simplified in this case. From the point of view of physics this is due to the fact that the δ-function potentials in the N-particle Hamiltonian (I.1.11) are now infinitely strong and the bosons cannot penetrate one another, so that the wave function should be equal to zero if the bosons' coordinates coincide. In this respect impenetrable bosons are rather similar to free fermions (see Appendix I.1). In this chapter correlation functions of impenetrable bosons are considered and their representations as Fredholm determinants of linear integral operators are given. The (very essential) simplification with respect to the general case is due to the fact that now all the auxiliary quantum fields can be set equal to zero (see IX.6), so that the kernels of these operators do not contain quantum fields. The determinant representation for the time-dependent correlator is easy to obtain in this case. Temperature correlation functions are considered as well as zero temperature ones.

In section 1 the model of the impenetrable Bose gas is defined. In section 2, the representation for the generating functional of currents

$$\langle \exp\{\alpha Q_1(x)\} \rangle$$

for impenetrable bosons is obtained from the corresponding representation obtained in Chapter XI for the general case $c < \infty$. In section 3, the representation for the equal-time correlator of two fields is given; this is obtained from the corresponding representation in Chapter XII. It appears to be equivalent to the well known representation of this correlator as the first Fredholm minor of a linear integral operator, obtained by A. Lenard in 1964 [7], [8]. In sections 4–6 the time-dependent zero- and nonzero-temperature correlators of two fields of impenetrable bosons are expressed as Fredholm minors; hence the generalization of the Lenard formula for this case is given. These results were obtained in [6]. In section 4 the form factor of the local Bose field is represented as a determinant. In section 5 we show that correlation functions (being represented in terms of form factors) can also be written down as a determinant. In section 6 the thermodynamic limit is considered and the Fredholm determinant representation is finally obtained for time-dependent correlation functions of the local field. Both zero- and nonzero-temperature correlation functions are considered. In the following chapters this will be used for the explicit evaluation of time- and temperature-dependent correlation functions of the local Bose fields. In section 7 the multipoint correlator of fields is considered; the determinant representation in this case was obtained by N.A. Slavnov in paper [4]. For the equal-time correlator this representation gives the formula obtained earlier [7], [8].

XIII.1 Impenetrable bosons

As has already been stated, impenetrable bosons are described by the second-quantized Hamiltonian

$$H = \int dx \left(\partial_x \Psi^\dagger \, \partial_x \Psi + c \Psi^\dagger \Psi^\dagger \Psi \Psi - h \Psi^\dagger \Psi \right) \tag{1.1}$$

in the limiting case $c = +\infty$; h is a chemical potential. The canonical Heisenberg fields

$$\Psi^\dagger(x, t) = \exp\{iHt\} \, \Psi^\dagger(x) \exp\{-iHt\}$$
$$\Psi(x, t) = \exp\{iHt\} \, \Psi(x) \exp\{-iHt\} \tag{1.2}$$

satisfy commutation relations

$$[\Psi(x, t), \, \Psi^\dagger(y, t)] = \delta(x - y),$$
$$[\Psi(x, t), \, \Psi(y, t)] = [\Psi^\dagger(x, t), \, \Psi^\dagger(y, t)] = 0. \tag{1.3}$$

We consider the theory in a box of length L $(-L/2 < x < L/2)$, going to the thermodynamic limit $L \to \infty$.

The fields $\Psi^\dagger(x) = \Psi^\dagger(x, 0)$ generate the natural coordinate basis of quantum states in the Fock space over the bare vacuum $|0\rangle$:

$$\Psi(x)|0\rangle = 0, \qquad \langle 0|0\rangle = 1$$
$$\langle 0|\Psi^\dagger(x) = 0. \tag{1.4}$$

The Bethe eigenstates $|\Psi_N\rangle$ of the Hamiltonian,

$$H|\Psi_N\rangle = E|\Psi_N\rangle,$$

are easily obtained from (I.1.25) for $c \to +\infty$, after removing the constant factor:

$$|\Psi_N\rangle = \frac{1}{\sqrt{N!}} \int d^N z \chi_N(z_1, \ldots, z_N | \lambda_1, \ldots, \lambda_N) \, \Psi^\dagger(z_1) \cdots \Psi^\dagger(z_N)|0\rangle$$

$$(1.5)$$

with wave function χ_N,

$$\chi_N = \frac{1}{\sqrt{N!}} \prod_{N \geq j > k \geq 1} \epsilon(z_j - z_k) \sum_{\mathcal{P}} (-1)^{[\mathcal{P}]} \exp\left\{ i \sum_{n=1}^{N} z_n \lambda_{\mathcal{P}n} \right\}, \quad (1.6)$$

where $\mathcal{P}$ is the permutation $(1, 2, \ldots, N) \to (\mathcal{P}_1, \ldots, \mathcal{P}_N)$ and $\epsilon(x) = \text{sign}(x)$ is the sign function.

The operator Q for the number of particles,

$$Q = \int dx \, \Psi^\dagger(x)\Psi(x), \qquad [Q, H] = 0, \qquad (1.7)$$

is conserved.

The periodic boundary conditions imply the Bethe equations (I.2.2) which are now very simple:

$$\exp\{iL\lambda_j\} = (-1)^{N-1}, \qquad j = 1, \ldots, N. \qquad (1.8)$$

Under these conditions, the norm of the wave function is easily calculated directly from (1.5), (1.6) to be

$$\langle \Psi_N | \Psi_N \rangle = \|\chi_N\|^2 = L^N \qquad (1.9)$$

which is in complete correspondence with the general formula for squared norms of Bethe eigenfunctions (see section X.1).

The eigenstate $|\Psi_N\rangle$ (1.5) contains N particles with momenta λ_j ($j = 1, \ldots, N$) and energies $\varepsilon(\lambda_j)$,

$$\varepsilon(\lambda) = \lambda^2 - h, \qquad E = \sum_{j=1}^{N} \left(\lambda_j^2 - h \right). \qquad (1.10)$$

Let us denote the set $\{\lambda\}$ in the ground state by $\{\mu_j\}$. For convenience we suppose that the number of particles N_g in the ground state is even. It is evident that the corresponding $\{\mu_j\}$ fill symmetrically (with respect to $\mu = 0$) all the possible vacancies,

$$e^{iL\mu_j} = (-1)^{N_g - 1} = -1, \qquad \sum_{j=1}^{N_g} \mu_j = 0$$

$$(1.11)$$

$$\mu_j = \frac{2\pi}{L} \left(j - \frac{N_g + 1}{2} \right), \qquad j = 1, \ldots, N_g.$$

The equilibrium condition is:

$$\varepsilon(\mu_{N_g}) \le 0, \qquad \varepsilon(\mu_{N_g+1}) \ge 0.$$

In the thermodynamic limit

$$N \to \infty, \quad L \to \infty, \quad D = \frac{N}{L} \text{ fixed} \tag{1.12}$$

the set $\{\mu_j\}$ fills the Fermi zone $-q \le \mu \le q$, the Fermi momentum q being given by

$$q = \sqrt{h} \tag{1.13}$$

and the gas density D by

$$D = \frac{q}{\pi}. \tag{1.14}$$

The thermodynamics of the model (see section I.5) are considerably simplified as there is no "dressing", i.e., the operator $\widehat{K} = 0$ (as well as $\widehat{K}_T = 0$). In other words, the particles λ_j are free particles. In particular, the Yang-Yang energy (I.6.1) is equal to $\varepsilon(\lambda)$ (1.10). At temperatures $T > 0$, the thermal equilibrium distribution of particles is given by the Fermi weight

$$\vartheta(\lambda, h, T) = \frac{1}{1 + e^{\varepsilon(\lambda)/T}}, \qquad \varepsilon(\lambda) = \lambda^2 - h. \tag{1.15}$$

The gas density is

$$D = \frac{1}{2\pi} \int\limits_{-\infty}^{+\infty} \vartheta(\lambda, h, T) \, d\lambda \tag{1.16}$$

and the kinetic energy of the particles is

$$E_k = \frac{1}{2\pi} \int\limits_{-\infty}^{+\infty} \lambda^2 \vartheta(\lambda, h, T) \, d\lambda. \tag{1.17}$$

The chemical potential h determines the gas density as a monotonically increasing function of h. At zero temperature, $D = 0$ at $h = 0$ and $0 < D < \infty$ as $0 < h < \infty$. At nonzero temperature, $D = 0$ at $h = -\infty$, and $0 < D < \infty$ as $-\infty < h < \infty$. We shall see that even for $T > 0$ correlation functions behave differently depending on the sign of h (see Chapter XVI).

We have finished the brief description of the impenetrable Bose gas. In the next section the generating functions for correlation functions of currents are considered.

Let us recall the main thermodynamic properties of the model (see sections 5 to 8 of Chapter I, section 5 of Chapter XI or papers [1], [2],

[3]. Let us take one of the eigenfunctions $|\Omega_T\rangle$ which are present in the state of thermodynamic equilibrium and let us evaluate the mean value

$$\frac{\langle\Omega_T|\Psi^\dagger(x)\Psi(x)|\Omega_T\rangle}{\langle\Omega_T|\Omega_T\rangle}.$$

We shall discover below that this mean value depends only on $\rho_p(\lambda)$, $\rho_h(\lambda)$ and $\rho_t(\lambda)$, but not on the set of integers n_j (I.2.13) which specify this particular state. This means that this mean value is the same for all the eigenfunctions which are present in the state of thermodynamic equilibrium. A functional integral representation similar to those in section I.5 shows that this mean value is equal to the correlation function at finite temperature (for details see section I.8).

XIII.2 Generating functionals
of current correlation functions

Consider first the generating functional of current correlators at zero temperature. It is defined as (see the beginning of section XI.1)

$$\langle\exp\{\alpha Q_1(x)\}\rangle = \frac{\langle\Omega|\exp\{\alpha Q_1(x)\}|\Omega\rangle}{\langle\Omega|\Omega\rangle}, \qquad x > 0, \qquad (2.1)$$

i.e., it is the normalized mean value with respect to the ground state ("physical vacuum") of the model. Here $Q_1(x)$,

$$Q_1(x) = \int_0^x \Psi^\dagger(y)\Psi(y)\,dy, \qquad x > 0, \qquad (2.2)$$

is the operator for the number of particles in the interval $[0, x]$. Due to translation invariance and parity conservation the operator for the number of particles in the interval $[x_1, x_2]$,

$$Q(x_1, x_2) = \int_{x_1}^{x_2} \Psi^\dagger(y)\Psi(y)\,dy, \qquad x_2 > x_1,$$

depends only on $|x_2 - x_1|$ ($\equiv |x|$), so that we put $x > 0$. The determinant representation for this correlator has already been obtained (see XI.4.15). In our case the considerable simplification is due to the fact that all the auxiliary "momenta" and "coordinates" $\mathring{p}$, $\mathring{q}$ entering the fields $\mathring{\varphi}$ in the kernel $V(\lambda, \mu)$ (XI.4.13) commute at $c = \infty$ (see XI.4.18, XI.3.8, XI.3.9) since

$$h(\lambda, \mu) = \frac{\lambda - \mu + ic}{ic} \xrightarrow[c\to\infty]{} 1.$$

Thus the result of normal ordering in calculating the vacuum mean value in the auxiliary quantum space $(\widetilde{0}|\widehat{V}|0)$ is trivial: in the expression for

$V(\lambda, \mu)$ one can put all the auxiliary quantum fields $\overset{\circ}{\varphi}_a(\lambda)$ equal to zero. Also

$$t(\lambda, \mu) = \frac{g(\lambda, \mu)}{h(\lambda, \mu)} = \frac{(ic)^2}{(\lambda - \mu)(\lambda - \mu + ic)} \xrightarrow[c \to \infty]{} \frac{ic}{\lambda - \mu} \qquad (2.3)$$

and the operator $\widehat{K}$ (XI.4.8) is equal to zero at $c = \infty$ so that

$$\det\left(1 - \frac{1}{2\pi}\widehat{K}\right) = 1$$

in the denominator of (XI.4.15); consequently one has at $c = +\infty$

$$\langle \exp\{\alpha Q_1(x)\} \rangle = \det\left(1 + \frac{1}{2\pi}\widehat{V}_\infty\right) \qquad (2.4)$$

with the operator $\widehat{V}_\infty$ possessing kernel $V_\infty(\lambda, \mu)$,

$$V_\infty(\lambda, \mu) = 2\left(e^\alpha - 1\right)e^{-ix(\lambda-\mu)/2}\left(\frac{\sin\dfrac{x(\lambda - \mu)}{2}}{\lambda - \mu}\right). \qquad (2.5)$$

The similarity transformation

$$\widetilde{V}(\lambda, \mu) = e^{ix\lambda/2}V_\infty(\lambda, \mu)e^{-ix\mu/2} \qquad (2.6)$$

does not change the determinant. But $\widetilde{V}(\lambda, \mu)$ is the symmetric kernel

$$\widetilde{V}(\lambda, \mu) = 2\left(e^\alpha - 1\right)\left(\frac{\sin\dfrac{x(\lambda - \mu)}{2}}{\lambda - \mu}\right).$$

It is convenient to rewrite the answer in the form:

$$\langle \exp\{\alpha Q_1(x)\} \rangle = \det\left(1 - \gamma \widehat{V}\right)\Big|_{\gamma=(1-\exp\alpha)/\pi} \qquad (2.7)$$

where the integral linear operator $\widehat{V}$, acting on the interval $[-q, q]$, possesses the kernel $V(\lambda, \mu)$,

$$V(\lambda, \mu) = \frac{\sin\dfrac{x(\lambda - \mu)}{2}}{\lambda - \mu}. \qquad (2.8)$$

The kernel of the integral operator is defined in the standard way. To act with $\widehat{V}$ on some function $f(\mu)$ one should integrate:

$$\left(\widehat{V} f\right)(\lambda) = \int_{-q}^{q} V(\lambda, \mu)f(\mu)\, d\mu. \qquad (2.9)$$

There exists another derivation of the determinant representation (2.7). One can start with the wave function for the impenetrable Bose gas (1.6) and integrate it, or one can use fermionisation (see Appendix 1 of Chapter I).

This is the result for the correlator $\langle \exp\{\alpha Q_1(x)\}\rangle$ in the impenetrable Bose gas at zero temperature. It should be said that at $\alpha = -\infty$ ($\gamma = 1/\pi$) the correlator gives the "emptiness formation probability" $P(x)$ (XI.1.26):

$$P(x) = \langle \exp\{\alpha Q_1(x)\}\rangle \Big|_{\alpha=-\infty} = \det\left(1 - \gamma \widehat{V}\right)\Big|_{\gamma=1/\pi} \qquad (2.10)$$

i.e., the probability that there are no particles in the interval $[0, x]$. The long distance asymptotics of this probability $P(x)$ can be found in formula (9.11) of paper [4].

The normalized mean value $\langle j(x)\rangle$ of the operator $j(x)$ and the correlator $\langle j(x)j(0)\rangle$ of two currents are obtained from (2.7) (see XI.1.3):

$$\langle j(x)\rangle = \frac{\partial}{\partial x}\frac{\partial}{\partial \alpha}\langle \exp\{\alpha Q_1(x)\}\rangle \Big|_{\alpha=0\ (\gamma=0)} \qquad (2.11)$$

$$\langle :j(x)j(0):\rangle = \frac{1}{2}\frac{\partial^2}{\partial x^2}\frac{\partial^2}{\partial \alpha^2}\langle \exp\{\alpha Q_1(x)\}\rangle \Big|_{\alpha=0\ (\gamma=0)}, \qquad (2.12)$$

for $x > 0$ (see section XI.1). As $\det(1 - \gamma\widehat{V}) = 1$ at $\alpha = 0$ ($\gamma = 0$), one easily performs the differentiations with the results

$$\frac{\partial}{\partial \alpha}\langle \exp\{\alpha Q_1(x)\}\rangle \Big|_{\alpha=0} = \frac{1}{\pi}\,\mathrm{Sp}\,V = \langle Q_1(x)\rangle$$

$$= \frac{1}{\pi}\int_{-q}^{q}\frac{x}{2}\,d\lambda = x\frac{q}{\pi} = Dx,$$

$$\frac{\partial^2}{\partial \alpha^2}\langle \exp\{\alpha Q_1(x)\}\rangle \Big|_{\alpha=0} = \frac{1}{\pi^2}(\mathrm{Sp}\,V)^2 - \frac{1}{\pi^2}\,\mathrm{Sp}\left(V^2\right) + \frac{1}{\pi}\,\mathrm{Sp}\,V = \langle Q_1(x)^2\rangle$$

$$= \frac{q^2}{\pi^2}x^2 - \frac{1}{\pi^2}\int_{-q}^{q}d\lambda\int_{-q}^{q}d\mu\left(\frac{\sin\dfrac{x(\lambda-\mu)}{2}}{\lambda-\mu}\right)^2 + \frac{q}{\pi}x.$$

Now we take the x-derivatives with the result

$$\langle j(x)\rangle = D, \qquad D = \frac{q}{\pi}$$

$$\langle j(x)j(0)\rangle = D^2 - \left(\frac{\sin\pi Dx}{\pi x}\right)^2 \qquad (2.13)$$

reproducing the well known result.

At nonzero temperature the correlation function is defined in the standard way:

$$\langle \exp\{\alpha Q_1(x)\}\rangle_T = \frac{\text{tr}\left(e^{-H/T}\exp\{\alpha Q_1(x)\}\right)}{\text{tr}\,e^{-H/T}}.$$

In the thermodynamic limit (Chapter I, sections 5–8) not all eigenfunctions of the Hamiltonian contribute to this trace, but only those which enter in the state of thermodynamic equilibrium. (It is described by equations (I.5.24)–(I.5.26).) We shall denote these eigenfunctions by $|\Omega_T\rangle$. The number of such eigenfunctions goes to infinity as $\exp S$ (here S is the entropy). Calculations similar to those in section I.8 show that in the thermodynamic limit only $\frac{\langle\Omega_T|\exp\{\alpha Q_1(x)\}|\Omega_T\rangle}{\langle\Omega_T|\Omega_T\rangle}$ contributes to the trace $\text{tr}\exp\{\alpha Q_1(x)\}$. But the mean value $\frac{\langle\Omega_T|\exp\{\alpha Q_1(x)\}|\Omega_T\rangle}{\langle\Omega_T|\Omega_T\rangle}$ can be calculated in a way similar to the zero temperature case (see section XI.5). The remarkable result is that this mean value does not depend on the particular choice of $|\Omega_T\rangle$, as long as $|\Omega_T\rangle$ belongs to the state of thermodynamic equilibrium. Section I.8 shows that this independence means that

$$\langle\exp\{\alpha Q_1(x)\}\rangle_T = \frac{\text{tr}\left(e^{-H/T}\exp\{\alpha Q_1(x)\}\right)}{\text{tr}\,e^{-H/T}} = \frac{\langle\Omega_T|\exp\{\alpha Q_1(x)\}|\Omega_T\rangle}{\langle\Omega_T|\Omega_T\rangle}.$$

$$(2.14)$$

Here $|\Omega_T\rangle$ is any one of the eigenfunctions of the Hamiltonian contributing to the state of thermodynamic equilibrium. In practice the mean value

$$\frac{\langle\Omega_T|\exp\{\alpha Q_1(x)\}|\Omega_T\rangle}{\langle\Omega_T|\Omega_T\rangle}$$

depends only on the densities $\rho_p(\lambda)$, $\rho_h(\lambda)$, $\rho_t(\lambda)$ and $\varepsilon(\lambda)$, but they are the same for each $|\Omega_T\rangle$ from the stateof thermodynamic equilibrium. Now the mean value $\langle\Omega_T|\exp\{\alpha Q_1(x)\}|\Omega_T\rangle$ can be calculated in a way similar to that at zero temperature (see section XI.5 and papers [1], [2], [3]).

Putting $c = +\infty$ in the representation (XI.5.17) one obtains (as all the fields $\mathring{\varphi}_a$ can be put to zero) that the temperature-dependent correlation function $\langle\exp\{\alpha Q_1(x)\}\rangle_T$ for the impenetrable Bose gas is represented as

$$\langle\exp\{\alpha Q_1(x)\}\rangle_T = \frac{\text{tr}\left(e^{-H/T}\exp\{\alpha Q_1(x)\}\right)}{\text{tr}\,e^{-H/T}}$$

$$= \frac{\langle\Omega_T|\exp\{\alpha Q_1(x)\}|\Omega_T\rangle}{\langle\Omega_T|\Omega_T\rangle}$$

$$= \det\left(1 - \gamma\widehat{\widetilde{V}}_T\right)\Big|_{\gamma=(1-\exp\alpha)/\pi} \qquad (2.15)$$

where the operator $\widehat{\widetilde{V}}_T$ is obtained from $\widehat{V}$ (2.8), (2.9) by the change of integration measure (XI.5.18)

$$\int\limits_{-q}^{q} d\lambda \longrightarrow \int\limits_{-\infty}^{+\infty} d\lambda\, \vartheta(\lambda) \tag{2.16}$$

with the Fermi weight $\vartheta(\lambda)$ given as in (1.15):

$$\vartheta(\lambda) = (1 + \exp\{\varepsilon(\lambda)/T\})^{-1}, \qquad \varepsilon(\lambda) = \lambda^2 - h,$$

so that

$$\left(\widehat{\widetilde{V}}_T f\right)(\lambda) = \int\limits_{-\infty}^{+\infty} d\mu\, \widetilde{V}_T(\lambda,\mu) f(\mu)$$

$$\widetilde{V}_T(\lambda,\mu) = V(\lambda,\mu)\vartheta(\mu), \tag{2.17}$$

f being an arbitrary function, with the function $V(\lambda,\mu)$ given (for $-\infty < \lambda,\mu < +\infty$) by formula (2.8). For $T \to 0$, $\vartheta(\lambda) \to \theta(q-\lambda)\theta(\lambda+q)$, so that the operator $\widehat{\widetilde{V}}_T \to \widehat{V}$ and (2.15) turns into (2.7). The change of measure (I.8.5) is the same as for excitations.

It is convenient to symmetrize further the kernel $\widetilde{V}_T$, going to the operator $\widehat{V}_T$ with kernel $V_T(\lambda,\mu)$,

$$V_T(\lambda,\mu) = \sqrt{\vartheta(\lambda)}\, V(\lambda,\mu)\, \sqrt{\vartheta(\mu)}. \tag{2.18}$$

(This is a similarity transformation.) So one finally rewrites (2.15) as

$$\langle \exp\{\alpha Q_1(x)\}\rangle_T = \det\left(1 - \gamma \widehat{V}_T\right)\Big|_{\gamma=(1-\exp\alpha)/\pi} \tag{2.19}$$

where on the right hand side the Fredholm determinant of the operator $\widehat{V}_T$ is present:

$$\left(\widehat{V}_T f\right)(\lambda) = \int\limits_{-\infty}^{+\infty} d\mu\, V_T(\lambda,\mu) f(\mu)$$

$$V_T(\lambda,\mu) = (1 + \exp\{\varepsilon(\lambda)/T\})^{-1/2}\, \frac{\sin\dfrac{(\lambda-\mu)x}{2}}{\lambda-\mu}$$

$$\times (1 + \exp\{\varepsilon(\mu)/T\})^{-1/2}, \qquad \varepsilon(\lambda) = \lambda^2 - h. \tag{2.20}$$

For the emptiness formation probability $P(x)$ one has

$$P(x) = \langle\exp\{\alpha Q_1(x)\}\rangle_T\Big|_{\alpha=-\infty}$$

$$= \det\left(1 - \gamma\hat{V}_T\right)\Big|_{\gamma=1/\pi}. \qquad (2.21)$$

For the temperature-dependent normalized mean value of operator $j(x)$ one has

$$\langle j(x)\rangle_T = \frac{\partial}{\partial x}\frac{\partial}{\partial\alpha}\langle\exp\{\alpha Q_1(x)\}\rangle_T\Big|_{\alpha=0\,(\gamma=0)}$$

$$= \frac{1}{2\pi}\int\limits_{-\infty}^{+\infty}\vartheta(\lambda)\,d\lambda = D(h,T) \qquad (2.22)$$

(where D is the gas density in the state of thermodynamic equilibrium). For the two-point current correlation function the expression is also obtained in an explicit form:

$$\langle :j(x)j(0):\rangle_T = \frac{1}{2}\frac{\partial^2}{\partial x^2}\frac{\partial^2}{\partial\alpha^2}\langle\exp\{\alpha Q_1(x)\}\rangle_T\Big|_{\alpha=0\,(\gamma=0)}$$

$$= D^2(h,T) - \frac{1}{4\pi^2}\left(\int\limits_{-\infty}^{+\infty}\frac{\exp\{ix\lambda\}\,d\lambda}{1+\exp\{(\lambda^2-h)/T\}}\right)^2 \qquad (2.23)$$

$(x > 0)$ so that the current correlator is given in essence by the squared Fourier transform of the Fermi weight.

So, in the case of impenetrable bosons, the correlators $\langle\exp\{\alpha Q_1(x)\}\rangle_T$ and the emptiness formation probability are represented (for zero as well as nonzero temperature) as Fredholm determinants of the linear integral operators (2.7), (2.10), (2.19), (2.21). For the two-point current correlation function the known explicit expressions are reproduced (2.13), (2.23). Many space-, time- and and temperature-dependent current correlation functions are evaluated in [4].

XIII.3 Equal-time two-point
correlator of fields. Lenard's formula

We turn now to the correlator of two fields for the impenetrable Bose gas, defined for zero and nonzero temperature by

$$\langle\Psi^\dagger(x)\Psi(0)\rangle = \frac{\langle\Omega|\Psi^\dagger(x)\Psi(0)|\Omega\rangle}{\langle\Omega|\Omega\rangle} \qquad (T=0),\, x > 0$$

$$\langle\Psi^\dagger(x)\Psi(0)\rangle_T = \frac{\mathrm{tr}\left(e^{-H/T}\,\Psi^\dagger(x)\Psi(0)\right)}{\mathrm{tr}\,e^{-H/T}} = \frac{\langle\Omega_T|\Psi^\dagger(x)\Psi(0)|\Omega_T\rangle}{\langle\Omega_T|\Omega_T\rangle}.$$

Here $|\Omega_T\rangle$ is one of the eigenfunctions of the Hamiltonian which is present in the state of thermodynamic equilibrium, see the end of section 1 and section I.8.

Due to translation invariance the correlators

$$\langle \Psi^\dagger(x_1)\Psi(x_2)\rangle, \qquad \langle \Psi^\dagger(x_1)\Psi(x_2)\rangle_T$$

depend on the difference $x_1 - x_2$ and are easily expressed in terms of the correlators written above ($x = x_1 - x_2$). Due to parity conservation, the correlators depend only on $|x|$, so we also put $x > 0$. The correlators $\langle \Psi^\dagger(x)\Psi(0)\rangle$ for the one-dimensional Bose gas with finite coupling constant $c > 0$ were represented in section XII.5 as the vacuum mean values of the Fredholm determinants of linear integral operators $\widehat{W}$, $\widehat{W}_T$ (see (XII.5.2), (XII.5.9)) in the auxiliary Fock space where the operators $\overset{\circ}{\varphi}$, $\overset{\circ}{\Phi}$ act. As explained already in the previous section, in the case of impenetrable bosons ($c = +\infty$) all these fields can be put to zero, so that one obtains similarly to (2.7), (2.15) the following results. In the zero temperature case,

$$\langle \Psi^\dagger(x)\Psi(0)\rangle = \frac{\partial}{\partial \alpha} \det\left(1 - \frac{2}{\pi}\widehat{V} + \frac{\alpha}{2\pi}\widehat{A}\right)\bigg|_{\alpha=0} \qquad (3.1)$$

where $\widehat{V}$ is the same linear integral operator as for the current correlator (2.8) with kernel

$$V(\lambda, \mu) = \frac{\sin\dfrac{x(\lambda - \mu)}{2}}{\lambda - \mu} \qquad (3.2)$$

and the operator $\widehat{A}$ possesses the kernel

$$A(\lambda, \mu) = \exp\left\{-i\frac{x(\lambda + \mu)}{2}\right\}. \qquad (3.3)$$

For temperature $T > 0$ one has

$$\langle \Psi^\dagger(x)\Psi(0)\rangle_T = \frac{\partial}{\partial \alpha} \det\left(1 - \frac{2}{\pi}\widehat{\widetilde{V}}_T + \frac{\alpha}{2\pi}\widehat{\widetilde{A}}_T\right)\bigg|_{\alpha=0}$$

$$= \frac{\partial}{\partial \alpha} \det\left(1 - \frac{2}{\pi}\widehat{V}_T + \frac{\alpha}{2\pi}\widehat{A}_T\right)\bigg|_{\alpha=0} \qquad (3.4)$$

where $\widehat{\widetilde{V}}_T$, $\widehat{V}_T$ are the same as in (2.17), (2.20), possessing kernels

$$\widetilde{V}_T(\lambda, \mu) = V(\lambda, \mu)\vartheta(\mu) \qquad (3.5)$$

$$V_T(\lambda, \mu) = \sqrt{\vartheta(\lambda)}\, V(\lambda, \mu) \sqrt{\vartheta(\mu)}. \qquad (3.6)$$

The operators $\widehat{A}$, $\widehat{\widetilde{A}}_T$, $\widehat{A}_T$ are linear integral operators with kernels, correspondingly,

$$A(\lambda, \mu) = \exp\left\{-i\frac{x(\lambda + \mu)}{2}\right\} \tag{3.7}$$

$$\widetilde{A}_T(\lambda, \mu) = \exp\left\{-i\frac{x(\lambda + \mu)}{2}\right\} \vartheta(\mu) \tag{3.8}$$

$$A_T(\lambda, \mu) = \sqrt{\vartheta(\lambda)} \exp\left\{-i\frac{x(\lambda + \mu)}{2}\right\} \sqrt{\vartheta(\mu)}. \tag{3.9}$$

Recall that $\vartheta(\lambda)$ is the Fermi weight

$$\vartheta(\lambda) \equiv \vartheta(\lambda, h, T) = \frac{1}{1 + e^{(\lambda^2 - h)/T}}. \tag{3.10}$$

It is to be noted that taking explicitly the α-derivative one obtains, e.g., in (3.4)

$$\langle \Psi^\dagger(x)\Psi(0)\rangle = \frac{1}{4}\gamma \operatorname{Sp}\left[\left(1 - \gamma\widehat{V}\right)^{-1} \widehat{A}\right] \det\left(1 - \gamma\widehat{V}\right)\Big|_{\gamma = 2/\pi} \tag{3.11}$$

and, similarly,

$$\langle \Psi^\dagger(x)\Psi(0)\rangle_T = \frac{1}{4}\gamma \operatorname{Sp}\left[\left(1 - \gamma\widehat{V}_T\right)^{-1} \widehat{A}_T\right] \det\left(1 - \gamma\widehat{V}_T\right)\Big|_{\gamma = 2/\pi}. \tag{3.12}$$

It is to be mentioned that the same Fredholm determinants (but with different values of the parameter γ) enter the representations (2.7), (2.19) for

$$\left\langle \exp\left\{\alpha \int_0^x \Psi^\dagger(y)\Psi(y)\,dy\right\}\right\rangle, \qquad \gamma = \frac{1 - e^\alpha}{\pi},$$

and (3.11), (3.12) for $\langle \Psi^\dagger(x)\Psi(0)\rangle$ ($\gamma = 2/\pi$ which corresponds to $\alpha = i\pi$). This is closely related to the fermionization phenomenon (see Appendix I.1), which gives the following connection between the canonical Fermi field $\Psi_F(x)$ and the canonical Bose field $\Psi_B(x)$:

$$\Psi_B^\dagger(x) = \Psi_F^\dagger(x) \exp\left\{i\pi \int_{-\infty}^x \Psi_F^\dagger(z)\Psi_F(z)\,dz\right\},$$

$$\Psi_B^\dagger(x)\Psi_B(x) \cong \Psi_F^\dagger(x)\Psi_F(x).$$

In our case, $\Psi_B(x) = \Psi(x)$ and this relation shows that $\alpha = i\pi$ for the correlator $\langle \Psi^\dagger(x)\Psi(0)\rangle$. So the formula

$$\gamma = \frac{1 - e^\alpha}{\pi}$$

gives the correct answer for the value of γ for the field correlator.

Formulæ (3.11), (3.12) are in fact equivalent to the known formulæ [7], [8] for the equal-time correlators of impenetrable bosons obtained by A. Lenard in 1964. Lenard showed that

$$\langle \Psi^\dagger(x)\Psi(-x)\rangle_T = \frac{1}{\pi}\rho(x,-x)\det\left(1 - \frac{2}{\pi}\hat{\theta}_T\right), \qquad x > 0. \qquad (3.13)$$

Here $\hat{\theta}_T$ is a linear integral operator acting in coordinate space on the interval $[-x, x]$ with kernel equal to

$$\theta_T(\xi - \eta) = \frac{1}{2}\int\limits_{-\infty}^{+\infty} d\lambda\, e^{i(\xi-\eta)\lambda}\vartheta(\lambda) \qquad (3.14)$$

i.e., to the Fourier transform of the Fermi weight

$$\vartheta(\lambda) \equiv \vartheta(\lambda, h, T) = \frac{1}{1 + e^{(\lambda^2 - h)/T}}.$$

The kernel $\theta_0(\xi - \eta)$ for $T = 0$ is

$$\theta_0(\xi - \eta) = \frac{\sin q(\xi - \eta)}{\xi - \eta} \qquad (-x \le \xi, \eta \le x) \qquad (3.15)$$

where $q = \sqrt{h}$ is the Fermi momentum. The operator $\hat{\theta}_0$ acts on the interval $[-x, x]$.

The kernel $\rho(\xi, \eta)$ of the resolvent operator $\hat{\rho}$ is defined in the usual way

$$\rho_T(\xi, \eta) - \frac{2}{\pi}\int\limits_{-x}^{x} \theta_T(\xi - \xi')\rho_T(\xi', \eta)\, d\xi' = \theta_T(\xi - \eta). \qquad (3.16)$$

The value of the kernel $\rho_T(x, -x)$ at the boundary of the interval $[-x, x]$ enters (3.13), so that the right hand side of (3.13) is the special value of the first Fredholm minor of the operator $\hat{\theta}_T$.

Let us show the equivalence between, e.g., (3.12) and (3.13). Formula (3.12) is readily rewritten as

$$\langle \Psi^\dagger(x)\Psi(-x)\rangle_T = \frac{1}{2\pi}\det\left(1 - \frac{2}{\pi}\widehat{K}_T\right)$$

$$\times \int\limits_{-\infty}^{+\infty} d\lambda\, e^{-i\lambda x} f_-(\lambda)\vartheta(\lambda) \qquad (3.17)$$

where the operator $\widehat{K}_T$ acting in momentum space possesses the kernel $K_T(\lambda, \mu)$ given by

$$K_T(\lambda, \mu) = \sqrt{\vartheta(\lambda)}\,\frac{\sin x(\lambda - \mu)}{\lambda - \mu}\sqrt{\vartheta(\mu)}, \qquad -\infty \le \lambda, \mu \le +\infty. \qquad (3.18)$$

For $T = 0$ the kernel is

$$K(\lambda, \mu) = \frac{\sin x(\lambda - \mu)}{\lambda - \mu}, \qquad -q \le \lambda, \mu \le q. \qquad (3.19)$$

The functions $f_{\pm}(\lambda)$ are defined by the integral equations

$$f_{\pm}(\lambda) - \frac{2}{\pi} \int_{-\infty}^{+\infty} K_T(\lambda, \mu) f_{\pm}(\mu) \, d\mu = \sqrt{\vartheta(\lambda)} e^{\pm i\lambda x}. \qquad (3.20)$$

In writing (3.17) we used the fact that the correlator $\langle \Psi^\dagger(x_1)\Psi(x_2) \rangle$ is a function of $|x_1 - x_2|$.

One can show by means of Fourier transforms that (3.13) and (3.17) (and hence (3.12)) are indeed equivalent.

Consider the integral equation

$$\varphi(\xi) - \frac{2}{\pi} \int_{-x}^{x} \theta_T(\xi - \xi')\varphi(\xi') \, d\xi' = \Phi(\xi). \qquad (3.21)$$

Taking the Fourier transform

$$\varphi(\xi) = \int_{-\infty}^{+\infty} d\lambda \sqrt{\vartheta(\lambda)} e^{-i\lambda\xi} f(\lambda)$$

$$f(\lambda) = \frac{1}{2\pi\sqrt{\vartheta(\lambda)}} \int_{-\infty}^{+\infty} d\xi \, e^{i\lambda\xi} \varphi(\xi) \qquad (3.22)$$

$$F(\lambda) = \frac{1}{2\pi\sqrt{\vartheta(\lambda)}} \int_{-\infty}^{+\infty} d\xi \, e^{i\lambda\xi} \Phi(\xi)$$

one obtains the following equation for the function $f(\lambda)$:

$$f(\lambda) - \frac{2}{\pi} \int_{-\infty}^{+\infty} K_T(\lambda, \mu) f(\mu) \, d\mu = F(\lambda) \qquad (3.23)$$

where the kernel $K_T(\lambda, \mu)$ is given by (3.18). So the Fredholm determinants of the operators $\widehat{\theta}_T$ and $\widehat{K}_T$ are equal:

$$\det\left(1 - \frac{2}{\pi}\widehat{K}_T\right) = \det\left(1 - \frac{2}{\pi}\widehat{\theta}_T\right). \qquad (3.24)$$

Now let us show that also

$$\rho_T(x, -x) = \frac{1}{2} \int_{-\infty}^{+\infty} e^{-i\lambda x} \sqrt{\vartheta(\lambda)} f_{-}(\lambda) \, d\lambda. \qquad (3.25)$$

To show this, consider equation (3.16) for $\eta = -x$. The Fourier transform of $\rho_T(\xi, -x)$ is denoted by $r(\lambda)$ where

$$r(\lambda) = \frac{1}{2\pi\sqrt{\vartheta(\lambda)}} \int\limits_{-\infty}^{+\infty} d\xi \, e^{i\lambda\xi} \rho_T(\xi, -x). \tag{3.26}$$

The Fourier transform of (3.16) leads to the following equation for $r(\lambda)$:

$$r(\lambda) - \frac{2}{\pi} \int\limits_{-\infty}^{+\infty} K_T(\lambda - \mu) r(\mu) \, d\mu = \frac{1}{2}\sqrt{\vartheta(\lambda)} e^{-ix\lambda}. \tag{3.27}$$

Comparison with (3.20) shows that

$$r(\lambda) = \frac{1}{2} f_-(\lambda) \tag{3.28}$$

and taking the inverse Fourier transform one obtains exactly (3.25) which in turn proves the equivalence of our formulæ for the equal-time correlators to those obtained by Lenard.

So we have described the equal-time correlators of the impenetrable Bose gas in terms of Fredholm determinants of linear operators. In the following sections of this chapter the determinant representations for the time-dependent correlation function

$$\langle \Psi^\dagger(x_1, t_1) \Psi(x_2, t_2) \rangle$$

for zero temperature and for nonzero temperature are obtained, which are the generalization of Lenard's formula. One should also realize that we have already obtained a generalization of Lenard's formula for finite coupling (penetrable bosons), see (XII.5.9).

XIII.4 Form factors

Our aim in what follows is to obtain, as an analogue of the Lenard representation, a Fredholm minor of the time-dependent correlation function

$$\langle \Psi^\dagger(x_2, t_2) \Psi(x_1, t_1) \rangle$$

for the impenetrable Bose gas. It can be expressed in the standard way in terms of form factors, i.e., matrix elements of operators Ψ, $\Psi^\dagger$ between Bethe eigenfunctions. So we begin with the study of form factors. In this section the form factor between states with a finite number of particles is considered. In the following sections the mean value of the operator $\Psi(x_2, t_2)\Psi^\dagger(x_1, t_1)$ with respect to the ground state of the Hamiltonian in the case of a finite number N of particles is represented as a first Fredholm minor of the determinant of a $N \times N$-dimensional matrix. In the thermodynamic limit it becomes the first Fredholm minor of a linear integral operator. We study the form factor of the field $\Psi^\dagger(x)$ between Bethe eigenstates (1.5):

$$F(x) = \langle \Psi_{N+1}(\{\lambda\}) | \Psi^\dagger(x) | \Psi_N(\{\mu\}) \rangle. \tag{4.1}$$

Eigenstate $|\Psi_N(\{\mu\})\rangle$ contains N particles. If the number of momenta in set $\{\lambda\}$ were not equal to $N+1$ the form factor would be equal to zero due to the conservation of the number of particles (1.7). Momenta μ, λ satisfy the Bethe equations. For convenience we take $N = N_g$, the number of particles in the ground state, to be even. We denote the momentum of particles in the ground state by μ_j:

$$e^{iL\mu_j} = (-1)^{N-1} = -1,$$
$$\mu_j = \frac{2\pi}{L}\left(m_j - \frac{N+1}{2}\right), \qquad j = 1, \dots, N; \quad m_j \in \mathbf{Z} \tag{4.2}$$

$$e^{iL\lambda_j} = (-1)^N = +1,$$
$$\lambda_j = \frac{2\pi}{L} n_j, \qquad j = 1, \dots, N+1; \quad n_j \in \mathbf{Z} \tag{4.3}$$

so that λ_j would never coincide with any of the μ_k. The form factor $F(x)$ is readily expressed in terms of the eigenfunctions χ_{N+1}, χ_N (1.5), (1.6). Using the commutation relations (1.3), ($\Psi(x) = \Psi(x,0)$) and the definition of the Fock vacuum (1.4) one obtains

$$F(x) = \sqrt{N+1} \int d^N z\, \chi^*_{N+1}(z_1, \dots, z_N, x | \{\lambda\}) \chi_N(z_1, \dots, z_N | \{\mu\})$$
$$= \frac{1}{N!} \sum_{\mathcal{P}} \sum_{\mathcal{Q}} (-1)^{[\mathcal{P}]+[\mathcal{Q}]} \exp\{-ix\lambda_{\mathcal{P}(N+1)}\}$$
$$\times \int_{-L/2}^{L/2} \prod_{n=1}^{N} dz_n\, \epsilon(x - z_n) \exp\left\{-i \sum_{n=1}^{N} z_n \left(\lambda_{\mathcal{P}n} - \mu_{\mathcal{Q}n}\right)\right\}. \tag{4.4}$$

The sum here is taken over permutations $\mathcal{P}$ of $\lambda_1, \dots, \lambda_{N+1}$ and $\mathcal{Q}$ of $\mu_1, \dots, \mu_N$; $(-1)^{[\mathcal{P}]}$ and $(-1)^{[\mathcal{Q}]}$ are the signs of the permutations. To calculate the integrals one applies integration by parts. It is essential that contributions from the boundary terms vanish due to the Bethe equations (4.2), (4.3) and the fact that $\lambda_{\mathcal{P}n} - \mu_{\mathcal{Q}n}$ is never zero. As

$$\frac{d\epsilon(x - z_n)}{dz_n} = -2\delta(x - z_n),$$

the integral term is calculated explicitly with the result

$$F(x) = \frac{(2i)^N}{N!} \exp\left\{ ix \left[\sum_{j=1}^{N} \mu_j - \sum_{j=1}^{N+1} \lambda_j \right] \right\}$$

$$\times \sum_{\mathcal{P},\mathcal{Q}} (-1)^{[\mathcal{P}]+[\mathcal{Q}]} \prod_{j=1}^{N} (\lambda_{\mathcal{P}j} - \mu_{\mathcal{Q}j})^{-1}. \tag{4.5}$$

The eigenfunctions $\chi_{N+1}(\{\lambda\})$, $\chi_N(\{\mu\})$ are antisymmetric in all λ_j and all μ_k, correspondingly. This fact allows us to prove a formula which plays the basic role in the following considerations:

$$\frac{1}{N!} \sum_{\mathcal{P},\mathcal{Q}} (-1)^{[\mathcal{P}]+[\mathcal{Q}]} \prod_{j=1}^{N} \frac{1}{\lambda_{\mathcal{P}j} - \mu_{\mathcal{Q}j}}$$

$$= \left(1 + \frac{\partial}{\partial \alpha}\right) \det{}_N \left[\frac{1}{\lambda_j - \mu_k} - \frac{\alpha}{\lambda_{N+1} - \mu_k} \right]\Bigg|_{\alpha=0}. \tag{4.6}$$

On the right hand side we have the determinant of an $N \times N$ matrix, the (j,k)-th element of this matrix (j is the number of the row, k is the number of the column) being equal to the expression in square brackets.

To prove (4.6) let us first consider the sum in a case similar to the left hand side of (4.6), but where the number of λ's is equal to the number of μ's and equal to N. In this case card $\{\lambda\}$ = card $\{\mu\}$ = N :

$$\frac{1}{N!} \sum_{\mathcal{P},\mathcal{Q}} (-1)^{[\mathcal{P}]+[\mathcal{Q}]} \prod_{j=1}^{N} \frac{1}{\lambda_{\mathcal{P}j} - \mu_{\mathcal{Q}j}} = \det \frac{1}{\lambda_j - \mu_k}.$$

This follows from the definition of the determinant. Now let us prove (4.6) in the case where card $\{\lambda\}$ = $N+1$ and card $\{\mu\}$ = N. In each term on the left hand side one of the λ_j is missing so the left hand side can be represented as

$$\sum_{m=1}^{N+1} (-1)^{N+1-m} \det{}_N \Lambda^{(m)} \tag{4.7}$$

where the $N \times N$ matrix $\Lambda^{(m)}$ is obtained by removing the m-th row from the $(N+1) \times N$ matrix Λ with matrix elements

$$\Lambda_{lk} = \frac{1}{\lambda_l - \mu_k}; \qquad l = 1, \ldots, N+1; \quad k = 1, \ldots, N. \tag{4.8}$$

$(-1)^{N+1-m}$ is the sign of the permutation putting the m-th row of matrix Λ to the last place,

$$\{1, \ldots, m-1, m, m+1, \ldots, N+1\} \longrightarrow \{1, \ldots, m-1, m+1, \ldots, N+1, m\}.$$

On the other hand, let us differentiate the determinant in the right hand side of (4.6) row by row (then putting $\alpha = 0$). One obtains then for the right hand side an expression similar to (4.7),

$$\sum_{m=1}^{N+1} \det{}_N \widetilde{\Lambda}^{(m)} \tag{4.9}$$

where the $N \times N$ matrix $\Lambda^{(m)}$ $(m = 1, \ldots, N)$ is the matrix obtained by differentiating the m-th row of the matrix in the square brackets in (4.6) and does not contain λ_m; the m-th row of $\widetilde{\Lambda}^{(m)}$ is

$$\widetilde{\Lambda}_{mk}^{(m)} = -\frac{1}{\lambda_{N+1} - \mu_k} \qquad (m = 1, \ldots, N; \quad k = 1, \ldots, N). \tag{4.10}$$

The matrix $\widetilde{\Lambda}^{(N+1)}$ has elements

$$\widetilde{\Lambda}_{jk}^{(N+1)} = \frac{1}{\lambda_j - \mu_k} \tag{4.11}$$

and does not contain λ_{N+1}. It is now obvious that

$$\det{}_N \widetilde{\Lambda}^{(m)} = -(-1)^{N-m} \det{}_N \Lambda^{(m)} \tag{4.12}$$

where the minus sign is due to the minus sign in (4.10) and the factor $(-1)^{N-m}$ is the sign of the permutation

$$\{1, \ldots, m-1, m, m+1, \ldots, N\} \longrightarrow \{1, \ldots, m-1, m+1, \ldots, N, m\}.$$

It is now obvious that expressions (4.7) and (4.9) are equal. So the identity (4.6) is proved.

As the left hand side of (4.6) is an antisymmetric function of all λ_j one can write (4.6) as

$$\left(1 + \frac{\partial}{\partial \alpha}\right) \det{}_N \left(\frac{1}{\lambda_j - \mu_k} - \frac{\alpha}{\lambda_{N+1} - \mu_k}\right)\bigg|_{\alpha=0}$$

$$= \sum_{\mathcal{P}} (-1)^{\mathcal{P}} \prod_{j=1}^{N} \frac{1}{\lambda_{\mathcal{P}j} - \mu_j}. \tag{4.13}$$

The final result for the time-independent form-factor is

$$F(x) = (2i)^N \exp\left\{ix\left[\sum_{j=1}^{N} \mu_j - \sum_{j=1}^{N+1} \lambda_j\right]\right\}$$

$$\times \left(1 + \frac{\partial}{\partial \alpha}\right) \det{}_N \left(M_{jk}^{(\alpha)}\right)\bigg|_{\alpha=0} \tag{4.14}$$

$$M_{jk}^{(\alpha)} = \frac{1}{\lambda_j - \mu_k} - \frac{\alpha}{\lambda_{N+1} - \mu_k} \qquad (j, k = 1, \ldots, N); \tag{4.15}$$

the exponential factor is obviously due to translation invariance. It is also worth mentioning that $\det {}_N M^{(\alpha)}$ is in fact a linear function of α, which follows from the fact that the rank of the matrix $A_{ik} = \alpha(\lambda_{N+1} - \mu_k)^{-1}$ is equal to 1 (all the columns are proportional). So the time-independent form factor is presented as a determinant.

It is easy now to write the expression also for the time-dependent form factor. As the Heisenberg field $\Psi^\dagger(x, t)$ depends on time t in the canonical way (1.2), and the eigenvectors $|\Psi_{N+1}\rangle$, $|\Psi_N\rangle$ are eigenvectors of the Hamiltonian (1.1), with the following eigenvalues (1.10),

$$E_N = \sum_{j=1}^N (\mu_j^2 - h), \qquad E_{N+1} = \sum_{j=1}^{N+1} (\lambda_j^2 - h),$$

one obtains

$$F(x, t) \equiv \langle \Psi_{N+1}(\{\lambda\}) | \Psi^\dagger(x, t) | \Psi_N(\{\mu\}) \rangle$$

$$= \exp \left\{ -iht + it \left(\sum_{j=1}^{N+1} \lambda_j^2 - \sum_{j=1}^N \mu_j^2 \right) + ix \left(\sum_{j=1}^N \mu_j - \sum_{j=1}^{N+1} \lambda_j \right) \right\}$$

$$\times (2i)^N \left(1 + \frac{\partial}{\partial \alpha} \right) \det {}_N \left(M_{jk}^{(\alpha)} \right) \Big|_{\alpha=0} \qquad (4.16)$$

with the same matrix $M^{(\alpha)}$ (4.15).

It is also obvious that complex conjugation leads to the form factor of field $\Psi(x, t)$,

$$\langle \Psi_N(\{\mu\}) | \Psi(x, t) | \Psi_{N+1}(\{\lambda\}) \rangle = F^*(x, t). \qquad (4.17)$$

So the time-independent form factor is also expressed as a determinant. Representation (4.14) is valid even if $|\Psi_N(\{\mu\})\rangle$ is not the ground state.

XIII.5 Normalized mean value
for a finite number of particles

Consider now the normalized mean value of the operator

$$\Psi(x_2, t_2) \Psi^\dagger(x_1, t_1)$$

with respect to state $|\Psi_N(\{\mu\})\rangle$ (4.2),

$$\langle \Psi(x_2, t_2) \Psi^\dagger(x_1, t_1) \rangle_N = \frac{\langle \Psi_N(\{\mu\}) | \Psi(x_2, t_2) \Psi^\dagger(x_1, t_1) | \Psi_N(\{\mu\}) \rangle}{\langle \Psi_N(\{\mu\}) | \Psi_N(\{\mu\}) \rangle},$$

$$(5.1)$$

$(\langle \Psi_N | \Psi_N \rangle = L^N$, see (1.9)). Let us decompose it over the complete set of eigenstates $|\Psi_{N+1}(\{\lambda\})\rangle$, $\langle \Psi_{N+1}(\{\lambda\}) | \Psi_{N+1}(\{\lambda\}) \rangle = L^{N+1}$; only the states with number of particles equal to $N + 1$ contribute as the number of particles is conserved (1.7). Thus one obtains

$$\langle \Psi(x_2, t_2)\Psi^\dagger(x_1, t_1)\rangle_N = \frac{1}{L^{2N+1}} \sum_{\{\lambda\}} F^*(x_2, t_2)F(x_1, t_1) \qquad (5.2)$$

where $F(x_1, t_1)$, $F^*(x_2, t_2)$ are the form factors (4.16), (4.17). The sum is over all the different sets $\{\lambda\}$ (card $\{\lambda\} = N+1$) which are solutions of the Bethe equations (4.3).

Using (4.16), (4.17) one rewrites (5.2) as

$$\langle \Psi(x_2, t_2)\Psi^\dagger(x_1, t_1)\rangle_N$$

$$= \sum_{\{\lambda\}} \exp\left\{ iht_{21} + it_{21}\left[\sum_{j=1}^{N}\mu_j^2 - \sum_{j=1}^{N+1}\lambda_j^2\right] + ix_{12}\left[\sum_{j=1}^{N}\mu_j - \sum_{j=1}^{N+1}\lambda_j\right] \right\}$$

$$\times \frac{2^{2N}}{L^{2N+1}}\left(1+\frac{\partial}{\partial\alpha}\right)\det{}_N\left(M_{jk}^{(\alpha)}\right)\bigg|_{\alpha=0}\left(1+\frac{\partial}{\partial\beta}\right)\det{}_N\left(M_{jk}^{(\beta)}\right)\bigg|_{\beta=0}.$$

$$(5.3)$$

Here the notation

$$x_{12} = x_1 - x_2, \qquad t_{ab} = t_a - t_b \quad (a, b = 1, 2) \qquad (5.4)$$

is used; the matrix $M^{(\alpha)}$ is (for $M^{(\beta)}$ one should merely replace α by β)

$$M_{jk}^{(\alpha)} = \frac{1}{\lambda_j - \mu_k} - \alpha\frac{1}{\lambda_{N+1} - \mu_k}. \qquad (5.5)$$

Each of the two determinants in (5.3) is an antisymmetric function of all the λ_j (4.13). So one can replace, e.g., the second determinant by

$$(N+1)! \prod_{j=1}^{N} \frac{1}{\lambda_j - \mu_k},$$

thus obtaining

$$\langle \Psi(x_2, t_2)\Psi^\dagger(x_1, t_1)\rangle_N$$

$$= \exp\left\{ iht_{21} + it_{21}\sum_{j=1}^{N}\mu_j^2 + ix_{12}\sum_{j=1}^{N}\mu_j \right\}\frac{1}{L}\left(\frac{2}{L}\right)^{2N}$$

$$\times (N+1)! \sum_{\{\lambda\}}\left(\exp\left\{ ix_{21}\lambda_{N+1} + it_{12}\lambda_{N+1}^2 \right\} + \frac{\partial}{\partial\alpha} \right)$$

$$\times \det{}_N\left(\frac{\exp\left\{ it_{12}\lambda_j^2 + ix_{21}\lambda_j \right\}}{(\lambda_j - \mu_k)(\lambda_j - \mu_j)} \right.$$

$$\left. -\alpha\frac{\exp\left\{ it_{12}\lambda_j^2 + ix_{21}\lambda_j \right\}}{\lambda_j - \mu_j}\frac{\exp\left\{ it_{12}\lambda_{N+1}^2 + ix_{21}\lambda_{N+1} \right\}}{\lambda_{N+1} - \mu_k} \right)\bigg|_{\alpha=0}. \qquad (5.6)$$

Let us now replace the summation with respect to all sets $\{\lambda\}$ by independent summations in each individual λ_j:

$$(N+1)! \sum_{\{\lambda\}} = \sum_{\lambda_1} \sum_{\lambda_2} \cdots \sum_{\lambda_N} \sum_{\lambda_{N+1}} \qquad (5.7)$$

(if some of the λ_j coincide then the corresponding contribution is zero due to the antisymmetry of the form factors in λ). Here $\sum_{\lambda_j}$ on the right hand side denotes the sum over all the permitted values of λ_j, i.e., over all $m_j \in \mathbb{Z}$ ($\lambda_j = 2\pi m_j/L$), i.e.,

$$\sum_{\lambda_j} = \sum_{m_j=-\infty}^{\infty} .$$

Due to the fact (mentioned already) that the determinant in (5.6) is a linear function of α, the sum over λ_{N+1} in the term proportional to $\partial/\partial\alpha$ can be taken inside the determinant (in the coefficient at α of each matrix element). Then one notices that the variable λ_j enters only the j-th row of the determinant ($j = 1, \ldots, N$), so that the corresponding sum can be taken inside the determinant in each element of this row. Taking this into account, and multiplying the rows and columns of the determinant by the corresponding factors from the first exponential on the right hand side of (5.6), one arrives at the following answer for the mean value with respect to the N-particle eigenstate:

$$\langle \Psi(x_2, t_2)\Psi^\dagger(x_1, t_1)\rangle_N = \exp\{iht_{21}\} \left(\frac{1}{2\pi}\widetilde{G}(t_{12}, x_{12}) + \frac{\partial}{\partial\alpha} \right)$$

$$\times \det{}_N \left[\delta_{jk}\widetilde{\varepsilon}(\mu_k|t_{12}, x_{12}) \exp\{it_{21}\mu_k^2 - ix_{21}\mu_k\} \right.$$

$$+ \exp\{it_2\mu_j^2 - ix_2\mu_j - it_1\mu_k^2 + ix_1\mu_k\}$$

$$\times \left(\frac{2(1-\delta_{jk})}{\pi L(\mu_j - \mu_k)}[\widetilde{E}(\mu_j|t_{12}, x_{12}) - \widetilde{E}(\mu_k|t_{12}, x_{12})] \right.$$

$$\left. \left. - \frac{\alpha}{L\pi^2}\widetilde{E}(\mu_j|t_{12}, x_{12})\widetilde{E}(\mu_k|t_{12}, x_{12}) \right) \right]\Bigg|_{\alpha=0} .$$

$$(5.8)$$

Here the following notation for the sums that appear is introduced:

$$\frac{1}{L} \sum_{\lambda_{N+1}} \exp\{it\lambda_{N+1}^2 - ix\lambda_{N+1}\} \equiv \frac{1}{2\pi}\widetilde{G}(t, x) \qquad (5.9)$$

$$\frac{1}{L} \sum_{\lambda_j} \frac{\exp\{it\lambda_j^2 - ix\lambda_j\}}{\lambda_j - \mu_k} \equiv \frac{1}{2\pi}\widetilde{E}(\mu_k|t, x) \qquad (5.10)$$

$$\frac{4}{L^2} \sum_{\lambda_j} \frac{\exp\{it\lambda_j^2 - ix\lambda_j\}}{(\lambda_j - \mu_k)^2} \equiv \tilde{\varepsilon}(\mu_k|t,x), \tag{5.11}$$

and also the relation

$$\frac{1}{(\lambda_j - \mu_k)(\lambda_j - \mu_j)} = \left[\frac{1}{\lambda_j - \mu_j} - \frac{1}{\lambda_j - \mu_k}\right]\frac{1}{\mu_j - \mu_k}$$

was used.

Formula (5.8) gives the final answer for the correlator of two fields in the N-particle eigenstate. Our aim in the next section is to go to the thermodynamic limit and to obtain thus the time-dependent correlator of the impenetrable Bose gas at zero and nonzero temperature. It should be mentioned that the thermodynamic limits of sums (5.9)–(5.11) are calculated in Appendix 1.

XIII.6 Time-dependent correlation function for impenetrable bosons

To obtain the correlator most interesting from the physical point of view, the time-dependent correlation function of two fields in the thermodynamic limit, $N \rightarrow \infty$, $L \rightarrow \infty$, $D = N/L$ fixed, one should take the thermodynamic limit of expression (5.8) for the correlator for the N-particle eigenstate. This procedure was explained in sections I.3, I.5, I.8, X.4, XI.4, XI.5, so here we do not describe it in much detail. The only peculiar feature now is to calculate the thermodynamic limit of the sums $\tilde{G}, \tilde{E}, \tilde{\varepsilon}$ entering (5.8). This is done in Appendix 1. The result is:

$$\lim \tilde{G}(t,x) = G(t,x);$$

$$G(t,x) = \int_{-\infty}^{+\infty} \exp\{it\mu^2 - ix\mu\}\, d\mu, \tag{6.1}$$

$$\lim \tilde{E}(\lambda|t,x) = E(\lambda|t,x);$$

$$E(\lambda|t,x) = \fint_{-\infty}^{+\infty} \frac{\exp\{it\mu^2 - ix\mu\}}{\mu - \lambda}\, d\mu \tag{6.2}$$

(the slashed integral means the principal value of the integral),

$$\lim \tilde{\varepsilon}(\lambda|t,x) = \exp\{it\lambda^2 - ix\lambda\} + \frac{2}{\pi L}\frac{\partial}{\partial\lambda}E(\lambda|t,x). \tag{6.3}$$

After these formulæ are established, taking the thermodynamic limit in (5.8) is quite trivial, so we give little more than the answers.

The zero temperature correlator, which is the normalized mean value with respect to the ground state $|\Omega\rangle$ of the gas,

$$\langle \Psi(x_2, t_2) \Psi^\dagger(x_1, t_1) \rangle = \frac{\langle \Omega | \Psi(x_2, t_2) \Psi^\dagger(x_1, t_1) | \Omega \rangle}{\langle \Omega | \Omega \rangle} \qquad (6.4)$$

is obtained from (5.8) if one chooses for $|\Psi_N(\{\mu\})\rangle$ the ground state of impenetrable bosons with $N = N_g$, see (1.11), and then takes the limit. The answer is

$$\langle \Psi(x_2, t_2) \Psi^\dagger(x_1, t_1) \rangle$$
$$= \exp\{-iht_{12}\} \left(\frac{1}{2\pi} G(t_{12}, x_{12}) + \frac{\partial}{\partial \alpha} \right) \det\left(1 + \widehat{V}_0\right) \Big|_{\alpha=0}. \qquad (6.5)$$

Here $\widehat{V}_0$ is an integral operator. It acts on the interval $[-q, q]$. The action of $\widehat{V}_0$ on an arbitrary function $f(\mu)$ is given by the formula

$$\left(\widehat{V}_0 f\right)(\lambda) = \int_{-q}^{q} V_0(\lambda, \mu) f(\mu)\, d\mu. \qquad (6.6)$$

The kernel $V_0(\lambda, \mu)$ is equal to

$$V_0(\lambda, \mu) = \left[\frac{E(\lambda | t_{12}, x_{12}) - E(\mu | t_{12}, x_{12})}{\pi^2 (\lambda - \mu)} \right.$$
$$\left. - \frac{\alpha}{2\pi^3} E(\lambda | t_{12}, x_{12}) E(\mu | t_{12}, x_{12}) \right]$$
$$\times \exp\left\{ -\frac{i}{2} t_{12}(\lambda^2 + \mu^2) + \frac{i}{2} x_{12}(\lambda + \mu) \right\}, \qquad (6.7)$$

$t_{ab} = t_a - t_b$, $x_{ab} = x_a - x_b$; the functions E, G are given by (6.1) and (6.2). It is to be noted that the second term on the right hand side of (6.3) is naturally included in the first term in the square brackets in (6.7), being reproduced in the limit $\lambda \to \mu$. Expression (6.7) is the final answer for the time-dependent zero temperature correlation function of the impenetrable Bose gas.

For the equal-time case, $t_1 = t_2$, the corresponding representation (3.1) is, of course, reproduced. Indeed, in this case

$$\frac{1}{2\pi} G(0, x_{12}) = \delta(x_1 - x_2),$$

which gives the commutator

$$[\Psi(x_1), \Psi^\dagger(x_2)] = \delta(x_1 - x_2)$$

and

$$E(\mu | 0, x_{12}) = -i\pi \exp\{-i\mu x_{12}\} \qquad \text{for } x_{12} > 0$$

and the kernel $V_0(\lambda, \mu)$ is reduced to ($x_{12} \equiv x$)

$$V_0(\lambda, \mu) = -\frac{2}{\pi} \left(\frac{\sin \dfrac{x_{12}(\lambda - \mu)}{2}}{\lambda - \mu} \right) + \frac{\alpha}{2\pi} \exp\left\{ -i\frac{x_{12}(\lambda + \mu)}{2} \right\},$$

in complete agreement with (3.1), (3.2) and (3.3), which were shown to be equivalent to Lenard's formula.

We now turn to the temperature-dependent correlation function (see section I.8)

$$\langle \Psi(x_2, t_2) \Psi^\dagger(x_1, t_1) \rangle_T = \frac{\mathrm{tr}\left(e^{-H/T} \Psi(x_2, t_2) \Psi^\dagger(x_1, t_1) \right)}{\mathrm{tr}\, e^{-H/T}}.$$

According to section I.8 we should calculate the mean value

$$\frac{\langle \Omega_T | \Psi(x_2, t_2) \Psi^\dagger(x_1, t_1) | \Omega_T \rangle}{\langle \Omega_T | \Omega_T \rangle}.$$

Here $|\Omega_T\rangle$ is one of the eigenfunctions present in the state of thermal equilibrium. This mean value can be calculated similarly to the zero temperature case (6.4), (6.5). The answer is also similar. It differs only by the measure of integration:

$$\int_{-q}^{q} d\lambda \longrightarrow \int_{-\infty}^{+\infty} d\lambda\, \vartheta(\lambda), \quad \vartheta(\lambda) = \frac{1}{1 + e^{(\lambda^2 - h)/T}}.$$

In the thermodynamic limit one obtains the following representation:

$$\frac{\langle \Omega_T | \Psi(x_2, t_2) \Psi^\dagger(x_1, t_1) | \Omega_T \rangle}{\langle \Omega_T | \Omega_T \rangle}$$

$$= \exp\{-iht_{12}\} \left(\frac{1}{2\pi} G(t_{12}, x_{12}) + \frac{\partial}{\partial \alpha} \right) \det\left(1 + \widehat{V}_T \right) \Big|_{\alpha=0}$$

$$\tag{6.8}$$

where the operator $\widehat{V}_T$ acts now on the whole real axis, its kernel $V_T(\lambda, \mu)$ being equal to

$$V_T(\lambda, \mu) = \sqrt{\vartheta(\lambda)}\, V_0(\lambda, \mu)\, \sqrt{\vartheta(\mu)}$$

$$\vartheta(\mu) = \frac{1}{1 + e^{(\mu^2 - h)/T}}.$$

$$\tag{6.9}$$

Here $V_0(\lambda, \mu)$ is given by (6.7) (we have symmetrized the kernel, as for (3.6)). So we see that the mean value depends only on $\vartheta(\lambda)$ which is the same for all eigenfunctions of the Hamiltonian $|\Omega_T\rangle$ present in the state of thermal equilibrium. So the mean value (6.8) does not depend on the particular choice of $|\Omega_T\rangle$. Now we can use the results of section I.8 (see I.8.19):

$$\langle \Psi(x_2, t_2)\Psi^\dagger(x_1, t_1)\rangle_T = \frac{\mathrm{tr}\left(e^{-H/T}\,\Psi(x_2, t_2)\Psi^\dagger(x_1, t_1)\right)}{\mathrm{tr}\,e^{-H/T}}$$

$$= \exp\{-iht_{12}\}\left(\frac{1}{2\pi}G(t_{12}, x_{12}) + \frac{\partial}{\partial\alpha}\right)$$

$$\times \det\left(1 + \widehat{V}_T\right)\bigg|_{\alpha=0}. \qquad (6.10)$$

Finally, let us give the representation for the correlator in Euclidean time, $\tau = it$. It is defined as ($\Psi(x, \tau) = e^{H\tau}\Psi(x)e^{-H\tau}$):

$$\langle \Psi(x_2, \tau_2)\Psi^\dagger(x_1, \tau_1)\rangle_T$$

$$= \frac{\mathrm{tr}\left(e^{-H/T}\,\Psi(x_2, \tau_2)\Psi^\dagger(x_1, \tau_1)\right)}{\mathrm{tr}\,e^{-H/T}}$$

$$= \exp\{h\tau_{12}\}\left(\frac{1}{2\pi}G(\tau_{12}, x_{12}) + \frac{\partial}{\partial\alpha}\right)\det\left(1 + \widehat{V}_\tau\right)\bigg|_{\alpha=0} \qquad (6.11)$$

with the kernel V_τ given by

$$V_\tau(\lambda, \mu) = V_T(\lambda, \mu)\bigg|_{t_{12}=-i\tau_{12}}$$

$$= \sqrt{\vartheta(\lambda)\,\vartheta(\mu)} \times \exp\left\{\frac{1}{2}\tau_{21}(\lambda^2 + \mu^2) - \frac{i}{2}x_{21}(\lambda + \mu)\right\}$$

$$\times \left[\frac{E(\lambda|\tau_{12}, x_{12}) - E(\mu|\tau_{12}, x_{12})}{\pi^2(\lambda - \mu)}\right.$$

$$\left. - \frac{\alpha}{2\pi^3}E(\lambda|\tau_{12}, x_{12})E(\mu|\tau_{12}, x_{12})\right] \qquad (6.12)$$

with

$$G(\tau_{12}, x_{12}) = \int d\mu \,\exp\{-\tau_{21}\mu^2 - ix_{12}\mu\}$$

$$E(\lambda|\tau_{12}, x_{12}) = \fint \frac{d\mu}{\mu - \lambda} \exp\{-\tau_{21}\mu^2 - ix_{12}\mu\}. \qquad (6.13)$$

So the representations for the time-dependent correlators, which generalize Lenard's formulæ for the equal-time correlators, are given.

XIII.7 Multifield correlation function for impenetrable bosons

First we shall consider the equal-time correlation function

$$\langle \Psi^\dagger(z_1^+)\Psi^\dagger(z_2^+)\cdots\Psi^\dagger(z_N^+)\Psi(z_1^-)\Psi(z_2^-)\cdots\Psi(z_N^-)\rangle. \qquad (7.1)$$

It can be considered similarly to the two-field correlation function. Actually, the simplest way to obtain the determinant representation is to take Lenard's formula [7], [8] and to make a Fourier transformation (similar to that for the two-point case treated in section 3, see also paper [5]). We can consider temperature correlation functions at once, because zero temperature is just a particular case. Explicit formulæ for the many-field equal-time correlation function (7.1) in the form of determinants of Fredholm integral operator are presented in section 4 of Chapter XIV (see formulæ (XIV.4.14) and (XIV.4.15)).

Now let us discuss many-field time-dependent correlators for the impenetrable Bose gas. To obtain the Fredholm determinant representation one should go through the same steps as for the two-point correlation function. The most important step is to obtain the determinant representation for the form factor (4.14). The summation of form factors is quite similar to that in section 5. The thermodynamic limit can be evaluated as in section 6. To write down an explicit determinant formula for the correlation function it is convenient to choose the following ordering:

$$\langle \Psi(x_{2N}, t_{2N})\Psi^\dagger(x_{2N-1}, t_{2N-1}) \cdots \Psi(x_4, t_4)\Psi^\dagger(x_3, t_3)\Psi(x_2, t_2)\Psi^\dagger(x_1, t_1)\rangle$$

$$\equiv \langle \prod_{k=1}^{N} \Psi(x_{2k}, t_{2k})\Psi^\dagger(x_{2k-1}, t_{2k-1})\rangle. \qquad (7.2)$$

First let us consider the case of zero temperature. For this case we shall express (7.2) in terms of the determinant of the Fredholm integral operator K, which acts on the interval $[-q, q]$. The kernel $K(\lambda_0, \lambda_N)$ of this integral operator is given by

$$K(\lambda_0, \lambda_N)$$

$$= \int_{-\infty}^{+\infty} d\lambda_1 \int_{-\infty}^{+\infty} d\lambda_2 \cdots \int_{-\infty}^{+\infty} d\lambda_N \prod_{m=1}^{N} \Big[\delta(\lambda_m - \lambda_{m-1}) + K_m(\lambda_m, \lambda_{m-1})$$

$$\times \exp i\{t_{2m}\lambda_m^2 - t_{2m-1}\lambda_{m-1}^2 + x_{2m-1}\lambda_{m-1} - x_{2m}\lambda_m\}\Big]. \qquad (7.3)$$

Here

$$K_m(\lambda_m, \lambda_{m-1}) = \frac{E_m(\lambda_m) - E_m(\lambda_{m-1})}{\pi^2(\lambda_m - \lambda_{m-1})} - \frac{\alpha_m}{2\pi^2} E_m(\lambda_m)E_m(\lambda_{m-1})$$

$$(7.4)$$

and the function $E_m(\lambda)$ is defined by

$$E_m(\lambda) = \fint \frac{d\mu}{\mu - \lambda} \exp\{i\mu^2(t_{2m-1} - t_{2m}) - i\mu(x_{2m-1} - x_{2m})\}. \qquad (7.5)$$

$\delta(\lambda - \mu)$ in (7.3) is Dirac's δ-function.

The correlation function (7.2) can be related to the determinant of integral operator (7.3) by the following formula:

$$\left\langle \prod_{k=1}^{N} \Psi(x_{2k}, t_{2k}) \Psi^{\dagger}(x_{2k-1}, t_{2k-1}) \right\rangle = \exp\left\{ ih \sum_{k=1}^{N} (t_{2k} - t_{2k-1}) \right\}$$

$$\times \prod_{k=1}^{N} \left(\frac{1}{2\pi} G(t_{2k-1} - t_{2k}, x_{2k-1} - x_{2k}) + \frac{\partial}{\partial \alpha_k} \right) \det K \bigg|_{\alpha_j=0}.$$

$$(7.6)$$

Here the function $G(t, x)$ is defined by

$$G(t, x) = \int_{-\infty}^{+\infty} d\mu \, e^{it\mu^2 - ix\mu}. \tag{7.7}$$

In section 1, Chapter XIV, it is explained that integral operators with kernels like (7.4) form an infinite dimensional group. This makes it possible to represent the kernel of integral operator $K(\lambda_0, \lambda_N)$ (7.3) in the form

$$K(\lambda, \mu) = \delta(\lambda - \mu) + \frac{1}{\lambda - \mu} \sum_{j=1}^{2N} e_j(\lambda) \tilde{e}_j(\mu) \tag{7.8}$$

where functions $e_j(\lambda)$ and $\tilde{e}_j(\mu)$ can be explicitly evaluated using the rules of section XIV.1. In Chapter XIV it is explained that the representation (7.8) leads to a differential equation for the quantum correlation function. It is interesting to mention that finite temperature does not spoil the representation (7.8).

The usual calculations show that for temperature correlation functions (see section I.8)

$$\left\langle \prod_{k=1}^{N} \Psi(x_{2k}, t_{2k}) \Psi^{\dagger}(x_{2k-1}, t_{2k-1}) \right\rangle_T$$

representation (7.6) is still valid but integral operator $\widehat{K}$ (7.3) should be replaced by $\widehat{K}_T$. The kernel of $\widehat{K}_T$ is equal to

$$K_T(\lambda, \mu) = \delta(\lambda - \mu) + \sqrt{\vartheta(\lambda)\vartheta(\mu)} \left[K(\lambda, \mu) - \delta(\lambda - \mu) \right]. \tag{7.9}$$

Here

$$\vartheta(\lambda) = \frac{1}{1 + e^{\varepsilon(\lambda)/T}} \tag{7.10}$$

and $\varepsilon(\lambda) = \lambda^2 - h$.

All of this shows that many-field correlation functions can be perfectly embedded in our approach.

For impenetrable bosons, the correlation function of many currents $j(x,t)$, which depends on space, time and temperature, can be found in [5].

Conclusion

In this chapter we have discussed determinant representations for the quantum correlation functions of impenetrable bosons. The Fredholm determinant for the time-independent field correlation function was obtained in papers [7], [8]. The Fredholm determinant representation for the time-dependent correlation function was found in [6]. In the following chapters these determinant representations will be used to write down the differential equations for quantum correlation functions. Finally, the asymptotics will be explicitly evaluated, which is especially interesting for time- and temperature-dependent correlation functions.

Appendix XIII.1:
Temperature correlations

In this Appendix we will follow the presentation in papers [1], [2], [3].

Let us consider correlation functions at finite temperature. To be specific, let us consider the field correlation function

$$\langle \Psi^\dagger(x)\Psi(0)\rangle_T = \frac{\text{tr}\left(e^{-H/T}\,\Psi^\dagger(x)\Psi(0)\right)}{\text{tr}\,e^{-H/T}}. \qquad (A.1.1)$$

Below we shall write down a functional integral representation for this correlation function similar to (I.5.14). First, let us write

$$\text{tr}\left[e^{-H/T}\Psi^\dagger(x)\Psi(0)\right] = \frac{1}{N!}\sum_{n_1,\dots,n_N} e^{-E_n/T}\frac{\langle\Psi_N|\Psi^\dagger(x)\Psi(0)|\Psi_N\rangle}{\langle\Psi_N|\Psi_N\rangle}. \qquad (A.1.2)$$

Here $|\Psi_N\rangle$ is the complete set of all eigenfunctions of the Hamiltonian H. This representation is similar to the representation (I.5.11) for $\text{tr}\,e^{-H/T}$. Now one can write a functional integral representation for (A.1.2) similar to (I.5.11):

$$\text{tr}\left[e^{-H/T}\Psi^\dagger(x)\Psi(0)\right] =$$

$$\text{const}\int D\left(\frac{\rho_t(\lambda)}{\rho_p(\lambda)}\right)\delta\left(L\int\rho_p(\lambda)\,d\lambda - D\right)\frac{\langle\Psi_N|\Psi^\dagger(x)\Psi(0)|\Psi_N\rangle}{\langle\Psi_N|\Psi_N\rangle}e^{S-E/T}. \qquad (A.1.3)$$

The mean value

$$\frac{\langle\Psi_N|\Psi^\dagger(x)\Psi(0)|\Psi_N\rangle}{\langle\Psi_N|\Psi_N\rangle}$$

has a finite thermodynamic limit. On the contrary, in the thermodynamic limit $(L \to \infty)$ the entropy S and energy E are both proportional to the length of the box L. This shows that the stationary point of (A.1.3) coincides with the stationary point of (I.5.14) and is given by equations (I.5.24)–(I.5.26) (this is the state of thermodynamic equilibrium). In

311

addition to this, the calculations of Chapter XIII show that if we take one of the eigenfunctions $|\Omega_T\rangle$ (which is present in the state of thermodynamic equilibrium) then the mean value

$$\frac{\langle\Omega_T|\Psi^\dagger(x)\Psi(0)|\Omega_T\rangle}{\langle\Omega_T|\Omega_T\rangle} \tag{A.1.4}$$

depends only on $\varepsilon(\lambda)$, $\rho_p(\lambda)$, $\rho_h(\lambda)$ and $\rho_t(\lambda)$, but not on the particular choice of the set of integer numbers n_j (I.2.13) which specify the eigenfunction. But the values $\varepsilon(\lambda)$, $\rho_p(\lambda)$, $\rho_h(\lambda)$, $\rho_t(\lambda)$ are the same for all eigenfunctions of the Hamiltonian that are present in the state of thermodynamic equilibrium. So, finally, we have

$$\langle\Psi^\dagger(x)\Psi(0)\rangle_T = \frac{\mathrm{tr}\left(e^{-H/T}\,\Psi^\dagger(x)\Psi(0)\right)}{\mathrm{tr}\,e^{-H/T}}$$

$$= \frac{\langle\Omega_T|\Psi^\dagger(x)\Psi(0)|\Omega_T\rangle}{\langle\Omega_T|\Omega_T\rangle}. \tag{A.1.5}$$

Here $|\Omega_T\rangle$ is one of the eigenfunctions which are present in the state of thermodynamic equilibrium and the right hand side of (A.1.5) does not depend on the particular choice of $|\Omega_T\rangle$. This is very similar to what has happened with excitations at finite temperature (see section I.8). Let us also recall that we know how many different eigenfunctions of the Hamiltonian are present in the state of thermodynamic equilibrium. This number is related to the entropy S. It is $\exp S$ where

$$S = \frac{L}{T}\int d\lambda\,[\rho_t(\lambda)\ln\rho_t(\lambda) - \rho_p(\lambda)\ln\rho_p(\lambda) - \rho_h(\lambda)\ln\rho_h(\lambda)]. \tag{A.1.6}$$

So the number of eigenfunctions in the state of thermodynamic equilibrium goes to infinity in the thermodynamic limit, but this is not the complete set of eigenfunctions. We have considered correlation functions of fields. But formulæ like (A.1.5) are also correct for any other correlation functions. Take, for example, the current correlation function

$$\langle j(x,t)j(0,0)\rangle_T = \frac{\mathrm{tr}\left(e^{-H/T}\,j(x,t)j(0,0)\right)}{\mathrm{tr}\,e^{-H/T}} = \frac{\langle\Omega_T|j(x,t)j(0,0)|\Omega_T\rangle}{\langle\Omega_T|\Omega_T\rangle} \tag{A.1.7}$$

or that for $\exp\{\alpha Q_1(x)\}$:

$$\frac{\mathrm{tr}\left(e^{-H/T}\exp\alpha Q_1(x)\right)}{\mathrm{tr}\,e^{-H/T}} = \frac{\langle\Omega_T|\exp\alpha Q_1(x)|\Omega_T\rangle}{\langle\Omega_T|\Omega_T\rangle}. \tag{A.1.8}$$

Here

$$Q_1(x) = \int\limits_0^x \Psi^\dagger(y)\Psi(y)\,dy. \tag{A.1.9}$$

Appendix XIII.2: Thermodynamic limit of singular sums

Here the behavior in the thermodynamic limit of sums of the type (5.9)–(5.11) is studied.

Consider first the nonsingular sum

$$S_0 = \frac{1}{L} \sum_{\lambda_j} g(\lambda_j) = \frac{1}{L} \sum_{n_j=-\infty}^{+\infty} g(\lambda_j) \tag{A.2.1}$$

where the sum is over all the $\lambda_j = 2\pi n_j/L$, $n_j \in \mathbf{Z}$ and $g(\lambda)$ is a smooth decreasing function. For example,

$$g(\lambda) = \exp\{it\lambda^2 - ix\lambda\}$$

with the standard regularization $t \to t + i0$. It is obvious that

$$\lim S_0 = \frac{1}{2\pi} \int_{-\infty}^{+\infty} g(\lambda)\, d\lambda, \tag{A.2.2}$$

(A.2.1) being in fact the integral sum for this integral. So (6.1) is true.

Consider now the "singular" sum of the type (5.10),

$$S_1 = \frac{1}{L} \sum_{\lambda_j} \frac{g(\lambda_j)}{\lambda_j - \mu_k} \tag{A.2.3}$$

with given $\mu_k = 2\pi(m_k + \frac{1}{2})/L$, $m_k \in \mathbf{Z}$. One transforms (A.2.3) as

$$S_1 = \frac{1}{L} \sum_{\lambda_j} \frac{g(\lambda_j) - g(\mu_k)}{\lambda_j - \mu_k} + \frac{1}{2\pi} g(\mu_k) \sum_{n_j=-\infty}^{\infty} \frac{1}{n_j - \frac{1}{2}}$$

$$\longrightarrow \frac{1}{2\pi} \int \frac{d\lambda}{\lambda - \mu_k} [g(\lambda) - g(\mu_k)] = \frac{1}{2\pi} \fint \frac{d\lambda}{\lambda - \mu_k} g(\lambda) \tag{A.2.4}$$

313

where the formulæ

$$\sum_{n_j=-\infty}^{\infty} \left(\frac{1}{n_j - \frac{1}{2}} \right) = 0, \qquad \fint \frac{d\lambda}{\lambda - \mu_k} = 0 \qquad (A.2.5)$$

were used. So the thermodynamic limit of S_1 (A.2.3) is

$$\lim S_1 = \frac{1}{2\pi} \fint \frac{d\lambda}{\lambda - \mu_k} g(\lambda), \qquad (A.2.6)$$

the particular case of this relation being (6.2).

Consider now an even more singular sum,

$$S_2 = \frac{1}{L^2} \sum_{\lambda_j} \frac{g(\lambda_j)}{(\lambda_j - \mu_k)^2}. \qquad (A.2.7)$$

Transform it in a similar way,

$$S_2 = \frac{g(\mu_k)}{L^2} \sum_{\lambda_j} \frac{1}{(\lambda_j - \mu_k)^2} + \frac{1}{L^2} \sum_{\lambda_j} \frac{g(\lambda_j) - g(\mu_k)}{(\lambda_j - \mu_k)^2}. \qquad (A.2.8)$$

The first term is calculated as

$$\frac{g(\mu_k)}{L^2} \sum_{\lambda_j} \frac{1}{(\lambda_j - \mu_k)^2} = \frac{1}{4\pi^2} g(\mu_k) \sum_{n_j=-\infty}^{\infty} \frac{1}{\left(n_j - \frac{1}{2} \right)^2}$$

$$= \frac{1}{4} g(\mu_k). \qquad (A.2.9)$$

For the second term which is a sum of the type S_1 (A.2.3) one uses (A.2.6) to get

$$\frac{1}{L^2} \sum_{\lambda_j} \frac{g(\lambda_j) - g(\mu_k)}{(\lambda_j - \mu_k)^2} \longrightarrow \frac{1}{2\pi L} \fint d\lambda \frac{g(\lambda) - g(\mu_k)}{(\lambda - \mu_k)^2}. \qquad (A.2.10)$$

Using (A.2.5) one can obtain

$$\frac{1}{2\pi L} \fint d\lambda \frac{g(\lambda) - g(\mu_k)}{(\lambda - \mu_k)^2} = \frac{1}{2\pi L} \frac{\partial}{\partial \mu_k} \fint d\lambda \frac{g(\lambda) - g(\mu_k)}{\lambda - \mu_k}$$

$$= \frac{1}{2\pi L} \frac{\partial}{\partial \mu_k} \fint d\lambda \frac{g(\lambda)}{\lambda - \mu_k}. \qquad (A.2.11)$$

Finally, one has for S_2 (A.2.7):

$$\lim S_2 = \frac{1}{4} g(\mu_k) + \frac{1}{2\pi L} \frac{\partial}{\partial \mu_k} \fint d\lambda \frac{g(\lambda)}{\lambda - \mu_k}. \qquad (A.2.12)$$

Relation (6.3) is an obvious particular case of this formula.

XIV

Differential Equations
for Correlation Functions

Introduction

The natural language for the description of correlation functions in exactly solvable models is the language of differential equations [11]. The determinant representation of correlation functions in the previous chapter can be used to obtain these differential equations. An ordinary differential equation (of the Painlevè type) for the equal-time, zero temperature correlation function of the impenetrable Bose gas was obtained in [8]. In that paper (starting from Lenard's formula for the two-point zero temperature equal-time correlator of impenetrable bosons) it was shown that the correlator is described by an ordinary differential equation (reducible to the Painlevè trascendent), and multipoint zero temperature equal-time correlators were also described as solutions of differential equations. The considerations in that paper were based on the spectral isomonodromic deformation method.

Our approach here is different [4], [5], [6]. We begin by considering the logarithm of the determinant of the linear integral operator entering the representation for the correlator as a τ-function for some integrable nonlinear evolutionary system. Then the linear integral equations playing the role of the Gel'fand-Levitan-Marchenko equations for this integrable system are constructed. The differential equations for correlators are derived, in fact, from these integral equations (it is worth mentioning that the idea of obtaining the differential equations from the integral ones can be found in [1], [9], [10]). This approach permits one to include time- and temperature-dependent correlators for impenetrable bosons in the same framework. It is remarkable that space- and time-dependent correlation functions are derived by partial differential equations which almost coincide with the original classical nonlinear Schrödinger equation, the only difference being that complex involution is dropped. Moreover,

315

the Bose gas with finite coupling constant $0 < c < +\infty$ (penetrable bosons) can also be treated.

The contents of this chapter are as follows. In section 1, the class of "integrable" linear integral equations is introduced. It is to be emphasized that all the linear integral operators entering the representations of correlation functions belong to this class. The properties of integrable linear integral operators are crucial in deriving differential equations for correlators. In sections 2 and 3 differential equations for the equal-time temperature correlator of two fields for impenetrable bosons are constructed. The correlator at zero temperature is obtained as a particular case. In section 4, the multipoint equal-time correlators for the impenetrable Bose gas at nonzero temperature are considered. The time-dependent two-point correlator of fields in the impenetrable Bose gas is treated in section 5, the temperature-dependent correlator in this case being considered in section 6. Finally, in section 7 the differential equations for the simplest equal-time correlator (the "emptiness formation probability") are derived for a penetrable Bose gas ($0 < c < \infty$). In this case the Bose gas is not equivalent to free fermions.

It is to be said that the results of this sections give, in principle, a complete description of the correlators. One can obtain results interesting from a physical point of view (see Appendices 1 and 2), for the short distance and low density expansions. The most important thing is, of course, to obtain the large time and long distance asymptotics. To do this, one uses the matrix Riemann problem related to the corresponding integrable differential equations. These matrix Riemann problems for the correlator considered are given in Chapter XV. The large time and long distance asymptotics are obtained in Chapter XVI. The present chapter is closely related to Chapter V (especially section V.1).

XIV.1 Integrable linear integral operators

Our aim in this chapter is to use the representation of correlation functions as Fredholm determinants of linear integral operators (see Chapters XI–XIII) to derive differential equations for the former. All these Fredholm integral operators belong to a special class: the kernel of each of them $V(\lambda, \mu)$ can be represented in the form

$$V(\lambda, \mu) = \frac{1}{\lambda - \mu} \sum_{j=1}^{N} e_j(\lambda) E_j(\mu). \tag{1.1}$$

Sometimes we shall call such operators "integrable" integral operators. Here the $e_j(\lambda)$ are N linearly independent functions (they are also supposed to be continuous and integrable, rapidly decaying at infinity). The functions $E_j(\lambda)$ have similar properties. In all examples related to correlation functions, the kernel (1.1) is nonsingular. This means that

$$\sum_{j=1}^{N} e_j(\lambda) E_j(\lambda) = 0. \tag{1.2}$$

But first, we shall study a more general situation, when the condition (1.2) is not necessarily satisfied and the integral operator (1.1) is singular. Suppose we act with the integral operator on the function $\varphi(\mu)$:

$$\left(\widehat{V}\varphi\right)(\lambda) = \fint_{-\infty}^{\infty} V(\lambda,\mu)\varphi(\mu)\,d\mu. \tag{1.3}$$

Here we shall understand the integral as the principal value. We will assume that the operator $1 + \widehat{V}$ is not degenerate.

It is interesting to mention that the product of two such operators has the same form. Let us take two different integrable operators, $\widehat{V}^{(1)}$ and $\widehat{V}^{(2)}$,

$$V^{(a)}(\lambda,\mu) = \frac{1}{\lambda-\mu}\sum_{j=1}^{N_a} e_j^{(a)}(\lambda) E_j^{(a)}(\mu), \qquad a = 1, 2. \tag{1.4}$$

Let us show that their product is also an "integrable" operator:

$$V^{(3)}(\lambda,\mu) = \fint d\nu\, V^{(1)}(\lambda,\nu) V^{(2)}(\nu,\mu)$$

$$= \frac{1}{\lambda-\mu}\left\{ \sum_{j=1}^{N_1} e_j^{(1)}(\lambda) E_j^{(3)}(\mu) + \sum_{k=1}^{N_2} e_k^{(3)}(\lambda) E_k^{(2)}(\mu) \right\} \tag{1.5}$$

with

$$E_j^{(3)}(\mu) = \fint d\nu\, E_j^{(1)}(\nu) V^{(2)}(\nu,\mu),$$

$$e_j^{(3)}(\lambda) = \fint d\nu\, V^{(1)}(\lambda,\nu) e_j^{(2)}(\nu) \tag{1.6}$$

which becomes obvious if one uses the relation

$$\frac{1}{(\lambda-\nu)(\nu-\mu)} = \frac{1}{\lambda-\mu}\left\{ \frac{1}{\lambda-\nu} + \frac{1}{\nu-\mu} \right\}. \tag{1.7}$$

All integrals here are understood in the sense of the principal value. But if (1.2) is valid for each operator then it is also valid for the product.

This is a very important property. In particular, it results in the following universal construction for the resolvent operator $\widehat{R}$ (with kernel $R(\lambda,\mu)$). Let us define the resolvent as usual:

$$(1+\widehat{V})(1-\widehat{R}) = 1; \qquad (1+\widehat{V})\widehat{R} = \widehat{V} \tag{1.8}$$

$$R(\lambda,\mu) + \fint d\nu\, V(\lambda,\nu) R(\nu,\mu) = V(\lambda,\mu). \tag{1.9}$$

To write down the resolvent more explicitly, one introduces functions $f_j^L(\lambda)$ and $f_j^R(\lambda)$ by the following integral equations:

$$f_j^L(\lambda) + \fint V(\lambda, \mu) f_j^L(\mu)\, d\mu = e_j(\lambda) \qquad (j = 1, \ldots, N) \qquad (1.10)$$

$$f_j^R(\lambda) + \fint f_j^R(\mu) V(\mu, \lambda)\, d\mu = E_j(\lambda) \qquad (j = 1, \ldots, N). \qquad (1.11)$$

These relations can also be rewritten in the form

$$f_j^L(\lambda) = e_j(\lambda) - \fint R(\lambda, \mu) e_j(\mu)\, d\mu \qquad (1.12)$$

$$f_j^R(\lambda) = E_j(\lambda) - \fint E_j(\mu) R(\mu, \lambda)\, d\mu. \qquad (1.13)$$

Theorem 1. *The resolvent $\widehat{R}$ of the operator $\widehat{V}$ (1.1) is given in terms of the functions f_j^L, f_j^R by*

$$R(\lambda, \mu) = \frac{1}{\lambda - \mu} \sum_{j=1}^{N} f_j^L(\lambda) f_j^R(\mu). \qquad (1.14)$$

Proof: Equation (1.9) defining the resolvent kernel can be rewritten as

$$(\lambda - \mu) R(\lambda, \mu) + \sum_{j=1}^{N} e_j(\lambda) \fint E_j(\nu) R(\nu, \mu)\, d\nu$$

$$+ \fint V(\lambda, \nu) \left[(\nu - \mu) R(\nu, \mu) \right] d\nu = \sum_{j=1}^{N} e_j(\lambda) E_j(\mu)$$

$$(1.15)$$

after multiplication of equation (1.9) by $(\lambda - \mu)$. Let us now move the second term on the left hand side to the right hand side and use (1.13); then we get

$$(\lambda - \mu) R(\lambda, \mu) + \fint V(\lambda, \nu) \left[(\nu - \mu) R(\nu, \mu) \right] d\nu = \sum_{j=1}^{N} e_j(\lambda) f_j^R(\mu). \qquad (1.16)$$

Here we have used (1.13). We can now act on this equation with the integral operator $1 - \widehat{R}$ from the left and get

$$(\lambda - \mu) R(\lambda, \mu) = \sum_{j=1}^{N} f_j^L(\lambda) f_j^R(\mu) \qquad (1.17)$$

by means of (1.8) and (1.12). In this way the theorem is proved. (All integrals are, as usual, to be understood in the principal value sense.)

If we consider the regular subgroup (1.2) then equation (1.9) shows that the function $R(\lambda, \mu)$ has no singularities at $\lambda = \mu$. This means that the f's satisfy the identity

$$\sum_{j=1}^{N} f_j^L(\lambda) f_j^R(\lambda) = 0. \tag{1.18}$$

Thus we have shown that the product of two operators of the form (1.1) has the same form, and the inverse operator has the same form. These properties define a group. Later we shall relate the Riemann problem to each of these operators.

XIV.2 Differential equations for temperature-dependent equal-time two-point correlators for the impenetrable Bose gas

We begin by deriving differential equations in the case of temperature-dependent equal-time two-point correlators for the impenetrable Bose gas considered earlier in sections XIII.2, XIII.3. This simplest case possesses in fact all the features of more general cases dealt with later. So in this case the necessary calculations are given in considerable detail.

We consider now the correlator of two fields

$$\langle \Psi^\dagger(x_1)\Psi(x_2)\rangle_T = \frac{\mathrm{tr}\left(e^{-H/T}\,\Psi^\dagger(x_1)\Psi(x_2)\right)}{\mathrm{tr}\,e^{-H/T}} \tag{2.1}$$

(see section I.8) and the generating functional for current correlators,

$$\langle \exp\{\alpha Q_1(x_1, x_2)\}\rangle_T = \frac{\mathrm{tr}\left(e^{-H/T}\exp\{\alpha Q_1(x_1, x_2)\}\right)}{\mathrm{tr}\,e^{-H/T}} \tag{2.2}$$

where $Q_1(x_1, x_2)$ is the operator for the number of particles in the interval $[x_2, x_1]$:

$$Q_1(x_1, x_2) = \int_{x_2}^{x_1} \Psi^\dagger(y)\Psi(y)\,dy. \tag{2.3}$$

The value of the correlator (2.1) at $\alpha = -\infty$ is the simplest two-point correlator $P(x_2, x_1)$ giving the probability that there are no particles of the gas in the interval $[x_2, x_1]$ (the "emptiness formation probability"):

$$P(x_1, x_2) = \langle \exp\{\alpha Q_1(x_1, x_2)\}\rangle_T\Big|_{\alpha=-\infty}. \tag{2.4}$$

The correlators considered here depend on two space points, x_1 and x_2. Due to translation invariance and parity conservation, the dependence is on the modulus, $|x_1 - x_2|$, of the coordinate difference only. So in what follows we suppose $x_1 > x_2$. Our starting point is the representations (XIII.3.4), (XIII.2.19) and (XIII.2.20) for the correlators (2.1), (2.2). The correlators depend on three variables: $|x_1 - x_2|$, the chemical potential h and the temperature T. It is easily seen that after introducing new variables, the scaled distance x and scaled chemical potential β,

$$x = \frac{1}{2}(x_1 - x_2)\sqrt{T}, \qquad \beta = \frac{h}{T}, \tag{2.5}$$

the explicit dependence on temperature of the correlators becomes trivial and the representations (XIII.3.12), (XIII.2.19) and (XIII.2.20) can be rewritten, after the corresponding change $\lambda \to \lambda/\sqrt{T}$ of the spectral parameter, as (see Appendix 4)

$$\langle \Psi^\dagger(x_1)\Psi(x_2)\rangle_T = \sqrt{T}g(x,\beta), \tag{2.6}$$

$$g(x,\beta) = \frac{1}{4}B_{++}(x,\beta,\gamma)\Delta(x,\beta,\gamma)\Big|_{\gamma=2/\pi}\ ;$$

$$\langle \exp\{\alpha Q_1(x_1,x_2)\}\rangle_T = \Delta(x,\beta,\gamma)\Big|_{\gamma=(1-\exp\alpha)/\pi}\ ; \tag{2.7}$$

$$P(x_1,x_2) = \langle \exp\{\alpha Q_1(x_1,x_2)\}\rangle_T\Big|_{\alpha=-\infty}$$

$$= \Delta(x,\beta,\gamma)\Big|_{\gamma=1/\pi}. \tag{2.8}$$

Let us explain the notation in (2.6)–(2.8). All the correlators are expressed in terms of a linear integral operator $\widehat{K}$ acting on any function f of the spectral parameter λ as

$$\left(\widehat{K}f\right)(\lambda) = \int\limits_{-\infty}^{+\infty} K(\lambda,\mu)f(\mu)\,d\mu \tag{2.9}$$

where $f(\mu)$ is an arbitrary function and the kernel $K(\lambda,\mu)$ is given by

$$K(\lambda,\mu) = \sqrt{\vartheta(\lambda)}\,\frac{\sin x(\lambda-\mu)}{\lambda-\mu}\,\sqrt{\vartheta(\mu)}. \tag{2.10}$$

It should be emphasized that the kernel of the operator $\widehat{K}$ can be put into the form (1.1):

$$K(\lambda,\mu) = \frac{e_+(\lambda)e_-(\mu) - e_-(\lambda)e_+(\lambda)}{2i(\lambda-\mu)} \tag{2.11}$$

where the functions e_+, e_- are introduced:

$$e_\pm(\lambda) = \sqrt{\vartheta(\lambda)}\, e^{\pm i\lambda x} \qquad (2.12)$$

so that the operator $\widehat{K}$ is integrable in the sense of section 1. The operator (2.11) is not singular, so condition (1.2) is valid. It is also symmetric:

$$K(\lambda, \mu) = K(\mu, \lambda).$$

This means that the sets of functions $\{e_j(\lambda)\}$ and $\{E_j(\lambda)\}$ are almost the same as e_+, e_-:

$$e_1 = e_+, \quad e_2 = e_-, \quad E_1 = \frac{e_-}{2i}, \quad E_2 = -\frac{e_+}{2i}.$$

The function $\vartheta(\lambda)$ here is the rescaled Fermi weight (XIII.1.15)

$$\vartheta(\lambda) = \frac{1}{1 + e^{\lambda^2 - \beta}}. \qquad (2.13)$$

($\vartheta(\lambda)$ is also a function of β and functions $e_\pm(\lambda)$ depend on x, β; this dependence will not be as a rule written down explicitly.)

The function Δ in (2.6)–(2.8) is the Fredholm determinant of the operator $\widehat{K}$,

$$\Delta(x, \beta, \gamma) = \det\left(1 - \gamma\widehat{K}\right), \qquad (2.14)$$

taken at the corresponding values ($\gamma = 2/\pi$ or $\gamma = (1 - e^\alpha)/\pi$) of the parameter γ.

To describe the coefficient B_{++} of Δ in (2.7) more explicitly it is natural to introduce functions $f_\pm(\lambda, x, \beta, \gamma)$ defined as the solutions of the linear integral equations

$$f_\pm(\lambda) - \gamma \int\limits_{-\infty}^{+\infty} K(\lambda, \mu) f_\pm(\mu)\, d\mu = e_\pm(\lambda). \qquad (2.15)$$

These are just the functions (1.10) and (1.11) in terms of which the resolvent of the integrable operator $\widehat{K}$ can be constructed. Of importance will be the 2×2 matrix of "potentials" $B_{lm}(x, \beta, \gamma)$ ($l = +, -; m = +, -$), the coefficient B_{++} being one of them,

$$B_{lm}(x, \beta, \gamma) \equiv \gamma \int\limits_{-\infty}^{+\infty} e_l(\lambda) f_m(\lambda)\, d\lambda. \qquad (2.16)$$

It is easily seen that the matrix B_{lm} is a real symmetrical matrix with two independent matrix elements B_{++} and B_{+-}:

$$B_{++}^* = B_{++} = B_{--}; \qquad B_{+-}^* = B_{+-} = B_{-+}. \qquad (2.17)$$

So the coefficient B_{++} in (2.6) is defined as:

$$B_{++}(x, \beta, \gamma) \equiv \gamma \int\limits_{-\infty}^{+\infty} e_+(\lambda) f_+(\lambda) \, d\lambda$$

$$= \gamma \int\limits_{-\infty}^{+\infty} e_-(\lambda) f_-(\lambda) \, d\lambda. \qquad (2.18)$$

So all the quantities entering the right hand sides of (2.6)–(2.8) are defined in terms of the linear integral operator $\widehat{K}$. The "integrability" (see section 1) of these operators allows one to prove the following properties of the two-point correlators under consideration:

Theorem 2. *The two-component vector function*

$$\vec{F}(\lambda) = \begin{pmatrix} f_+(\lambda) \\ f_-(\lambda) \end{pmatrix}$$

(where the functions $f_\pm$ are defined by the integral equations (2.15)) satisfies the following differential equations:

$$\partial_x \vec{F}(\lambda) = (i\lambda\sigma_3 + B_{++}\sigma_1)\vec{F}(\lambda) \qquad (2.19)$$

$$(2\lambda\partial_\beta + \partial_\lambda)\vec{F}(\lambda) = [ix\sigma_3 - i(\partial_\beta B_{+-})\sigma_3 - (\partial_\beta B_{++})\sigma_2]\vec{F}(\lambda). \quad (2.20)$$

Here σ_a (a=1, 2, 3) are the Pauli matrices.

   The proof of these relations is given in the next section.
   Relations (2.19), (2.20) can be regarded as a Lax (zero curvature) representation for some classical nonlinear partial differential evolution equation for the potentials B_{++}, B_{+-} (2.16), (2.17). The compatibility condition for equations (2.19), (2.20)

$$[\partial_x - i\lambda\sigma_3 - B_{++}\sigma_1, 2\lambda\partial_\beta + \partial_\lambda - ix\sigma_3 + i\partial_\beta B_{+-}\sigma_3 + \partial_\beta B_{++}\sigma_2] = 0$$
$$(2.21)$$

immediately results in the following statement:

Theorem 3. *The potentials $B_{++}(x, \beta, \gamma)$, $B_{+-}(x, \beta, \gamma)$ satisfy the following system of partial differential equations in x, β valid for any γ:*

$$\partial_\beta B_{+-} = x + \frac{1}{2}\frac{\partial_x\partial_\beta B_{++}}{B_{++}} \qquad (2.22)$$

$$\partial_x B_{+-} = B_{++}^2. \qquad (2.23)$$

The function B_{++} itself satisfies the nonlinear equation

$$\partial_\beta B_{++}^2 = 1 + \frac{1}{2}\frac{\partial}{\partial x}\left(\frac{\partial_x\partial_\beta B_{++}}{B_{++}}\right). \tag{2.24}$$

The "initial conditions" are given by the following asymptotics at $x \to 0$ with β fixed:

$$B_{++}(x,\beta,\gamma) = \gamma d(\beta) + [\gamma d(\beta)]^2 x + O\left(x^2\right),$$

$$B_{+-}(x,\beta,\gamma) = \gamma d(\beta) + [\gamma d(\beta)]^2 x + O\left(x^2\right), \tag{2.25}$$

$$d(\beta) = \int\limits_{-\infty}^{+\infty} \vartheta(\lambda)\, d\lambda$$

and by the requirement that

$$\lim_{\beta\to-\infty} B_{++}(x,\beta,\gamma) = \lim_{\beta\to-\infty} B_{+-}(x,\beta,\gamma) = 0. \tag{2.26}$$

Proof: As equation (2.21) is valid for any λ, after some calculations one obtains equation (2.22) and also the equation

$$\partial_\beta\partial_x B_{+-} = \partial_\beta B_{++}^2 \tag{2.27}$$

which, due to (2.26), is equivalent to equation (2.23). Equation (2.24) is obtained if one differentiates (2.22) with respect to x and (2.23) with respect to β and then equates the right hand sides. Equations (2.25) and (2.26) follow immediately from results in Appendices 1 and 2. So we have constructed a completely integrable system of nonlinear partial differential equations for the potentials B.

We now turn to the Fredholm determinant

$$\Delta = \det\left(1 - \gamma\widehat{K}\right)$$

(2.14). It will be convenient to introduce the function σ which is the logarithm of the determinant:

$$\sigma(x,\beta,\gamma) \equiv \ln\Delta(x,\beta,\gamma) = \ln\det\left(1 - \gamma\widehat{K}\right). \tag{2.28}$$

The function σ possesses the following properties:

Theorem 4. *The partial derivatives of $\sigma(x,\beta,\gamma)$ with respect to x and β are given, for any γ, by*

$$\partial_x\sigma = -B_{+-}, \qquad \partial_x^2\sigma = -B_{++}^2 \tag{2.29}$$

$$\partial_\beta\sigma = -x\partial_\beta B_{+-} + \frac{1}{2}(\partial_\beta B_{+-})^2 - \frac{1}{2}(\partial_\beta B_{++})^2. \tag{2.30}$$

The function $\sigma(x, \beta, \gamma)$ at any γ satisfies the partial differential equation

$$(\partial_\beta \partial_x^2 \sigma)^2 = -4(\partial_x^2 \sigma)[2x \partial_\beta \partial_x \sigma + (\partial_\beta \partial_x \sigma)^2 - 2\partial_\beta \sigma] \qquad (2.31)$$

with "initial conditions"

$$\sigma = -\gamma d(\beta)x - [\gamma d(\beta)]^2 \frac{x^2}{2} + O\left(x^3\right), \qquad x \to 0$$

$$d(\beta) = \int\limits_{-\infty}^{+\infty} \vartheta(\lambda)\, d\lambda \qquad (2.32)$$

$$\lim_{\beta \to -\infty} \sigma(x, \beta, \gamma) = 0. \qquad (2.33)$$

Proof: Equations (2.29), (2.30) are proved in the next section. To obtain equation (2.31) it is sufficient to substitute the expressions $B_{+-} = -\partial_x \sigma$ (2.29) and $B_{++} = (-\partial_x^2 \sigma)^{1/2}$ (from (2.23) it follows that $\partial_x^2 \sigma = -\partial_x B_{+-} = B_{++}^2$) into equation (2.30). It is obvious that the kernel $K(\lambda, \mu)$ (2.10) of $\widehat{K}$ is equal to zero at $x = 0$, and at $\beta = -\infty$. Also, $e_\pm(\lambda) \to 0$ for $\beta \to -\infty$. So (2.33) for $\sigma = \ln \det \left(1 - \gamma \widehat{K}\right)$ is true and relation (2.32) follows from (2.29), (2.25).

So the potentials B and function $\sigma = \ln \Delta$ entering the representations for the correlators satisfy nonlinear partial differential equations which are completely integrable. The derivation of most of these results is given in the next section. The main idea is to consider integral equation (2.15) as a Gel'fand-Levitan-Marchenko-type equation for a classical integrable evolution system, the function $\vartheta(\lambda)$ (2.13) playing the role of "scattering data" (reflection coefficient) in these equations. As shown in the next section, theorems 2–4 are true not only for the function $\vartheta(\lambda)$ given for impenetrable bosons in (2.13) but also if $\vartheta(\lambda)$ is an arbitrary function of the difference $\lambda^2 - \beta$, decreasing as $\lambda^2 - \beta \to +\infty$, i.e., satisfying the requirements

$$(2\lambda\partial_\beta + \partial_\lambda)\vartheta(\lambda) = 0,$$

$$\vartheta\Big|_{\lambda^2 - \beta \to +\infty} = 0. \qquad (2.34)$$

It should be mentioned that the function $d(\beta)$ entering the "initial conditions" (2.25), (2.32) is in one-to-one correspondence with the function $\vartheta(\lambda)$ possessing properties (2.34). It follows from the reversibility of the formula

$$d(\beta) = \int\limits_{-\infty}^{+\infty} \vartheta(\lambda^2 - \beta)\, d\lambda \qquad (2.35)$$

(we have indicated the dependence of the function ϑ on its arguments λ, β explicitly). Indeed, taking a Fourier transform one has

$$\tilde{d}(k) = \int\limits_{-\infty}^{+\infty} e^{ik\beta} d(\beta)\, d\beta = \sqrt{\frac{i\pi}{k}} \int\limits_{-\infty}^{+\infty} e^{-iku} \vartheta(u)\, du,$$

so that the Fourier transform of the function $\vartheta(u)$ $(u = \lambda^2 - \beta)$ is restored uniquely from the function $\tilde{d}(k)$.

It should be mentioned that the correlators considered can be expressed entirely in terms of the potentials B. Indeed, it is easy to see that the operator $\widehat{K} \to 0$ as $x \to 0$, so that

$$\sigma(x = 0, \beta, \gamma) = \ln \det \left(1 - \gamma\widehat{K}\right)\Big|_{x=0} = 0.$$

Equation (2.29), $\partial_x \sigma(x, \beta, \gamma) = -B_{+-}(x, \beta, \gamma)$, then allows one to obtain σ by integrating B_{+-} over x so that

$$\Delta(x, \beta, \gamma) = \exp\{\sigma(x, \beta, \gamma)\} = \exp\left\{-\int\limits_0^x B_{+-}(y, \beta, \gamma)\, dy\right\}. \qquad (2.36)$$

Putting this expression into representations (2.6), (2.7) one obtains representations of the correlators in terms of the potentials B.

To conclude this section let us consider the correlators at zero temperature, $T = 0$. In this case one can make an independent derivation starting directly from representations (XIII.2.7), (XIII.3.1) for the correlators $\langle \exp\{\alpha Q_1(x)\}\rangle$ and $\langle \Psi^\dagger(x_1)\Psi(x_2)\rangle$ at zero temperature. The case of $T > 0$ having been considered already, the results for $T = 0$ can be obtained using the limiting procedure $T \to 0$. At zero temperature, the correlators depend on two variables, namely, the chemical potential h $(q = \sqrt{h}$ is the Fermi momentum) and the distance $|x_1 - x_2|$. It is easily seen from (XIII.2.7), (XIII.3.1) that the essential dependence is on the product $|x_1 - x_2|q$ only. So we introduce a new variable, ξ,

$$\xi = \frac{1}{2}q|x_1 - x_2| = x\beta^{1/2} \qquad (2.37)$$

(recall that $x = \frac{1}{2}|x_1 - x_2|\sqrt{T}$, $\beta = h/T$ (2.5) are the variables used in the nonzero temperature case). In the limit $T \to 0$ one evidently has for the Fredholm determinant $\Delta = \det\left(1 - \gamma\widehat{K}\right)$ (2.14)

$$\Delta(x, \beta, \gamma) = \Delta^{(0)}(\xi, \gamma) \qquad (T = 0). \qquad (2.38)$$

The partial differential equations for $\sigma = \ln\Delta$ in the limit $T = 0$ turn into ordinary differential equations in the variable ξ for $\sigma^{(0)} \equiv \ln\Delta^{(0)}$. It is convenient to introduce a new function σ_0 by

$$\sigma_0(\xi, \gamma) \equiv \left(\xi \sigma^{(0)}\right)' = \xi \partial_\xi \ln \Delta^{(0)}(\xi, \gamma). \tag{2.39}$$

The ordinary differential equation for σ_0 can be obtained directly from equation (2.31)):

$$(\xi \sigma_0'')^2 = -4(\xi \sigma_0' - \sigma_0)[4\xi \sigma_0' - 4\sigma_0 + (\sigma_0')^2. \tag{2.40}$$

The initial conditions are obtained from (2.32):

$$
\begin{aligned}
\sigma^{(0)} &= -2\gamma\xi - 2\gamma^2\xi^2 + O\left(\xi^3\right), \quad \text{or} \\
\sigma_0 &= \xi(\sigma^{(0)})' = -2\gamma\xi - 4\gamma^2\xi^2 + O\left(\xi^3\right).
\end{aligned}
\tag{2.41}
$$

This reproduces the results of [8], where the zero temperature, equal-time correlation function of fields was described in terms of an ordinary differential equation. One does similar calculations to obtain differential equations for the zero temperature limit of B_{++}, B_{+-}. They are very straightforward.

It is interesting to mention that the same equation (2.24) drives autocorrelation in the Ising model (for critical transverse magnetic field), see [7].

XIV.3 Correlators as a completely integrable system

Our main idea is to consider the integral equations (2.15)

$$f_\pm(\lambda) - \gamma \int\limits_{-\infty}^{+\infty} K(\lambda, \mu) f_\pm(\mu) \, d\mu = e_\pm(\lambda) \tag{3.1}$$

with "plane waves" $e_\pm(\lambda)$,

$$e_\pm(\lambda) = \sqrt{\vartheta(\lambda)} e^{\pm i\lambda x} \tag{3.2}$$

and kernel

$$K(\lambda, \mu) = \frac{e_+(\lambda)e_-(\mu) - e_-(\lambda)e_+(\mu)}{2i(\lambda - \mu)}, \tag{3.3}$$

$$K(\lambda, \mu) = K(\mu, \lambda) = K^*(\lambda, \mu)$$

as Gel'fand-Levitan-Marchenko-type equations for some nonlinear evolution system in two space-time dimensions. The function σ,

$$\sigma(x, \beta, \gamma) = \ln \Delta(x, \beta, \gamma) = \ln \det \left(1 - \gamma \widehat{K}\right), \tag{3.4}$$

plays the role of τ-function for this integrable system. The function $\vartheta(\lambda)$ with the properties

$$(2\lambda\partial_\beta + \partial_\lambda)\vartheta(\lambda) = 0; \qquad \vartheta(\lambda)\Big|_{\lambda^2 - \beta = +\infty} = 0 \qquad (3.5)$$

plays the role of "scattering data"; in our case of impenetrable bosons $\vartheta(\lambda)$ is given by

$$\vartheta(\lambda) = \frac{1}{1 + e^{\lambda^2 - \beta}}. \qquad (3.6)$$

It is to be emphasized, however, that all the considerations in this section are valid also for any function $\vartheta(\lambda)$ with properties (3.5).

It is also worth mentioning that the idea of obtaining differential equations for the solution of integral equations has been used in various ways in papers [1], [9], [10].

The operator $\widehat{K}$ (2.11) being integrable in the sense of section 1, the resolvent kernel $R(\lambda, \mu)$ of the resolvent operator $\widehat{R}$,

$$(1 - \gamma\widehat{K})(1 + \gamma\widehat{R}) = 1; \qquad (1 - \gamma\widehat{K})\widehat{R} = \widehat{K},$$

$$R(\lambda, \mu) - \gamma \int_{-\infty}^{+\infty} K(\lambda, \lambda')R(\lambda', \mu)\, d\lambda' = K(\lambda, \mu), \qquad (3.7)$$

is readily expressed, due to Theorem 1, in terms of functions $f_\pm(\lambda)$ (3.1) as

$$R(\lambda, \mu) = \frac{f_+(\lambda)f_-(\mu) - f_-(\lambda)f_+(\mu)}{2i(\lambda - \mu)}. \qquad (3.8)$$

We begin by obtaining the zero curvature representation (2.19), (2.20) involving the potentials B

$$B_{lm}(x, \beta, \gamma) \equiv \gamma \int_{-\infty}^{+\infty} e_l(\lambda)f_m(\lambda)\, d\lambda, \qquad l = +, -; \; m = +, -$$

$$B_{++} = B_{++}^* = B_{--}, \qquad B_{+-} = B_{+-}^* = B_{-+} \qquad (3.9)$$

from the basic equations (3.1). To obtain the "L-operator" (2.19), let us differentiate integral equations (3.1) with respect to x. The result, in obvious notation, is

$$\left[(1 - \gamma\widehat{K})\partial_x f_\pm\right](\lambda) - \gamma\left[(\partial_x \widehat{K})f_\pm\right](\lambda) = \pm i\lambda e_\pm(\lambda) \qquad (3.10)$$

where $\partial_x \widehat{K}$ denotes the integral operator with kernel equal to $\partial_x K(\lambda, \mu)$. It is crucial that this kernel is proportional to the sum of two one-dimensional projectors:

$$\partial_x K(\lambda, \mu) = \frac{1}{2} \left[e_+(\lambda) e_-(\mu) + e_-(\lambda) e_+(\mu) \right]. \tag{3.11}$$

Apply now the operator $(1 - \gamma \widehat{K})^{-1}$ to both sides of (3.10). The result of its action on the first term on the left hand side is evidently $\partial_x f_\pm(\lambda)$. The action on the second term is easily calculated if one represents the kernel of $\partial_x \widehat{K}$ in the form (3.11) and then uses the formula

$$\left[(1 - \gamma \widehat{K})^{-1} e_\pm \right] (\lambda) = f_\pm(\lambda)$$

which is obvious from (3.1), so that

$$-\gamma \left[(1 - \gamma \widehat{K})^{-1} (\partial_x \widehat{K}) f_\pm \right] (\lambda) = -\frac{1}{2} f_+(\lambda) B_{-,\pm} - \frac{1}{2} f_-(\lambda) B_{+,\pm}.$$

To calculate the action on the right hand side one uses the relation (3.7)

$$(1 - \gamma \widehat{K})^{-1} = (1 + \gamma \widehat{R})$$

and the explicit formula (3.8) for the resolvent kernel with the result

$$\pm i \int\limits_{-\infty}^{+\infty} [\delta(\lambda - \mu) + \gamma R(\lambda, \mu)] \mu e_\pm(\mu) \, d\mu$$

$$= \pm i \lambda f_\pm(\lambda) \mp i\gamma \int\limits_{-\infty}^{+\infty} [(\lambda - \mu) R(\lambda, \mu)] e_\pm(\mu) \, d\mu$$

$$= \pm i \lambda f_\pm(\lambda) \mp \frac{1}{2} f_+(\lambda) B_{-,\pm} - \pm \frac{1}{2} f_-(\lambda) B_{+,\pm}.$$

Taking all this into account one obtains the relations

$$\partial_x f_+(\lambda) = i\lambda f_+(\lambda) + B_{++} f_-(\lambda)$$
$$\partial_x f_-(\lambda) = -i\lambda f_-(\lambda) + B_{--} f_+(\lambda). \tag{3.12}$$

As $B_{--} = B_{++}$ (see (3.9)) this is exactly the L-operator (2.19) in Theorem 2 of the previous section.

The derivation of the M-operator (2.20) is based essentially on properties (3.5) of the function $\vartheta(\lambda)$ and is therefore more cumbersome. Let us give this derivation. Our aim is to act with the operator $2\lambda \partial_\beta + \partial_\lambda$ on both sides of equation (3.1). Let us calculate first the derivative with respect to β. The kernel $K(\lambda, \mu)$ (3.3) satisfies the relation

$$\partial_\beta K(\lambda, \mu) = \frac{\partial_\beta \vartheta(\lambda)}{2\vartheta(\lambda)} K(\lambda, \mu) + K(\lambda, \mu) \frac{\partial_\beta \vartheta(\mu)}{2\vartheta(\mu)} \tag{3.13}$$

so that differentiating (3.1) with respect to β one has

$$\left[(1 - \gamma \widehat{K}) \partial_\beta f_\pm \right] (\lambda) = \left[(1 + \gamma \widehat{K}) \left(\frac{\partial_\beta \vartheta}{2\vartheta} f_\pm \right) \right] (\lambda) \tag{3.14}$$

and now acting with the operator $(1 - \gamma\widehat{K})^{-1} = (1 + \gamma\widehat{R})$ on both sides of (3.14) one arrives at the relation

$$\partial_\beta f_\pm = \frac{\partial_\beta \vartheta(\lambda)}{2\vartheta(\lambda)} f_\pm(\lambda)$$

$$+ 2\gamma \int\limits_{-\infty}^{+\infty} R(\lambda,\mu) \frac{\partial_\beta \vartheta(\mu)}{2\vartheta(\mu)} f_\pm(\mu)\, d\mu \qquad (3.15)$$

(notice that $(1 - \gamma\widehat{K})^{-1}\widehat{K} = \widehat{R}$). Taking into account the explicit formula (3.8) for the resolvent kernel one rewrites (3.15) as

$$2\lambda\partial_\beta f_\pm(\lambda) = 2\lambda\frac{\partial_\beta \vartheta(\lambda)}{2\vartheta(\lambda)} f_\pm(\lambda)$$

$$+ \frac{2\gamma}{i} \int\limits_{-\infty}^{+\infty} [f_+(\lambda)f_-(\mu) - f_-(\lambda)f_+(\mu)] \frac{\partial_\beta \vartheta(\mu)}{2\vartheta(\mu)} f_\pm(\mu)\, d\mu$$

$$+ 2\gamma \int\limits_{-\infty}^{+\infty} R(\lambda,\mu)\mu \frac{\partial_\beta \vartheta(\mu)}{\vartheta(\mu)} f_\pm(\mu)\, d\mu. \qquad (3.16)$$

Let us now discuss the derivative with respect to λ. Differentiating (3.1) and performing integration by parts one has

$$\partial_\lambda f_\pm(\lambda) = \frac{\partial_\lambda \vartheta(\lambda)}{2\vartheta(\lambda)} f_\pm(\lambda) \pm ix f_\pm(\lambda)$$

$$+ 2\gamma \int\limits_{-\infty}^{+\infty} R(\lambda,\mu)\frac{\partial_\mu \vartheta(\mu)}{2\vartheta(\mu)} f_\pm(\mu)\, d\mu. \qquad (3.17)$$

Adding up equations (3.16) and (3.17) and taking into account (3.5),

$$(2\lambda\partial_\beta + \partial_\lambda)\vartheta(\lambda) = 0,$$

one obtains that

$$(2\lambda\partial_\beta + \partial_\lambda)f_\pm(\lambda) = \pm ix f_\pm(\lambda) - if_+(\lambda)u_{-,\pm}(x,\beta) + if_-(\lambda)u_{+,\pm}(x,\beta).$$
$$(3.18)$$

Here the potentials u_{lm} ($l = +, -$; $m = +, -$) are introduced:

$$u_{lm}(x,\beta) = \gamma \int\limits_{-\infty}^{+\infty} f_l(\mu)f_m(\mu)\frac{\partial_\beta \vartheta(\mu)}{\vartheta(\mu)}\, d\mu. \qquad (3.19)$$

Let us now prove that

$$u_{lm} = \partial_\beta B_{lm} = \gamma\partial_\beta \int\limits_{-\infty}^{+\infty} e_l(\mu)f_m(\mu)\, d\mu \qquad (3.20)$$

(for the potentials B see (3.9)). Indeed, using the explicit form (3.2) for e_l, formula (3.14) and equation (3.1) one calculates

$$\partial_\beta B_{lm} = \gamma \int\limits_{-\infty}^{+\infty} \frac{\partial_\beta \vartheta(\mu)}{2\vartheta(\mu)} e_l(\mu) f_m(\mu)\, d\mu$$

$$+ \gamma \int\limits_{-\infty}^{+\infty} f_l(\lambda)[1 - \gamma\widehat{K}](\lambda, \mu)\partial_\beta f_m(\mu)\, d\lambda\, d\mu$$

$$= \gamma \int\limits_{-\infty}^{+\infty} \frac{\partial_\beta \vartheta(\mu)}{2\vartheta(\mu)}[1 - \gamma\widehat{K}](\mu, \lambda) f_l(\lambda) f_m(\mu)\, d\lambda\, d\mu$$

$$+ \gamma \int\limits_{-\infty}^{+\infty} f_l(\lambda)[1 + \gamma\widehat{K}](\lambda, \mu)\frac{\partial_\beta \vartheta(\mu)}{2\vartheta(\mu)} f_m(\mu)\, d\lambda\, d\mu$$

$$= \gamma \int\limits_{-\infty}^{+\infty} f_l(\lambda) f_m(\lambda)\frac{\partial_\beta \vartheta(\lambda)}{\vartheta(\lambda)}\, d\lambda = u_{lm}, \tag{3.21}$$

so that (3.20) is proved. Putting (3.20) into (3.18) one has

$$(2\lambda\partial_\beta + \partial_\lambda)f_+(\lambda) = ixf_+(\lambda) - i(\partial_\beta B_{+-})f_+(\lambda) + i(\partial_\beta B_{++})f_-(\lambda);$$
$$(2\lambda\partial_\beta + \partial_\lambda)f_-(\lambda) = -ixf_-(\lambda) + i(\partial_\beta B_{+-})f_-(\lambda) - i(\partial_\beta B_{++})f_+(\lambda);$$
$$\tag{3.22}$$

(it is taken into account that $B_{+-} = B_{-+}$, $B_{++} = B_{--}$, see (3.9)). Equations (3.22) give, of course, exactly the M-operator (2.20). So the relation (2.20) is also proved. As explained in section 2, equations (2.22)–(2.24) for the potentials are the compatibility conditions for linear differential equations (3.12), (3.22).

We turn now to the function

$$\sigma(x, \beta, \gamma) = \ln \Delta(x, \beta, \gamma) = \ln \det \left(1 - \gamma\widehat{K}\right) \tag{3.23}$$

which plays the role of the τ-function for the integrable nonlinear system. Let us first calculate its derivatives with respect to x and β. To obtain the derivative with respect to x is rather simple:

$$\partial_x \sigma = \partial_x \ln \det \left(1 - \gamma\widehat{K}\right) = \partial_x \operatorname{Sp} \ln(1 - \gamma\widehat{K})$$
$$= -\gamma \operatorname{Sp}\left[(1 - \gamma\widehat{K})^{-1}\partial_x\widehat{K}\right]. \tag{3.24}$$

Using now formula (3.11) for the kernel of the operator $\widehat{K}$ and the relations following from (3.1)

$$\left[(1 - \gamma\widehat{K})^{-1}e_\pm\right](\lambda) = f_\pm(\lambda),$$

one easily calculates the trace in (3.24),

$$
\mathrm{Sp}\left[(1 - \gamma\widehat{K})^{-1}\partial_x\widehat{K}\right] = \frac{1}{2}\int\limits_{-\infty}^{+\infty} f_+(\lambda)e_-(\lambda)\,d\lambda + \frac{1}{2}\int\limits_{-\infty}^{+\infty} f_-(\lambda)e_+(\lambda)\,d\lambda
$$

$$
= \frac{1}{2\gamma}\left(B_{-+} + B_{+-}\right).
$$

As the potentials B_{+-} and B_{-+} are equal, see (3.9), one obtains for $\partial_x\sigma$

$$
\partial_x\sigma = -B_{+-} \tag{3.25}
$$

which is exactly equation (2.29).

The expression for the derivative of σ with respect to β is not obtained so straightforwardly. Let us give the corresponding calculations. One begins with the obvious formula,

$$
\partial_\beta\sigma = \partial_\beta \ln\det\left(1 - \gamma\widehat{K}\right) = \partial_\beta\,\mathrm{Sp}\ln(1 - \gamma\widehat{K})
$$

$$
= -\gamma\,\mathrm{Sp}\left[(1 - \gamma\widehat{K})^{-1}\partial_\beta\widehat{K}\right].
$$

Using representation (3.13) for the kernel of $\partial_\beta\widehat{K}$ and, following from (3.7), the relation $(1 - \gamma\widehat{K})^{-1}\widehat{K} = \widehat{R}$, one represents the trace in the formula above in the form

$$
\partial_\beta\sigma = -\gamma\int\limits_{-\infty}^{+\infty} R(\lambda,\lambda)\frac{\partial_\beta\vartheta(\lambda)}{\vartheta(\lambda)}\,d\lambda \tag{3.26}
$$

which by means of (3.8) can be written as

$$
\partial_\beta\sigma = \frac{i\gamma}{2}\int\limits_{-\infty}^{+\infty} d\lambda\,\frac{\partial_\beta\vartheta(\lambda)}{\vartheta(\lambda)}\left[f_-(\lambda)\partial_\lambda f_+(\lambda) - f_+(\lambda)\partial_\lambda f_-(\lambda)\right]. \tag{3.27}
$$

Now taking into account expression (3.17) for $\partial_\lambda f_\pm$, one transforms the right hand side into

$$
\partial_\beta\sigma = -\gamma x\int\limits_{-\infty}^{+\infty} f_+(\lambda)f_-(\lambda)\frac{\partial_\beta\vartheta(\lambda)}{\vartheta(\lambda)}\,d\lambda
$$

$$
+ \frac{i\gamma^2}{2}\int\limits_{-\infty}^{+\infty} d\lambda\int\limits_{-\infty}^{+\infty} d\mu\left[f_-(\lambda)f_+(\mu)R(\lambda,\mu)\frac{\partial_\mu\vartheta(\mu)}{\vartheta(\mu)}\frac{\partial_\beta\vartheta(\lambda)}{\vartheta(\lambda)}\right.
$$

$$
\left. - f_+(\lambda)f_-(\mu)R(\lambda,\mu)\frac{\partial_\mu\vartheta(\mu)}{\vartheta(\mu)}\frac{\partial_\beta\vartheta(\lambda)}{\vartheta(\lambda)}\right].
$$

$$
\tag{3.28}
$$

The first integral here is equal to $-x\partial_\beta B_{+-}$, see (3.9), (3.19), (3.20). Now making a change of variables, $\lambda \leftrightarrow \mu$, in the second term inside the square brackets in (3.28) and using the symmetry of the resolvent kernel, $R(\lambda, \mu) = R(\mu, \lambda)$, one arrives at

$$\partial_\beta \sigma = -x\partial_\beta B_{+-} + \frac{i\gamma^2}{2} \int_{-\infty}^{+\infty} d\lambda \int_{-\infty}^{+\infty} d\mu \, f_-(\lambda) f_+(\mu) R(\lambda, \mu)$$

$$\times \left[\frac{\partial_\mu \vartheta(\mu)}{\vartheta(\mu)} \frac{\partial_\beta \vartheta(\lambda)}{\vartheta(\lambda)} - \frac{\partial_\lambda \vartheta(\lambda)}{\vartheta(\lambda)} \frac{\partial_\beta \vartheta(\mu)}{\vartheta(\mu)} \right].$$

$$(3.29)$$

By means of (3.5) one can transform the derivatives $\partial_\lambda \vartheta(\lambda)$ and $\partial_\mu \vartheta(\mu)$ into $\partial_\beta \vartheta(\lambda)$, $\partial_\beta \vartheta(\mu)$, e.g.,

$$\partial_\lambda \vartheta(\lambda) = -2\lambda \partial_\beta \vartheta(\lambda).$$

Thus

$$\partial_\beta \sigma = -x\partial_\beta B_{+-} + i\gamma^2 \int_{-\infty}^{+\infty} d\lambda \int_{-\infty}^{+\infty} d\mu \, f_-(\lambda) f_+(\mu) R(\lambda, \mu)$$

$$\times (\lambda - \mu) \frac{\partial_\beta \vartheta(\mu)}{\vartheta(\mu)} \frac{\partial_\beta \vartheta(\lambda)}{\vartheta(\lambda)} \qquad (3.30)$$

and now making use of formula (3.8) for the resolvent kernel one sees that the factor $(\lambda - \mu)$ in the integrand cancels the denominator of the resolvent kernel:

$$\partial_\beta \sigma = -x\partial_\beta B_{+-} + \frac{\gamma^2}{2} \int_{-\infty}^{+\infty} d\lambda \int_{-\infty}^{+\infty} d\mu \, f_-(\lambda) f_+(\mu)$$

$$\times [f_+(\lambda) f_-(\mu) - f_-(\lambda) f_+(\mu)] \frac{\partial_\beta \vartheta(\mu)}{\vartheta(\mu)} \frac{\partial_\beta \vartheta(\lambda)}{\vartheta(\lambda)}$$

$$(3.31)$$

Now the calculation of the two-dimensional integral is reduced to the calculations of one-dimensional integrals. Using again (3.9), (3.19) and (3.20) one obtains finally for $\partial_\beta \sigma$

$$\partial_\beta \sigma = -x\partial_\beta B_{+-} + \frac{1}{2}(\partial_\beta B_{+-})^2 - \frac{1}{2}(\partial_\beta B_{++})^2 \qquad (3.32)$$

which is exactly equation (2.30). It was shown in section 2 that equations (3.25) and (3.32) result in the nonlinear differential equation (2.31) for σ.

So Theorems 2–4 of section 2 are proved, except for the "initial conditions" (2.25), (2.26), (2.32), (2.33). In Appendix 1 the expansion of correlators considered at small distance $x \to 0$ is constructed; in particular, formulæ (2.25), (2.32) are proved there. In Appendix 2 where the low density expansion of the correlators is given, formulæ (2.26), (2.33) are established.

XIV.4 Differential equations for temperature-dependent equal-time multifield correlators for the impenetrable Bose gas

The determinant representation for temperature-dependent equal-time multifield correlators can be obtained in a way similar to the corresponding two-point correlation function (see comments in section XIII.7). It is a version of Lenard's formula. Introducing normalized distance and chemical potential

$$x \equiv z\sqrt{T}; \qquad \beta = \frac{h}{T} \tag{4.1}$$

as in the two-point case in section 2 one has, for the multipoint correlator,

$$\langle \Psi^\dagger(z_1^+) \cdots \Psi^\dagger(z_N^+)\Psi(z_1^-) \cdots \Psi(z_N^-)\rangle_T = T^{N/2} G_N(\{x^+\}, \{x^-\}, \beta) \tag{4.2}$$

where

$$G_N(\{x^+\}, \{x^-\}, \beta) \equiv G_N(x_1^+, \ldots, x_N^+; x_1^-, \ldots, x_N^-; \beta)$$

depends on the normalized variables.

As for the two-point case considered in section 2 one can represent the multipoint correlator in terms of the linear operator $\widehat{K}$,

$$\left(\widehat{K} f\right)(\lambda) = \int\limits_{-\infty}^{+\infty} K(\lambda, \mu)f(\mu)\,d\mu,$$

where now the kernel $K(\lambda, \mu)$ is given as

$$K(\lambda, \mu) = \frac{1}{2i(\lambda - \mu)} \sum_{m=1}^{2N} (-1)^m e_m^+(\lambda)e_m^-(\mu). \tag{4.3}$$

The "plane waves" $e_m^\pm(\lambda)$ here are

$$e_m^\pm(\lambda) = \sqrt{\vartheta(\lambda)}e^{\pm i\lambda x_m} \qquad (m = 1, \ldots, 2N). \tag{4.4}$$

(for x_m see (4.13)). It is easily seen that

$$\sum_{m=1}^{2N} (-1)^m e_m^+(\lambda)e_m^-(\lambda) = 0 \tag{4.5}$$

so that the operator $\widehat{K}$ is a nonsingular "integrable" operator in the sense of section 1. It is then quite natural to introduce functions $f_m^+(\lambda)$,

$f_m^-(\lambda)$ (playing the role of the functions f^L, f^R, see (1.10), (1.11)) as the solutions of the linear integral equations

$$f_m^+(\lambda) - \gamma \int_{-\infty}^{+\infty} K(\lambda, \mu) f_m^+(\mu) \, d\mu = e_m^+(\lambda) \qquad (4.6a)$$

$$f_m^-(\lambda) - \gamma \int_{-\infty}^{+\infty} K(\mu, \lambda) f_m^-(\mu) \, d\mu = e_m^-(\lambda). \qquad (4.6b)$$

The normalized chemical potential $\beta = h/T$, which enters only through the function

$$\vartheta(\lambda) = \frac{1}{1 + e^{\lambda^2 - \beta}} \qquad (4.7)$$

with properties

$$(2\lambda \partial_\beta + \partial_\lambda)\vartheta(\lambda) = 0, \qquad \vartheta(\lambda)\Big|_{\lambda^2 - \beta = +\infty} = 0, \qquad (4.8)$$

plays the role of a parameter in these integral equations.

Theorem 1 (1.14) of section 1 allows one to write the kernel $R(\lambda, \mu)$ of the resolvent operator $\widehat{R}$,

$$(1 - \gamma \widehat{K})\widehat{R} = \widehat{K}, \qquad (4.9)$$

in terms of the functions $f_m^\pm$:

$$R(\lambda, \mu) = \frac{1}{2i(\lambda - \mu)} \sum_{m=1}^{2N} (-1)^m f_m^+(\lambda) f_m^-(\mu). \qquad (4.10)$$

Due to the property (see (1.18))

$$\sum_{l=1}^{2N} (-1)^l f_l^+(\lambda) f_l^-(\lambda) = 0 \qquad (4.11)$$

there is no pole at $\lambda = \mu$ in the kernel and the operator $\widehat{R}$ is also a nonsingular integrable operator.

It is also convenient to define "potentials" V_{lm} similar to the potentials B_{++}, B_{+-} in the two-point case,

$$V_{lm}(\{x\}, \beta, \gamma) \equiv (-1)^l \gamma \int_{-\infty}^{+\infty} e_l^-(\mu) f_m^+(\mu) \, d\mu. \qquad (4.12)$$

We turn now to the multipoint correlator (4.2). Let us introduce variables x_j $(j = 1, \ldots, 2N)$ which are the variables $x_1^+, \ldots, x_N^+, x_1^-, \ldots, x_N^-$

taken in increasing order:

$$\{x_j;\ j = 1, \ldots, 2N\} = \{x_k^+;\ k = 1, \ldots, N\} \bigcup \{x_l^-;\ l = 1, \ldots, N\},$$

$$x_1 \leq x_2 \leq x_3 \leq \cdots \leq x_{2N}. \tag{4.13}$$

Then the representation of the correlator $G_N(\{x^+\}, \{x^-\}, \beta)$ in terms of the linear operator $\widehat{K}$ is

$$G_N(\{x^+\}, \{x^-\}, \beta) = \left(\frac{1}{4}\right)^N (-1)^{[(N+1)/2]}$$

$$\times \prod_{j<k} \text{sign}(x_k^+ - x_j^+)\,\text{sign}(x_k^- - x_j^-)$$

$$\times \det{}_N [V_{lm}(\{x\}, \beta, \gamma)] \Delta(\{x\}, \beta, \gamma)\Big|_{\gamma=2/\pi}. \tag{4.14}$$

Here $[(N+1)/2] = N/2$ for N even and $(N+1)/2$ for N odd; $\det_N V_{lm}$ is the determinant of the $N \times N$ matrix with matrix elements V_{lm} (4.12) $(1 \leq l \leq N;\ N+1 \leq m \leq 2N)$ and Δ is the Fredholm determinant of $\widehat{K}$,

$$\Delta(\{x\}, \beta, \gamma) = \det\left(1 - \gamma\widehat{K}\right). \tag{4.15}$$

Our aim in what follows is to obtain integrable partial differential equations for the potential V_{lm} and the Fredholm determinant Δ.

Let us first analyze the operator $\widehat{K}$ and properties of the functions $f_m^\pm$. It is quite obvious that

$$K^*(\lambda, \mu) = K(\mu, \lambda) \qquad (\text{Im}\,\mu = \text{Im}\,\lambda = 0),$$
$$K(-\lambda, -\mu) = K(\mu, \lambda) \tag{4.16}$$

and hence

$$f_m^-(\lambda) = \left(f_m^+(\lambda)\right)^*, \qquad \text{Im}\,\lambda = 0$$
$$f_m^-(-\lambda) = f_m^+(\lambda). \tag{4.17}$$

It results in the fact that the potentials V_{lm} are real and symmetric "up to the sign",

$$V_{lm}^* = V_{lm}, \qquad V_{lm} = (-1)^{l+m} V_{ml}. \tag{4.18}$$

The resolvent kernel $R(\lambda, \mu)$ has already been defined in terms of the functions f_m^+ (4.10): it possesses property (4.11) as does the kernel $K(\lambda, \mu)$ (4.5). The symmetry properties of $R(\lambda, \mu)$ are also similar:

$$R^*(\lambda, \mu) = R(\mu, \lambda) \qquad (\text{Im}\,\mu = \text{Im}\,\lambda = 0),$$
$$R(-\lambda, -\mu) = R(\mu, \lambda). \tag{4.19}$$

We turn now to deriving partial differential equations for the potentials V_{lm}. The first step is to obtain a Lax representation. It is done in complete analogy with the two-point case considered in much detail in section 3, so here we give the answers. Differentiating the linear integral equations (4.6a) for the functions $f_m^+(\lambda)$ one obtains a set of L-operators (we put $\partial_k \equiv \partial/\partial x_k$):

$$\partial_k f_m^+(\lambda) = \delta_{km}\left\{i\lambda f_m^+(\lambda) - \frac{1}{2}\sum_{l=1}^{2N} f_l^+(\lambda)V_{lm}\right\}$$

$$+ \frac{1}{2}f_k^+(\lambda)V_{km} \qquad (k, m = 1, \ldots, 2N), \qquad (4.20)$$

Applying to (4.6a) the operator $(2\lambda\partial_\beta + \partial_\lambda)$ one gets the M-operator,

$$(2\lambda\partial_\beta + \partial_\lambda)f_m^+(\lambda) = ix_m f_m^+(\lambda) - i\sum_{l=1}^{2N} f_l^+(\lambda)\partial_\beta V_{lm}$$

$$(m = 1, \ldots, 2N). \qquad (4.21)$$

Equations (4.20), (4.21) give a Lax representation for the multidimensional integrable classical system describing multipoint correlators (x_1, ..., x_{2N} are coordinates and β corresponds to "time").

Demanding now the compatibility of these equations one arrives at a system of partial differential equations for the potentials. First, we write down equations without the "time" derivative:

$$\partial_k V_{lm} = \frac{1}{2}V_{lk}V_{km} \qquad (k \neq m, l) \qquad (4.22)$$

$$(\partial_m + \partial_j)V_{jm} = -\frac{1}{2}\sum_{l \neq m, j}^{2N} V_{jl}V_{lm}, \quad m \neq j \qquad (4.23)$$

$$\partial_m V_{mm} = -\frac{1}{2}\sum_{l \neq m}^{2N} V_{ml}V_{lm} \qquad (4.24)$$

(in deriving (4.22) the fact is taken into account that the potentials go to zero as $\beta \to -\infty$).

The equations containing the β-derivative are

$$\partial_\beta(\partial_l - \partial_m)V_{lm} + (x_l - x_m)V_{lm} + (\partial_\beta V_{mm} - \partial_\beta V_{ll})V_{lm}$$

$$+ \frac{1}{2}\sum_{p \neq l, m}(V_{lp}\partial_\beta V_{pm} - V_{pm}\partial_\beta V_{lp}) = 0. \qquad (4.25)$$

Also, one easily obtains the following simple consequences:

$$\left(\sum_{l=1}^{2N}\partial_l\right)V_{jm} = 0; \qquad \sum_{m=1}^{2N} V_{mm} = 0. \qquad (4.26)$$

So all the equations for the potentials V_{lm} are given.

Now let us derive an equation for the function σ:

$$\sigma(\{x\}, \beta, \gamma) = \ln \det \left(1 - \gamma \widehat{K}\right) = \ln \Delta(\{x\}, \beta, \gamma). \qquad (4.27)$$

First one calculates, as in the two-point case in section 2, the differential $d\sigma$ of the function σ with respect to the variables $\{x\}, \beta$. The calculations are similar to those made already in section 3 in the case $N = 1$ and are given in Appendix 3. The result is:

$$\partial_m \sigma = -\frac{1}{2} V_{mm} \qquad (4.28)$$

$$\partial_\beta \sigma = -\frac{1}{2} \sum_{m=1}^{2N} x_m \partial_\beta V_{mm} + \frac{1}{4} \sum_{m,l=1}^{2N} (\partial_\beta V_{lm})(\partial_\beta V_{ml}). \qquad (4.29)$$

Using now equations (4.18), (4.22) and (4.28) one expresses the potentials entering the right hand side of (4.29) in terms of derivatives of the function σ,

$$V_{km}^2 = 2(-1)^{k+m} \partial_k V_{mm} = 4(-1)^{k+m+1} \partial_k \partial_m \sigma$$
$$V_{mm} = -2 \partial_m \sigma. \qquad (4.30)$$

Putting these expressions into the right-hand side of (4.29) one arrives at the following equation for the function $\sigma(\{x\}, \beta, \gamma)$, which is valid for any γ:

$$\partial_\beta \sigma = \sum_{m=1}^{2N} x_m \partial_\beta \partial_m \sigma + \sum_{m=1}^{2N} (\partial_\beta \partial_m \sigma)^2$$
$$- \frac{1}{4} \sum_{k \neq m}^{2N} (\partial_k \partial_m \sigma)^{-1} (\partial_\beta \partial_k \partial_m \sigma)^2. \qquad (4.31)$$

It is readily shown also that

$$\sum_{m=1}^{2N} \partial_m \sigma = 0. \qquad (4.32)$$

Considering the equality of mixed derivatives, $\partial_\beta \partial_k \sigma = \partial_k \partial_\beta \sigma$, one obtains from (4.28), (4.29) "conserved currents":

$$\partial_\beta \left(\sum_{m=1}^{2N} (x_m - x_k) V_{mk} V_{km} \right) = \partial_k \left(\sum_{m,l}^{2N} (\partial_\beta V_{lm})(\partial_\beta V_{ml}) \right). \qquad (4.33)$$

To conclude this section let us discuss the zero temperature limit of equation (4.31) for σ. In this limit the functions $\Delta = \det \left(1 - \gamma \widehat{K}\right)$

and $\sigma = \ln \Delta$ depend on $2N$ variables τ_j only;

$$\tau_j = x_j \sqrt{\beta} = z_j \sqrt{h}, \qquad T \to 0. \tag{4.34}$$

Recalculating the derivatives in (4.31) one comes to the following equation for σ at zero temperature, $T = 0$:

$$\sum_{i,m}^{2N} \tau_i \tau_m \partial_i \partial_m \sigma + \frac{1}{2} \sum_m^{2N} \left(\partial_m \sigma + \sum_i \tau_i \partial_i \partial_m \sigma \right)^2$$

$$= \frac{1}{2} \sum_{k \neq m}^{2N} (\partial_m \partial_k \sigma)^{-1} \left[\partial_m \partial_k \sigma + \frac{1}{2} \sum_i \tau_i \partial_i \partial_k \partial_m \sigma \right]^2. \tag{4.35}$$

Here $\partial_i \equiv \partial/\partial\tau_i$. Also, of course, $\sum_i \partial_i \sigma = 0$.

XIV.5 Differential equations for the time-dependent correlator of fields for the impenetrable Bose gas at zero temperature

Our aim here is to derive partial differential equations for the time-dependent correlation function of two fields for the impenetrable Bose gas. One begins by rewriting the representation (XIII.6.5) in a form similar to (2.6) (see Appendix 4). Introducing distance x and time t by

$$x \equiv \frac{1}{2}(x_1 - x_2), \qquad t \equiv \frac{1}{2}(t_2 - t_1) \tag{5.1}$$

one rewrites (XIII.6.5) as

$$\langle \Psi(x_2, t_2) \Psi^\dagger(x_1, t_1) \rangle = -\frac{1}{2\pi} \exp\{2iht\} \, b_{++} \det\left(1 + \hat{V}\right). \tag{5.2}$$

Let us explain the notation. The linear integral operator $\hat{V}$ acts on arbitrary functions φ on the interval $[-q, q]$ as follows:

$$\left(\hat{V}\varphi\right)(\lambda) = \int_{-q}^{q} V(\lambda, \mu)\varphi(\mu) \, d\mu$$

where the Fermi momentum q is defined by the chemical potential h: $q = \sqrt{h}$. The kernel $V(\lambda, \mu)$ of $\hat{V}$ is given by

$$V(\lambda, \mu) = \frac{e_+(\lambda)e_-(\mu) - e_-(\lambda)e_+(\mu)}{\lambda - \mu}, \tag{5.3}$$

the functions $e_\pm(\lambda)$ being given by

$$e_-(\lambda) = \frac{1}{\pi} \exp\{it\lambda^2 + ix\lambda\}$$

$$e_+(\lambda) = e_-(\lambda)E(\lambda). \tag{5.4}$$

The function $E(\lambda) \equiv E(\lambda|x,t)$ (and also the function $G \equiv G(x,t)$, which is later used) are defined similarly to (XIII.6.1), (XIII.6.2):

$$G \equiv G(x,t) \equiv \int\limits_{-\infty}^{+\infty} d\mu \, \exp\{-2it\mu^2 - 2ix\mu\} \qquad (5.5)$$

$$E(\lambda) \equiv E(\lambda|x,t) \equiv \fint\limits_{-\infty}^{+\infty} \frac{d\mu}{\mu - \lambda} \, \exp\{-2it\mu^2 - 2ix\mu\}. \qquad (5.6)$$

It is to be mentioned that due to the factor $1/2$ in (5.1) the definitions of these functions here differ slightly from those used in section XIII.6.

The operator $\widehat{V}$ is an "integrable" operator in the sense of section 1. So it is natural to introduce functions $f_\pm(\lambda)$ (see (1.10)). As the operator $\widehat{V}$ is symmetric,

$$\mathcal{V}(\lambda, \mu) = \mathcal{V}(\mu, \lambda), \qquad (5.7)$$

the functions f^L and f^R in (1.10) and (1.11) coincide up to their signs. So one has

$$f_\pm(\lambda) + \int\limits_{-q}^{q} \mathcal{V}(\lambda, \mu) f_\pm(\mu) \, d\mu = e_\pm(\lambda). \qquad (5.8)$$

The "potential" $b_{++} \equiv b_{++}(x,t)$ entering (5.2) is defined as

$$b_{++} = B_{++} - G \qquad (5.9)$$

where B_{++} is one of the potentials B_{lm}

$$B_{lm} = \int\limits_{-q}^{q} e_l(\mu) f_m(\mu) \, d\mu, \qquad l, m = +, -. \qquad (5.10)$$

So we have defined all the quantities entering the right hand side of representation (5.2) for the correlator.

In what follows the potentials

$$C_{lm} = \int\limits_{-q}^{q} \mu e_l(\mu) f_m(\mu) \, d\mu, \qquad l, m = +, - \qquad (5.11)$$

will also be used. It is worth mentioning that these new potentials can be expressed in terms of the potentials B (see below).

Our first aim is to obtain the Lax representation similar to the corresponding formulæ (2.19), (2.20) in the equal-time case. This representation generates the nonlinear evolution partial differential equations for the potentials B. Before doing this, however, let us establish some important properties of the functions introduced. Our first remark is that the operator $\widehat{\mathcal{V}}$ being integrable, one easily constructs the kernel $\mathcal{R}(\lambda, \mu)$ of the resolvent operator $\widehat{\mathcal{R}}$,

$$(1 + \widehat{\mathcal{V}})(1 - \widehat{\mathcal{R}}) = 1, \qquad (1 + \widehat{\mathcal{V}})\widehat{\mathcal{R}} = \widehat{\mathcal{V}},$$

$$\mathcal{R}(\lambda, \mu) + \int_{-q}^{q} \mathcal{V}(\lambda, \nu)\mathcal{R}(\nu, \mu)\, d\nu = \mathcal{V}(\lambda, \mu). \tag{5.12}$$

It can be expressed in terms of the functions $f_{\pm}(\lambda)$ (5.8) (see (1.14)):

$$\mathcal{R}(\lambda, \mu) = \frac{f_+(\lambda)f_-(\mu) - f_-(\lambda)f_+(\mu)}{\lambda - \mu}, \tag{5.13}$$

$$\mathcal{R}(\lambda, \mu) = \mathcal{R}(\mu, \lambda). \tag{5.14}$$

It is easily seen now that the following relation is valid:

$$B_{+-}(x, t) = B_{-+}(x, t) \tag{5.15}$$

so that only the potentials B_{++}, B_{--} and B_{+-} are independent. However, these potentials are now complex and $B_{++} \neq B_{--}$, contrary to the equal-time case.

Turn now to the functions $e_{\pm}(\lambda)$ (5.4). Due to the obvious relation

$$\partial_x E(\lambda) = -2iG - 2i\lambda E(\lambda) \tag{5.16}$$

for the function $E(\lambda)$ entering the definition of $e_+(\lambda)$ one easily calculates

$$\partial_x e_+(\lambda) = -i\lambda e_+(\lambda) - 2iGe_-(\lambda)$$
$$\partial_x e_-(\lambda) = i\lambda e_-(\lambda). \tag{5.17}$$

Due to the relation

$$\partial_t E(\lambda) = -2i\lambda^2 E - 2i\lambda G + \partial_x G \tag{5.18}$$

one also has

$$\partial_t e_+(\lambda) = -i\lambda^2 e_+(\lambda) - 2i\lambda Ge_-(\lambda) + e_-(\lambda)\partial_x G$$
$$\partial_t e_-(\lambda) = i\lambda^2 e_-(\lambda). \tag{5.19}$$

These relations result in the following formulæ for the kernel of $\widehat{\mathcal{V}}$ which are of primary importance in the derivation of the Lax representation:

$$\partial_x \mathcal{V}(\lambda, \mu) = -i\left[e_+(\lambda)e_-(\mu) + e_-(\lambda)e_+(\mu)\right] \tag{5.20}$$

$$\partial_t \mathcal{V}(\lambda, \mu) = -i(\lambda + \mu)\left[e_+(\lambda)e_-(\mu) + e_-(\lambda)e_+(\mu)\right]$$
$$- 2iGe_-(\lambda)e_-(\mu). \tag{5.21}$$

So the derivatives of $\widehat{\mathcal{V}}$ with respect to x and t are proportional to the sum of one-dimensional projectors in the space where $\widehat{\mathcal{V}}$ acts.

We introduce now, together with the functions $e_\pm(\lambda)$ (5.4), functions $e_\pm^{(1)}(\lambda)$, $e_\pm^{(2)}(\lambda)$,

$$e_\pm^{(1)}(\lambda) = \lambda e_\pm(\lambda), \qquad e_\pm^{(2)}(\lambda) = \lambda^2 e_\pm(\lambda). \tag{5.22}$$

It is not difficult to calculate the action of the operator $(1+\widehat{\mathcal{V}})^{-1} = (1-\widehat{\mathcal{R}})$ on these functions:

$$\left[(1+\widehat{\mathcal{V}})^{-1}e_\pm\right](\lambda) = f_\pm(\lambda); \tag{5.23}$$

$$\left[(1+\widehat{\mathcal{V}})^{-1}e_+^{(1)}\right](\lambda) = \lambda f_+(\lambda) + B_{+-}f_+(\lambda) - B_{++}f_-(\lambda);$$

$$\left[(1+\widehat{\mathcal{V}})^{-1}e_-^{(1)}\right](\lambda) = \lambda f_-(\lambda) + B_{--}f_+(\lambda) - B_{-+}f_-(\lambda); \tag{5.24}$$

$$\left[(1+\widehat{\mathcal{V}})^{-1}e_+^{(2)}\right](\lambda) = \lambda^2 f_+(\lambda) + \lambda B_{+-}f_+(\lambda) - \lambda B_{++}f_-(\lambda)$$
$$+ C_{+-}f_+(\lambda) - C_{++}f_-(\lambda);$$

$$\left[(1+\widehat{\mathcal{V}})^{-1}e_-^{(2)}\right](\lambda) = \lambda^2 f_-(\lambda) + \lambda B_{--}f_+(\lambda) - \lambda B_{-+}f_-(\lambda) \tag{5.25}$$
$$+ C_{--}f_+(\lambda) - C_{-+}f_-(\lambda).$$

To conclude this preliminary part, we list also the important relations of the potentials C (5.11) to the potentials B (5.10) proved in Appendix 5:

$$C_{++} = \frac{i}{2}\partial_x B_{++} - 2GB_{+-} + B_{+-}B_{++};$$

$$C_{--} = \frac{1}{2i}\partial_x B_{--} - B_{+-}B_{--}; \tag{5.26}$$

$$C_{+-} - C_{-+} = B_{+-}^2 - B_{++}B_{--}. \tag{5.27}$$

Now everything is ready to prove the following theorem:

Theorem 5. *The two-component function*

$$\vec{F}(\lambda) = \begin{pmatrix} f_+(\lambda) \\ f_-(\lambda) \end{pmatrix} \tag{5.28}$$

(where the functions $f_\pm(\lambda)$ are defined by the integral equations (5.8)) satisfies, for any λ, the following equations:

$$(\partial_x + i\lambda\sigma_3 - 2iQ)\vec{F} = 0 \tag{5.29}$$
$$(\partial_t + i\lambda^2\sigma_3 - 2i\lambda Q - iV)\vec{F} = 0 \tag{5.30}$$

where the 2×2 matrices Q and V (λ-independent) are expressed in terms of the potentials B by

$$Q = \begin{pmatrix} 0 & b_{++} \\ B_{--} & 0 \end{pmatrix} \tag{5.31}$$

$$V = \begin{pmatrix} 2b_{++}B_{--} & i\partial_x b_{++} \\ -i\partial_x B_{--} & -2B_{--}b_{++} \end{pmatrix}, \qquad b_{++} = B_{++} - G. \tag{5.32}$$

The matrix V can also be represented as

$$V = i\partial_x U, \qquad U = \begin{pmatrix} -B_{+-} & b_{++} \\ -B_{--} & B_{+-} \end{pmatrix}. \tag{5.33}$$

The relations (5.29) and (5.30) can be regarded as a Lax representation for some classical nonlinear evolution partial differential equation for the potentials b_{++}, B_{++}.

Proof: Let us first prove equation (5.29) which is the simpler. Differentiating with respect to x equation (5.8) for $f_\pm(\lambda)$ one obtains, in an obvious notation,

$$[(1 + \widehat{V})\partial_x f_\pm](\lambda) = -[(\partial_x \widehat{V})f_\pm](\lambda) + \partial_x e_\pm(\lambda) \tag{5.34}$$

which by virtue of (5.17) is rewritten as

$$\partial_x f_+(\lambda) = -[(1 + \widehat{V})^{-1}(\partial_x \widehat{V})f_+](\lambda) - i[(1 + \widehat{V})^{-1}e_+^{(1)}](\lambda)$$
$$- 2iGf_-(\lambda) \tag{5.35}$$
$$\partial_x f_-(\lambda) = -[(1 + \widehat{V})^{-1}(\partial_x \widehat{V})f_-](\lambda) + i[(1 + \widehat{V})^{-1}e_-^{(1)}](\lambda). \tag{5.36}$$

The action of $(1 + \widehat{V})^{-1}$ on $\partial_x \widehat{V}$ in the first terms in the right hand sides here is easily calculated due to the fact that $\partial_x \widehat{V}$ is proportional to the sum of one-dimensional projectors as given by (5.20) (recall that $[(1 + \widehat{V})^{-1}e_\pm](\lambda) = f_\pm(\lambda)$). The action of $(1 + \widehat{V})^{-1}$ on the functions $e_\pm^{(1)}$ is already known (see (5.24)). So one easily calculates the right hand sides:

$$\partial_x f_+(\lambda) = -i\lambda f_+(\lambda) + 2iB_{++}f_-(\lambda) - 2iGf_-(\lambda)$$
$$= -i\lambda f_+(\lambda) + 2ib_{++}f_-(\lambda) \tag{5.37}$$
$$\partial_x f_-(\lambda) = i\lambda f_-(\lambda) + 2iB_{--}f_+(\lambda). \tag{5.38}$$

If we write this in matrix form we get (5.29).

Turn now to equation (5.30). Differentiating the equations for the functions $f_\pm(\lambda)$ one obtains similarly to (5.34):

$$[(1 + \widehat{V})\partial_t f_\pm](\lambda) = -[(\partial_t \widehat{V})f_\pm](\lambda) + \partial_t e_\pm(\lambda) \tag{5.39}$$

which is, due to (5.19), equivalent to the following equalities

$$\partial_t f_+(\lambda) = -[(1+\widehat{\mathcal{V}})^{-1}(\partial_t \widehat{\mathcal{V}})f_+](\lambda) - i[(1+\widehat{\mathcal{V}})^{-1}e_+^{(2)}](\lambda)$$
$$- 2iG[(1+\widehat{\mathcal{V}})^{-1}e_-^{(1)}](\lambda) + \partial_x G f_-(\lambda) \qquad (5.40)$$

$$\partial_t f_-(\lambda) = -[(1+\widehat{\mathcal{V}})^{-1}(\partial_t \widehat{\mathcal{V}})f_-](\lambda) + i[(1+\widehat{\mathcal{V}})^{-1}e_-^{(2)}](\lambda). \qquad (5.41)$$

Taking into account that the kernel of $\partial_t \widehat{\mathcal{V}}$ can, due to (5.21), be represented as

$$\partial_t \mathcal{V}(\lambda, \mu) = -ie_+^{(1)}(\lambda)e_-(\mu) - ie_-^{(1)}(\lambda)e_+(\mu)$$
$$- ie_+(\lambda)e_-^{(1)}(\mu) - ie_-(\lambda)e_+^{(1)}(\mu)$$
$$- 2iGe_-(\lambda)e_-(\mu), \qquad (5.42)$$

one calculates the kernel of $(1+\widehat{\mathcal{V}})^{-1}(\partial_t \widehat{\mathcal{V}})$ using (5.23), (5.24). The action of $(1+\widehat{\mathcal{V}})^{-1}(\partial_t \widehat{\mathcal{V}})$ on the functions $f_\pm$ is then easily calculated in an explicit form. The action of $(1+\widehat{\mathcal{V}})^{-1}$ on the other terms on the right hand sides of (5.40), (5.41) is also calculated in explicit form using (5.24), (5.25). So the right hand sides of (5.40), (5.41) are expressed in terms of the potentials B and C. It happens that only C_{++}, C_{--} and $C_{+-} - C_{-+}$ enter these expressions. Using (5.26), (5.27) one obtains the following expressions where only the potentials $b_{++} = B_{++} - G$, B_{--} enter the right hand sides:

$$\partial_t f_+ = -i\lambda^2 f_+ + 2i\lambda b_{++} f_- - (\partial_x b_{++})f_- + 2iB_{--}b_{++}f_+,$$
$$\partial_t f_- = i\lambda^2 f_- + 2i\lambda B_{--} f_+ + (\partial_x B_{--})f_+ - 2iB_{--}b_{++}f_-. \qquad (5.43)$$

Rewriting this in matrix form we get (5.30).

The only thing which remains is to establish equality (5.33). It is equivalent to the relation

$$\partial_x B_{+-} = 2iB_{--}b_{++}. \qquad (5.44)$$

Using (5.17) and (5.38) one has, from the definition of the potential B_{+-},

$$\partial_x B_{+-} = \partial_x \int_{-q}^{q} e_+(\lambda)f_-(\lambda)\, d\lambda$$

$$= -2iGB_{--} - i\int_{-q}^{q} \lambda e_+(\lambda)f_-(\lambda)\, d\lambda$$

$$+ \int_{-q}^{q} e_+(\lambda)[i\lambda f_-(\lambda) + 2iB_{--}f_+(\lambda)]\, d\lambda$$

$$= 2iB_{--}b_{++}, \qquad (5.45)$$

so that (5.44) holds true. The theorem is thus proved.

Relations (5.29), (5.30) are valid for any λ and can be regarded as a Lax representation for some nonlinear partial differential evolution equation for the potentials B_{++}, B_{--}. The compatibility condition for (5.29), (5.30),

$$[\partial_x + i\lambda\sigma_3 - 2iQ, \partial_t + i\lambda^2\sigma_3 - 2i\lambda Q - iV] = 0 \qquad (5.46)$$

results in the following equations for the potentials:

Theorem 6. *The potentials b_{++}, B_{--} satisfy the following system of partial differential equations, which represent the nonlinear Schrödinger equation (however, it is to be remembered that the complex conjugation involution is absent in the real-time case):*

$$
\begin{aligned}
i\partial_t b_{++} &= -\frac{1}{2}\partial_x^2 b_{++} - 4b_{++}^2 B_{--} \\
i\partial_t B_{--} &= \frac{1}{2}\partial_x^2 B_{--} + 4B_{--}^2 b_{++}.
\end{aligned}
\qquad (5.47)
$$

Initial data for these equations can be extracted from the equal-time ($T = 0$) correlator (see section 2 and paper [8]).

The proof of this theorem is rather straightforward. One has merely to calculate the left hand side of the compatibility condition (5.46).

So, we have obtained the description of the potentials b_{++}, B_{--} entering the representation of the correlator. Now our aim is to relate the Fredholm determinant $\Delta = \det\left(1 + \widehat{V}\right)$ to these potentials. Again, as in section 2, it is convenient to introduce the function σ, which is the logarithm of the determinant and plays the role of τ-function for the integrable system

$$\sigma = \ln\det\left(1 + \widehat{V}\right) = \sigma(x, t). \qquad (5.48)$$

The function σ at $t = 0$ was completely described in section 2 (see (2.40), (2.41)) as a solution of the Painlevè type equation. Derivatives of σ with respect to x and t can be expressed in terms of the potentials B, C (5.10), (5.11):

Theorem 7. *The derivatives of the function $\sigma = \ln\det\left(1 + \widehat{V}\right)$ are expressed in terms of the potentials B, C (5.10), (5.11) by the following formulæ:*

$$\partial_x\sigma = -2iB_{+-}; \qquad \partial_x^2\sigma = 4b_{++}B_{--} \qquad (5.49)$$
$$\partial_t\sigma = -2iGB_{--} - 2i(C_{+-} + C_{-+}). \qquad (5.50)$$

All the potentials entering the right hand sides here can be expressed in terms of B_{++}, B_{--}, *namely*

$$\partial_x(C_{+-} + C_{-+}) = (B_{++} - 2G)\partial_x B_{--} - B_{--}\partial_x B_{++} \quad (5.51)$$

$$\partial_t B_{+-} = b_{++}\partial_x B_{--} - B_{--}\partial_x b_{++}. \quad (5.52)$$

At $t = 0$ *the function* σ *was described completely in section 2 (see (2.40), (2.41)) as a solution to a Painlevè type ordinary differential equation.*

Proof: The proof of the relation $\partial_x \sigma = -2iB_{+-}$ almost literally repeats the proof of formula (2.29) given in detail in section 3. The second relation in (5.49), $\partial_x^2 \sigma = 4b_{++}B_{--}$, becomes obvious if one takes into account the equality (5.44), $\partial_x B_{+-} = 2ib_{++}B_{--}$. To obtain (5.50) one writes

$$\partial_t \sigma = \partial_t \ln \det \left(1 + \widehat{V}\right) = \partial_t \operatorname{Sp} \ln(1 + \widehat{V})$$

$$= \operatorname{Sp}\left\{(1 + \widehat{V})^{-1}(\partial_t \widehat{V})\right\}. \quad (5.53)$$

Now one should use formulæ (5.21) and (5.42) for the kernel of $\partial_t \widehat{V}$ and formulæ (5.23), (5.24) describing the action of $(1+\widehat{V})^{-1}$ on the functions $e_{\pm}(\lambda)$, $e_{\pm}^{(1)}(\lambda) = \lambda e_{\pm}(\lambda)$. After taking the trace one uses the definitions (5.10), (5.11) of the potentials B, C to obtain exactly the right hand side of (5.50).

Finally, the relations (5.51), (5.52) are established in a way very similar to that used in the proof of (5.44). Thus the theorem is proved.

So, we have obtained a set of equations which completely describe the correlation function

$$\langle \Psi(x_2, t_2)\Psi^\dagger(x_1, t_1)\rangle_{T=0}.$$

XIV.6 Temperature- and time-dependent correlation function of two fields for the impenetrable Bose gas

Consider now the temperature-dependent correlation function

$$\langle \Psi(x_2, t_2)\Psi^\dagger(x_1, t_1)\rangle_T$$

(XIII.6.8). We begin again by putting the determinant representation (XIII.6.7) and (XIII.6.9) for this correlator into a form similar to (5.2). The correlator (XIII.6.8) depends on four variables, i.e., on the modulus of the difference in position, $|x_1 - x_2|$, on the difference $t_2 - t_1$, on the chemical potential h and on the temperature T. The essential dependence is, however, on only three parameters. Let us introduce the normalized distance x, time t and chemical potential β

$$x \equiv \frac{1}{2}|x_1 - x_2|\sqrt{T}, \qquad t \equiv \frac{1}{2}(t_2 - t_1)T,$$

$$\beta \equiv \frac{h}{T}.$$

(6.1)

(It should be emphasized that the variables x, t here differ by a factor of $\sqrt{T}$ from (5.1).) Using a procedure similar to that used in sections 2, 5 it is not difficult to represent the correlator in the form:

$$\langle \Psi(x_2, t_2)\Psi^\dagger(x_1, t_1)\rangle_T = \sqrt{T}g(x, t, \beta),$$

$$g(x, t, \beta) = -\frac{1}{2\pi}\exp\{2it\beta\}\, b_{++}\det\left(1 + \widehat{V_T}\right)$$

(6.2)

where the Fredholm determinant of the linear integral operator $\widehat{V_T}$, acting on the whole real axis, enters the right hand side. The kernel $V_T(\lambda, \mu)$ is given by

$$V_T(\lambda, \mu) = \frac{e_+(\lambda)e_-(\mu) - e_-(\lambda)e_+(\mu)}{\lambda - \mu},$$

$$V_T(\mu, \lambda) = V_T(\lambda, \mu).$$

(6.3)

This formula is quite similar to (5.3) but the functions $e_\pm(\lambda)$ are different:

$$e_-(\lambda) = \frac{1}{\pi}\sqrt{\vartheta(\lambda)}\exp\{it\lambda^2 + i\lambda x\},$$

$$e_+(\lambda) = e_-(\lambda)E(\lambda),$$

$$\vartheta(\lambda) = \frac{1}{1 + \exp\{\lambda^2 - \beta\}}.$$

(6.4)

The functions $E(\lambda)$ and G (which is used later) are defined as in (5.5), (5.6):

$$G \equiv G(x, t) \equiv \int\limits_{-\infty}^{+\infty} d\mu \, \exp\{-2it\mu^2 - 2ix\mu\}$$

(6.5)

$$E(\lambda) \equiv E(\lambda|x, t) \equiv \fint\limits_{-\infty}^{+\infty} \frac{d\mu}{\mu - \lambda}\exp\{-2it\mu^2 - 2ix\mu\}.$$

(6.6)

The operator $\widehat{V_T}$ is an integrable operator. As usual, one introduces functions $f_\pm(\lambda)$ by

$$f_\pm(\lambda) + \int\limits_{-\infty}^{+\infty} V_T(\lambda, \mu)f_\pm(\mu)\, d\mu = e_\pm(\lambda).$$

(6.7)

The "potential" $b_{++} \equiv b_{++}(x,t)$ entering (6.2) is defined exactly as in (5.9),

$$b_{++} = B_{++} - G \tag{6.8}$$

where B_{++} is one of the potentials B_{lm},

$$B_{lm} = \int\limits_{-\infty}^{+\infty} e_l(\mu) f_m(\mu) \, d\mu, \qquad l,m = +,-. \tag{6.9}$$

So all the quantities entering the right hand side of (6.2) are defined in a quite similar way to the zero temperature case considered in section 5. The potentials C_{lm},

$$C_{lm} = \int\limits_{-\infty}^{+\infty} \mu e_l(\mu) f_m(\mu) \, d\mu, \qquad l,m = +,- \tag{6.10}$$

are also defined similarly.

Again, one constructs the kernel $\mathcal{R}(\lambda,\mu)$ of the resolvent operator $\widehat{\mathcal{R}}$,

$$(1 + \widehat{V_T})(1 - \widehat{\mathcal{R}}) = 1, \qquad (1 + \widehat{V_T})\widehat{\mathcal{R}} = \widehat{V_T},$$

$$\mathcal{R}(\lambda,\mu) + \int\limits_{-\infty}^{+\infty} V_T(\lambda,\nu)\mathcal{R}(\nu,\mu) \, d\nu = V_T(\lambda,\mu), \tag{6.11}$$

as

$$\mathcal{R}(\lambda,\mu) = \frac{f_+(\lambda)f_-(\mu) - f_-(\lambda)f_+(\mu)}{\lambda - \mu}, \tag{6.12}$$

$$\mathcal{R}(\lambda,\mu) = \mathcal{R}(\mu,\lambda). \tag{6.13}$$

The relation (5.15) is also valid

$$B_{+-}(x,t,\beta) = B_{-+}(x,t,\beta). \tag{6.14}$$

The first step is to obtain the Lax representation for the classical system of nonlinear evolution equations for the potentials and hence for the correlator. Before doing this, let us list the important properties of the functions $e_\pm(\lambda)$ and of the kernel $V_T(\lambda,\mu)$. One easily calculates for $E(\lambda)$ (6.6) that

$$\begin{aligned}
\partial_x E(\lambda) &= -2iG - 2i\lambda E(\lambda) \\
\partial_t E(\lambda) &= -2i\lambda^2 E - 2i\lambda G + \partial_x G \\
\partial_\beta E(\lambda) &= 0 \\
\partial_\lambda E(\lambda) &= -(4it\lambda + 2ix)E - 4itG.
\end{aligned} \tag{6.15}$$

It should be emphasized that the dependence on the chemical potential β enters only through the function $\vartheta(\lambda)$ (see (6.4)) which possesses the following important property:

$$(2\lambda\partial_\beta + \partial_\lambda)\vartheta(\lambda) = 0. \tag{6.16}$$

We also introduce functions $e_\pm^{(1)}(\lambda)$, $e_\pm^{(2)}(\lambda)$ exactly by formula (5.22)

$$e_\pm^{(1)}(\lambda) = \lambda e_\pm(\lambda), \qquad e_\pm^{(2)}(\lambda) = \lambda^2 e_\pm(\lambda). \tag{6.17}$$

The action of $(1 + \widehat{\mathcal{V}_T})^{-1}$ on these functions is easily seen to be given by formulæ (5.23)–(5.25):

$$\left[(1 + \widehat{\mathcal{V}_T})^{-1}e_\pm\right](\lambda) = f_\pm(\lambda); \tag{6.18}$$

$$\left[(1 + \widehat{\mathcal{V}_T})^{-1}e_\pm^{(1)}\right](\lambda) = \lambda f_\pm(\lambda) + B_{\pm,-}f_+(\lambda) - B_{\pm,+}f_-(\lambda); \tag{6.19}$$

$$\left[(1 + \widehat{\mathcal{V}_T})^{-1}e_\pm^{(2)}\right](\lambda) = \lambda^2 f_\pm(\lambda) + \lambda B_{\pm,-}f_+(\lambda) - \lambda B_{\pm,+}f_-(\lambda)$$
$$+ C_{\pm,-}f_+(\lambda) - C_{\pm,+}f_-(\lambda) \tag{6.20}$$

so that the right hand sides are expressed in terms of the functions $f_\pm(\lambda)$ and potentials B, C. Taking into account (6.15), (6.16) one calculates the derivatives of $e_\pm(\lambda)$:

$$\partial_x e_-(\lambda) = ie_-^{(1)}(\lambda)$$
$$\partial_x e_+(\lambda) = -ie_+^{(1)}(\lambda) - 2iGe_-(\lambda) \tag{6.21}$$

$$\partial_t e_-(\lambda) = ie_-^{(2)}(\lambda)$$
$$\partial_t e_+(\lambda) = -ie_+^{(2)}(\lambda) - 2iGe_-^{(1)}(\lambda) + e_-(\lambda)\partial_x G \tag{6.22}$$

$$\partial_\beta e_+(\lambda) = \frac{\partial_\beta\vartheta(\lambda)}{2\vartheta(\lambda)}e_+(\lambda); \qquad \partial_\beta e_-(\lambda) = \frac{\partial_\beta\vartheta(\lambda)}{2\vartheta(\lambda)}e_-(\lambda) \tag{6.23}$$

$$(2\lambda\partial_\beta + \partial_\lambda)e_-(\lambda) = 2ite_-^{(1)}(\lambda) + ixe_-(\lambda)$$
$$(2\lambda\partial_\beta + \partial_\lambda)e_+(\lambda) = -2ite_+^{(1)}(\lambda) - ixe_+(\lambda) - 4itGe_-(\lambda). \tag{6.24}$$

It is worth mentioning that formulæ (6.21), (6.22) are just the same as in the zero temperature case. The derivatives of the kernel $\mathcal{V}_T(\lambda, \mu)$ with respect to x and t are hence also the same

$$\partial_x \mathcal{V}_T(\lambda, \mu) = -i\left[e_+(\lambda)e_-(\mu) + e_-(\lambda)e_+(\mu)\right] \tag{6.25}$$
$$\partial_t \mathcal{V}_T(\lambda, \mu) = -ie_+^{(1)}(\lambda)e_-(\mu) - ie_+(\lambda)e_-^{(1)}(\mu)$$
$$- ie_-^{(1)}(\lambda)e_+(\mu) - ie_-(\lambda)e_+^{(1)}(\mu)$$
$$- 2iGe_-(\lambda)e_-(\mu). \tag{6.26}$$

For the derivatives with respect to β and λ one has

$$\partial_\beta \mathcal{V}_T(\lambda, \mu) = \frac{\partial_\beta \vartheta(\lambda)}{2\vartheta(\lambda)} \mathcal{V}_T(\lambda, \mu) + \mathcal{V}_T(\lambda, \mu) \frac{\partial_\beta \vartheta(\mu)}{2\vartheta(\mu)} \qquad (6.26a)$$

$$\partial_\lambda \mathcal{V}_T(\lambda, \mu) = \frac{\partial_\lambda \vartheta(\lambda)}{2\vartheta(\lambda)} \mathcal{V}_T(\lambda, \mu) + \mathcal{V}_T(\lambda, \mu) \frac{\partial_\mu \vartheta(\mu)}{2\vartheta(\mu)}$$
$$\qquad\qquad - \partial_\mu \mathcal{V}_T(\lambda, \mu) + 2t \partial_x \mathcal{V}_T(\lambda, \mu) \qquad (6.26b)$$

so that

$$(2\lambda \partial_\beta + \partial_\lambda)\mathcal{V}_T(\lambda, \mu) = -\partial_\mu \mathcal{V}_T(\lambda, \mu) + 2t \partial_x \mathcal{V}_T(\lambda, \mu)$$
$$\qquad\qquad + 2(\lambda - \mu)\mathcal{V}_T(\lambda, \mu)\frac{\partial_\beta \vartheta(\mu)}{2\vartheta(\mu)}. \qquad (6.27)$$

Equations (6.21), (6.22), (6.25), (6.26) are the same as in the zero temperature case, so one easily sees that the expressions for $\partial_x f_\pm$, $\partial_t f_\pm$ (i.e., L- and M-operators in the Lax representation) are also the same. It is also true for the expressions (5.26), (5.27) for the potentials C in terms of the potentials B; one has simply to repeat the corresponding proof given in Appendix 5 to obtain

$$C_{++} = \frac{i}{2}\partial_x B_{++} - 2GB_{+-} + B_{+-}B_{++},$$
$$\qquad\qquad\qquad\qquad\qquad\qquad\qquad\qquad (6.28)$$
$$C_{--} = -\frac{i}{2}\partial_x B_{--} - B_{+-}B_{--},$$

$$C_{+-} - C_{-+} = B_{+-}^2 - B_{++}B_{--}. \qquad (6.29)$$

However, now one can apply the operator $(2\lambda \partial_\beta + \partial_\lambda)$ to $f_\pm(\lambda)$ also to obtain the third equation in the Lax representation ("N-operator")

Theorem 8. *The two-component function* $\vec{F}(\lambda)$

$$\vec{F}(\lambda) = \begin{pmatrix} f_+(\lambda) \\ f_-(\lambda) \end{pmatrix}$$

(where the functions $f_\pm(\lambda)$ *are defined by the integral equations (6.7)) satisfies, for any* λ, *the following Lax representation:*

$$\widehat{\mathsf{L}}(\lambda)\vec{F} = 0, \qquad \widehat{\mathsf{M}}(\lambda)\vec{F} = 0, \qquad \widehat{\mathsf{N}}(\lambda)\vec{F} = 0. \qquad (6.30)$$

The operators $\widehat{\mathsf{L}}$ *and* $\widehat{\mathsf{M}}$ *are the same as in (5.29), (5.30):*

$$\widehat{\mathsf{L}}(\lambda) = \partial_x + i\lambda\sigma_3 - 2iQ \qquad (6.31)$$
$$\widehat{\mathsf{M}}(\lambda) = \partial_t + i\lambda^2\sigma_3 - 2i\lambda Q - iV. \qquad (6.32)$$

The operator $\widehat{N}(\lambda)$ is given by

$$\widehat{N}(\lambda) = 2\lambda\partial_\beta + \partial_\lambda + 2i\lambda t\sigma_3 + ix\sigma_3$$
$$- 4itQ - 2\partial_\beta U \tag{6.33}$$

where the 2×2 matrices Q, V and U are

$$Q = \begin{pmatrix} 0 & b_{++} \\ B_{--} & 0 \end{pmatrix}, \qquad b_{++} = B_{++} - G, \tag{6.34}$$

$$V = \begin{pmatrix} 2b_{++}B_{--} & i\partial_x b_{++} \\ -i\partial_x B_{--} & -2B_{--}b_{++} \end{pmatrix}, \tag{6.35}$$

$$U = \begin{pmatrix} -B_{+-} & b_{++} \\ -B_{--} & B_{+-} \end{pmatrix}. \tag{6.36}$$

As in the zero temperature case, the matrix V can be represented in the form

$$V = i\partial_x U. \tag{6.37}$$

Proof: It has already been mentioned that the first two equations in (6.30) with $\widehat{L}$ and $\widehat{M}$ operators (6.31), (6.32) are established exactly as in the zero temperature case considered in section 5. The proof of relation (6.37) is also the same. So we have to prove the third equation in (6.30). It is similar to the proof of equation (2.20). To prove that $\widehat{N}\vec{F} = 0$ (see (6.33)) let us apply the operator $(2\lambda\partial_\beta + \partial_\lambda)$ to both sides of the integral equation (6.7) for $f_\pm(\lambda)$:

$$(2\lambda\partial_\beta + \partial_\lambda)f_\pm(\lambda) + \int\limits_{-\infty}^{+\infty} [(2\lambda\partial_\beta + \partial_\lambda)V_T(\lambda,\mu)]\, f_\pm(\mu)\, d\mu$$

$$+ 2\lambda \int\limits_{-\infty}^{+\infty} V_T(\lambda,\mu)\partial_\beta f_\pm(\mu)\, d\mu = (2\lambda\partial_\beta + \partial_\lambda)e_\pm(\lambda)$$

which, by means of (6.27), is easily transformed to the following form:

$$(2\lambda\partial_\beta + \partial_\lambda)f_\pm(\lambda) + \int\limits_{-\infty}^{+\infty} V_T(\lambda,\mu)(2\mu\partial_\beta + \partial_\mu)f_\pm(\mu)\, d\mu$$

$$= -2t \int\limits_{-\infty}^{+\infty} [\partial_x V_T(\lambda,\mu)]f_\pm(\mu)\, d\mu + (2\lambda\partial_\beta + \partial_\lambda)e_\pm(\lambda)$$

$$- 2 \int\limits_{-\infty}^{+\infty} [(\lambda-\mu)V_T(\lambda,\mu)]\left(\frac{\partial_\beta\vartheta(\mu)}{2\vartheta(\mu)} + \partial_\beta\right) f_\pm(\mu)\, d\mu \tag{6.38}$$

(the term in the integrand containing the operator $-\partial_\mu \mathcal{V}_T(\lambda, \mu)$ from (6.27) has been integrated by parts).

The operators $\partial_x \mathcal{V}_T$ and $(\lambda - \mu)\mathcal{V}_T(\lambda, \mu)$ on the right hand side of (6.38) are represented as the sum of one-dimensional projectors, see (6.3) and (6.25). The term $(2\lambda\partial_\beta + \partial_\lambda)e_\pm(\lambda)$ is represented, by means of (6.24), as a linear combination of the functions $e_\pm^{(1)}(\lambda)$, $e_\pm(\lambda)$.

Acting now with the operator $(1 + \widehat{\mathcal{V}_T})^{-1}$ on both sides of (6.38), one obtains on the left hand side $(2\lambda\partial_\beta + \partial_\lambda)f_\pm(\lambda)$. Due to (6.18), (6.19) the right hand side is easily expressed in terms of the functions $f_\pm(\lambda)$, the potentials B_{lm} (6.9) and also the integrals (arising from the second integral in the right hand side)

$$
A_{lm} = \int_{-\infty}^{+\infty} e_l(\mu) \left(\frac{\partial_\beta \vartheta(\mu)}{2\vartheta(\mu)} + \partial_\beta \right) f_m(\mu)\, d\mu
$$

$$
= \partial_\beta B_{lm} + \int_{-\infty}^{+\infty} \left[\left(\frac{\partial_\beta \vartheta(\mu)}{2\vartheta(\mu)} - \partial_\beta \right) e_l(\mu) \right] f_m(\mu)\, d\mu \qquad (6.39)
$$

$$
(l, m = +, -).
$$

Due to (6.23), the integral is equal to zero, so that

$$
A_{lm} = \partial_\beta B_{lm} = \partial_\beta \int_{-\infty}^{+\infty} e_l(\mu) f_m(\mu)\, d\mu. \qquad (6.40)
$$

Now, after elementary calculations, one arrives at the expressions for $(2\lambda\partial_\beta + \partial_\lambda)f_\pm(\lambda)$ in terms of the potentials B_{lm}, which are easily seen to be equivalent to the third equation in (6.30) with the operator $\widehat{\mathsf{N}}$ given by (6.33). The theorem is thus proved.

All three operators $\widehat{\mathsf{L}}$, $\widehat{\mathsf{M}}$, $\widehat{\mathsf{N}}$ should commute for arbitrary values of the spectral parameter λ,

$$
[\widehat{\mathsf{L}}(\lambda), \widehat{\mathsf{M}}(\lambda)] = [\widehat{\mathsf{L}}(\lambda), \widehat{\mathsf{N}}(\lambda)] = 0,
$$

$$
[\widehat{\mathsf{M}}(\lambda), \widehat{\mathsf{N}}(\lambda)] = 0. \qquad (6.41)
$$

These equations are the compatibility conditions for (6.30). Writing down equations (6.41) explicitly and then putting the coefficient at each power of λ equal to zero, one obtains nonlinear differential equations for the potentials. The calculations are obvious, though rather extensive. We will give the results.

It is convenient to introduce the notation

$$g_- \equiv e^{-2it\beta} B_{--}; \qquad g_+ \equiv e^{2it\beta} b_{++}, \tag{6.42}$$

$$n \equiv g_+ g_- = b_{++} B_{--}; \qquad p \equiv g_- \partial_x g_+ - g_+ \partial_x g_- \tag{6.43}$$

and write down first the equations without the β-derivative for the potentials b_{++}, B_{--}. This is the separated nonlinear Schrödinger equation, just as in the zero temperature case:

$$- i\partial_t g_+ = 2\beta g_+ + \frac{1}{2}\partial_x^2 g_+ + 4g_+^2 g_- \tag{6.44}$$

$$i\partial_t g_- = 2\beta g_- + \frac{1}{2}\partial_x^2 g_- + 4g_-^2 g_+ \tag{6.45}$$

$$- 2i\partial_t n = \partial_x p. \tag{6.46}$$

The equations containing the β-derivative are

$$\frac{\partial_\beta \partial_x g_+}{g_+} = \frac{\partial_\beta \partial_x g_-}{g_-} \equiv \varphi(x, t, \beta) \tag{6.47}$$

(the function φ is defined here) and

$$- i\partial_t \varphi + 4\partial_\beta p = 0 \tag{6.48}$$

$$\partial_x \varphi + 8\partial_\beta n + 2 = 0. \tag{6.49}$$

This is a complete set of equations for the potentials b_{++} (B_{++}) and B_{--}. Initial data at $t = 0$ for them can be extracted from the equal-time correlator

$$g(x, t = 0, \beta) = g(x, \beta)$$

completely described in section 2, where the equal-time correlator

$$\langle \Psi^\dagger(x_2)\Psi(x_1)\rangle_T$$

was considered.

Other potentials B, C are defined in terms of b_{++} (B_{++}) and B_{--}. Namely,

$$\partial_x B_{+-} = 2in, \qquad \partial_t B_{+-} = -p \tag{6.50}$$

$$\partial_\beta B_{+-} = -\frac{ix}{2} - \frac{i\varphi}{4} \tag{6.51}$$

and

$$C_{++} = \frac{i}{2}\partial_x B_{++} + B_{++}B_{+-} - 2GB_{+-} \tag{6.52}$$

$$C_{--} = -\frac{i}{2}\partial_x B_{--} - B_{+-}B_{--} \tag{6.53}$$

$$\partial_x(C_{+-} + C_{-+}) = (B_{++} - 2G)\partial_x B_{--} - B_{--}\partial_x B_{++} \tag{6.54}$$

$$C_{+-} - C_{-+} = B_{+-}^2 - B_{++}B_{--}. \tag{6.55}$$

So the set of equations for the potentials is obtained. It is worth mentioning that it is not difficult to obtain a closed equation for the potential g_+ that enters the definition of the correlator directly. Indeed, equation (6.44) can be used to express

$$g_- = \frac{1}{8}(-2i\partial_t g_+ - 4\beta g_+ - \partial_x^2 g_+)g_+^{-2} \tag{6.56}$$

and then one can substitute this expression into the remaining equations (6.45)–(6.49).

To describe the correlator (6.2) one has also to describe the Fredholm determinant $\det\left(1 + \widehat{V_T}\right)$. As usual, it is convenient to introduce the function $\sigma(x, t, \beta)$:

$$\sigma(x, t, \beta) = \ln \det \left(1 + \widehat{V_T}\right). \tag{6.57}$$

At $t = 0$, the function σ was completely described in section 2. The derivatives of σ can be expressed in terms of the potentials. The following expressions are valid:

Theorem 9. *The derivatives of the logarithm of the Fredholm determinant of $\widehat{V_T}$ are given by:*

$$\partial_x \sigma = -2iB_{+-} \tag{6.58}$$

$$\partial_t \sigma = -2iGB_{--} - 2i(C_{+-} + C_{-+}) \tag{6.59}$$

$$\partial_\beta \sigma = -2it\partial_\beta(C_{+-} + C_{-+}) - 2ix\partial_\beta B_{+-}$$
$$- 2itB_{--}\partial_\beta B_{++} + 2it(B_{++} - 2G)\partial_\beta B_{--}$$
$$+ 2(\partial_\beta B_{++})(\partial_\beta B_{--}) - 2(\partial_\beta B_{+-})^2. \tag{6.60}$$

Proof: The first two relations, (6.58) and (6.59), are proved exactly as are the corresponding formulæ (5.49), (5.50); these derivatives are expressed in terms of the potentials in the same way as in the temperature-dependent and zero temperature cases. The proof of formula (6.60) for $\partial_\beta \sigma$ is rather similar to the proof of formula (2.30) in the equal-time case given in section 3, though one has to do more calculations. One begins with the relation

$$\partial_\beta \sigma = \partial_\beta \operatorname{Sp} \ln(1 + \widehat{V_T}) = \operatorname{Sp}\left\{(1 + \widehat{V_T})^{-1}(\partial_\beta \widehat{V_T})\right\}$$

$$= 2 \int\limits_{-\infty}^{+\infty} d\lambda \, \frac{\partial_\beta \vartheta(\lambda)}{2\vartheta(\lambda)} \mathcal{R}(\lambda, \lambda) \tag{6.61}$$

which, by means of expression (6.12) for the resolvent kernel, is readily put into the form corresponding to (3.27),

$$\partial_\beta \sigma = 2 \int\limits_{-\infty}^{+\infty} d\lambda \, \frac{\partial_\beta \vartheta(\lambda)}{2\vartheta(\lambda)} [f_-(\lambda)\partial_\lambda f_+(\lambda) - f_+(\lambda)\partial_\lambda f_-(\lambda)]. \qquad (6.62)$$

It is not difficult to obtain the expression for $\partial_\lambda f_\pm(\lambda)$ analogous to (3.17),

$$\partial_\lambda f_\pm(\lambda) = \frac{\partial_\lambda \vartheta(\lambda)}{2\vartheta(\lambda)} f_\pm(\lambda) \mp ix f_\pm(\lambda) + 2t\partial_x f_\pm(\lambda)$$

$$- 2 \int\limits_{-\infty}^{+\infty} \mathcal{R}(\lambda,\mu) \frac{\partial_\mu \vartheta(\mu)}{2\vartheta(\mu)} f_\pm(\mu)\, d\mu \qquad (6.63)$$

using relation (6.26) and the formula

$$\partial_\lambda e_\pm(\lambda) = \frac{\partial_\lambda \vartheta(\lambda)}{2\vartheta(\lambda)} e_\pm(\lambda) + 2t\partial_x e_\pm(\lambda) \mp ix e_\pm(\lambda). \qquad (6.64)$$

Now one substitutes expressions (6.63) for $\partial_\lambda f_\pm(\lambda)$ into (6.62) and performs calculations quite analogous to those made in section 3 (see (3.28)–(3.31)), arriving at expression (6.60) for $\partial_\beta \sigma$. The theorem is thus proved.

So the correlation function $\langle \Psi^\dagger(x_2,t_2)\Psi(x_1,t_1)\rangle_T$ is defined completely.

XIV.7 Integro-differential equation
for finite coupling constant

In this section the integro-differential equation which drives the simplest two-point correlator at finite coupling, $0 < c < \infty$, is derived. This correlator is the "emptiness formation probability" (XI.1.17):

$$P(x_2, x_1) = \frac{\langle \Omega| \exp\{\alpha Q(x_2,x_1)\}|\Omega\rangle}{\langle \Omega|\Omega\rangle}\Bigg|_{\alpha=-\infty} ;$$

$$Q(x_2, x_1) = \int\limits_{x_1}^{x_2} \Psi^\dagger(z)\Psi(z)\, dx. \qquad (7.1)$$

The quantity $P(x_2, x_1)$ gives the probability that there are no particles of the gas in the interval $[x_1, x_2]$; we consider the case of zero temperature. Introducing the variable x,

$$x \equiv \frac{1}{2}|x_2 - x_1|, \qquad (7.2)$$

one rewrites the representation (XI.1.26) for the correlator as

$$P(x_2, x_1) \equiv P(x) = \frac{(0|\det\left(1 + \widehat{V}\right)|0)}{\det\left(1 - \dfrac{1}{2\pi}\widehat{K}\right)} \tag{7.3}$$

where both operators $\widehat{V}$ and $\widehat{K}$ act on the interval $[-q, q]$, q being the Fermi momentum (see Chapter I, and especially sections I.3, I.4):

$$\left(\widehat{V} f\right)(\lambda) = \int\limits_{-q}^{q} V(\lambda, \mu) f(\mu) \, d\mu;$$

$$\left(\widehat{K} f\right)(\lambda) = \int\limits_{-q}^{q} K(\lambda, \mu) f(\mu) \, d\mu. \tag{7.4}$$

This is the action of our operators on some function $f(\mu)$.

The kernels $K(\lambda, \mu)$ and $V(\lambda, \mu)$ are given by

$$K(\lambda, \mu) = \frac{2c}{(\lambda - \mu)^2 + c^2} \tag{7.5}$$

$$V(\lambda, \mu) = -\frac{c}{2\pi}\left(\frac{\exp\{ix(\lambda - \mu) + \dfrac{1}{2}[\varphi(\lambda) - \varphi(\mu)]\}}{(\lambda - \mu)(\lambda - \mu + ic)}\right.$$

$$\left. + \frac{\exp\{-ix(\lambda - \mu) - \dfrac{1}{2}[\varphi(\lambda) - \varphi(\mu)]\}}{(\lambda - \mu)(\lambda - \mu - ic)}\right) \tag{7.6}$$

where $\varphi(\lambda)$ is the dual quantum field, commuting, however, at different values of the spectral parameter

$$[\varphi(\lambda), \, \varphi(\mu)] = 0. \tag{7.7}$$

The vectors $|0)$ and $(0|$ in (7.3) are Fock vacuum and dual Fock vacuum in the auxiliary quantum space (see the details in section XI.1, formula (XI.1.24)). It is essential here that before performing the averaging in the auxiliary Fock space, $(0|\cdots|0)$, in (7.3), the quantum field $\varphi(\lambda)$ (due to property (7.7)) can be treated as an ordinary c-number valued function.

We begin by representing the kernel $V(\lambda, \mu)$ of $\widehat{V}$ in the form

$$V(\lambda, \mu) = \frac{i}{\lambda - \mu} \int\limits_{0}^{\infty} ds \, e^{-sc} \left[e_+(\lambda|s)e_-(\mu|s) - e_-(\lambda|s)e_+(\mu|s)\right] \tag{7.8}$$

where the functions $e_\pm(\lambda|s)$ are given by

$$e_\pm(\lambda|s) = \sqrt{\frac{c}{2\pi}} \exp\left\{\pm\frac{1}{2}[ix\lambda + is\lambda + \varphi(\lambda)]\right\}. \tag{7.9}$$

The equality of the right hand sides of (7.8) and (7.6) is verified by elementary integration.

One should emphasize that the representation (7.8) shows that the operator $\widehat{V}$ is an integrable operator in the sense of section 1. Indeed, its kernel $V(\lambda, \mu)$ is represented in a form similar to (1.1), with the sum replaced by an integral over s,

$$\sum_{j=1}^{N} \longrightarrow \int_{0}^{\infty} ds, \tag{7.10}$$

so that the procedure for deriving partial differential integrable equations for the correlators used in the previous sections of this chapter can be generalized. First, we define the resolvent operator $\widehat{R}$ with kernel $R(\lambda, \mu)$ in the standard way:

$$(1 + \widehat{V})(1 - \widehat{R}) = 1, \qquad (1 + \widehat{V})\widehat{R} = \widehat{V},$$

$$R(\lambda, \mu) + \int_{-q}^{q} V(\lambda, \nu)R(\nu, \mu)\, d\nu = V(\lambda, \mu). \tag{7.11}$$

Due to the symmetry of $V(\lambda, \mu)$,

$$V(\lambda, \mu) = V(\mu, \lambda), \tag{7.12}$$

the resolvent is also symmetric,

$$R(\lambda, \mu) = R(\mu, \lambda). \tag{7.13}$$

To write down an explicit expression for the resolvent kernel one introduces, as usual, functions $f_{\pm}(\lambda|s)$ defined by the linear integral equations

$$f_{\pm}(\lambda|s) + \int_{-q}^{q} V(\lambda, \mu)f_{\pm}(\mu|s)\, d\mu = e_{\pm}(\lambda|s) \tag{7.14}$$

(the parameter s "enumerates" the functions $f_{\pm}$). The standard arguments result in the following representation for the resolvent kernel:

$$R(\lambda, \mu) = \frac{i}{\lambda - \mu} \int_{0}^{\infty} ds\, e^{-sc} \left[f_{+}(\lambda|s)f_{-}(\mu|s) - f_{-}(\lambda|s)f_{+}(\mu|s) \right], \tag{7.15}$$

i.e., the resolvent kernel is expressed in terms of the functions $f_{\pm}(\lambda|s)$ in the same way as the kernel of $\widehat{V}$ (7.8) is expressed in terms of $e_{\pm}(\lambda|s)$. It is worth mentioning that the resolvent kernel has no pole at $\lambda = \mu$, $R(\lambda, \lambda) < +\infty$.

Define now potentials $B_{lm}(s,t)$,

$$B_{lm}(s,t) = \int_{-q}^{q} e_l(\mu|s) f_m(\mu|t)\, d\mu, \qquad l, m = +, - \qquad (7.16)$$

which are now the kernel of a 2×2 matrix integral operator. Due to (7.12), (7.13) this kernel is symmetric,

$$B_{lm}(s,t) = B_{ml}(t,s). \qquad (7.17)$$

Again, one considers equations (7.14) as equations of the Gel'fand-Levitan-Marchenko type for some nonlinear evolution equation.

To obtain the corresponding L-operator (which is now an integral operator, i.e., the "infinite-dimensional matrix") one differentiates equations (7.14) with respect to x. Introducing the two-component vector $\vec{F}(\lambda|t)$,

$$\vec{F}(\lambda|t) = \begin{pmatrix} f_+(\lambda|t) \\ f_-(\lambda|t) \end{pmatrix} \qquad (7.18)$$

and using the relation

$$\partial_x e_\pm(\lambda|s) = \pm i\lambda e_\pm(\lambda|s) \qquad (7.19)$$

one obtains, generalizing in an obvious way the corresponding calculations of section 3, the following result:

$$\partial_x \vec{F}(\lambda|t) = i\lambda\sigma_3 \vec{F}(\lambda|t)$$
$$+ 2\int_0^\infty ds\, e^{-sc} Q(t,s) \vec{F}(\lambda|s) \qquad (7.20)$$

where $Q(t,s)$ is now a matrix integral kernel:

$$Q(t,s) = \begin{pmatrix} 0 & B_{++}(s,t) \\ B_{--}(s,t) & 0 \end{pmatrix} \qquad (7.21)$$
$$Q(t,s) = Q(s,t).$$

Now let us differentiate integral equations (7.14) with respect to q,

$$\partial_q f_\pm(\lambda|t) + \int_{-q}^{q} V(\lambda,\mu) \partial_q f_\pm(\mu|t)\, d\mu$$
$$+ V(\lambda,q) f_\pm(q|t) + V(\lambda,-q) f_\pm(-q,t) = 0 \qquad (7.22)$$

or, acting with $(1+\widehat{V})^{-1}$,

$$\partial_q f_\pm(\lambda|t) + R(\lambda,q) f_\pm(q,t) + R(\lambda,-q) f_\pm(-q,t) = 0. \qquad (7.23)$$

This can be rewritten in the form

$$\partial_q \vec{F}(\lambda|t) + i \int_0^\infty ds\, e^{-sc} U(t,s) \vec{F}(\lambda|s) = 0 \tag{7.24}$$

where the 2×2 matrix $U(t,s)$ is given by

$$U(t,s) = \frac{A_+(s,t)}{\lambda - q} + \frac{A_-(s,t)}{\lambda + q} \tag{7.25}$$

$$A_+(s,t) = \begin{pmatrix} f_-(q|s)f_+(q|t) & -f_+(q|s)f_+(q|t) \\ f_-(q|s)f_-(q|t) & -f_+(q|s)f_-(q|t) \end{pmatrix} \tag{7.26}$$

$$A_-(s,t) = \begin{pmatrix} f_-(-q|s)f_+(-q|t) & -f_+(-q|s)f_+(-q|t) \\ f_-(-q|s)f_-(-q|t) & -f_+(-q|s)f_-(-q|t) \end{pmatrix}. \tag{7.27}$$

The compatibility conditions for (7.20), (7.24)

$$\partial_x \partial_q \vec{F}(\lambda|t) = \partial_q \partial_x \vec{F}(\lambda|t),$$

should be valid for arbitrary λ. Omitting the straightforward calculations we give the resulting set of equations for the functions B, f:

$$\partial_q B_{++}(s,t) = f_+(q|s)f_+(q|t) + f_+(-q|s)f_+(-q|t) \tag{7.28}$$

$$\partial_q B_{--}(s,t) = f_-(q|s)f_-(q|t) + f_-(-q|s)f_-(-q|t) \tag{7.29}$$

$$\partial_x f_+(q|t) = iqf_+(q|t) + 2\int_0^\infty ds\, e^{-sc} B_{++}(t,s)f_-(q|s) \tag{7.30}$$

$$\partial_x f_-(q|t) = -iqf_-(q|t) + 2\int_0^\infty ds\, e^{-sc} B_{--}(t,s)f_+(q|s) \tag{7.31}$$

$$\partial_x f_+(-q|t) = -iqf_+(-q|t) + 2\int_0^\infty ds\, e^{-sc} B_{++}(t,s)f_-(-q|s) \tag{7.32}$$

$$\partial_x f_-(-q|t) = iqf_-(-q|t) + 2\int_0^\infty ds\, e^{-sc} B_{--}(t,s)f_+(-q|s). \tag{7.33}$$

Equations (7.28)–(7.33) are the complete set of six equations for the six unknown functions

$$B_{++}(s,t), \quad B_{--}(s,t), \quad f_+(q|s), \quad f_-(q|s), \quad f_+(-q|s), \quad f_-(-q|s). \tag{7.34}$$

These are the integro-differential equations which drive the correlation function at finite coupling constant. To show this one has to express the

Fredholm determinant $\det\left(1+\widehat{V}\right)$ entering representation (7.3) for the correlator in terms of the functions (7.34). Let us do this.

It is again convenient to consider the logarithm σ of the Fredholm determinant

$$\sigma(x,q) = \ln \det\left(1+\widehat{V}\right). \tag{7.35}$$

The derivatives of σ with respect to x and q can be expressed in terms of the functions (7.34). To obtain the derivative with respect to x one has to make only obvious modifications to the proof given in the previous cases considered earlier in this chapter with the result

$$\partial_x \sigma = \partial_x \ln \det\left(1+\widehat{V}\right) = \text{Sp} \ln[(1+\widehat{V})^{-1}\partial_x \widehat{V}]$$

$$= -2\int_0^\infty ds\, e^{-sc} B_{+-}(s,s). \tag{7.36}$$

As the potentials B_{+-} do not enter the set of six equations (7.28)–(7.33), one has also to express them in terms of the functions entering this set; this is readily done by differentiating equations (7.16) with respect to x and q:

$$\partial_x B_{+-}(s,s) = 2\int_0^\infty dt\, e^{-ts} B_{++}(s,t) B_{--}(t,s). \tag{7.37}$$

This is an analogue of (2.23), while (7.36) is similar to (2.29).

$$\partial_q B_{+-}(s,s) = f_+(q|s) f_-(q|s) + f_+(-q|s) f_-(-q|s). \tag{7.38}$$

So the x-derivative of σ is expressed in terms of functions entering the basic set of integro-differential equations.

Let us now calculate the derivative of σ with respect to q,

$$\partial_q \sigma = \partial_q \ln \det\left(1+\widehat{V}\right) = \text{Sp} \ln[(1+\widehat{V})^{-1}\partial_q \widehat{V}]. \tag{7.39}$$

The dependence of $\widehat{V}$ on q enters only through the integration limits, so $\partial_q \widehat{V}$ acts on a function $f(\lambda)$ in the following way:

$$\left((\partial_q \widehat{V})f\right)(\lambda) = \partial_q \int_{-q}^{q} V(\lambda,\mu) f(\mu)\, d\mu$$

$$= V(\lambda|q) f(q) + V(\lambda|-q) f(-q). \tag{7.40}$$

As $(1+\widehat{V})^{-1}\widehat{V} = \widehat{R}$, one has

$$\partial_q \sigma = R(q,q) + R(-q,-q). \tag{7.41}$$

Consider, e.g., the quantity $R(q, q)$; it follows from (7.15) that

$$R(q,q) = i \int_0^\infty ds\, e^{-sc} \left(f_-(q|s) \partial_\lambda f_+(\lambda|s) \Big|_{\lambda=q} - f_+(q|s) \partial_\lambda f_-(\lambda|s) \Big|_{\lambda=q} \right).$$

$$(7.42)$$

The "complete" derivative $\partial_q f_\pm(q|s)$ can be represented as

$$\partial_q f_\pm(q|s) = \partial_q f_\pm(\lambda|s) \Big|_{\lambda=q} + \partial_\lambda f_\pm(\lambda|s) \Big|_{\lambda=q} \qquad (7.43)$$

so that

$$R(q,q) = i \int_0^\infty ds\, e^{-sc} \left[f_-(q|s) \partial_q f_+(q|s) - f_+(q|s) \partial_q f_-(q|s) \right]$$

$$- i \int_0^\infty ds\, e^{-sc} \left[f_-(q|s) \partial_q f_+(\lambda|s) - f_+(q|s) \partial_q f_-(\lambda|s) \right] \Big|_{\lambda=q}.$$

Now using formula (7.23) for the functions $\partial_q f_\pm(\lambda|s)$ entering the second integral one sees that the terms containing $R(\lambda, q)$ in (7.23) do not contribute (they are cancelled after putting $\lambda = q$). Finally, one gets:

$$R(q,q) = i \int_0^\infty ds\, e^{-sc} \left[f_-(q|s) \partial_q f_+(q|s) - f_+(q|s) \partial_q f_-(q|s) \right]$$

$$+ \frac{1}{2q} \left\{ \int_0^\infty ds\, e^{-sc} \left[f_+(q|s) f_-(-q|s) - f_-(q|s) f_+(-q|s) \right] \right\}^2. \quad (7.44)$$

Similarly,

$$R(-q,-q) = i \int_0^\infty ds\, e^{-sc} \left[f_+(q|s) \partial_q f_-(-q|s) - f_-(-q|s) \partial_q f_+(-q|s) \right]$$

$$+ \frac{1}{2q} \left\{ \int_0^\infty ds\, e^{-sc} \left[f_+(q|s) f_-(-q|s) - f_-(q|s) f_+(-q|s) \right] \right\}^2. \quad (7.45)$$

So we have expressed $\partial_q \ln \det \left(1 + \widehat{V} \right)$ in terms of solutions of the system (7.28)–(7.33). It is also to be mentioned that

$$\sigma \Big|_{q=0} = \ln \det \left(1 + \widehat{V} \right) \Big|_{q=0} = 0. \qquad (7.46)$$

This shows that our approach is applicable not only to the impenetrable Bose gas (free fermions), but also to a Bose gas with any finite coupling (interacting fermions). It is also interesting that N.A. Slavnov [15] wrote down the differential equation which describes the field correlation function $\langle \Psi(x,t)\Psi^\dagger(0,0)\rangle$ at finite coupling c. This is a completely integrable integro-differential equation. It resembles an integro-differential version of the classical NS equation (the isotopic index becomes continuous).

Conclusion

In this chapter we have explained how to write down differential equations for quantum correlation functions. We started from the determinant (Fredholm) representation for quantum correlation functions. In section 1 we explained that integral operators appearing in the representation for correlation functions are of a very special form (we called them "integrable"). This helped us to write down differential equations. We started with the integral equation and obtained two (or more) linear differential equations for its solutions (they play the role of the Lax representation). Compatibility conditions for these linear differential equations lead to nonlinear partial differential equations (completely integrable) which drive the correlation function. Separate calculations show that logarithmic derivatives of the correlation function (the Fredholm determinant) can be expressed in terms of solutions of the nonlinear partial differential equations mentioned above. It is interesting to mention that for temperature correlation functions we found a differential equation which contains temperature derivatives as well as space and time derivatives. In the next two chapters we shall solve differential equations for quantum correlation functions by means of the Riemann-Hilbert problem and evaluate the asymptotics of correlation functions.

A differential equation of the Painlevé type for the equal-time zero temperature correlation function for the impenetrable Bose gas was constructed in [8]. Before that differential equations for correlation functions in the Ising model were written down in [11]. Partial differential equations for time- and temperature-dependent correlation functions for the impenetrable Bose gas were constructed in [4], [5], [6]. Integro-differential equations for the penetrable Bose gas (NS equation, $0 < c < \infty$) were constructed in [6], [15].

Equation (2.24) is related to the Maxwell-Bloch equation [2]. The Lax representation (2.20) contains $\partial/\partial\lambda$; in relation to this we should mention the papers [12], [13]. In those papers the symmetry of soliton equations was studied and $\partial/\partial\lambda$ appeared as one of the generators. The symmetry was extended to Virasoro algebra. We would like to emphasize the roles of A. Shabat and V. Zakharov in the development of the theory of nonlinear partial differential equations and their applications.

The idea of obtaining differential equations for solutions of integral equations was that of Krein ([9], [10]). It was later developed and applied to the solution of nonlinear differential equations. M.J. Ablowitz and A.S. Fokas [1] demonstrated how to obtain the Korteweg-de Vries equation starting from the Gel'fand-Levitan integral equation. The Bäcklund transformation from this point of view was constructed in [3] and [14].

Appendix XIV.1: Short distance asymptotics of temperature-dependent equal-time two-point correlators for the impenetrable Bose gas

In this Appendix the expansions for small distance $(x_1 \to x_2)$ of the correlators

$$\langle \Psi^\dagger(x_1)\Psi(x_2)\rangle_T$$

and

$$\langle \exp\rangle_T\{\alpha Q_1(x_1, x_2)\}$$

are constructed. Due to (2.6), (2.7), (2.36) these correlators may be represented in terms of the potentials B_{++}, B_{+-} (2.16), (2.17) (see also (2.28)–(2.30)). So we begin by constructing the short distance expansions for the potentials. We use them, in particular, to establish the "initial conditions" (2.25), (2.32) for the partial differential equations for the potentials and the function σ.

It is easily seen that the potentials B_{+-}, B_{++} (2.16) can be represented as

$$B_{+-}(x, \beta, \gamma) = \int\limits_{-\infty}^{+\infty} s(\lambda)\, d\lambda \tag{A.1.1}$$

$$B_{++}(x, \beta, \gamma) = \int\limits_{-\infty}^{+\infty} e^{2i\lambda x} s(\lambda)\, d\lambda \tag{A.1.2}$$

where the following linear integral equation for $s(\lambda)$ is obtained from the integral equations (2.15) for the functions $f_\pm(\lambda)$:

$$s(\lambda) - \gamma\vartheta(\lambda) \int\limits_{-\infty}^{+\infty} \frac{1 - e^{-2ix(\lambda-\mu)}}{2i(\lambda - \mu)} s(\mu)\, d\mu = \gamma\vartheta(\lambda). \tag{A.1.3}$$

The kernel of this equation being small for x small, one can expand the solution $s(\lambda)$ in a Taylor series in x,

$$s(\lambda) = \sum_{k=0}^{\infty} s_k(\lambda) x^k \tag{A.1.4}$$

where the $s_k(\lambda)$ are also functions of β and γ. The following recursion relation is readily obtained for s_k:

$$s_m(\lambda) = s_0(\lambda) \sum_{k=0}^{m-1} \frac{(2i)^{m-k-1}}{(m-k)!} \int_{-\infty}^{+\infty} (\mu - \lambda)^{m-k-1} s_k(\mu) \, d\mu;$$

$$s_0(\lambda) = \gamma \vartheta(\lambda) \tag{A.1.5}$$

which allows us to calculate any s_m in an explicit but somewhat lengthy form. So the complete short distance expansions for B_{+-} (A.1.1), B_{++} (A.1.2) as well as for σ,

$$\sigma = \ln \det \left(1 - \gamma \widehat{K}\right) = -\int_0^x B_{+-}(y, \beta, \gamma) \, dy \tag{A.1.6}$$

and g,

$$g(x, \beta, \gamma) = \frac{1}{4} B_{++}(x, \beta, \gamma) e^{\sigma(x, \beta, \gamma)}, \tag{A.1.7}$$

enter (2.6), (2.7) all of which are obtained. The coefficients in these expansions are expressed in terms of the functions $\beta_l(\beta, \gamma)$ which are the moments of $\vartheta(\lambda)$ (the Fermi weight):

$$\beta_l(\beta, \gamma) = \gamma \int_{-\infty}^{+\infty} \lambda^l \vartheta(\lambda) \, d\lambda \qquad (\beta_{2k+1} \equiv 0) \tag{A.1.8}$$

Let us give the first few terms:

$$B_{++} = \beta_0 + \beta_0^2 x + \left(\beta_0^3 - 2\beta_2\right) x^2$$
$$+ \left(\beta_0^4 - \frac{4}{3}\beta_0 \beta_2\right) x^3 + O\left(x^4\right) \tag{A.1.9}$$

$$B_{+-} = \beta_0 + \beta_0^2 x + \beta_0^3 x^2 + \left(\beta_0^4 - \frac{4}{3}\beta_0 \beta_2\right) x^3$$
$$+ \left(\beta_0^5 - \frac{5}{3}\beta_0^2 \beta_2\right) x^4 + O\left(x^5\right) \tag{A.1.10}$$

$$\sigma = -\beta_0 x - \frac{1}{2}\beta_0^2 x^2 - \frac{1}{3}\beta_0^3 x^3$$
$$- \left(\frac{\beta_0^4}{4} - \frac{\beta_0 \beta_2}{3}\right) x^4 - \left(\frac{\beta_0^5}{5} - \frac{\beta_0^2 \beta_2}{3}\right) x^5 + O\left(x^6\right) \tag{A.1.11}$$

$$g = \frac{1}{4}\beta_0 \left(1 - \frac{2\beta_2}{\beta_0} x^2 + \frac{2}{3}\beta_2 x^3 \right) + O\left(x^4 \right). \tag{A.1.12}$$

Taking into account that in (2.25), (2.32) the notation

$$d(\beta) = \frac{1}{\gamma}\beta_0(\beta, \gamma)$$

is used, one sees that the first three terms of (A.1.9), (A.1.10) give just (2.25) and similarly equation (A.1.11) results in (2.32).

Equation (A.1.12) is readily rewritten for the correlator (2.6) as (see also (2.5)):

$$\langle \Psi^\dagger(x_1)\Psi(x_2)\rangle_T$$

$$= \sqrt{T} g\left(\frac{x_1 - x_2}{2}\sqrt{T}, \frac{h}{T}, \gamma \right)\Bigg|_{\gamma = 2/\pi}$$

$$= D\left[1 - \frac{1}{2}\frac{E}{D}(x_1 - x_2)^2 + \frac{1}{3}E(x_1 - x_2)^3 \right]$$

$$+ O(x_1 - x_2)^4,\ x_1 > x_2. \tag{A.1.13}$$

Here

$$D = \frac{1}{2\pi} \int\limits_{-\infty}^{+\infty} \frac{d\lambda}{1 + e^{(\lambda^2 - h)/T}}$$

is the gas density and

$$E = \frac{1}{2\pi} \int\limits_{-\infty}^{+\infty} \frac{\lambda^2\, d\lambda}{1 + e^{(\lambda^2 - h)/T}}$$

is the kinetic energy density.

Thus we have obtained the short distance expansion of the two-point correlators of the impenetrable Bose gas.

Appendix XIV.2: Low density expansion

In this Appendix the expansions of the temperature-dependent equal-time two-point correlators (2.6), (2.7) of the impenetrable Bose gas are constructed. Recall that the gas density D is given by (XIII.1.16); taking into account that the rescaled Fermi weight $\vartheta(\lambda)$ (2.13) is used in this section, one has

$$D = \frac{\sqrt{T}}{2\pi} \int\limits_{-\infty}^{+\infty} \vartheta(\lambda)\, d\lambda, \qquad \vartheta(\lambda) = \frac{1}{1 + e^{\lambda^2 - \beta}}. \tag{A.2.1}$$

So $D \to 0$ as $\beta \to -\infty$. It is convenient here to introduce a new variable, ζ,

$$\zeta = -e^{\beta}, \qquad \zeta \xrightarrow[\beta \to -\infty]{} 0. \tag{A.2.2}$$

It is shown below that the functions B_{++}, B_{+-}, σ and the two-point correlator

$$g = \frac{1}{4} B_{++} e^{\sigma} \bigg|_{\gamma = 2/\pi}$$

can be expanded into absolutely convergent Taylor series in ζ for $\zeta \to 0$:

$$
\begin{aligned}
B_{++} &= \sum_{k=1}^{\infty} b_k(x)\zeta^k \\
B_{+-} &= \sum_{k=1}^{\infty} c_k(x)\zeta^k \\
\sigma &= \sum_{k=1}^{\infty} \sigma_k(x)\zeta^k = -\int\limits_0^x B_{+-}(y, \beta, \gamma)\, dy \\
g &= \sum_{k=1}^{\infty} g_k(x)\zeta^k = \frac{1}{4} B_{++}(x, \beta, \gamma) e^{\sigma(x, \beta, \gamma)} \bigg|_{\gamma = 2/\pi}.
\end{aligned}
\tag{A.2.3}
$$

The explicit formulæ for the coefficients in these series as well as the lower bound of the convergence radius are given below. It is to be mentioned that formulæ (A.2.3) result, in particular, in formulæ (2.26), (2.33).

Our starting point is again the representations (A.1.1), (A.1.2) for the potentials B_{+-}, B_{++} and integral equation (A.1.3) for the function $s(\lambda)$. Putting into (A.1.3) $s(\lambda)$ in the form

$$s(\lambda) = \sum_{k=1}^{\infty} \zeta^k s_k(\lambda, x) \tag{A.2.4}$$

and representing the Fermi weight $\vartheta(\lambda)$ (A.2.1) as

$$\vartheta(\lambda) = -\sum_{k=1}^{\infty} \zeta^k \exp\{-k\lambda^2\} \tag{A.2.5}$$

one obtains the following recursive relations for s_k:

$$s_1(\lambda) = -\gamma \exp\{-\lambda^2\} \tag{A.2.6}$$

$$s_k(\lambda, x) = \exp\{-\lambda^2\} s_{k-1}(\lambda, x)$$
$$- \gamma \exp\{-\lambda^2\} \int_{-\infty}^{+\infty} \frac{1 - \exp\{-2i(\lambda - \mu)x\}}{2i(\lambda - \mu)} s_{k-1}(\mu, x)\, d\mu$$

$$(k \geq 2). \tag{A.2.7}$$

The function $s_k(\lambda)$ is obtained from $s_{k-1}(\lambda)$ by the action of two operators $\mathbf{O_1}$ and $\mathbf{O_2}$. The operator $\mathbf{O_1}$ is multiplication with respect to $\exp(-\lambda^2)$. The integral operator $\mathbf{O_2}$ has the following kernel:

$$\delta(\lambda - \mu) - \gamma \frac{[1 - \exp(2ix(\mu - \lambda))]}{2i(\lambda - \mu)} \tag{A.2.8}$$

This operator can be diagonalized by means of a Fourier transform. Its eigenvalues can be equal to one of the following numbers:

$$1, \qquad 1 - \pi\gamma = \exp \alpha \tag{A.2.9}$$

Eigenvalues of the operator $\mathbf{O_1}$ belong to the interval $[0,1]$.

Now let us try to estimate the behaviour of s_k as $k \to \infty$. It will be different, depending on γ.

(i) If $|1 - \pi\gamma| < 1$ then s_k does not increase with k, and the series (A.2.4) is convergent for $\zeta < 1$.

(ii) If $|1 - \pi\gamma| \geq 1$ then the series will be convergent for negative β such that

$$\exp(\beta)|(1 - \pi\gamma)| \geq 1.$$

One can consider s_k as a function of λ. One can prove that

$$|s_k(\lambda)| \le c_k \exp(-\lambda^2);$$

here c_k is some constant.

Let us give the first term of the expansions for σ and g:

$$\sigma = -\gamma\sqrt{\pi}\, x e^\beta \tag{A.2.10}$$

$$g(x,\beta,\gamma)\Big|_{\gamma=2/\pi} = \frac{e^{-x^2}}{2\pi} \int\limits_{-\infty}^{+\infty} \vartheta(\lambda)\, d\lambda + O\left(e^{2\beta}\right) \tag{A.2.11}$$

or (see formula (XIV. 2.6))

$$\langle \Psi^\dagger(x_1)\Psi(x_2)\rangle_T = D \exp\left\{-\frac{T}{4}(x_1 - x_2)^2\right\} \tag{A.2.12}$$

$$(h \ll -(x_1 - x_2)^2 T^2).$$

In conclusion, let us emphasize once more that for $\beta \to -\infty$ (and x fixed) the functions B_{+-}, B_{++} can be expanded in convergent series in integer powers of e^β, the same being true, of course, also for the functions σ and g.

Appendix XIV.3: Logarithmic derivatives of correlation function

Here the differential of the function $\sigma = \ln \det \left(1 - \gamma \widehat{K}\right)$ in the multi-point equal-time case is calculated. First consider the derivatives with respect to the variables x_m,

$$\partial_m \sigma = \partial_m \ln \det \left(1 - \gamma \widehat{K}\right) = \partial_m \operatorname{Sp} \ln(1 - \gamma \widehat{K})$$
$$= -\gamma \operatorname{Sp}\{(1 - \gamma \widehat{K})^{-1} \partial_m \widehat{K}\}. \tag{A.3.1}$$

From definition (4.3) of the kernel of $\widehat{K}$ one has for its derivative

$$\partial_m K(\lambda, \mu) = \frac{(-1)^m}{2} e_m^+(\lambda) e_m^-(\mu), \qquad m = 1, \dots, 2N \tag{A.3.2}$$

i.e., it is proportional to a one-dimensional projector. So the action of the operator $(1 - \gamma \widehat{K})^{-1}$ on the operator $\partial_m \widehat{K}$ is easy to calculate. One takes into account the integral equations (4.6),

$$\left[(1 - \gamma \widehat{K})^{-1} e_m^+\right](\lambda) = f_m^+(\lambda).$$

Taking now the trace in (A.3.1) one has, due to definition (4.12) of the potentials V,

$$\partial_m \sigma = -\frac{1}{2} V_{mm} \tag{A.3.3}$$

which is exactly equation (4.28).

To calculate the derivative with respect to β is somewhat more difficult. The calculations necessary are similar to those given in section 3 in the case $N = 1$. One begins from the obvious expression, analogous to (3.26),

(3.27), for $\delta_\beta\sigma$:

$$\partial_\beta\sigma = -\gamma \int\limits_{-\infty}^{+\infty} \frac{\partial_\beta\vartheta(\lambda)}{\vartheta(\lambda)} R(\lambda, \lambda)\, d\lambda$$

$$= \frac{i\gamma}{2} \sum_{m=1}^{2N} (-1)^m \int\limits_{-\infty}^{+\infty} d\lambda\, \frac{\partial_\beta\vartheta(\lambda)}{\vartheta(\lambda)} f_m^-(\lambda)\, \partial_\lambda f_m^+(\lambda). \qquad (A.3.4)$$

Here the expression for the diagonal of the resolvent kernel

$$R(\lambda, \lambda) = \frac{1}{2i} \sum_{m=1}^{2N} (-1)^m f_m^-(\lambda)\, \partial_\lambda f_m^+(\lambda), \qquad (A.3.5)$$

is used, which follows from (4.10), (4.11). Direct calculations give the following representation, similar to (3.17); for the function $\delta_\lambda f_m^p(\lambda)$:

$$\partial_\lambda f_m^+(\lambda) = ix_m f_m^+(\lambda) + \frac{\partial_\lambda\vartheta(\lambda)}{2\vartheta(\lambda)} f_m^+(\lambda)$$

$$+ \gamma \int\limits_{-\infty}^{+\infty} R(\lambda, \mu)\frac{\partial_\mu\vartheta(\mu)}{\vartheta(\mu)} f_m^+(\mu)\, d\mu. \qquad (A.3.6)$$

Substituting (A.3.6) into (A.3.4) one has

$$\partial_\beta\sigma = \frac{i\gamma}{2} \int\limits_{-\infty}^{+\infty} \frac{\partial_\beta\vartheta(\lambda)}{\vartheta(\lambda)} \sum_{m=1}^{2N} (-1)^m f_m^-(\lambda)$$

$$\times \left[ix_m f_m^+(\lambda) + \frac{\partial_\lambda\vartheta(\lambda)}{2\vartheta(\lambda)} f_m^+(\lambda) \right.$$

$$\left. + \gamma \int\limits_{-\infty}^{+\infty} R(\lambda, \mu)\frac{\partial_\mu\vartheta(\mu)}{\vartheta(\mu)} f_m^+(\mu)\, d\mu \right] d\lambda. \qquad (A.3.7)$$

The second term in the square brackets can be omitted due to the property (4.11)

$$\sum_{m=1}^{2N} (-1)^m f_m^+(\lambda) f_m^-(\lambda) = 0.$$

The contribution corresponding to the first term can be transformed into

$$-\frac{\gamma}{2} \sum_{m=1}^{2N} (-1)^m x_m \int\limits_{-\infty}^{+\infty} \frac{\partial_\beta\vartheta(\lambda)}{\vartheta(\lambda)} f_m^-(\lambda) f_m^+(\lambda)\, d\lambda$$

$$= -\frac{1}{2} \sum_{m=1}^{2N} x_m\, \partial_\beta V_{mm}. \qquad (A.3.8)$$

To do this one uses relations similar to (3.15) and (3.19), (3.20), namely:

$$\partial_\beta f_m^+(\lambda) = \frac{\partial_\beta \vartheta(\lambda)}{2\vartheta(\lambda)} f_m^+(\lambda) + 2\gamma \int\limits_{-\infty}^{+\infty} R(\lambda,\mu) \frac{\partial_\beta \vartheta(\mu)}{2\vartheta(\mu)} f_m^+(\mu)\, d\mu \quad (A.3.9)$$

and

$$\partial_\beta V_{lm} = \partial_\beta \left[(-1)^l \gamma \int\limits_{-\infty}^{+\infty} e_l^-(\lambda) f_m^+(\lambda)\, d\lambda \right]$$

$$= (-1)^l \gamma \int\limits_{-\infty}^{+\infty} f_l^-(\lambda) f_m^+(\lambda) \frac{\partial_\beta \vartheta(\lambda)}{\vartheta(\lambda)}\, d\lambda. \qquad (A.3.10)$$

The remaining (third) term inside the square brackets in (A.3.7) generates the following contribution to $\partial_\beta \sigma$ (one uses (4.10) for the resolvent kernel):

$$\frac{\gamma^2}{4} \sum_{m,l=1}^{2N} (-1)^{m+l} \int\limits_{-\infty}^{+\infty} d\lambda\, d\mu\, f_m^-(\lambda) f_l^+(\lambda) f_l^-(\mu) f_m^+(\mu)$$

$$\times \frac{1}{\lambda-\mu} \frac{\partial_\beta \vartheta(\lambda)}{\vartheta(\lambda)} \frac{\partial_\mu \vartheta(\mu)}{\vartheta(\mu)}. \qquad (A.3.11)$$

Due to the property of the function ϑ,

$$\partial_\mu \vartheta(\mu) = -2\mu \partial_\beta \vartheta(\mu), \qquad (A.3.12)$$

one can transform (A.3.11) into

$$-\frac{\gamma^2}{4} \sum_{m,l=1}^{2N} (-1)^{m+l} \int\limits_{-\infty}^{+\infty} d\lambda\, d\mu\, f_m^-(\lambda) f_l^+(\lambda) f_l^-(\mu) f_m^+(\mu)$$

$$\times \frac{(\mu+\lambda)+(\mu-\lambda)}{\lambda-\mu} \frac{\partial_\beta \vartheta(\lambda)}{\vartheta(\lambda)} \frac{\partial_\beta \vartheta(\mu)}{\vartheta(\mu)}. \qquad (A.3.13)$$

It is now obvious that the term proportional to $(\mu+\lambda)$ in the integrand does give zero contribution to the integral due to the symmetry properties. In the remaining term, the factor $(\mu-\lambda)$ cancels the denominator and the two-fold integral is factorized into a product of two simple integrals. Now (A.3.13) is easily transformed into

$$\frac{1}{4} \sum_{m,l=1}^{2N} (\partial_\beta V_{lm})(\partial_\beta V_{ml}). \qquad (A.3.14)$$

Here the relation (A.3.10) has been also used. Summing up the contributions (A.3.8) and (A.3.14) one obtains exactly equation (4.29):

$$\partial_\beta \sigma = -\frac{1}{2} \sum_{m=1}^{2N} x_m \partial_\beta V_{mm} + \frac{1}{4} \sum_{m,l=1}^{2N} (\partial_\beta V_{lm})(\partial_\beta V_{ml}). \qquad (A.3.15)$$

Appendix XIV.4: Evaluation of Fredholm minor

Our aim here is to derive the representation (5.2) for the time-dependent correlator of two fields of impenetrable bosons at zero temperature from representation (XIII.6.5). In the variables (5.1),

$$x \equiv \frac{x_1 - x_2}{2}, \quad t \equiv \frac{t_2 - t_1}{2}$$

one rewrites (XIII.6.5) as

$$\langle \Psi(x_2, t_2)\Psi^\dagger(x_1, t_1)\rangle$$
$$= \frac{1}{2\pi} \exp\{2iht\} \left(G + 2\pi \frac{\partial}{\partial \alpha} \right) \det \left(1 + \widehat{V}^{(\alpha)} \right) \Big|_{\alpha=0} \quad \text{(A.4.1)}$$

where the operator $\widehat{V}^{(\alpha)}$ is obtained from $\widehat{V}_0$ (XIII.6.7) to be

$$V^{(\alpha)}(\lambda, \mu) = \frac{e_+(\lambda)e_-(\mu) - e_-(\lambda)e_+(\mu)}{\lambda - \mu} - \frac{\alpha}{2\pi}e_+(\lambda)e_+(\mu) \quad \text{(A.4.2)}$$

with the functions $e_\pm(\lambda)$ given by (5.4) and G and E (entering the expression for e_+) defined now by (5.5) and (5.6).

Let us apply the derivative in (A.4.1) explicitly:

$$\frac{\partial}{\partial \alpha} \det \left(1 + \widehat{V}^{(\alpha)} \right)$$
$$= -\frac{1}{2\pi} \det \left(1 + \widehat{V} \right) \operatorname{Sp} \left[(1 + \widehat{V})^{-1} \widehat{A} \right] \quad \text{(A.4.3)}$$

where $\widehat{A}$ is a one-dimensional projector with kernel

$$\mathcal{A}(\lambda, \mu) = e_+(\lambda)e_+(\mu). \quad \text{(A.4.4)}$$

This fact and the definition (5.8) of $f_+(\lambda)$,

$$\left[(1 + \widehat{V})^{-1}e_+ \right](\lambda) = f_+(\lambda), \quad \text{(A.4.5)}$$

allow one to take the trace in (A.4.3) with the result

$$\mathrm{Sp}\left[(1+\widehat{V})^{-1}\widehat{A}\right] = \int_{-q}^{q} f_+(\lambda)e_+(\lambda)\,d\lambda = B_{++} \tag{A.4.6}$$

where the potential B_{++} was defined in (5.10). Taking into account the definition (5.9), $b_{++} = B_{++} - G$, it is now easily seen that (A.4.1) is indeed equivalent to (5.2). To prove formula (2.6) one should do similar calculations or just put $t_1 = t_2 = 0$ above.

Appendix XIV.5:
Expression for C potentials

We prove here the relations (5.26), (5.27).

Taking explicitly the derivative of the potential B_{++} with respect to x and using (5.17), (5.37) (it is to be emphasized that in the derivation of (5.37) equations (5.26), (5.27) were not used) one has

$$\partial_x B_{++} = \partial_x \int_{-q}^{q} e_+(\lambda) f_+(\lambda) \, d\lambda$$

$$= -iC_{++} - 2iGB_{-+} - iC_{++} + 2iB_{+-}B_{++} - 2iGB_{+-}.$$

As $B_{-+} = B_{+-}$, this is equivalent to the first of the relations (5.26). The second of them is established in a similar way.

To obtain (5.27) one writes for the potentials C_{+-}, C_{-+}

$$C_{+-} = \int_{-q}^{q} \lambda e_+(\lambda) f_-(\lambda) \, d\lambda$$

$$= \int_{-q}^{q} \lambda e_-(\mu) \left[(1 - \widehat{R})\right] (\mu, \lambda) \, e_+(\lambda) \, d\lambda \, d\mu \qquad (A.5.1)$$

$$C_{-+} = \int_{-q}^{q} \lambda e_-(\lambda) f_+(\lambda) \, d\lambda$$

$$= \int_{-q}^{q} \lambda e_-(\lambda) \left[(1 - \widehat{R})\right] (\lambda, \mu) \, e_+(\mu) \, d\lambda \, d\mu. \qquad (A.5.2)$$

Here the obvious relations $f_\pm(\lambda) = [(1 - \widehat{R})e_\pm](\lambda)$ were used; it is to be mentioned that the resolvent kernel in the case considered is symmetric, $R(\lambda, \mu) = R(\mu, \lambda)$. Exchanging λ and μ in the second integrand in (A.5.1)

375

and subtracting leads to

$$C_{+-} - C_{-+} = \int\limits_{-q}^{q} (\lambda - \mu) R(\lambda, \mu) e_-(\lambda) e_+(\mu) \, d\lambda \, d\mu$$

$$= -B_{--}B_{++} + B_{-+}B_{+-} \qquad (A.5.3)$$

(we used the explicit formula (5.13) for the resolvent kernel). Taking into account that $B_{+-} = B_{-+}$, one sees that (A.5.3) is exactly the same as (5.27).

It should be noted that equations (5.26) are in fact among the compatibility conditions of the Lax representation.

XV

The Matrix Riemann-Hilbert Problem for Correlation Functions

Introduction

In the previous chapters correlation functions were represented as determinants of a special form of Fredholm integral operator. First, a differential equation was obtained for the time-independent zero temperature field correlation function for the impenetrable Bose gas (it depends only on one variable—distance). By means of an isomonodromic deformation technique A. Jimbo, T. Miwa, Y. Mori and M. Sato [9] wrote down the ordinary differential equation (of Painlevè type). In the previous chapter differential equations for correlation functions in more general situations (with time, temperature and for finite coupling) were constructed. A.R. Its came up with the idea of applying the matrix Riemann problem for the description of these correlation functions (Fredholm determinants). This permits us to write down partial differential equations (completely integrable) for correlation functions and evaluate the asymptotics of correlation functions [3], [4], [5], [7]. Below we follow this line.

Integrable nonlinear partial differential equations for correlation functions of the one-dimensional nonrelativistic Bose gas with point-like repulsion between gas particles were given in section XIV. So quantum correlation functions can be expressed in terms of the solutions of classical nonlinear integrable evolution systems. These classical systems have been investigated already by means of the classical inverse scattering method [1]. Our approach allows one, in particular, to solve the particularly interesting problem of calculating the asymptotics of correlators at large time and distance; this is done in the next section.

The standard way, now, of investigating nonlinear differential equations in the framework of the inverse scattering method is the matrix Riemann-Hilbert problem (RP) formalism ([10], [11], [12], [13]). Detailed considerations of the corresponding questions are given in [1]. In this introduction the matrix RPs for classical integrable systems describing the correlators are constructed. This chapter is closely related to Chapter V (see section V.1).

Let us give now necessary facts about the matrix RP. The standard formulation of the problem is as follows. One has to find the $N \times N$ matrix-valued function $\chi(\lambda)$ ($\det \chi(\lambda) \neq 0$ for any λ) which is holomorphic for $\text{Im}\,\lambda > 0$ and for $\text{Im}\,\lambda < 0$, and whose boundary values $\chi_\pm(\lambda)$,

$$\chi_\pm(\lambda) = \lim_{\epsilon \to 0} \chi(\lambda \pm i\epsilon), \qquad \epsilon > 0, \tag{0.1}$$

at the real axis are related by

$$\chi_-(\lambda) = \chi_+(\lambda)G(\lambda), \qquad \lambda \in \mathbb{R}; \tag{0.2}$$

$\chi(\lambda)$ is equal to the unit matrix I at $\lambda = \infty$,

$$\chi(\infty) = I. \tag{0.3}$$

The smoothly varying matrix $G(\lambda)$ is a given $N \times N$ matrix defined only for real λ:

$$G(\lambda) \to I, \quad \lambda \to \pm\infty, \text{ and } \det G(\lambda) \neq 0. \tag{0.4}$$

The classical result of paper [2] is that this problem is uniquely solvable if one of the matrices

$$\frac{1}{2}[G(\lambda) + G^\dagger(\lambda)] \quad \text{or} \quad \frac{1}{2i}[G(\lambda) - G^\dagger(\lambda)]$$

(the dagger here means hermitian conjugation) is positive (or negative) definite:

$$\frac{1}{2}[G(\lambda) + G^\dagger(\lambda)] > 0 \quad \text{or} \quad \frac{1}{2i}[G(\lambda) - G^\dagger(\lambda)] > 0. \tag{0.5}$$

The investigation of the matrix RP can be reduced to the investigation of the following system of linear singular integral equations for the function $\chi_+(\lambda)$:

$$\chi_+(\lambda) = I + \frac{1}{2\pi i} \int\limits_{-\infty}^{+\infty} \frac{\chi_+(\mu)[I - G(\mu)]}{\mu - \lambda - i0}\, d\mu, \qquad \lambda \in \mathbb{R}. \tag{0.6}$$

The singular integral matrix equation (0.6) is equivalent to matrix RP (0.2), (0.3).

Indeed, let $\chi(\lambda)$ be the solution of this matrix RP. Then equation (0.6) can be rewritten as

$$\chi_+(\lambda) = I + \frac{1}{2\pi i} \int\limits_{-\infty}^{+\infty} \frac{\chi_+(\mu) - I}{\mu - \lambda - i0}\, d\mu$$

$$- \frac{1}{2\pi i} \int\limits_{-\infty}^{+\infty} \frac{\chi_-(\mu) - I}{\mu - \lambda - i0}\, d\mu. \tag{0.7}$$

The integration contour in the second integral here can be shifted into the lower half-plane of μ (the function $(\chi_-(\mu) - I)$ is holomorphic there and there is no pole of the integrand in the lower half-plane). So the value of the second integral in (0.7) is equal to zero. The first integral is easily calculated by shifting the contour into the upper half-plane. It is equal to the residue at pole $\mu = \lambda + i0$:

$$\frac{1}{2\pi i} \int\limits_{-\infty}^{+\infty} \frac{\chi_+(\mu) - I}{\mu - \lambda - i0} \, d\mu = \chi_+(\lambda) - I.$$

So equation (0.6) is satisfied if $\chi(\lambda)$ is the solution of the matrix RP.

On the other hand, suppose that the function $\chi_+(\lambda)$ is the solution of equation (0.6). Define a matrix function $\chi(\lambda)$ for complex λ by

$$\chi(\lambda) = I + \frac{1}{2\pi i} \int\limits_{-\infty}^{+\infty} \frac{\chi_+(\mu)[I - G(\mu)]}{\mu - \lambda} \, d\mu. \tag{0.8}$$

The function $\chi(\lambda)$ thus constructed is holomorphic for $\operatorname{Im}\lambda > 0$ and $\operatorname{Im}\lambda < 0$. Its boundary values

$$\chi_\pm(\lambda) = \chi(\lambda \pm i0), \quad \lambda \in \mathbb{R},$$

are related:

$$\chi_+(\lambda) - \chi_-(\lambda) = \frac{1}{2\pi i} \int\limits_{-\infty}^{+\infty} \left(\frac{1}{\mu - \lambda - i0} - \frac{1}{\mu - \lambda + i0} \right)$$

$$\times \, \chi_+(\mu)[I - G(\mu)] \, d\mu$$

$$= \chi_+(\lambda)[I - G(\lambda)]. \tag{0.9}$$

Here the well known formula

$$\frac{1}{\mu - \lambda - i0} - \frac{1}{\mu - \lambda + i0} = 2\pi i \delta(\lambda - \mu)$$

was used. Equation (0.9) means that relation (0.2) is satisfied for $\chi(\lambda)$ (0.8); the validity of (0.3) is also evident. So equation (0.6) is indeed equivalent to the corresponding matrix RP for $\chi(\lambda)$ (0.8). In our examples it will often happen that $\det G(\lambda) = 1$ in (0.2). This will imply that $\det \chi(\lambda) = 1$.

The further contents of this chapter are as follows. In sections 1 and 2 the matrix Riemann problem for the temperature-dependent, equal-time two-point correlators for the impenetrable Bose gas is presented. In sections 3 and 4 the matrix Riemann problem in the time-dependent case is considered. In section 5 the equal-time multipoint correlators of fields of impenetrable bosons are described in terms of the corresponding matrix problem. Finally, in section 6 the integral-operator-valued Riemann-Hilbert problem for the simplest correlator of penetrable bosons with finite coupling constant $0 < c < +\infty$ is given.

The results presented in this chapter were obtained in papers [3], [4], [5], [6], [7]. We would like to emphasize the role of A.R. Its in obtaining them. We recommend the reader to look through Appendix XV1.1 where a complete solution of the scalar Riemann problem is presented.

XV.1 The matrix Riemann-Hilbert problem for temperature-dependent equal-time two-point correlators for the impenetrable Bose gas

The results of section XIV.2 show that the investigation of the temperature-dependent equal-time correlators for the impenetrable Bose gas,

$$\langle \Psi^\dagger(x_1)\Psi(x_2)\rangle_T \tag{1.1}$$

and

$$\langle \exp\{\alpha Q_1(x_1, x_2)\}\rangle_T, \qquad Q_1(x_1, x_2) = \int_{x_2}^{x_1} \Psi^\dagger(y)\Psi(y)\, dy \tag{1.2}$$

can be reduced to studying the integrable nonlinear partial differential equations (XIV.2.22)–(XIV.2.24) for the potentials B_{lm} ($l, m = +, -$) (XIV.2.16). The reader is recommended to recall the contents of section XIV.2 since the notation of that section will be used in what follows. The investigation of the classical integrable system for these correlators can be carried out by means of the classical inverse scattering method. In this section the basic object of this method for the integrable system of section XIV.2 is constructed, i.e., the matrix Riemann problem (RP). In this connection, the main object of our analysis, the integral equations (XIV.2.15) for the functions $f_\pm(\lambda)$, will be shown to be equivalent to simple algebraic reductions of the corresponding Gel'fand-Levitan-Marchenko equations (or, more exactly, to singular integral equations of the type (0.7)) for the special type of "initial data" parametrized by functions $\vartheta(\lambda)$ with the properties

$$(2\lambda\partial_\beta + \partial_\lambda)\vartheta(\lambda) = 0, \qquad \vartheta(\lambda) \xrightarrow[\lambda\to\infty]{} 0. \tag{1.3}$$

For impenetrable bosons the function $\vartheta(\lambda)$ is given by the Fermi weight (XIV.2.13),

$$\vartheta(\lambda) = \frac{1}{1 + e^{\lambda^2 - \beta}}. \tag{1.4}$$

It is to be remembered that the normalized chemical potential β and distance x (see (XIV.2.5)),

$$x = \frac{1}{2}(x_1 - x_2)\sqrt{T} > 0, \qquad \beta = \frac{h}{T}, \tag{1.5}$$

are used. It should be mentioned that the constructions of section XIV.2 (as well as of the present section) are valid for all initial data with properties (1.3), the "arbitrary function" $\vartheta(\lambda)$ with these properties playing the role of reflection coefficient in the inverse scattering method.

Consider in the complex λ-plane the following matrix RP. One has to find a 2×2 matrix-valued function $\chi(\lambda)$ (with determinant not equal to zero for any λ, $\det \chi(\lambda) \neq 0$) which is holomorphic for $\operatorname{Im} \lambda > 0$ and for $\operatorname{Im} \lambda < 0$; the boundary values at the real axis are related by

$$\chi_-(\lambda) = \chi_+(\lambda)G(\lambda); \qquad \chi_\pm(\lambda) = \lim_{\epsilon \to 0^+} \chi(\lambda \pm i\epsilon) \tag{1.6a}$$

and $\chi(\lambda)$ is equal to the unit matrix I for $\lambda = \infty$,

$$\chi(\infty) = I. \tag{1.6b}$$

The conjugation matrix $G(\lambda)$ here is given by

$$G(\lambda) = \begin{pmatrix} 1 + \pi\gamma e_+(\lambda)e_-(\lambda) & -\pi\gamma e_+^2(\lambda) \\ \pi\gamma e_-^2(\lambda) & 1 - \pi\gamma e_+(\lambda)e_-(\lambda) \end{pmatrix} \tag{1.7}$$

where the "plane waves" $e_\pm(\lambda)$ are given by (XIV.2.12):

$$e_\pm(\lambda) = \sqrt{\vartheta(\lambda)}\, e^{\pm i\lambda x}. \tag{1.8}$$

The functions χ, G, $e_\pm$ depend, of course, also on x, β and γ but this dependence is as a rule suppressed in our notation.

Let us now discuss the solvability of the matrix RP (1.6). It is related to the generating functional for the current correlators (1.2) ($\gamma = (1 - e^\alpha)/\pi$, see (XIV.2.7)). For $\gamma = 1/\pi$ ($\alpha = -\infty$) one obtains the emptiness formation probability (XIV.2.8). In what follows we will consider the region

$$0 < \gamma \leq \frac{1}{\pi}, \quad -\infty < \beta < +\infty, \quad x > 0 \tag{1.9}$$

for these correlators. Using a result of Gokhberg and Krein (see (0.5)) one easily checks that the matrix RP (1.6) with G given by (1.7) is uniquely solvable in region (1.9).

Another case interesting from the physical point of view is $\gamma = 2/\pi$. The matrix RP is related to the equal-time correlator of two fields at this value of γ, see (XIV.2.6). In this case the situation regarding solvability is more complicated. For negative chemical potentials ($\beta < 0$) condition

(0.5) is still valid for any $x > 0$, so that the matrix RP is uniquely solvable in the region

$$\gamma = \frac{2}{\pi}, \quad \beta < 0, \quad x > 0. \tag{1.10}$$

However, for positive chemical potentials, i.e., in the region

$$\gamma = \frac{2}{\pi}, \quad \beta > 0, \quad x > 0 \tag{1.11}$$

the sufficiency conditions (0.5) of the solvability are not fulfilled. This results in the possibility of the existence (for $\beta > 0$) of a set $X \in \mathbb{R}$ such that for $x \in X$ the matrix RP (1.6) possesses particular indices and hence is not uniquely solvable for these x. It should be noted that the set X is at most countable,

$$X = \{x_n\} \tag{1.12}$$

which is the corollary of the compactness of the operator $\widehat{K}$ (XIV.2.9), (XIV.2.10) (see Appendix 1). In what follows in this section it is supposed that the matrix RP is solvable, i.e., $x \neq x_n$.

Let us study the properties of the solution of matrix RP (1.6) as a function of x, β. First of all, it should be noted that the function $\chi(\lambda)$ can be expanded for $|\lambda| \to \infty$ as

$$\chi(\lambda) = 1 + \frac{\Psi_1}{\lambda} + \frac{\Psi_2}{\lambda^2} + \cdots \tag{1.13}$$

where the 2×2 matrix-valued functions $\Psi_i(x, \beta)$ (which are also functions of γ) do not depend on λ. Due to the symmetry properties of G (1.7)

$$G(\lambda) = \sigma_1 \, G^{-1}(-\lambda) \, \sigma_1 \tag{1.14}$$

$$G(\lambda) = G^*(-\lambda) \tag{1.15}$$

one concludes that the coefficient Ψ_1 in expansion (1.13) can be written as

$$\Psi_1 = -\frac{1}{2}B_{++}\sigma_2 - \frac{i}{2}B_{+-}\sigma_3 = -\frac{i}{2}\begin{pmatrix} B_{+-} & -B_{++} \\ B_{++} & -B_{+-} \end{pmatrix} \tag{1.16}$$

where, as usual,

$$\sigma_1 = \begin{pmatrix} 0 & 1 \\ 1 & 0 \end{pmatrix}, \quad \sigma_2 = \begin{pmatrix} 0 & -i \\ i & 0 \end{pmatrix}, \quad \sigma_3 = \begin{pmatrix} 1 & 0 \\ 0 & -1 \end{pmatrix}$$

are the Pauli matrices. The functions $B_{++}(x, \beta, \gamma)$, $B_{+-}(x, \beta, \gamma)$ in (1.16) are real-valued scalar functions of x, β, γ (it will be shown later that they are just the potentials B_{lm} (XIV.2.16)).

Our aim now is to show that the function $\Psi(\lambda)$,

$$\Psi(\lambda) \equiv \chi(\lambda) \, e^{i\lambda x \sigma_3} \tag{1.17}$$

is a matrix solution of a linear system (the Lax or zero-curvature representation) (XIV.2.19), (XIV.2.20). It is done in the way standard in the classical inverse scattering method (see, e.g., [1]). One rewrites the conjugation condition (1.6a) of the matrix RP for $\Psi(\lambda)$ as

$$\Psi_-(\lambda) = \Psi_+(\lambda)\, G_0(\lambda) \qquad (\text{Im}\,\lambda = 0) \tag{1.18}$$

where the matrix $G_0(\lambda)$ is

$$\begin{aligned} G_0(\lambda) &= e^{-i\lambda x \sigma_3}\, G(\lambda)\, e^{i\lambda x \sigma_3} \\ &= \begin{pmatrix} 1 + \pi\gamma\vartheta(\lambda) & -\pi\gamma\vartheta(\lambda) \\ \pi\gamma\vartheta(\lambda) & 1 - \pi\gamma\vartheta(\lambda) \end{pmatrix}. \end{aligned} \tag{1.19}$$

This matrix does not depend on x,

$$\partial_x G_0(\lambda) = 0. \tag{1.20}$$

As the dependence on β is through the function $\vartheta(\lambda)$ with properties (1.3) one also has

$$(2\lambda\partial_\beta + \partial_\lambda)G_0(\lambda) = 0. \tag{1.21}$$

This means that the derivatives of $\Psi(\lambda)$ should satisfy the same Riemann problem as $\Psi(\lambda)$ (the behaviour for large λ will be different),

$$\begin{aligned} \left(\partial_x \Psi(\lambda)\right)_- &= \left(\partial_x \Psi(\lambda)\right)_+ G_0(\lambda), \\ \left((2\lambda\partial_\beta + \partial_\lambda)\Psi(\lambda)\right)_- &= \left((2\lambda\partial_\beta + \partial_\lambda)\Psi(\lambda)\right)_+ G_0(\lambda), \end{aligned}$$

so that the logarithmic derivatives,

$$\left(\partial_x \Psi\right)\Psi^{-1}(\lambda) \equiv F_1(\lambda) \tag{1.22}$$

$$\left((2\lambda\partial_\beta + \partial_\lambda)\Psi\right)\Psi^{-1}(\lambda) \equiv F_2(\lambda) \tag{1.23}$$

can be analytically continued through the real axis—they have no cuts at the real axis and are holomorphic in the whole complex plane (for any $\lambda \in \mathbb{C}$). It is easily seen from (1.13) and (1.17) that the asymptotics of the logarithmic derivatives for large λ are given by:

$$F_1(\lambda) = i\lambda\sigma_3 + O(1), \qquad F_2(\lambda) = O(1), \quad \lambda \to \infty.$$

Thus one concludes, using Liouville's theorem, that $F_1(\lambda)$ and $F_2(\lambda)$ are polynomials in λ of the first and zeroth degree, respectively,

$$F_1(\lambda) = i\lambda\sigma_3 + U_0, \qquad F_2(\lambda) = V_0,$$

and the matrices $U_0(x, \beta, \gamma)$ and $V_0(x, \beta, \gamma)$ (which do not depend on λ) are easily calculated in terms of the first matrix coefficient $\Psi_1(\lambda)$ in the expansion (1.13):

$$U_0 = i[\Psi_1, \sigma_3], \qquad V_0 = i\sigma_3 x + 2\partial_\beta \Psi_1.$$

Using now the parametrization (1.16) of Ψ_1 in terms of real scalar potentials B_{++}, B_{+-} one calculates U_0, V_0 explicitly and rewrites (1.22), (1.23) as the following linear differential equations, valid for any γ and any $\lambda \in \mathbb{C}$:

$$\partial_x \Psi(\lambda) = \left(i\lambda\sigma_3 + B_{++}\sigma_1 \right)\Psi(\lambda) \tag{1.24}$$

$$(2\lambda\partial_\beta + \partial_\lambda)\Psi(\lambda) = \left(ix\sigma_3 - i\partial_\beta B_{+-}\sigma_3 - \partial_\beta B_{++}\sigma_2 \right)\Psi(\lambda) \tag{1.25}$$

which is exactly the zero-curvature representation (XIV.2.19), (XIV.2.20) written now, however, for a matrix-valued function $\Psi(\lambda)$ of dimension 2×2. The immediate corollary is, of course, equations (XIV.2.22)–(XIV.2.24) for the potentials B_{++}, B_{+-}:

$$\partial_\beta B_{+-} = x + \frac{1}{2}\frac{\partial_x\partial_\beta B_{++}}{B_{++}} \tag{1.26a}$$

$$\partial_x B_{+-} = B_{++}^2 \tag{1.26b}$$

(one takes into account that $B_{++} \to 0$ as $\beta \to -\infty$) and

$$\partial_\beta B_{++}^2 = 1 + \frac{1}{2}\frac{\partial}{\partial x}\frac{\partial_x\partial_\beta B_{++}}{B_{++}}. \tag{1.27}$$

In the above considerations the explicit form (1.4) of $\vartheta(\lambda)$ was not used. It was only necessary that it should possess properties (1.3). This means that the matrix RP under consideration may be used to construct solutions of equations (1.25), (1.26) that depend on a functional parameter. The choice (1.4) for $\vartheta(\lambda)$ leads to solutions (potentials B_{++}, B_{+-}) which were defined earlier, in (XIV.2.16), the correlation functions being defined in section XIV.2 in terms of these potentials. Let us prove this statement.

To do this one considers the system of singular integral equations (see (0.6)) equivalent to the matrix RP (1.6) for the 2×2 matrix $\chi(\lambda)$:

$$\chi_+(\lambda) = I + \frac{1}{2\pi i}\int\limits_{-\infty}^{+\infty}\frac{\chi_+(\mu)[I - G(\mu)]}{\mu - \lambda - i0}\,d\mu \qquad (\text{Im }\lambda = 0). \tag{1.28}$$

The special property

$$\det[I - G(\lambda)] = 0 \tag{1.29}$$

of the conjugation matrix $G(\lambda)$ (1.7) allows one to reduce (1.28) to two scalar equations. Indeed, let us put

$$\widehat{\chi}(\lambda) \equiv \chi_+(\lambda)\,E(\lambda) \tag{1.30}$$

where the triangular matrix $E(\lambda)$ is defined by

$$E(\lambda) = \begin{pmatrix} 1 & e_+(\lambda) \\ 0 & e_-(\lambda) \end{pmatrix}. \tag{1.31}$$

In terms of the function $\widehat{\chi}(\lambda)$ the system (1.28) is readily rewritten as

$$\widehat{\chi}(\lambda) = E(\lambda) + \frac{1}{2\pi i} \int\limits_{-\infty}^{+\infty} \frac{\widehat{\chi}(\mu)\widehat{G}(\mu, \lambda)}{\mu - \lambda - i0} \, d\mu \tag{1.32}$$

with the matrix $\widehat{G}(\mu, \lambda)$ given by

$$\widehat{G}(\mu, \lambda) = E^{-1}(\mu)[I - G(\mu)]E(\lambda)$$

$$= \begin{pmatrix} 0 & 0 \\ -\pi\gamma e_-(\mu) & \pi\gamma \left[e_+(\mu)e_-(\lambda) - e_+(\lambda)e_-(\mu) \right] \end{pmatrix}. \tag{1.33}$$

Due to the zero first line of this matrix, equations (1.32) for the matrix elements $\widehat{\chi}_{11}(\lambda)$, $\widehat{\chi}_{21}(\lambda)$ are representations for these functions in terms of $\widehat{\chi}_{12}$, $\widehat{\chi}_{22}$:

$$\widehat{\chi}_{11}(\lambda) = 1 - \frac{\gamma}{2i} \int\limits_{-\infty}^{+\infty} \frac{\widehat{\chi}_{12}(\mu)e_-(\mu)}{\mu - \lambda - i0} \, d\mu \tag{1.34}$$

$$\widehat{\chi}_{21}(\lambda) = -\frac{\gamma}{2i} \int\limits_{-\infty}^{+\infty} \frac{\widehat{\chi}_{22}(\mu)e_-(\mu)}{\mu - \lambda - i0} \, d\mu. \tag{1.35}$$

The remaining two equations in (1.32) are the following two scalar equations for $\widehat{\chi}_{12}(\lambda)$, $\widehat{\chi}_{22}(\lambda)$:

$$\widehat{\chi}_{12}(\lambda) = e_+(\lambda) + \gamma \int\limits_{-\infty}^{+\infty} K(\lambda, \mu)\widehat{\chi}_{12}(\mu) \, d\mu$$

$$\widehat{\chi}_{22}(\lambda) = e_-(\lambda) + \gamma \int\limits_{-\infty}^{+\infty} K(\lambda, \mu)\widehat{\chi}_{22}(\mu) \, d\mu \tag{1.36}$$

with kernel

$$K(\lambda, \mu) = \frac{e_+(\lambda)e_-(\mu) - e_-(\lambda)e_+(\mu)}{2i(\lambda - \mu)}$$

$$= \sqrt{\vartheta(\lambda)} \, \frac{\sin x(\lambda - \mu)}{\lambda - \mu} \sqrt{\vartheta(\mu)} \tag{1.37}$$

which is just the same (XIV.2.10), (XIV.2.11) as in integral equations (XIV.2.15),

$$f_\pm(\lambda) - \gamma \int\limits_{-\infty}^{+\infty} K(\lambda, \mu) f_\pm(\mu)\, d\mu = e_\pm(\lambda) \qquad (1.38)$$

for the functions $f_\pm(\lambda)$. So one concludes that

$$\begin{aligned}
\hat\chi_{12}(\lambda) &= f_+(\lambda), \\
\hat\chi_{22}(\lambda) &= f_-(\lambda).
\end{aligned} \qquad (1.39)$$

So the reduction of the system of singular integral equations (1.28) of the matrix RP formulation of the inverse scattering method to the scalar integral equations (1.38) of section XIV.2 is realized. Thus these equations for $f_\pm(\lambda)$ are indeed Gel'fand–Levitan–Marchenko-type equations for the integrable classical nonlinear system describing the equal-time two-point correlators; this argument was implied in section XIV.2 from the beginning in our derivation of integrable partial differential equations for correlators.

Turn now to relations (1.34), (1.35). Taking into account that the matrix elements $\hat\chi_{j1}(\lambda) = \chi_{j1}^+(\lambda)$ ($j = 1, 2$) (which is obvious from (1.30)) and also the identification (1.39) (see also (0.8)) one obtains

$$\begin{aligned}
\chi_{11}(\lambda) &= 1 - \frac{\gamma}{2i} \int\limits_{-\infty}^{+\infty} \frac{f_+(\mu) e_-(\mu)}{\mu - \lambda}\, d\mu, \\
\chi_{21}(\lambda) &= -\frac{\gamma}{2i} \int\limits_{-\infty}^{+\infty} \frac{f_-(\mu) e_-(\mu)}{\mu - \lambda}\, d\mu,
\end{aligned} \qquad (1.40)$$

$$\operatorname{Im} \lambda > 0.$$

Expanding these expressions in $1/\lambda$ for $\lambda \to \infty$, one obtains for the corresponding matrix elements of Ψ_1 in (1.13):

$$\begin{aligned}
(\Psi_1)_{11} &= \frac{\gamma}{2i} \int\limits_{-\infty}^{+\infty} f_+(\mu) e_-(\mu)\, d\mu \\
(\Psi_1)_{21} &= \frac{\gamma}{2i} \int\limits_{-\infty}^{+\infty} f_-(\mu) e_-(\mu)\, d\mu.
\end{aligned} \qquad (1.41)$$

Recall now that the potentials B_{lm} were defined in section XIV.2 by the formulæ:

$$B_{lm}(x, \beta, \gamma) \equiv \gamma \int\limits_{-\infty}^{+\infty} e_l(\lambda) f_m(\lambda) \, d\lambda; \qquad l, m = +, - \tag{1.42}$$

$$B_{lm}^* = B_{lm}, \quad B_{lm} = B_{ml}, \quad B_{++} = B_{--}$$

(see (XIV.2.16)–(XIV.2.18)). Comparing now (1.16), (1.41) and (1.42) one sees that the functions B_{++}, B_{+-} introduced in (1.16) are indeed the same potentials (1.42) as were introduced in section XIV.2.

So the equivalence between the systems of integral equations (1.38) and (1.28) (the latter system being equivalent to the matrix Riemann problem of the classical inverse scattering method) is established. Equations (1.38) can be thus interpreted as Gel'fand–Levitan–Marchenko equations for the nonlinear system (1.25), (1.26). In turn, this makes it possible to consider the Fredholm determinant logarithm,

$$\sigma = \ln \det \left(1 - \gamma \widehat{K} \right),$$

as a τ-function for this nonlinear system.

In conclusion let us make the following remark. It is not difficult to verify using (1.39), (1.30) that the following equation is valid:

$$\begin{pmatrix} f_+(\lambda) \\ f_-(\lambda) \end{pmatrix} = \chi_+(\lambda) \begin{pmatrix} e_+(\lambda) \\ e_-(\lambda) \end{pmatrix} \qquad (\text{Im } \lambda = 0). \tag{1.43}$$

XV.2 Representation of equal-time two-point correlators in terms of the matrix Riemann-Hilbert problem

It was shown in the previous section that the potentials B_{++}, B_{+-} (in terms of which the correlators (1.1), (1.2) were expressed in section XIV.2) are naturally obtained from the solution of the matrix Riemann problem, see (1.16), as a coefficient in the expansion (1.13) of the solution. The aim of this section is to represent the potentials in terms of matrix RP in a form convenient for the calculation of the long distance asymptotics of the correlation functions (which is done later in Chapter XVI).

We begin with the generating functional (1.2) for equal-time current correlators. The result of section XIV.2 is the representation

$$\langle \exp\{\alpha Q_1(x_1, x_2)\} \rangle_T = \exp\{\sigma(x, \beta, \gamma)\} \Big|_{\gamma = (1 - \exp \alpha)/\pi} \tag{2.1}$$

where x and β are the rescaled distance and chemical potential (1.5). The logarithm, σ, of the Fredholm determinant of the operator $\widehat{K}$ (1.37),

$$\sigma(x, \beta, \gamma) = \ln \det \left(1 - \gamma \widehat{K} \right) \tag{2.2}$$

is expressed in terms of the potentials B_{++}, B_{+-} (see (XIV.2.29), (XIV.2.30)):

$$\partial_x \sigma = -B_{+-},$$

$$\partial_\beta \sigma = -x \partial_\beta B_{+-} + \frac{1}{2}(\partial_\beta B_{+-})^2 - \frac{1}{2}(\partial_\beta B_{++})^2. \tag{2.3}$$

As $\sigma(x = 0, \beta, \gamma) = 0$ one has (see (XIV.2.36))

$$\sigma(x, \beta, \gamma) = -\int_0^x B_{+-}(y, \beta, \gamma)\, dy. \tag{2.4}$$

So the correlator (2.1) is expressed in terms of potentials which are related to the solution of the matrix RP (1.6) by means of (1.13), (1.16). This matrix RP is uniquely solvable in the region (1.9)

$$-\infty < \alpha \leq 0, \quad 0 \leq \gamma \leq \frac{1}{\pi}, \quad -\infty < \beta < +\infty, \quad x > 0. \tag{2.5}$$

We will consider correlator (2.1) in this region of parameter space.

It is convenient to transform the initial matrix RP to the matrix RP that has diagonal elements of the new conjugation matrix equal to one. To this end, let us define scalar functions $\alpha(\lambda)$ and $\beta(\lambda)$ by means of the following explicit formulæ:

$$\alpha(\lambda) = \exp\left\{-\frac{1}{2\pi i}\int_{-\infty}^{+\infty}\frac{d\mu}{\mu - \lambda}\ln\left(\frac{1 - \pi\gamma + e^{\mu^2 - \beta}}{1 + e^{\mu^2 - \beta}}\right)\right\} \tag{2.6}$$

$$\beta(\lambda) = \exp\left\{-\frac{1}{2\pi i}\int_{-\infty}^{+\infty}\frac{d\mu}{\mu - \lambda}\ln\left(\frac{1 + \pi\gamma + e^{\mu^2 - \beta}}{1 + e^{\mu^2 - \beta}}\right)\right\}. \tag{2.7}$$

It is to be mentioned that for $\gamma \leq 1/\pi$, which is the case now (see (2.5)), the expressions inside the logarithm are positive, the logarithms being defined to be real. The functions $\alpha(\lambda)$ and $\beta(\lambda)$ are holomorphic for $\text{Im}\,\lambda > 0$ and for $\text{Im}\,\lambda < 0$, being the solutions of the scalar Riemann problems (see Appendix 1 of Chapter XVI):

$$\alpha_-(\lambda) = \alpha_+(\lambda)g_\alpha(\lambda), \qquad \text{Im}\,\lambda = 0 \tag{2.8a}$$

$$\alpha(\infty) = 1 \tag{2.8b}$$

$$g_\alpha(\lambda) = 1 - \pi\gamma e_+(\lambda)e_-(\lambda)$$

$$= \frac{1 - \pi\gamma + \exp\{\lambda^2 - \beta\}}{1 + \exp\{\lambda^2 - \beta\}} \tag{2.8c}$$

and

$$\beta_-(\lambda) = \beta_+(\lambda)g_\beta(\lambda), \qquad \text{Im}\,\lambda = 0 \tag{2.9a}$$

$$\beta(\infty) = 1 \tag{2.9b}$$

$$g_\beta(\lambda) = 1 + \pi\gamma e_+(\lambda)e_-(\lambda)$$

$$= \frac{1 + \pi\gamma + \exp\{\lambda^2 - \beta\}}{1 + \exp\{\lambda^2 - \beta\}} \tag{2.9c}$$

(the function β should not be confused with the chemical potential divided by temperature, $\beta = h/T$). Going from the matrix function $\chi(\lambda)$ in (1.6) to the matrix-valued function $\Phi(\lambda)$,

$$\chi(\lambda) \equiv \Phi(\lambda) \begin{pmatrix} \beta(\lambda) & 0 \\ 0 & \alpha(\lambda) \end{pmatrix}, \tag{2.10}$$

(which does not change the value at $\lambda = \infty$ (1.6b)) one rewrites the matrix RP (1.6) for $\Phi(\lambda)$:

$$\Phi_-(\lambda) = \Phi_+(\lambda)G_\Phi(\lambda), \qquad \text{Im}\,\lambda = 0 \tag{2.11a}$$

$$\Phi(\infty) = 1 \tag{2.11b}$$

where the new conjugation matrix is

$$G_\Phi(\lambda) = \begin{pmatrix} 1 & b^*(\lambda)e^{2i\lambda x} \\ -b(\lambda)e^{-2i\lambda x} & 1 \end{pmatrix}, \qquad \text{Im}\,\lambda = 0 \tag{2.11c}$$

with the "reflection coefficient" $b(\lambda)$ given by

$$b(\lambda) \equiv -\frac{\pi\gamma\alpha_+(\lambda)\vartheta(\lambda)}{\beta_-(\lambda)}. \tag{2.12}$$

Here obvious symmetry properties of the functions α, β are taken into account:

$$\alpha^*(\lambda^*) = \alpha(-\lambda), \qquad \beta^*(\lambda^*) = \beta(-\lambda) \tag{2.13}$$

$$\alpha^{-1}(\lambda) = \alpha(-\lambda), \qquad \beta^{-1}(\lambda) = \beta(-\lambda) \tag{2.14}$$

It is worth mentioning that the L-operator (1.24) in the zero-curvature representation for the potentials B is in fact the L-operator of the classical nonlinear Schrödinger equation (see (V.1.2), (V.1.14), (V.1.15), (V.1.16)). The reader is also recommended to compare the conjugation matrix $G_\Phi(\lambda)$ (2.11c) with the corresponding conjugation matrix $G(\lambda, x)$ of the NS equation as given, e.g., in Chapter II of the text-book [1], to convince himself that obtaining the potential $B_{++}(x, \beta, \gamma)$ is equivalent to reconstructing the potential in the NS equation from the reflection coefficient $b(\lambda)$.

In terms of the solution to the matrix RP (2.11) the potentials B_{++}, B_{+-} are given as

$$B_{+-} = 2i \lim_{\lambda \to \infty} \lambda \left[\Phi_{11}(\lambda) - \beta^{-1}(\lambda) \right],$$

$$B_{++} = -2i \lim_{\lambda \to \infty} \lambda\Phi_{12}(\lambda). \tag{2.15}$$

These formulæ are obtained in the obvious way from (1.13), (1.16) and (2.10). So we have represented the correlator (2.1) in terms of the solutions to the matrix RP (2.11). This representation will be the starting point in the calculation of the long distance asymptotics in the next chapter.

We turn now to the equal-time correlator of two fields (1.1). It was represented in section XIV.2 as:

$$\langle \Psi^\dagger(x_1)\Psi(x_2)\rangle_T = \sqrt{T}g(x,\beta);$$
$$g(x,\beta) = \frac{1}{4}B_{++}(x,\beta,\gamma)e^{\sigma(x,\beta,\gamma)}\Big|_{\gamma=2/\pi} \tag{2.16}$$

i.e., due to (2.3), (2.4) it is also given in terms of the potentials B. Now one has to distinguish between two cases. If the chemical potential β is negative (i.e., one considers the case (1.10)) the matrix Riemann problem (1.6) has a unique solution. So in the case

$$\gamma = \frac{2}{\pi}, \quad \beta < 0, \quad x > 0 \tag{2.17}$$

one can go to the matrix RP (2.11) introducing functions $\alpha(\lambda)$, $\beta(\lambda)$ by means of formulæ (2.6), (2.7), the corresponding scalar Riemann problems (2.8), (2.9) being again uniquely solvable. Potentials B_{++} and B_{+-} in this case are given by formulæ (2.15).

In the case of positive chemical potential (1.11)

$$\gamma = \frac{2}{\pi}, \quad \beta > 0, \quad x > 0 \tag{2.18}$$

the situation is more complicated. It was already mentioned that in this case the sufficient conditions (0.5) are not fulfilled, so that there is a possibility of the existence of a countable set $\{x_n\} \equiv X$ (1.12) for which the matrix RP (1.6) is not uniquely solvable. The transition from RP (1.6) to RP (2.11) cannot be made. The reason is that the function $g_\alpha(\lambda)$ (2.8c) at $\gamma = 2/\pi$

$$g_\alpha(\lambda) = 1 - \pi\gamma e_+(\lambda)e_-(\lambda)\Big|_{\gamma=2/\pi} = \frac{e^{\lambda^2-\beta}-1}{e^{\lambda^2-\beta}+1} \tag{2.19}$$

possesses zeros on the real axis at points $\pm\lambda_1$, where

$$\lambda_1 = \sqrt{\beta} > 0 \tag{2.20}$$

so that the scalar RP (2.8) for the function $\alpha(\lambda)$ certainly has an index (see Appendix 1 to Chapter XV1, formula (A.1.2)) and is not uniquely solvable. To treat this case we regularize the matrix RP as follows.

Let $x \notin X$ (see (1.12)). Then the matrix RP (1.6) is uniquely solvable. The conjugation matrix $G(\lambda)$ (1.7) can be continued analytically in λ

into some vicinity of the real axis (up to the nearest zero of $e^{\lambda^2+\beta}+1$), so that the conjugation condition (1.6a) can be changed for the equivalent conjugation condition on the contour Γ shown in Figure XV.1, where we suppose that $0 < \epsilon < \text{Im } \sqrt{\beta + i\pi}$. This change of integration contour can be made also in the initial linear integral equations (1.38), the logarithm σ of the Fredholm determinant (2.2) and the potentials B_{++}, B_{+-} (1.42) obviously remaining the same.

Going to contour Γ regularizes the scalar RP (2.8) for the function $\alpha(\lambda)$,

$$\alpha_-(\lambda) = \alpha_+(\lambda)\frac{e^{\lambda^2-\beta}-1}{e^{\lambda^2-\beta}+1}, \qquad \lambda \in \Gamma \qquad (2.21a)$$

$$\alpha(\infty) = 1 \qquad (2.21b)$$

and does not affect the solvability of the scalar RP for $\beta(\lambda)$,

$$\beta_-(\lambda) = \beta_+(\lambda)\frac{e^{\lambda^2-\beta}+3}{e^{\lambda^2-\beta}+1}, \qquad \lambda \in \Gamma \qquad (2.22a)$$

$$\beta(\infty) = 1. \qquad (2.22b)$$

The corresponding explicit formulæ for the functions α, β differ from (2.6), (2.7) with $\gamma = 2/\pi$ in the integration contour only,

$$\alpha(\lambda) = \exp\left\{-\frac{1}{2\pi i}\int_\Gamma \frac{d\mu}{\mu - \lambda} \ln\left(\frac{e^{\mu^2-\beta}-1}{e^{\mu^2-\beta}+1}\right)\right\} \qquad (2.23)$$

$$\beta(\lambda) = \exp\left\{-\frac{1}{2\pi i}\int_\Gamma \frac{d\mu}{\mu - \lambda} \ln\left(\frac{e^{\mu^2-\beta}+3}{e^{\mu^2-\beta}+1}\right)\right\}. \qquad (2.24)$$

The logarithm branch in (2.23) is fixed so that

$$\ln\frac{e^{\lambda^2-\beta}-1}{e^{\lambda^2-\beta}+1} \longrightarrow 0 \text{ for } \lambda \to +\infty. \qquad (2.25)$$

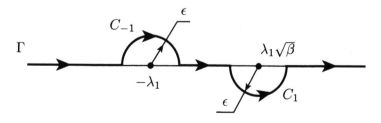

Figure XV.1

Going as previously from the matrix $\chi(\lambda)$ to $\Phi(\lambda)$,

$$\chi(\lambda) = \Phi(\lambda) \begin{pmatrix} \beta(\lambda) & 0 \\ 0 & \alpha(\lambda) \end{pmatrix} \qquad (2.26)$$

one obtains a matrix RP with ones on the diagonal,

$$\Phi_-(\lambda) = \Phi_+(\lambda) G_\Phi(\lambda), \qquad \lambda \in \Gamma \qquad (2.27a)$$

$$\Phi(\infty) = 1 \qquad (2.27b)$$

$$G_\Phi(\lambda) = \begin{pmatrix} 1 & c(\lambda)e^{2i\lambda x} \\ -b(\lambda)e^{-2i\lambda x} & 1 \end{pmatrix}, \qquad (2.27c)$$

where now

$$c(\lambda) = -\frac{2\vartheta(\lambda)\beta_+(\lambda)}{\alpha_-(\lambda)}, \qquad b(\lambda) = -\frac{2\vartheta(\lambda)\alpha_+(\lambda)}{\beta_-(\lambda)}. \qquad (2.28)$$

It should be mentioned that in going to the contour Γ one no longer has the complex conjugation involutions (2.13) but relations (2.14) are still valid, so that

$$c(\lambda) = b(-\lambda). \qquad (2.29)$$

It will be seen in the next chapter that for the long distance $(x \to +\infty)$ asymptotical analysis of the solution of the matrix RP (2.27) one has to take into account the zeros $\pm\lambda_1 = \pm\sqrt{\beta}$ (2.20) of $e^{\lambda^2 - \beta} - 1$ in an explicit way. To do this it is useful to perform another transformation of the RP (2.27). First, let us go to the function $\widetilde{\Phi}(\lambda)$,

$$\widetilde{\Phi}(\lambda) = \Phi(\lambda) \begin{pmatrix} \lambda + \lambda_1 & 0 \\ 0 & \lambda - \lambda_1 \end{pmatrix}, \qquad \lambda_1 = \sqrt{\beta} \qquad (2.30)$$

so that RP (2.27) is transformed to the following matrix RP with zeros for $\widetilde{\Phi}(\lambda)$:

$$\widetilde{\Phi}_-(\lambda) = \widetilde{\Phi}_+(\lambda) \widetilde{G}_\Phi(\lambda), \qquad \lambda \in \Gamma \qquad (2.31a)$$

$$\widetilde{\Phi}(\lambda) \sim \lambda I, \qquad \lambda \to \infty \qquad (2.31b)$$

$$\widetilde{\Phi}(-\lambda_1) \begin{pmatrix} 1 \\ 0 \end{pmatrix} = 0, \qquad \widetilde{\Phi}(+\lambda_1) \begin{pmatrix} 0 \\ 1 \end{pmatrix} = 0 \qquad (2.31c)$$

with the conjugation matrix

$$\widetilde{G}_\Phi(\lambda) = \begin{pmatrix} 1 & c(\lambda)\dfrac{\lambda - \lambda_1}{\lambda + \lambda_1} e^{2i\lambda x} \\ -b(\lambda)\dfrac{\lambda + \lambda_1}{\lambda - \lambda_1} e^{-2i\lambda x} & 1 \end{pmatrix}. \qquad (2.31d)$$

Secondly, to treat RP (2.31) it is convenient to use the "dressing soliton method" known in the theory of integrable systems [1]. Namely, one represents the solution of RP (2.31) in the form

$$\widetilde{\Phi}(\lambda) = P(\lambda)\overset{\circ}{\Phi}(\lambda) \tag{2.32}$$

where $\overset{\circ}{\Phi}(\lambda)$ is the solution of the corresponding regular Riemann problem and $P(\lambda)$ is the matrix Blaschke-Potapov factor,

$$P(\lambda) = \lambda I + P. \tag{2.33}$$

The matrix P on the right hand side here does not depend on λ and is to be defined from the following system of equations:

$$\begin{aligned}
(-\lambda_1 I + P)\, \overset{\circ}{\Phi}(-\lambda_1) \begin{pmatrix} 1 \\ 0 \end{pmatrix} &= 0, \\
(\lambda_1 I + P)\, \overset{\circ}{\Phi}(\lambda_1) \begin{pmatrix} 0 \\ 1 \end{pmatrix} &= 0.
\end{aligned} \tag{2.34}$$

The regular matrix problem for the matrix-valued function $\overset{\circ}{\Phi}(\lambda)$ is

$$\begin{aligned}
\overset{\circ}{\Phi}_-(\lambda) &= \overset{\circ}{\Phi}_+(\lambda)\widetilde{G}_\Phi(\lambda), &(2.35a) \\
\overset{\circ}{\Phi}(\infty) &= 1 &(2.35b)
\end{aligned}$$

where $\widetilde{G}_\Phi(\lambda)$ is given by (2.31d).

It is to be mentioned also that due to relations (2.14), (2.29) one easily sees that the matrix elements $\overset{\circ}{\Phi}_{21}(\lambda)$ and $\overset{\circ}{\Phi}_{22}(\lambda)$ are proportional to $\overset{\circ}{\Phi}_{12}(-\lambda)$, $\overset{\circ}{\Phi}_{11}(-\lambda)$:

$$\begin{aligned}
\overset{\circ}{\Phi}_{21}(\lambda) &= \overset{\circ}{\Phi}_{12}(-\lambda)\alpha^{-1}(\lambda)\beta^{-1}(\lambda) \\
\overset{\circ}{\Phi}_{22}(\lambda) &= \overset{\circ}{\Phi}_{11}(-\lambda)\alpha^{-1}(\lambda)\beta^{-1}(\lambda).
\end{aligned} \tag{2.36}$$

Taking this into account one has for the matrix P, after solving equations (2.34),

$$P = \frac{1}{d}\left(a\sigma_3 + e\sigma_2\right) \tag{2.37}$$

$$d \equiv \overset{\circ}{\Phi}_{22}(\lambda_1)\overset{\circ}{\Phi}_{11}(-\lambda_1) - \overset{\circ}{\Phi}_{12}(\lambda_1)\overset{\circ}{\Phi}_{21}(-\lambda_1) \tag{2.38}$$

$$a \equiv \lambda_1\left(\overset{\circ}{\Phi}_{11}(-\lambda_1)\overset{\circ}{\Phi}_{22}(\lambda_1) + \overset{\circ}{\Phi}_{12}(\lambda_1)\overset{\circ}{\Phi}_{21}(-\lambda_1)\right) \tag{2.39}$$

$$e \equiv -2i\lambda_1\overset{\circ}{\Phi}_{11}(-\lambda_1)\overset{\circ}{\Phi}_{12}(\lambda_1). \tag{2.40}$$

Matrix RP (2.35) will be the starting point for the asymptotic analysis of the equal-time correlator of two fields in the next chapter. In conclusion, let us discuss how the potentials B_{++}, B_{+-} are expressed in terms of the solution $\overset{\circ}{\Phi}(\lambda)$ of matrix RP (2.35). Summing up all the transformations (2.32), (2.30), (2.26) one sees that $\chi(\lambda)$ and $\overset{\circ}{\Phi}(\lambda)$ are related by

$$\chi(\lambda) = (\lambda I + P)\,\overset{\circ}{\Phi}(\lambda)\begin{pmatrix} (\lambda + \lambda_1)^{-1} & 0 \\ 0 & (\lambda - \lambda_1)^{-1} \end{pmatrix}\begin{pmatrix} \beta(\lambda) & 0 \\ 0 & \alpha(\lambda) \end{pmatrix}.$$
$$(2.41)$$

The function $\overset{\circ}{\Phi}(\lambda)$ for $\lambda \to \infty$ has an asymptotic expansion similar to (1.13),

$$\overset{\circ}{\Phi}(\lambda) = 1 + \frac{\overset{\circ}{\Psi}_1}{\lambda} + \cdots, \qquad \lambda \to \infty. \tag{2.42}$$

The 2×2 matrix $\Psi_1(x, \beta)$ (recall that $\gamma = 2/\pi$) in expansion (1.13) can be expressed by means of (2.41) as

$$\Psi_1 = P + \overset{\circ}{\Psi}_1 - \lambda_1\sigma_3 + \begin{pmatrix} \beta_0 & 0 \\ 0 & \alpha_0 \end{pmatrix} \tag{2.43}$$

where α_0 and β_0 are the coefficients of the corresponding expansions

$$\alpha(\lambda) = 1 + \frac{\alpha_0}{\lambda} + \cdots, \qquad \beta(\lambda) = 1 + \frac{\beta_0}{\lambda} + \cdots \tag{2.44}$$

of the functions $\alpha(\lambda)$, $\beta(\lambda)$, given explicitly by (2.23), (2.24).

As has already been discussed, the potentials B_{++}, B_{+-} are not changed by going to the contour Γ. So the left hand side of (2.43) can be expressed in terms of the potentials by means of relation (1.16). One thus obtains the following expressions for the potentials:

$$B_{+-} = 2ia_0 d^{-1} + \overset{\circ}{B}_{+-} \tag{2.45}$$

$$B_{++} = -2ed^{-1} + \overset{\circ}{B}_{++} \tag{2.46}$$

where

$$a_0 = a - \lambda_0 d = 2\lambda_1\overset{\circ}{\Phi}_{12}(\lambda_1)\overset{\circ}{\Phi}_{21}(-\lambda_1) \tag{2.47}$$

and the natural notation

$$\overset{\circ}{B}_{+-} = 2i\left[\left(\overset{\circ}{\Psi}_1\right)_{11} + \beta_0\right] \tag{2.48}$$

$$\overset{\circ}{B}_{++} = -2i\left(\overset{\circ}{\Psi}_1\right)_{12} \tag{2.49}$$

is introduced (this should be compared with (2.15)).

Regular matrix RP (2.35) is associated with a linear system similar to (1.24), (1.25) where the quantities $\overset{\circ}{B}$ should be taken as the potentials. In particular, the differential equation in x similar to (1.24) is written for $\overset{\circ}{\Phi}(\lambda)$ as

$$\partial_x\overset{\circ}{\Phi}(\lambda) = i\lambda[\sigma_3, \overset{\circ}{\Phi}(\lambda)] + \overset{\circ}{B}_{++}\sigma_1\overset{\circ}{\Phi}(\lambda). \tag{2.50}$$

Differentiating the expression (2.38) for the function d with respect to x and using (2.50) for $\partial_x\overset{\circ}{\Phi}_{lm}(\pm\lambda_1)$ $(l, m = 1, 2)$ one arrives at the following important identity:

$$\frac{i}{2}\partial_x d = a_0.$$
(2.51)

This allows one to rewrite (2.45) as

$$B_{+-} = -\partial_x \ln d + \mathring{B}_{+-}.$$
(2.52)

As the x-derivative of

$$\sigma(x, \beta, \gamma) = \ln \det \left(1 - \gamma \widehat{K}\right)\Big|_{\gamma = 2/\pi}$$

is given by $\partial_x \sigma = -B_{+-}$ (see (2.3)), one has the following representation for σ:

$$\sigma = \sigma_0 + \ln d; \qquad \partial_x \sigma_0 = -\mathring{B}_{+-}.$$
(2.53)

So the potentials B_{++}, B_{+-} and the logarithm of the Fredholm determinant, σ, are expressed by formulæ (2.45), (2.46) and (2.53) in terms of the solutions to the regular matrix RP (2.35) in the case of positive chemical potential $\beta > 0$ (2.18). These representations will be of primary importance in calculating the long distance asymptotics of the correlator in this case.

XV.3 The matrix Riemann-Hilbert problem for the temperature- and time-dependent correlation function of two fields for the impenetrable Bose gas

The aim of this section is to construct the matrix Riemann problem for the inverse scattering method in the case of the time-dependent temperature correlation function

$$\langle \Psi(x_2, t_2) \Psi^\dagger(x_1, t_1) \rangle_T$$
(3.1)

which is related to the completely integrable nonlinear system of partial differential equations in scaled distance x, time t and chemical potential β,

$$x = \frac{1}{2}(x_1 - x_2)\sqrt{T} > 0, \quad t = \frac{1}{2}(t_2 - t_1) > 0, \quad \beta = \frac{h}{T},$$
(3.2)

given in section XIV.6. It is done in an analogous way to the construction of the matrix RP for equal-time correlators in sections 1 and 2.

Consider the matrix RP for a 2×2 matrix-valued function $\chi(\lambda)$ which is holomorphic for $\text{Im}\,\lambda > 0$ and $\text{Im}\,\lambda < 0$, the boundary values on the real axis

$$\chi_\pm(\lambda) = \lim \chi(\lambda \pm i\epsilon), \quad \epsilon \to 0^+,$$

being related by

$$\chi_-(\lambda) = \chi_+(\lambda)G(\lambda), \quad \operatorname{Im}\lambda = 0 \tag{3.3a}$$

and $\chi(\lambda)$ being equal to the unit matrix I at $\lambda = \infty$

$$\chi(\infty) = I. \tag{3.3b}$$

The conjugation matrix $G(\lambda)$ here is quite similar to (1.7),

$$G(\lambda) = I + 2\pi i \begin{pmatrix} -e_+(\lambda)e_-(\lambda) & e_+^2(\lambda) \\ -e_-^2(\lambda) & e_+(\lambda)e_-(\lambda) \end{pmatrix} \tag{3.3c}$$

but functions $e_\pm(\lambda)$ in the time-dependent case are given by (XIV.6.4):

$$e_-(\lambda) = \frac{1}{\pi}\sqrt{\vartheta(\lambda)}\exp\{it\lambda^2 + i\lambda x\},$$
$$e_+(\lambda) = e_-(\lambda)E(\lambda). \tag{3.4}$$

The Fermi weight

$$\vartheta(\lambda) = \frac{1}{1 + e^{\lambda^2 - \beta}} \tag{3.5}$$

possesses the important properties

$$(2\lambda\partial_\beta + \partial_\lambda)\vartheta(\lambda) = 0, \quad \vartheta(\lambda)\Big|_{\lambda^2 - \beta = +\infty} = 0 \tag{3.6}$$

all the constructions below being valid in fact not only for $\vartheta(\lambda)$ (3.5) but also for arbitrary $\vartheta(\lambda)$ having properties (3.6). The function $E(\lambda)$ in (3.4) and also the function G, which will be used later, are defined in (XIV.6.5) and (XIV.6.6):

$$G \equiv G(x, t) \equiv \int\limits_{-\infty}^{+\infty} d\mu \, \exp\{-2it\mu^2 - 2ix\mu\} \tag{3.7}$$

$$E(\lambda) \equiv E(\lambda|x, t) \equiv \fint\limits_{-\infty}^{+\infty} \frac{d\mu}{\mu - \lambda} \exp\{-2it\mu^2 - 2ix\mu\}. \tag{3.8}$$

The solution $\chi(\lambda)$ and the matrix $G(\lambda)$ depend also on the parameters x, t, β but this dependence is as a rule suppressed in the notation.

In complete analogy with the equal-time case of section 1 one can show that the standard singular integral equations (0.6) for matrix RP (3.3) are equivalent to integral equations (XIV.6.7),

$$f_\pm(\lambda) + \int\limits_{-\infty}^{+\infty} V_T(\lambda, \mu)f_\pm(\mu)\,d\mu = e_\pm(\lambda) \tag{3.9}$$

$$\mathcal{V}_T(\lambda, \mu) = \frac{e_+(\lambda)e_-(\mu) - e_-(\lambda)e_+(\mu)}{\lambda - \mu} \tag{3.10}$$

which were the basic ones in constructing in section XIV.6 the integrable dynamical system for the correlator. Also, the potentials B_{lm} (XIV.6.9) and C_{lm} (XIV.6.10),

$$B_{lm} = \int\limits_{-\infty}^{+\infty} e_l(\mu)f_m(\mu)\,d\mu \qquad (l, m = +, -) \tag{3.11}$$

$$C_{lm} = \int\limits_{-\infty}^{+\infty} \mu e_l(\mu)f_m(\mu)\,d\mu \qquad (l, m = +, -) \tag{3.12}$$

are given by the corresponding matrix elements of the matrix coefficients $\Psi_1(x, t, \beta)$, $\Psi_2(x, t, \beta)$ in the expansion for $\lambda \to \infty$ of $\chi(\lambda)$ in powers of $1/\lambda$:

$$\chi(\lambda) = 1 + \frac{\Psi_1}{\lambda} + \frac{\Psi_2}{\lambda^2} + \cdots \tag{3.13}$$

$$\Psi_1 \equiv \begin{pmatrix} -B_{-+} & B_{++} \\ -B_{--} & B_{+-} \end{pmatrix} \tag{3.14}$$

$$\Psi_1 \equiv \begin{pmatrix} -C_{-+} & C_{++} \\ -C_{--} & C_{+-} \end{pmatrix}. \tag{3.15}$$

As $\det G(\lambda) = 1$ we have that

$$\det \chi(\lambda) = 1, \tag{3.16}$$

implying relation (XIV.6.14)

$$B_{+-}(x, t, \beta) = B_{-+}(x, t, \beta). \tag{3.17}$$

Complex conjugation involution (1.15) is not, however, valid so that the potentials B, C are not real.

The reduction of the singular integral equations (0.6) to the nonsingular integral equations (3.9) as well as the proof of the equality of the potentials B_{lm}, C_{lm} defined by (3.11), (3.12) to the corresponding potentials defined in (3.13)–(3.15) repeats almost exactly the corresponding derivation in section 1 (see (1.28)–(1.41)); only minor alterations are needed. So we leave this to the reader.

Let us also demonstrate that the matrix RP (3.3) is indeed related to the integrable system of section XIV.6 for correlator (3.1), reproducing the Lax representation (XIV.6.30) (and thus equations (XIV.6.44)–(XIV.6.55)) for the potentials. It is convenient to represent the solution $\chi(\lambda)$ of RP (3.3) in the form

$$\chi(\lambda) = \tilde{\chi}(\lambda)\chi_0(\lambda) \tag{3.18}$$

with triangular matrix $\chi_0(\lambda)$ given by

$$\chi_0(\lambda) = \begin{pmatrix} 1 & -\int\limits_{-\infty}^{+\infty} \dfrac{\exp\{-2it\mu^2 - 2ix\mu\}}{\mu - \lambda}\, d\mu \\ 0 & 1 \end{pmatrix}. \qquad (3.19)$$

It is easily seen that the matrix function $\widetilde{\chi}(\lambda)$ is the solution to the following matrix RP:

$$\widetilde{\chi}_-(\lambda) = \widetilde{\chi}_+(\lambda)\widetilde{G}(\lambda), \quad \operatorname{Im}\lambda = 0 \qquad (3.20a)$$

$$\widetilde{\chi}(\infty) = I, \qquad (3.20b)$$

$$\widetilde{G}(\lambda) = \begin{pmatrix} 1 - 2\vartheta(\lambda) & 2\pi i\,(\vartheta(\lambda) - 1)\, e^{-2it\lambda^2 - 2ix\lambda} \\ -\dfrac{2i}{\pi}\vartheta(\lambda)e^{2it\lambda^2 + 2ix\lambda} & 1 - 2\vartheta(\lambda) \end{pmatrix}.$$

$$(3.20c)$$

The advantage of RP (3.20) (as compared to the original problem (3.3)) is the simple dependence of $\widetilde{G}(\lambda)$ on the variables x, t.

In particular, for the function $\widetilde{\chi}_e(\lambda)$,

$$\widetilde{\chi}_e(\lambda) \equiv \widetilde{\chi}(\lambda) \exp\{-i(t\lambda^2 + x\lambda)\sigma_3\} \qquad (3.21)$$

the conjugation matrix $\widetilde{G}_e(\lambda)$,

$$\begin{aligned} \widetilde{\chi}_{e,-}(\lambda) &= \widetilde{\chi}_{e,+}(\lambda)\widetilde{G}_e(\lambda), \\ \widetilde{G}_e(\lambda) &= \exp\{i(t\lambda^2 + x\lambda)\sigma_3\}\,\widetilde{G}(\lambda)\,\exp\{-i(t\lambda^2 + x\lambda)\sigma_3\} \end{aligned} \qquad (3.22)$$

does not depend on x or t and the dependence in β and λ enters only through the function $\vartheta(\lambda)$, with properties (3.6), so that

$$\partial_x\widetilde{G}_e(\lambda) = 0, \quad \partial_t\widetilde{G}_e(\lambda) = 0, \quad (2\lambda\partial_\beta + \partial_\lambda)\widetilde{G}_e(\lambda) = 0. \qquad (3.23)$$

Thus it is established in the standard way (see (1.17)–(1.25)) that matrix-valued function $\widetilde{\chi}_e(\lambda)$ is a matrix solution of the linear system (the Lax representation) (XIV.6.30),

$$\widehat{\mathsf{L}}(\lambda)\widetilde{\chi}_e(\lambda) = 0, \qquad \widehat{\mathsf{M}}(\lambda)\widetilde{\chi}_e(\lambda) = 0, \qquad \widehat{\mathsf{N}}(\lambda)\widetilde{\chi}_e(\lambda) = 0 \qquad (3.24)$$

with operators $\widehat{\mathsf{L}}$, $\widehat{\mathsf{M}}$ and $\widehat{\mathsf{N}}$ given in (XIV.6.31)–(XIV.6.33).

It should be mentioned also that the correlator (3.1) is expressed in section XIV.6 in terms of potentials B_{+-}, B_{++}; the potential b_{++} (XIV.6.8),

$$b_{++} = B_{++} - G, \qquad (3.25)$$

simply related to B_{++}, is also often used. It is not difficult to obtain from (3.13), (3.14), (3.18) and (3.19) that these potentials are related to the matrix function $\widetilde{\chi}(\lambda)$ by

$$
\begin{aligned}
b_{++} &= \lim_{\lambda \to \infty} \lambda \widetilde{\chi}_{12}(\lambda), \\
B_{+-} &= - \lim_{\lambda \to \infty} \lambda \widetilde{\chi}_{21}(\lambda).
\end{aligned}
\tag{3.26}
$$

So we have constructed the matrix Riemann problem of the classical inverse scattering method for the integrable system in three dimensions (variables x, t, β) describing the correlator (3.1). All the results in this section were obtained assuming the uniqueness of solutions of this matrix RP. Later we shall prove this uniqueness for large x and t. It should be mentioned that in the case of positive chemical potential $\beta > 0$ regularization of the matrix RP (similar to the equal-time case) is needed. This is carried out in the next section.

XV.4 Representation of the temperature- and time-dependent two-field correlator for impenetrable bosons in terms of the matrix Riemann-Hilbert problem

It was shown in the previous section that the potentials B, C (in terms of which the correlator (3.1) was expressed in section XIV.6) can be obtained from the solution of the matrix RP. The aim of this section (similar in this respect to that of section 2) is to represent the matrix RP (3.20) (equivalent to original RP (3.3)) in a form convenient for calculating the large time and distance asymptotics in Chapter XVI. It will be seen that there exists a considerable difference between the equal-time and the time-dependent cases. The reason is that in the latter case one has to take into account properly the stationary point on the real axis

$$
\lambda_0 = -\frac{x}{2t}; \quad x > 0,\ t > 0,\ \lambda_0 < 0
\tag{4.1}
$$

of the phase

$$
\theta(x, t, \lambda) \equiv t\lambda^2 + x\lambda
\tag{4.2}
$$

in the exponent in the conjugation matrix $\widetilde{G}(\lambda)$ (3.20c) of the matrix RP (3.20), this being known in the classical theory of solitons to be the critical property for asymptotic analysis. We consider x and t to be positive (see the discussion in section XIV.6). The considerations are essentially different (as in section 2) in the cases $\beta < 0$ and $\beta > 0$.

Consider first the case of negative chemical potential. It is convenient to introduce a function $\Phi(\lambda)$,

$$
\Phi(\lambda) = \widetilde{\chi}(\lambda) \exp\{-\sigma_3 \ln \alpha(\lambda)\}
\tag{4.3}
$$

where $\alpha(\lambda)$ is defined exactly as in (2.6), at $\gamma = 2/\pi$:

$$
\alpha(\lambda) = \exp\left\{-\frac{1}{2\pi i} \int_{-\infty}^{+\infty} \frac{d\mu}{\mu - \lambda} \ln \varphi(\mu^2, \beta)\right\}, \qquad \beta < 0
\tag{4.4}
$$

and where the function φ is

$$\varphi(\mu^2, \beta) = \frac{e^{\mu^2 - \beta} - 1}{e^{\mu^2 - \beta} + 1}. \tag{4.5}$$

The function $\alpha(\lambda)$ is, in the case $\beta < 0$, the solution of the well defined scalar RP

$$\alpha_-(\lambda) = \alpha_+(\lambda)[1 - 2\vartheta(\lambda)] = \alpha_+(\lambda)\varphi(\lambda^2, \beta), \qquad \operatorname{Im}\lambda = 0 \tag{4.6a}$$

$$\alpha(\infty) = 1, \tag{4.6b}$$

since the function $\varphi(\lambda^2, \beta)$ possesses no zeros (or poles) on the real axis for $\beta < 0$.

The matrix RP is rewritten for $\Phi(\lambda)$ (4.3) as

$$\Phi_-(\lambda) = \Phi_+(\lambda)G_\Phi(\lambda, \{p, q\}), \qquad \operatorname{Im}\lambda = 0 \tag{4.7a}$$

$$\Phi(\infty) = 1 \tag{4.7b}$$

$$G_\Phi(\lambda, \{p, q\}) = \begin{pmatrix} 1 & p(\lambda)e^{-2i\theta} \\ q(\lambda)e^{2i\theta} & 1 + p(\lambda)q(\lambda) \end{pmatrix} \tag{4.7c}$$

where $\theta = \theta(x, t, \lambda) = \lambda^2 t + \lambda x$ (4.2) and the functions $p(\lambda)$ and $q(\lambda)$ are given by

$$p(\lambda) = -2\pi i [\alpha_-(\lambda)]^2 \frac{e^{\lambda^2 - \beta}}{e^{\lambda^2 - \beta} - 1}$$

$$q(\lambda) = -\frac{2i}{\pi}[\alpha_+(\lambda)]^{-2} \frac{1}{e^{\lambda^2 - \beta} - 1}. \tag{4.8}$$

Considering now the matrix coefficient Φ_1, Φ_2 of $\Phi(\lambda)$ in the expansion

$$\Phi(\lambda) = 1 + \frac{\Phi_1}{\lambda} + \frac{\Phi_2}{\lambda^2} + \cdots, \qquad \lambda \to \infty \tag{4.9}$$

one obtains using (3.13)–(3.15), (3.18) and (4.3) expressions for the potentials B, C in terms of Φ_1, Φ_2. We write down explicitly the expressions for the potentials that we shall need:

$$B_{+-} = B_{-+} = -\alpha_0 - (\Phi_1)_{11}$$

$$b_{++} = B_{++} - G = (\Phi_1)_{12}; \quad B_{--} = -(\Phi_1)_{21} \tag{4.10}$$

$$C_{+-} + C_{-+} + B_{--}G = (\Phi_2)_{22} - (\Phi_2)_{11}.$$

Here α_0 is the coefficient of $1/\lambda$ in the expansion

$$\alpha(\lambda) = 1 + \frac{\alpha_0}{\lambda} + \cdots \tag{4.11}$$

of the function $\alpha(\lambda)$. Explicitly:

$$\alpha_0 = \frac{1}{2\pi i} \int\limits_{-\infty}^{+\infty} d\mu \, \ln \varphi(\mu^2, \beta). \tag{4.12}$$

So the potentials are expressed in terms of the solution to the matrix RP (4.7) in the case $\beta < 0$.

Let us now turn to the case of positive chemical potential, $\beta > 0$. The transition to matrix RP (4.7) in this case cannot be realized directly since now the function $1 - e^{\lambda^2 - \beta}$ possesses zeros at $\pm\lambda_1$,

$$\lambda_1 = \sqrt{\beta} \tag{4.13}$$

on the real axis and the scalar RP (4.6) for the function $\alpha(\lambda)$ possesses indices. As in the equal-time case (section 2) we pass to the equivalent regularized formulation of the matrix RP, changing conjugation condition (3.20a) for the function $\widetilde{\chi}(\lambda)$ on the real axis for the conjugation condition on contour Γ shown in Figure XV.1. Besides the condition

$$0 < \epsilon < \text{Im} \sqrt{\beta + i\pi} \tag{4.14}$$

(used already in section 2) we demand also that the stationary point λ_0 and "critical points" $\pm\lambda_1$ (4.13) do not coincide, i.e.,

$$||\lambda_0| - \sqrt{\beta}| > \epsilon. \tag{4.15}$$

As was discussed in section 2, this change of integration contour does not change the potentials B, C and the Fredholm determinant, Δ. We shall go now to the matrix RP for $\Phi(\lambda)$ defined again by formula (4.3),

$$\Phi(\lambda) = \widetilde{\chi}(\lambda) \exp\{-\sigma_3 \ln \alpha(\lambda)\} \tag{4.16}$$

but with the function $\alpha(\lambda)$ given by

$$\alpha(\lambda) = \exp\left\{-\frac{1}{2\pi i} \int\limits_{\Gamma} \frac{d\mu}{\mu - \lambda} \ln \varphi(\mu^2, \beta)\right\}, \qquad \beta > 0. \tag{4.17}$$

This function $\alpha(\lambda)$ is now the solution to the well defined scalar RP on the contour Γ:

$$\alpha_-(\lambda) = \alpha_+(\lambda)[1 - 2\vartheta(\lambda)] = \alpha_+(\lambda)\varphi(\lambda^2, \beta), \quad \lambda \in \Gamma, \tag{4.18a}$$
$$\alpha(\infty) = 1. \tag{4.18b}$$

The matrix RP for $\Phi(\lambda)$ (4.16) is also formally given by formulæ (4.7)–(4.8), but with the conjugation contour changed to contour Γ and with the function $\alpha(\lambda)$ given by (4.17).

To use a matrix RP for calculating the asymptotics it is useful to make further transformations, similar to those made in section 2 in the equal-time case. However, now, in the case $\beta > 0$, one should distinguish between the two subcases, that is, between the "space-like"

$$\lambda_0 < -\sqrt{\beta}, \quad \beta > 0 \tag{4.19}$$

and "time-like"

$$-\sqrt{\beta} < \lambda_0 < 0, \quad \beta > 0 \tag{4.20}$$

regions.

Consider first the space-like region (4.19). Here, in complete correspondence with the equal-time case (2.30), (2.32) one represents the function $\Phi(\lambda)$ as

$$\Phi(\lambda) = (\lambda I + P) \, \mathring{\Phi}(\lambda) \begin{pmatrix} (\lambda - \lambda_1)^{-1} & 0 \\ 0 & (\lambda + \lambda_1)^{-1} \end{pmatrix}. \tag{4.21}$$

The function $\mathring{\Phi}(\lambda)$ here solves the matrix RP obtained from RP (4.7) (however, with conjugation contour Γ and function $\alpha(\lambda)$ given by (4.17)!) by changing functions p, q (given by (4.8) with $\alpha(\lambda)$ taken from (4.17)) for functions p_1, q_1:

$$\mathring{\Phi}_-(\lambda) = \mathring{\Phi}_+(\lambda) G_s(\lambda), \quad \lambda \in \Gamma \tag{4.22a}$$

$$\mathring{\Phi}(\infty) = I \tag{4.22b}$$

$$G_s(\lambda) \equiv G_\Phi(\lambda, \{p_1, q_1\}) \tag{4.22c}$$

where the functions $p_1(\lambda)$ and $q_1(\lambda)$ are given by

$$\begin{aligned} p_1(\lambda) &= p(\lambda) \frac{\lambda + \lambda_1}{\lambda - \lambda_1} \\ q_1(\lambda) &= q(\lambda) \frac{\lambda - \lambda_1}{\lambda + \lambda_1}. \end{aligned} \tag{4.23}$$

The matrix P is defined by the equations

$$\begin{aligned} (\lambda_1 I + P) \, \mathring{\Phi}(\lambda_1) \begin{pmatrix} 1 \\ 0 \end{pmatrix} &= 0, \\ (-\lambda_1 I + P) \, \mathring{\Phi}(-\lambda_1) \begin{pmatrix} 0 \\ 1 \end{pmatrix} &= 0 \end{aligned} \tag{4.24}$$

so that one obtains equations very similar to (2.37):

$$\begin{aligned} P_{11} &= -P_{22} = -\frac{\lambda_1}{d} \left(\mathring{\Phi}_{11}(\lambda_1) \mathring{\Phi}_{22}(-\lambda_1) + \mathring{\Phi}_{21}(\lambda_1) \mathring{\Phi}_{12}(-\lambda_1) \right) \\ P_{12} &= \frac{2\lambda_1}{d} \mathring{\Phi}_{11}(\lambda_1) \mathring{\Phi}_{12}(-\lambda_1) \\ P_{21} &= -\frac{2\lambda_1}{d} \mathring{\Phi}_{21}(\lambda_1) \mathring{\Phi}_{22}(-\lambda_1) \\ d &= \mathring{\Phi}_{22}(-\lambda_1) \mathring{\Phi}_{11}(\lambda_1) - \mathring{\Phi}_{12}(-\lambda_1) \mathring{\Phi}_{21}(\lambda_1) \\ & (\lambda_0 < -\sqrt{\beta}). \end{aligned} \tag{4.25}$$

The potentials B, C are again given in terms of the matrix $\Phi(\lambda)$ by formulæ (4.10), where it is presumed that contour Γ is used for $\Phi(\lambda)$, $\alpha(\lambda)$, so that one has the following expression for α_0:

$$\alpha_0 = \frac{1}{2\pi i} \int_\Gamma d\mu \, \ln \varphi(\mu^2, \beta)$$

$$= \frac{1}{2\pi i} \int_{-\infty}^{+\infty} d\mu \, \ln|\varphi(\mu^2, \beta)| - \lambda_1. \tag{4.26}$$

In terms of the function $\mathring{\Phi}(\lambda)$ expressions (4.10) are readily rewritten as

$$B_{+-} = -\alpha_0 - \lambda_1 - P_{11} + \mathring{B}_{+-}$$
$$b_{++} = B_{++} - G = P_{12} + \mathring{b}_{++}$$
$$B_{--} = -P_{21} + \mathring{B}_{--} \tag{4.27}$$
$$C_{+-} + C_{-+} + B_{--}G = P_{21}\mathring{b}_{++} + P_{12}\mathring{B}_{--} + \mathring{C}_{+-} + \mathring{C}_{-+}$$
$$(\lambda_0 < -\sqrt{\beta}).$$

The potentials $\mathring{B}$, $\mathring{b}$, $\mathring{C}$ are defined by the expansion of the solutions of the regularized RP:

$$\mathring{\Phi}(\lambda) = 1 + \frac{\mathring{\Phi}_1}{\lambda} + \frac{\mathring{\Phi}_2}{\lambda^2} + \cdots, \qquad \lambda \to \infty$$

$$\mathring{\Phi}_1 = \begin{pmatrix} -\mathring{B}_{+-} & \mathring{b}_{++} \\ -\mathring{B}_{--} & \mathring{B}_{+-} \end{pmatrix}, \quad \mathring{\Phi}_2 = \begin{pmatrix} -\mathring{C}_{-+} & \mathring{C}_{++} \\ -\mathring{C}_{--} & \mathring{C}_{+-} \end{pmatrix} \tag{4.28}$$

$$(\lambda_0 < -\sqrt{\beta}).$$

So in the space-like case the potentials B, C are expressed in terms of the solution $\mathring{\Phi}(\lambda)$ of the regularized RP (4.22).

Let us turn now to the time-like case (4.20). In this case one uses a different representation for $\Phi(\lambda)$:

$$\Phi(\lambda) = (\lambda I + P) \, \mathring{\Phi}(\lambda) \begin{pmatrix} (\lambda^2 - \lambda_1^2)^{-1} & 0 \\ 0 & 1 \end{pmatrix} \tag{4.29}$$

$$(-\sqrt{\beta} < \lambda_0 < 0).$$

Now the matrix $\mathring{\Phi}(\lambda)$ solves the matrix RP obtained from (4.7) (with conjugation contour Γ and function $\alpha(\lambda)$ (4.17)) by changing functions p, q for p_2, q_2 and also by changing the behavior for $\lambda \to \infty$:

$$\mathring{\Phi}_-(\lambda) = \mathring{\Phi}_+(\lambda)G_t(\lambda), \qquad \lambda \in \Gamma \tag{4.30a}$$
$$\mathring{\Phi}(\lambda) \sim \lambda^{\sigma_3}, \qquad \lambda \to \infty, \; \lambda \notin \Gamma \tag{4.30b}$$
$$G_t(\lambda) \equiv G_\Phi(\lambda, \{p_2, q_2\}) \tag{4.30c}$$

with functions p_2, q_2 given by

$$
\begin{aligned}
p_2(\lambda) &= p(\lambda)(\lambda^2 - \lambda_1^2)^{-1} \\
q_2(\lambda) &= q(\lambda)(\lambda^2 - \lambda_1^2).
\end{aligned}
\tag{4.31}
$$

The matrix P in (4.29) is defined by the equations

$$
(\pm\lambda_1 I + P)\,\mathring{\Phi}(\pm\lambda_1)\begin{pmatrix} 1 \\ 0 \end{pmatrix} = 0
\tag{4.32}
$$

so that one obtains for its matrix elements

$$
\begin{aligned}
P_{11} &= -P_{22} = -\frac{\lambda_1}{d}\left(\mathring{\Phi}_{11}(\lambda_1)\mathring{\Phi}_{21}(-\lambda_1) + \mathring{\Phi}_{11}(-\lambda_1)\mathring{\Phi}_{21}(\lambda_1)\right) \\
P_{12} &= \frac{2\lambda_1}{d}\mathring{\Phi}_{11}(\lambda_1)\mathring{\Phi}_{11}(-\lambda_1) \\
P_{21} &= -\frac{2\lambda_1}{d}\mathring{\Phi}_{21}(\lambda_1)\mathring{\Phi}_{21}(-\lambda_1) \\
d &\equiv \mathring{\Phi}_{21}(-\lambda_1)\mathring{\Phi}_{11}(\lambda_1) - \mathring{\Phi}_{11}(-\lambda_1)\mathring{\Phi}_{21}(\lambda_1) \\
&(-\sqrt{\beta} < \lambda_0 < 0).
\end{aligned}
\tag{4.33}
$$

Similarly to (4.27) one obtains for the potentials

$$
\begin{aligned}
B_{+-} &= -\alpha_0 - P_{11} + \mathring{B}_{+-} \\
b_{++} &= P_{12} + \mathring{b}_{++} \\
B_{--} &= -P_{21} + \mathring{B}_{--} \\
C_{+-} + C_{-+} + B_{--}G &= P_{21}\mathring{b}_{++} + P_{12}\mathring{B}_{--} - \lambda_1^2 + \mathring{C}_{+-} + \mathring{C}_{-+} \\
&(-\sqrt{\beta} < \lambda_0 < 0),
\end{aligned}
\tag{4.34}
$$

where now the potentials $\mathring{B}$, $\mathring{C}$ are defined from the expansion in $1/\lambda$ for $\lambda \to \infty$ of the function $\mathring{\Phi}(\lambda)$ as follows:

$$
\mathring{\Phi}(\lambda) = \left(1 + \frac{\mathring{\Phi}_1}{\lambda} + \frac{\mathring{\Phi}_2}{\lambda^2} + \cdots\right)\lambda^{\sigma_3}, \qquad \lambda \to \infty
$$

$$
\mathring{\Phi}_1 = \begin{pmatrix} -\mathring{B}_{+-} & \mathring{b}_{++} \\ -\mathring{B}_{--} & \mathring{B}_{+-} \end{pmatrix}, \qquad \mathring{\Phi}_2 = \begin{pmatrix} -\mathring{C}_{-+} & \mathring{C}_{++} \\ -\mathring{C}_{--} & \mathring{C}_{+-} \end{pmatrix}
\tag{4.35}
$$

$$
(-\sqrt{\beta} < \lambda_0 < 0).
$$

So in the time-like case the potentials B, C may also be expressed in terms of the solution to the regularized RP (4.30).

Equations (4.27), (4.34) for the potentials can be somewhat simplified, similarly as was done in section 2 for (2.45). To do this one notices that the function $\mathring{\Phi}_e(\lambda)$,

$$
\mathring{\Phi}_e(\lambda) = \mathring{\Phi}(\lambda)\exp\{-i\sigma_3\theta(x,t,\lambda)\},
\tag{4.36}
$$

in both the time-like and space-like cases satisfies the two equations

$$\mathring{L}(\lambda)\mathring{\Phi}_e(\lambda) = 0, \quad \mathring{M}(\lambda)\mathring{\Phi}_e(\lambda) = 0 \tag{4.37}$$

of the linear system (3.24) (XIV.6.30) where the operators $\mathring{L}$, $\mathring{M}$ are obtained by changing the potentials b_{++}, B_{--} in expressions (XIV.6.31), (XIV.6.32) for the operators $\hat{L}$, $\hat{M}$ for $\mathring{b}_{++}$, $\mathring{B}_{--}$. Using equations (4.37) one may check by direct calculation that the following identities hold true (similar to (2.52) for the equal-time case). In the space-like ($\lambda_0 < -\sqrt{\beta}$) case one has

$$\lambda_1 + P_{11} = -\frac{i}{2}\partial_x \ln d,$$

$$P_{21}\mathring{b}_{++} + P_{12}\mathring{B}_{--} = \frac{i}{2}\partial_t \ln d \tag{4.38}$$

$$(\lambda_0 < -\sqrt{\beta})$$

and in the time-like case

$$P_{11} = -\frac{i}{2}\partial_x \ln d$$

$$P_{21}\mathring{b}_{++} + P_{12}\mathring{B}_{--} - \lambda_1^2 = \frac{i}{2}\partial_t \ln d. \tag{4.39}$$

$$(-\sqrt{\beta} < \lambda_0 < 0).$$

Hence expressions (4.27), (4.34) can be written in the same form in both space- and time-like regions:

$$b_{++} = P_{12} + \mathring{b}_{++}$$

$$B_{--} = -P_{21} + \mathring{B}_{--}$$

$$B_{+-} = -\alpha_0 + \frac{i}{2}\partial_x \ln d + \mathring{B}_{+-} \tag{4.40}$$

$$C_{+-} + C_{-+} + B_{--}G = \frac{i}{2}\partial_t \ln d + \mathring{C}_{+-} + \mathring{C}_{-+}$$

$$(\beta > 0).$$

It should be remembered, of course, that the definitions of the quantities on the right hand side here depend on whether the region is space- or time-like (the quantity α_0 (4.26) remains the same).

Finally, let us turn to the main object of our interest, i.e., the correlator (3.1)

$$\langle \Psi(x_2, t_2)\Psi^\dagger(x_1, t_1)\rangle_T = \sqrt{T}g(x, t, \beta), \tag{4.41}$$

$$g(x, t, \beta) = -\frac{1}{2\pi}\exp\{2it\beta\}b_{++}\exp\{\sigma(x, t, \beta)\}$$

where $\sigma(x, t, \beta)$ is the logarithm of the Fredholm determinant (see

(XIV.6.2), (XIV.6.57)). As (see (XIV.6.58), (XIV.6.59))

$$\partial_x \sigma = -2iB_{+-} \tag{4.42}$$

$$\partial_t \sigma = -2i(C_{+-} + C_{-+} + GB_{--}) \tag{4.43}$$

one obtains the following representation for the correlator in the case $\beta > 0$:

$$g(x, t, \beta) = -\frac{1}{2\pi} \exp\{2it\beta\} \exp\{\sigma_0\} \left(m + d\mathring{b}_{++}\right) \quad (\beta > 0) \tag{4.44}$$

where

$$m = P_{12}d \tag{4.45}$$

and the function $\sigma_0(x, t, \beta)$ is defined by the equality

$$\sigma(x, t, \beta) \equiv \ln d(x, t, \beta) + \sigma_0(x, t, \beta) \tag{4.46}$$

so that

$$\partial_x \sigma_0 = 2i\alpha_0 - 2i\mathring{B}_{+-}, \tag{4.47}$$

$$\partial_t \sigma_0 = -2i\left(\mathring{C}_{+-} + \mathring{C}_{-+}\right). \tag{4.48}$$

This representation will be used to calculate the large time and distance asymptotics of the correlator in case of positive chemical potential $\beta > 0$. In the case $\beta < 0$, where no regularization of the matrix RP is needed, one may use directly the original representation (4.41).

XV.5 Matrix problem for the multifield equal-time correlator for the impenetrable Bose gas

Here the matrix Riemann problem for the multipoint equal-time temperature-dependent correlators for impenetrable bosons is presented. The multidimensional integrable system of partial differential equations for these correlators was constructed in section XVI.4. It is supposed here that the reader is acquainted with the contents of that section.

The matrix Riemann problem in the case of the multipoint correlator is obtained as a natural generalization of the matrix RP in the two-point case considered in much detail in section 1. Let us consider the $2N$-point correlator (XIV.4.2)

$$\langle \Psi^\dagger(z_1^+) \cdots \Psi^\dagger(z_N^+) \Psi(z_1^-) \cdots \Psi(z_N^-) \rangle_T = T^{N/2} G_N(\{x^+\}, \{x^-\}, \beta) \tag{5.1}$$

where

$$x_j^\pm \equiv z_j^\pm \sqrt{T}, \qquad \beta = \frac{h}{T} \tag{5.2}$$

are normalized distances and chemical potential. The variables x_j,

$$\{x_j; \ j = 1, \ldots, 2N\} = \{x_k^+; \ k = 1, \ldots, N\} \bigcup \{x_l^-; \ l = 1, \ldots, N\},$$
$$x_1 \leq x_2 \leq x_3 \leq \cdots \leq x_{2N},$$

$$(5.3)$$

were introduced in (XIV.4.13) and the determinant representation of the correlator G_N (5.1) as a function of x_j and β was given in (XIV.4.14).

Consider the following matrix RP. One has to find a $2N \times 2N$ ($N \geq 2$) matrix-valued function $\chi(\lambda)$ which is holomorphic for $\text{Im} \, \lambda > 0$ and for $\text{Im} \, \lambda < 0$ and

$$\chi_-(\lambda) = \chi_+(\lambda) G(\lambda, \{p, q\}), \qquad \text{Im} \, \lambda = 0 \qquad (5.4a)$$
$$\chi(\infty) = I. \qquad (5.4b)$$

Here I is the $2N \times 2N$ unit matrix and the conjugation matrix $G(\lambda)$ is given by ($\gamma = 2/\pi$)

$$G_{mn}(\lambda) = \delta_{mn} + (-1)^n \pi \gamma e_m^+(\lambda) e_n^-(\lambda), \quad m, n = 1, \ldots, 2N. \qquad (5.5)$$

The matrices $\chi(\lambda)$, $G(\lambda)$ depend also on the variables x_n, β and γ (for correlator (5.1) $\gamma = 2/\pi$, see (XIV.4.14)). The "plane waves" $e^\pm(\lambda)$ are

$$e_m^\pm(\lambda) = \sqrt{\vartheta(\lambda)} e^{\pm i \lambda x_m}. \qquad (5.6)$$

The function $\vartheta(\lambda)$,

$$\vartheta(\lambda) = \frac{1}{1 + e^{\lambda^2 - \beta}} \qquad (5.7)$$

possesses the properties

$$(2\lambda \partial_\beta + \partial_\lambda) \vartheta(\lambda) = 0; \qquad \vartheta(\lambda) \Big|_{\lambda^2 - \beta \to +\infty} = 0. \qquad (5.8)$$

First, let us demonstrate that (5.4) is indeed the matrix RP of the inverse scattering method for the integrable nonlinear system (XIV.4.22)–(XIV.4.25). As in the two-point case, one begins by obtaining the Lax representation (XIV.4.20), (XIV.4.21). The analogue of the function $\Psi(\lambda)$ (1.17) is now

$$\Psi(\lambda) = \chi(\lambda) e^{i\lambda J} \qquad (5.9)$$

where J is a diagonal $2N \times 2N$ matrix given by

$$J_{mn} = \delta_{mn} x_n \qquad (5.10)$$

(no summation over n is implied). The conjugation matrix $G_0(\lambda)$ of the matrix RP for $\Psi(\lambda)$ (to be compared with (1.19)) is now

$$G_0(\lambda) = e^{-iJ\lambda} G(\lambda) e^{iJ\lambda} \qquad (5.11)$$

with matrix elements

$$(G_0(\lambda))_{mn} = \delta_{mn} + (-1)^n \pi \gamma \vartheta(\lambda). \tag{5.12}$$

The matrix $G_0(\lambda)$ does not depend on x,

$$\partial_n G_0(\lambda) = 0, \quad \partial_n \equiv \frac{\partial}{\partial x_n}. \tag{5.13}$$

The dependence on chemical potential β enters only through the function $\vartheta(\lambda)$; due to (5.8)

$$(2\lambda \partial_\beta + \partial_\lambda) G_0(\lambda) = 0. \tag{5.14}$$

Using the standard arguments (see (1.20)–(1.25)) one arrives at the Lax representation

$$\partial_n \Psi(\lambda) = (i\lambda J_n + U_n) \Psi(\lambda) \tag{5.15}$$

$$(2\lambda \partial_\beta + \partial_\lambda)\Psi(\lambda) = V_0 \Psi(\lambda) \tag{5.16}$$

with λ-independent matrices J_n, U_n and V_0 given by

$$(J_n)_{ml} = \delta_{nm}\delta_{ml} \tag{5.17}$$

$$U_n = i[\Psi_1, J_n], \quad V_0 = iJ + 2\partial_\beta \Psi_1. \tag{5.18}$$

Here $\Psi_1(x, \beta, \gamma)$ is the coefficient of $1/\lambda$ in the expansion

$$\Psi(\lambda) = \left(1 + \frac{\Psi_1}{\lambda} + \cdots\right) e^{i\lambda J}, \quad \lambda \to \infty. \tag{5.19}$$

Introducing the notation

$$V_{lm} \equiv 2i\,(\Psi_1)_{ml} \tag{5.20}$$

one represents the matrix elements of U_n, V_0 in terms of "potentials" V_{lm}:

$$\begin{aligned}(U_n)_{lm} &= -\frac{1}{2}\delta_{nl}V_{ml} + \frac{1}{2}V_{nl}\delta_{nm} \\ (V_0)_{lm} &= ix_l\delta_{lm} - i\partial_\beta V_{ml}\end{aligned} \tag{5.21}$$

(no summation over repeated indices).

So we have reproduced the Lax representation (XIV.4.20), (XIV.4.21). Calculations similar to those in previous sections permit us to reproduce all the results of section XIV.4. Details of the calculations are presented in the appendix of paper [7], see also [8]. So we have shown that the problem of the evaluation of multifield correlation functions can also be solved by means of the Riemann-Hilbert problem.

XV.6 The operator Riemann-Hilbert problem
for a correlator with finite coupling constant

Here, we consider the simplest equal-time correlator for the Bose gas with finite coupling constant $(0 < c < \infty)$ (which corresponds to interacting fermions), i.e., the "emptiness formation probability" (XIV.7.1)–(XIV.7.3) at zero temperature:

$$P(x_2, x_1) = \frac{\langle \Omega | \exp \left\{ \alpha \int_{x_1}^{x_2} \Psi^\dagger(z) \Psi(z) \, dz \right\} | \Omega \rangle}{\langle \Omega | \Omega \rangle} \Bigg|_{\alpha = -\infty} \qquad (6.1)$$

$$= P(x), \qquad x \equiv \frac{1}{2} |x_1 - x_2|.$$

In section XIV.7 the integrable nonlinear "integro-differential" system (XIV.7.28)–(XIV.7.33) describing this correlator was given. In the present section the Riemann-Hilbert (RP) problem corresponding to this integrable nonlinear system is presented. The latter is a system of integro-differential equations, which is why the corresponding RP is an integral-operator-valued problem ($\chi(\lambda)$ is now an integral operator, but not a matrix as before). The constructions of this section are quite natural generalizations of the corresponding constructions for the correlators of impenetrable bosons $(c = +\infty)$ given in the previous sections of this chapter, see particularly section 1. It is, however, to be mentioned that here we consider the case of zero temperature; the generalization to the nonzero temperature case is also obvious.

Consider the integral-operator-valued function $G(\lambda)$ for real λ

$$\widehat{G}(\lambda) = \widehat{I} + \theta(q^2 - \lambda^2)\widehat{g}(\lambda), \quad \lambda \in \mathbb{R} \qquad (6.2)$$

where $\widehat{I}$ is the unit operator with matrix kernel

$$I(s, s') = \begin{pmatrix} \delta(s - s') & 0 \\ 0 & \delta(s - s') \end{pmatrix} e^{sc} \qquad (6.3)$$

and the matrix kernel $g(\lambda|s, s')$ of integral operator $\widehat{g}(\lambda)$ is given by

$$g(\lambda|s, s') = 2\pi \begin{pmatrix} e_+(\lambda|s)e_-(\lambda|s') & -e_+(\lambda|s)e_+(\lambda|s') \\ e_-(\lambda|s)e_-(\lambda|s') & -e_-(\lambda|s)e_+(\lambda|s') \end{pmatrix} \qquad (6.4)$$

with the functions $e_\pm(\lambda|s)$ defined by (XIV.7.9),

$$e_\pm(\lambda|s) = \sqrt{\frac{c}{2\pi}} \exp \left\{ ix\lambda + is\lambda + \frac{\varphi(\lambda)}{2} \right\} \qquad (6.5)$$

($\varphi(\lambda)$ is the auxiliary quantum field, see (XI.1.24)). The function $\theta(\lambda^2 - q^2)$ in (6.2) is the Heaviside function (Im $\lambda = 0$)

$$\theta(\lambda^2 - q^2) = 0 \ \text{ if } \ \lambda^2 - q^2 < 0; \qquad \theta(\lambda^2 - q^2) = 1 \ \text{ if } \ \lambda^2 - q^2 > 0.$$

All the operators are integral operators in $L_2(\mathbb{R}^+, e^{-sc})$ acting on functions $f(s)$ defined on the positive semiaxis of s with weight e^{-sc}:

$$\left(\widehat{G} f\right)(s) = \int\limits_0^\infty ds'\, e^{-s'c} G(s, s') f(s').$$

The operator $\widehat{G}(\lambda)$ can be considered as a "conjugation matrix" for an infinite-dimensional Riemann-Hilbert problem: that of constructing an integral-operator-valued function $\chi(\lambda)$ with the following properties (remember that $\chi(\lambda)$ is now not a matrix, it is an integral operator):

(1) $\chi(\lambda)$ is holomorphic in λ for $\mathrm{Im}\,\lambda > 0$ and $\mathrm{Im}\,\lambda < 0$;

(2) the boundary values

$$\chi_\pm(\lambda) = \lim_{\epsilon \to 0^+} \chi(\lambda \pm i\epsilon), \quad \mathrm{Im}\,\lambda = 0$$

on the real axis are related by

$$\chi_-(\lambda) = \chi_+(\lambda)\widehat{G}(\lambda); \tag{6.6a}$$

(3) for $\lambda \to \infty$, $\lambda \in \mathbb{C}$, $\chi(\lambda)$ is equal to the unit operator

$$\chi(\infty) = \widehat{I}. \tag{6.6b}$$

In terms of the corresponding kernels these properties can be rewritten in the following way:

(1) The kernel $\chi(\lambda|s, s')$ is an holomorphic function of λ in the complex half-planes $\mathrm{Im}\,\lambda > 0$ and $\mathrm{Im}\,\lambda < 0$.

(2) The boundary values on the real λ axis are related by

$$\chi_-(\lambda|s, s') = \int\limits_0^\infty ds''\, e^{-s''c} \chi_+(\lambda|s, s'') G(\lambda|s'', s)$$

$$= \chi_+(\lambda|s, s') + \theta(q^2 - \lambda^2) \int\limits_0^\infty ds''\, e^{-s''c} \chi_+(\lambda|s, s'') g(\lambda|s'', s'). \tag{6.7a}$$

(3)

$$\chi(\infty|s, s') = I(s, s') = \begin{pmatrix} 1 & 0 \\ 0 & 1 \end{pmatrix} \delta(s - s') e^{sc}. \tag{6.7b}$$

It is to be noted that for $\lambda \to \infty$ the operator function $\chi(\lambda)$ is expanded in powers of $1/\lambda$,

$$\chi(\lambda) = \widehat{I} + \frac{\Psi_1}{\lambda} + \frac{\Psi_2}{\lambda^2} + \cdots;$$

$$\chi(\lambda|s, s') = I(s, s') + \frac{\Psi_1(s, s')}{\lambda} + \frac{\Psi_2(s, s')}{\lambda^2} + \cdots. \tag{6.8}$$

Suppose that the Riemann-Hilbert problem has a unique solution. Then we show that the function $\Psi(\lambda)$,

$$\Psi(\lambda) = \chi(\lambda)E(\lambda),$$
$$E(\lambda|s, s') = \delta(s - s') \exp\{i\lambda(x + s')\sigma_3\}e^{sc} \qquad (6.9)$$

is the integral-operator-valued solution of the linear system (Lax representation) (XIV.7.20), (XIV.7.24). We begin by reproducing the "x-equation" (XIV.7.20). Applying the standard arguments, used already in section 1, based on the Liouville theorem and on the x-independence of the conjugation operator $\widehat{G}_0(\lambda)$ for $\Psi(\lambda)$,

$$\Psi_-(\lambda) = \Psi_+(\lambda)\widehat{G}_0(\lambda),$$
$$\widehat{G}_0(\lambda) = E^{-1}(\lambda)\,\widehat{G}(\lambda)\,E(\lambda) \qquad (6.10)$$

one obtains that the "logarithmic derivative"

$$(\partial_x\Psi(\lambda))\,\Psi^{-1}(\lambda)$$

is analytic on the whole complex λ-plane and is a first order polynomial in λ,

$$(\partial_x\Psi(\lambda))\,\Psi^{-1}(\lambda) = i\lambda\widehat{I}_0 + U_0 \qquad (6.11)$$

with kernels I_0, U_0 (λ-independent) easily calculated by means of (6.9), (6.8):

$$I_0(s, s') = \delta(s - s')e^{sc}\sigma_3, \qquad (6.12)$$
$$U_0(s, s') = i[\Psi_1(s, s'), \sigma_3]. \qquad (6.13)$$

Introducing the following notation for matrix elements of the 2×2 matrix Ψ_1:

$$\left[\Psi_1(s, s')\right]_{12} \equiv iB_{++}(s, s')$$
$$\left[\Psi_1(s, s')\right]_{21} \equiv -iB_{--}(s, s') \qquad (6.14)$$

one calculates

$$U_0(s, s') = 2Q(s, s'),$$
$$Q(s, s') \equiv \begin{pmatrix} 0 & B_{++}(s, s') \\ B_{--}(s, s') & 0 \end{pmatrix} \qquad (6.15)$$

so that (6.11) is rewritten in the form

$$\partial_x\Psi(\lambda|s, s') = i\lambda\sigma_3\Psi(\lambda|s, s')$$
$$+ 2\int_0^\infty e^{-s''c}Q(s, s'')\Psi(\lambda|s'', s')\,ds''. \qquad (6.16)$$

Thus equation (XIV.7.20) is reproduced for the function $\Psi(\lambda|s, s')$. It is to be emphasized that here the potentials B_{++}, B_{--} are introduced as matrix elements of the matrix Ψ_1 (6.14) and not by formulæ (XIV.7.16). In what follows we prove that both definitions coincide and that we have the symmetry

$$B_{++}(s, s') = B_{++}(s', s), \quad B_{--}(s, s') = B_{--}(s', s).$$

But before doing this let us reproduce also equation (XIV.7.24) for the q-derivative of Ψ.

To do this it is convenient to consider the integral-operator-valued function $\Psi_0(\lambda)$ with kernel given explicitly by

$$\Psi_0(\lambda|s, s') = \left[I(s, s') + \frac{ic}{2\pi} \ln \frac{\lambda + q}{\lambda - q} \begin{pmatrix} 0 & 1 \\ 0 & 0 \end{pmatrix} \right] \tag{6.17}$$
$$\times \begin{pmatrix} e^{-\varphi(\lambda)/2} & 0 \\ -e^{-\varphi(\lambda)/2} & e^{\varphi(\lambda)/2} \end{pmatrix}.$$

It is easily verified that for all real λ the following conjugation condition holds:

$$\Psi_{0,-}(\lambda) = \Psi_{0,+}(\lambda) \widehat{G}_0(\lambda), \quad \operatorname{Im} \lambda = 0 \tag{6.18}$$

with the same conjugation matrix $\widehat{G}_0(\lambda)$ (6.10) as for $\Psi(\lambda)$. So one can represent the integral operator $\Psi(\lambda)$ in the neighbourhood of the interval $[-q, q]$ in the form

$$\Psi(\lambda) = \widehat{\Psi}(\lambda) \Psi_0(\lambda) \tag{6.19}$$

where $\widehat{\Psi}(\lambda)$ is single-valued and analytic in that neighbourhood. Note that the representation (6.19) is just the infinite-dimensional analogue of formulæ in paper [9]. Just as was done in that paper one concludes from (6.19) that

$$\left(i\partial_q \Psi(\lambda) \right) \Psi^{-1}(\lambda) = \frac{A_+}{\lambda - q} + \frac{A_-}{\lambda + q} \tag{6.20}$$

where

$$A_{\pm} = i \lim_{\lambda \to \pm q} (\lambda \mp q) \widehat{\Psi}(\lambda) \, \partial_q \Psi_0(\lambda) \Psi_0^{-1}(\lambda) \widehat{\Psi}^{-1}(\lambda). \tag{6.21}$$

Note that the kernels of the integral operators $\Psi_0^{-1}(\lambda)$ and $\Psi^{-1}(\lambda)$ are given by

$$\Psi_0^{-1}(\lambda|s, s') = \begin{pmatrix} e^{\varphi(\lambda)/2} & 0 \\ e^{-\varphi(\lambda)/2} & e^{-\varphi(\lambda)/2} \end{pmatrix}$$
$$\times \left[I(s, s') - \frac{ic}{2\pi} \ln \frac{\lambda + q}{\lambda - q} \begin{pmatrix} 0 & 1 \\ 0 & 0 \end{pmatrix} \right] \tag{6.22}$$

$$\Psi^{-1}(\lambda|s,s') = \begin{pmatrix} \Psi_{22}(\lambda|s',s) & -\Psi_{12}(\lambda|s',s) \\ -\Psi_{21}(\lambda|s',s) & \Psi_{11}(\lambda|s',s) \end{pmatrix}. \tag{6.23}$$

Equality (6.22) is obvious; equality (6.23) follows from the identity

$$\widehat{G}^{-1}(\lambda) = \widehat{I} - \theta(q^2 - \lambda^2)\widehat{g}(\lambda). \tag{6.24}$$

In turn, this identity is a direct corollary of the important relation

$$\widehat{g}^2(\lambda) = 0. \tag{6.25}$$

Note also that one has from (6.22):

$$\left[(\partial_q \Psi_0(\lambda))\,\Psi_0^{-1}(\lambda)\right](s,s') = \frac{ic}{2\pi}\left(\frac{1}{\lambda - q} + \frac{1}{\lambda + q}\right)\begin{pmatrix} 0 & 1 \\ 0 & 0 \end{pmatrix}. \tag{6.26}$$

Returning now to integral operators $A_\pm(\lambda)$ (6.21) one obtains from (6.22), (6.23), (6.26) that

$$
\begin{aligned}
A_\pm(s,s') &= -\frac{c}{2\pi}\int_0^\infty\int_0^\infty ds''ds''' \, e^{-(s''+s''')c}\,\widehat{\Psi}(\pm q|s',s'') \\
&\quad \times \begin{pmatrix} 0 & 1 \\ 0 & 0 \end{pmatrix}\widehat{\Psi}^{-1}(\pm q|s''',s') \\
&= -\frac{c}{2\pi}\int_0^\infty ds'' \, e^{-s''c}\,\widehat{\Psi}(\pm q|s,s'')\begin{pmatrix} 0 & 1 \\ 0 & 0 \end{pmatrix} \\
&\quad \times \int_0^\infty ds''' \, e^{-s'''c}\,\widehat{\Psi}^{-1}(\pm q|s''',s') \\
&= -\frac{c}{2\pi}\int_0^\infty ds'' \, e^{-s''c}\,\Psi(\pm q|s,s'')\begin{pmatrix} e^{\varphi(\pm q)/2} & 0 \\ e^{-\varphi(\pm q)/2} & 0 \end{pmatrix} \\
&\quad \times \begin{pmatrix} -e^{-\varphi(\pm q)/2} & e^{\varphi(\pm q)/2} \\ 0 & 0 \end{pmatrix}\int_0^\infty ds''' \, e^{-s'''c}\,\Psi^{-1}(\pm q|s''',s').
\end{aligned}
\tag{6.27}
$$

We introduce the notation

$$f_+(\lambda|s) \equiv \int_0^\infty ds' \, e^{-s'c}\left[\chi_{11}(\lambda|s,s')e_+(\lambda|s') + \chi_{12}(\lambda|s,s')e_-(\lambda|s')\right]$$

$$f_-(\lambda|s) \equiv \int_0^\infty ds' \, e^{-s'c}\left[\chi_{21}(\lambda|s,s')e_+(\lambda|s') + \chi_{22}(\lambda|s,s')e_-(\lambda|s')\right].$$

$$\tag{6.28a}$$

Using the explicit form of the conjugation matrix $\widehat{G}(\lambda)$ (6.2) and relations (6.6a), (6.7a) it is not difficult to see that the left hand sides here possess in fact no cuts along the real λ axis, so that, for $\operatorname{Im} \lambda = 0$, one can also write

$$f_+(\lambda|s) \equiv \int_0^\infty ds'\, e^{-s'c} \left[\chi_{11,+}(\lambda|s,s')e_+(\lambda|s') + \chi_{12,+}(\lambda|s,s')e_-(\lambda|s')\right]$$

$$f_-(\lambda|s) \equiv \int_0^\infty ds'\, e^{-s'c} \left[\chi_{21,+}(\lambda|s,s')e_+(\lambda|s') + \chi_{22,+}(\lambda|s,s')e_-(\lambda|s')\right].$$

$$(6.28b)$$

In this sense these formulæ are completely analogous to (1.43).

Now one obtains from (6.27) the following formulæ for $A_\pm$:

$$A_\pm(s,s') = \begin{pmatrix} f_+(\pm q|s)f_-(\pm q|s') & -f_+(\pm q|s)f_+(\pm q|s') \\ f_-(\pm q|s)f_-(\pm q|s') & -f_-(\pm q|s)f_+(\pm q|s') \end{pmatrix} \qquad (6.29)$$

which is exactly (XIV.7.26), (XIV.7.27) (it is, of course, to be remembered that now the functions $f_\pm$ are defined by (6.28a) and not by (XIV.7.14); the equivalence of these definitions will be shown later). So equation (XIV.7.24) is also reproduced (see (6.20), (6.29)).

The reconstruction of the Lax representation (XIV.7.20), (XIV.7.24) from the Riemann-Hilbert problem (6.6) is finished. It is, of course, evident that the compatibility conditions for equations (6.16), (6.20) are the same equations (XIV.7.28)–(XIV.7.33) for the "potentials" B_{++}, B_{--} and the function $f_\pm(\pm q|s)$.

Our last task is to show that the formulæ (6.14), (6.28a) give the same potentials B and functions $f_\pm(\lambda)$ as (XIV.7.14) and (XIV.7.16),

$$f_\pm(\lambda|s) + \int_{-q}^{q} \mathcal{V}(\lambda,\mu)f_\pm(\mu|s)\, d\mu = e_\pm(\lambda|s) \qquad (6.30)$$

$$B_{lm}(s,s') = \int_{-q}^{q} e_l(\mu|s)f_m(\mu|s')\, d\mu, \quad l,m = +,- \qquad (6.31)$$

by which these quantities were defined in section XIV.7.

To do this let us consider the singular integral equation (see (0.6)) equivalent to RP (6.6),

$$\chi_+(\lambda) = \widehat{I} - \frac{1}{2\pi i} \int_{-q}^{q} \frac{\chi_+(\mu)\widehat{g}(\mu)}{\mu - \lambda - i0}\, d\mu. \qquad (6.32)$$

Set

$$\Phi_+(l) \equiv \chi_+(\lambda)T(\lambda) \qquad (6.33)$$

with

$$T(\lambda|s, s') = \tau(\lambda|s)\delta(s - s')e^{s'c} \qquad (6.34)$$

$$\tau(\lambda|s) \equiv \begin{pmatrix} 1 & e_+(\lambda|s) \\ 0 & e_-(\lambda|s) \end{pmatrix}.$$

Equations (6.32) are rewritten as

$$\Phi_+(\lambda) = T(\lambda) - \frac{1}{2\pi i} \int\limits_{-q}^{q} \frac{\Phi_+(\mu)T^{-1}(\mu)\widehat{g}(\mu)T(\lambda)}{\mu - \lambda - i0}\, d\mu. \qquad (6.35)$$

Note that

$$\left(T^{-1}(\mu)\widehat{g}(\mu)T(\lambda)\right)(\mu, \lambda|s, s')$$
$$= \tau^{-1}(\mu|s)g(\mu|s, s')\tau(\lambda|s') \qquad (6.36)$$
$$= 2\pi \begin{pmatrix} 0 & 0 \\ e_-(\mu|s') & e_-(\mu|s')e_+(\lambda|s') - e_+(\mu|s')e_-(\lambda|s') \end{pmatrix}.$$

So for matrix $(\Phi_+)_{ij}(s, s') \equiv \Phi_{ij}^+(s, s')$ of integral operator $\Phi_+(s, s')$ one obtains

$$\Phi_{11}^+(\lambda|s, s') = \delta(s - s')e^{s'c}$$
$$+ i \int\limits_{-q}^{q} \frac{d\mu}{\mu - \lambda - i0} \int\limits_{0}^{\infty} ds''\, e^{-s''c}\, \Phi_{12}^+(\mu|s, s'')e_-(\mu|s'),$$

$$\Phi_{12}^+(\lambda|s, s') = e_+(\lambda|s)\delta(s - s')e^{s'c}$$
$$+ i \int\limits_{-q}^{q} \frac{d\mu}{\mu - \lambda} \int\limits_{0}^{\infty} ds''\, e^{-s''c}\, \Phi_{12}^+(\mu|s, s'')$$
$$\times \left[e_-(\mu|s')e_+(\lambda|s') - e_+(\mu|s')e_-(\lambda|s')\right],$$

$$\Phi_{21}^+(\lambda|s, s') = i \int\limits_{-q}^{q} \frac{d\mu}{\mu - \lambda - i0} \int\limits_{0}^{\infty} ds''\, e^{-s''c}\, \Phi_{22}^+(\mu|s, s'')e_-(\mu|s'),$$

$$\Phi_{22}^+(\lambda|s, s') = e_-(\lambda|s)\delta(s - s')e^{s'c}$$
$$+ i \int\limits_{-q}^{q} \frac{d\mu}{\mu - \lambda} \int\limits_{0}^{\infty} ds''\, e^{-s''c}\, \Phi_{22}^+(\mu|s, s'')$$
$$\times \left[e_-(\mu|s')e_+(\lambda|s') - e_+(\mu|s')e_-(\lambda|s')\right].$$

$$(6.37)$$

Introduce now the functions

$$f_{ij}(\lambda|s) = \int_0^\infty ds'\, e^{-s'c}\, \Phi_{ij}^+(\lambda|s,s').$$ (6.38)

Then one obtains from (6.37) the following scalar integral equations:

$$f_{12}(\lambda|s) = e_+(\lambda|s) - \int_{-q}^q d\mu\, \mathcal{V}(\lambda,\mu) f_{12}(\mu|s),$$

(6.39)

$$f_{22}(\lambda|s) = e_-(\lambda|s) - \int_{-q}^q d\mu\, \mathcal{V}(\lambda,\mu) f_{22}(\mu|s),$$

where

$$\mathcal{V}(\lambda,\mu) = i\int_0^\infty ds\, e^{-sc}\left[\frac{e_+(\lambda|s)e_-(\mu|s) - e_+(\mu|s)e_-(\lambda|s)}{\lambda-\mu}\right]$$ (6.40)

is the basic integral linear operator (XIV.7.8) entering equations (6.30) for the functions $f_\pm(\lambda|s)$. Thus

$$f_{12}(\lambda|s) = f_+(\lambda|s); \quad f_{22}(\lambda|s) = f_-(\lambda|s).$$ (6.41)

Taking into account the definition (6.33) of the function $\Phi_+(\lambda)$, one sees that (6.41) coincides with (6.28a), i.e., that (6.30) and (6.28a), (6.28b) are equivalent. It is also clear now that indeed (6.28a) is analogous to (1.43).

To prove that the potentials B_{++}, B_{--} defined in (6.14) are indeed the same as were introduced in section XIV.7 by formulæ (XIV.7.16), one notices first that the matrix coefficient Ψ_1 in (6.8) can be expressed in the form (see (6.32))

$$\Psi_1(s,s') = \frac{1}{2\pi i}\int_{-q}^q d\mu \int_0^\infty ds''\, e^{-s''c}\chi_+(\mu|s,s'')g(\mu|s'',s').$$ (6.42)

Therefore one has for the corresponding matrix elements the following expressions in terms of the potentials B (6.31):

$$\left[\Psi_1(s,s')\right]_{11} = -i\int_{-q}^q d\mu\, e_-(\mu|s')f_+(\mu|s) = -iB_{-+}(s',s),$$

$$\left[\Psi_1(s,s')\right]_{12} = i\int_{-q}^q d\mu\, e_+(\mu|s')f_+(\mu|s) = iB_{++}(s',s),$$

$$\left[\Psi_1(s,s')\right]_{21} = -i \int_{-q}^{q} d\mu \, e_-(\mu|s')f_-(\mu|s) = -iB_{--}(s',s),$$

$$\left[\Psi_1(s,s')\right]_{22} = i \int_{-q}^{q} d\mu \, e_+(\mu|s')f_-(\mu|s) = +iB_{+-}(s',s).$$

$$(6.43)$$

It has already been shown that the potentials defined in (6.31) are symmetric (see (XIV.7.17)). In the framework of RP it is a direct corollary of the symmetry (6.40) resulting in the relation

$$B_{lm}(s,s') = B_{ml}(s',s). \tag{6.44}$$

(It is explained in more detail in Chapter XIV, section 7, formulæ (XIV.7.12), (XIV.7.17).) So all the potentials B are the same as in section XIV.7.

Thus the Riemann-Hilbert problem in the case of finite coupling constant

$$0 < c < \infty$$

is presented.

This shows that by our method we can calculate correlation functions for any Bethe Ansatz solvable model. We have done it for finite coupling, which means interacting fermions.

Conclusion

Here the Riemann-Hilbert problem was considered. It was used for the solution of classical completely integrable differential equations in [1], [10], [11], [12], [13]. A.R. Its came up with the idea of applying the Riemann-Hilbert problem to the solution of the differential equations that describe correlation functions. The Riemann problem was used for the solution of differential equations describing correlation functions in [3], [4], [5], [6], [7], [8]. In the next chapter the Riemann-Hilbert problem will be used for the evaluation of the asymptotics of correlation functions.

Appendix XV.1:
Degeneration points

It was shown in section 1 that the matrix Riemann problem (1.6) is equivalent to the nonsingular linear integral equations (1.38) for functions $f_\pm(\lambda)$,

$$f_\pm(\lambda) - \gamma \int\limits_{-\infty}^{+\infty} K(\lambda, \mu) f_\pm(\mu) \, d\mu = e_\pm(\lambda) \qquad (A.1.1)$$

with "plane waves"

$$e_\pm(\lambda) = \sqrt{\vartheta(\lambda)} \, e^{\pm i \lambda x}$$

and kernel

$$K(\lambda, \mu) = \frac{e_+(\lambda) e_-(\mu) - e_-(\lambda) e_+(\mu)}{2i(\lambda - \mu)}$$
$$= \sqrt{\vartheta(\lambda)} \, \frac{\sin x(\lambda - \mu)}{\lambda - \mu} \, \sqrt{\vartheta(\mu)}.$$

As condition (0.5) is fulfilled in regions (1.9), (1.10)

$$0 < \gamma < \frac{1}{\pi}, \quad -\infty < \beta < +\infty, \quad x > 0,$$
$$\gamma = \frac{2}{\pi}, \quad \beta < 0, \quad x > 0, \qquad (A.1.2)$$

the matrix RP under consideration has a unique solution. Hence, equations (A.1.1) also have a unique solution in regions (A.1.2). This means, in particular, that the Fredholm determinant

$$\Delta(x, \beta, \gamma) \equiv \det\left(1 - \gamma \widehat{K}\right) \qquad (A.1.3)$$

is not equal to zero if x, β, γ belong to regions (A.1.2).

Consider now the case $\beta > 0$ (1.11):

$$\gamma = \frac{2}{\pi}, \quad \beta > 0, \quad x > 0, \qquad (A.1.4)$$

where the sufficient condition (0.5) is not fulfilled. It is convenient to discuss the question of the solvability directly for the integral equations (A.1.1), equivalent to the matrix RP (1.6). These are Fredholm equations. For fixed γ and β the kernel $K(\lambda, \mu)$ in (A.1.1) is an analytical function of x uniformly bounded in x, λ in any finite strip

$$0 < \tilde{x}_0 < \operatorname{Re} x < \tilde{x}_1; \quad \operatorname{Im} x \le \epsilon.$$

It also satisfies the estimate

$$\int\limits_{-\infty}^{+\infty} \int\limits_{-\infty}^{+\infty} |K(\lambda, \mu)|^2 \, d\lambda \, d\mu < C\tilde{x}_1^2.$$

So the standard results in the theory of Fredholm integral equations show that in any finite region

$$0 < \tilde{x}_0 < x < \tilde{x}_1$$

there exists only a finite number of points x_k for which the Fredholm determinant

$$\det \left(1 - \gamma \widehat{K} \right)$$

is equal to zero (γ and β held fixed) and integral equations (A.1.1) are not uniquely solvable. The same is true for the matrix RP under consideration.

XVI

Asymptotics of Temperature-dependent Correlation Functions for the Impenetrable Bose Gas

Introduction

The first asymptotics formulæ for the impenetrable Bose gas were obtained by H. G. Vaidiya and C. A. Tracy [19] for the case of zero temperature and equal time. The large distance and time asymptotics of temperature correlation functions of the impenetrable Bose gas are given in this chapter. To calculate the asymptotics one uses the integrable nonlinear systems of differential equations for the correlators constructed in Chapter XIV, performing the asymptotical analysis by using matrix Riemann problems that correspond to the classical inverse scattering method; this was discussed in Chapter XV.

The asymptotics depend strongly on the region considered. The sign of the chemical potential h influences them strongly. The relevance of this parameter is especially clear at zero temperature. For $h < 0$ the density of the ground state of the gas is zero. For $h > 0$ the density of the ground state is positive. The temperature-dependent field correlator

$$\langle \Psi(x_2, t_2) \Psi^\dagger(x_1, t_1) \rangle_T$$

decays exponentially with respect to both space and time separations:

$$x = \frac{1}{2}|x_1 - x_2|\sqrt{T}, \quad t = \frac{1}{2}|t_2 - t_1|T.$$

Here T is the temperature. The important factor in the asymptotics of the field correlator

$$\langle \Psi(x_2, t_2) \Psi^\dagger(x_1, t_1) \rangle_T$$

is

$$\exp\left\{ \frac{1}{\pi} \int\limits_{-\infty}^{+\infty} d\mu \, |x + 2t\mu| \ln \left| \frac{e^{\mu^2 - \beta} - 1}{e^{\mu^2 - \beta} + 1} \right| \right\}$$

420

where $\beta = h/T$, and h is the chemical potential. The asymptotics are described more precisely in section 9.

In section 1 the asymptotics of the generating functional of currents

$$\langle \exp\{\alpha Q_1(x_1, x_2)\} \rangle$$

are evaluated for $\alpha < 0$, $|x_1 - x_2| \to \infty$. In section 2 the probability of the absence of particles (due to thermal fluctuations) on some space interval is evaluated. It illustrates for us the structure of the state of thermal equilibrium. In section 3 the long distance asymptotics of the field correlation functions are evaluated for negative chemical potential. In section 4 the asymptotics of the field correlator are evaluated for positive chemical potential. In section 5 we collect all the results for the asymptotics of equal-time temperature correlations. In section 6 we evaluate the asymptotics of the time-dependent temperature correlator of fields for negative chemical potential. In section 7 we calculate these asymptotics for $h > 0$, when x and $t \to \infty$ in space-like directions. In section 8 we consider $h > 0$ and time-like directions. Section 9 contains explicit formulæ describing the asymptotics of the field correlator

$$\langle \Psi(x_2, t_2) \Psi^\dagger(x_1, t_1) \rangle_T$$

at large space and time separations.

The results presented in this chapter were obtained in papers [6], [7], [8], [9], [12], [13]. Again the contribution of A.R. Its should be emphasized.

XVI.1 Generating functional of
currents for the impenetrable Bose gas

In this section we consider the temperature-dependent generating functional of currents for impenetrable bosons,

$$\langle \exp\{\alpha Q_1(x_1, x_2)\} \rangle_T = \frac{\mathrm{tr}\left(e^{-H/T} \exp\{\alpha Q_1(x_1, x_2)\}\right)}{\mathrm{tr}\, e^{-H/T}} \tag{1.1}$$

$$Q_1(x_1, x_2) = \int_{x_2}^{x_1} dz\, \Psi^\dagger(z)\Psi(z), \quad x_1 > x_2, \tag{1.2}$$

Q_1 is the operator for the number of particles in the interval $[x_2, x_1]$. This correlator was considered in detail in sections XIII.2, XIV.2, XV.1 and XV.2. The reader is recommended to look through the contents of those sections. The mean value (1.1) depends on $|x_1 - x_2|$.

Our starting point is the following representation for the correlator

$$\langle \exp\{\alpha Q_1(x_1, x_2)\} \rangle_T = \exp\{\sigma(x, \beta, \gamma)\} \Big|_{\gamma = (1 - \exp\alpha)/\pi} \tag{1.3}$$

where the function $\sigma(x, \beta, \gamma)$ (the logarithm of the Fredholm determinant, see (XIV.2.7)–(XIV.2.14)) depends on the rescaled distance x and chemical potential β,

$$x \equiv \frac{1}{2}|x_1 - x_2|\sqrt{T}, \qquad \beta \equiv \frac{h}{T}. \qquad (1.4)$$

Here the case of finite negative α is considered (the "emptiness formation probability" case, $\alpha = -\infty$ ($\gamma = 1/\pi$) possesses peculiar features and is thus considered separately in the next section). It is also supposed that $x > 0$ (it is not in fact a restriction, since the correlator (1.1) is an even function of x), so that

$$-\infty < \alpha < 0 \quad \left(\frac{1}{\pi} > \gamma > 0\right); \qquad x > 0. \qquad (1.5)$$

The function $\sigma(x, \beta, \gamma)$ was shown (XIV.2.32), (XIV.2.33) to be zero at $x = 0$ as well as at $\beta = -\infty$,

$$\sigma(0, \beta, \gamma) = \sigma(x, -\infty, \gamma) = 0. \qquad (1.6)$$

Its derivatives with respect to x and β are expressed in terms of potentials B_{++}, B_{+-} (XIV.2.29), (XIV.2.30) as

$$\partial_x \sigma = -B_{+-} \qquad (1.7)$$

$$\partial_\beta \sigma = -x\partial_\beta B_{+-} + \frac{1}{2}(\partial_\beta B_{+-})^2 - \frac{1}{2}(\partial_\beta B_{++})^2. \qquad (1.8)$$

In turn, the potentials $B(x, \beta, \gamma)$ are expressed in section XV.2 in terms of the matrix elements of the matrix-valued solution $\Phi(\lambda)$ of the matrix Riemann problem (RP) (XV.2.11) by means of formulæ (XV.2.15)

$$B_{+-} = 2i \lim_{\lambda \to \infty} \lambda \left[\Phi_{11}(\lambda) - \beta^{-1}(\lambda)\right], \qquad (1.9)$$

$$B_{++} = -2i \lim_{\lambda \to \infty} \lambda \Phi_{12}(\lambda). \qquad (1.10)$$

Recall that the scalar functions $\alpha(\lambda)$, $\beta(\lambda)$ solving scalar RP's (XV.2.6), (XV.2.7) are given explicitly by

$$\alpha(\lambda) = \exp\left\{-\frac{1}{2\pi i} \int_{-\infty}^{+\infty} \frac{d\mu}{\mu - \lambda} \ln\left(\frac{1 - \pi\gamma + e^{\mu^2 - \beta}}{1 + e^{\mu^2 - \beta}}\right)\right\} \qquad (1.11)$$

$$\beta(\lambda) = \exp\left\{-\frac{1}{2\pi i} \int_{-\infty}^{+\infty} \frac{d\mu}{\mu - \lambda} \ln\left(\frac{1 + \pi\gamma + e^{\mu^2 - \beta}}{1 + e^{\mu^2 - \beta}}\right)\right\}. \qquad (1.12)$$

So the asymptotic analysis for $x \to +\infty$ of the correlator is reduced to the asymptotic analysis of the solution of the matrix RP which is done by using the approach suggested in paper [15].

The standard system (XV.0.6) of singular integral equations equivalent to the matrix RP (XV.2.11) is easily seen to give the following system of two singular integral equations for the matrix elements Φ_{11}, Φ_{12}

$$\Phi_{11,+}(\lambda) = 1 - \frac{\gamma}{2i} \int_{-\infty}^{+\infty} \frac{\Phi_{12,+}(\mu)\alpha_-(\mu)e^{-2i\mu x}\,d\mu}{(\mu - \lambda - i0)\beta_-(\mu)(1 - \pi\gamma + e^{\mu^2-\beta})}; \quad (1.13)$$

$$\Phi_{12,+}(\lambda) = \frac{\gamma}{2i} \int_{-\infty}^{+\infty} \frac{\Phi_{11,+}(\mu)\beta_+(\mu)e^{2i\mu x}\,d\mu}{(\mu - \lambda - i0)\alpha_+(\mu)(1 - \pi\gamma + e^{\mu^2-\beta})}, \quad (1.14)$$

$$(\text{Im}\,\lambda = 0)$$

(in obtaining these equations one also uses the scalar RP's for $\alpha(\lambda)$, $\beta(\lambda)$). Directly from equations (1.13), (1.14) one can derive the following estimates for $\Phi_{11}(\lambda)$, $\Phi_{12}(\lambda)$:

$$|\Phi_{11}(\lambda, x, \beta, \gamma) - 1| \le \frac{C}{\lambda}, \qquad 0 < \gamma < \frac{1}{\pi}, \quad (1.15)$$

$$|\Phi_{12}(\lambda, x, \beta, \gamma)| \le \frac{D}{\lambda}, \qquad 0 < \gamma < \frac{1}{\pi}, \quad (1.16)$$

where $\text{Im}\,\lambda \ge 0$, $\beta \le \beta_0$, $x \ge x_0$ and the constants C, D depend, for fixed γ, on x_0, β_0 only. We note that one can interpret the Φ_{ik} as the Jost solutions for the Dirac operator L (see the remark after (XV.2.14)), so that these estimates can also be established by using the standard results [4] for the scattering theory for this operator.

Let us now take into account that the function $\Phi_{11}(\mu)\beta(\mu)/\alpha(\mu)$ is analytic in the upper half-plane and satisfies the estimates (1.15) (remember that $\Phi_{11}(\infty) = \alpha(\infty) = \beta(\infty) = 1$). Hence the integration contour in (1.14) can be shifted into the upper half-plane of μ and the integral can be represented as a sum of residues at first order poles at the points λ_k, $-\lambda_k^*$ $(k = 1, 2, \ldots)$ which are zeros of the function

$$1 - \pi\gamma + e^{\mu^2-\beta}$$

in the upper half-plane,

$$\lambda_k = [\ln(1 - \pi\gamma) + \beta + (2k - 1)i\pi]^{1/2}, \quad k = 1, 2, \ldots$$
$$\text{Im}\,\lambda_k > 0, \ \text{Re}\,\lambda_k > 0. \quad (1.17)$$

Thus the following representation is obtained for Φ_{12}:

$$\Phi_{12,+}(\lambda) = \frac{\pi\gamma\Phi_{11,+}(\lambda)\beta_+(\lambda)e^{2i\lambda x}}{\left(1 - \pi\gamma + e^{\lambda^2-\beta}\right)\alpha_+(\lambda)} + S(\lambda) \quad (1.18)$$

where the first term on the right hand side is due to the pole at $\mu = \lambda + i0$ and the function $S(\lambda)$ is given as a series

$$S(\lambda) = \sum_{k=1}^{\infty} [S(\lambda_k, \lambda) + S(-\lambda_k^*, \lambda)] \tag{1.19}$$

with

$$S(\lambda_k, \lambda) \equiv \frac{\pi \gamma \Phi_{11}(\lambda_k) \beta(\lambda_k) e^{2i\lambda_k x}}{2(\pi \gamma - 1) \lambda_k (\lambda_k - \lambda) \alpha(\lambda_k)}. \tag{1.20}$$

Series (1.19) is uniformly convergent for $\lambda \in \mathbb{R}$, $x \geq x_0$ and $\beta_1 \leq \beta \leq \beta_0$. This series is also asymptotical at $x \to \infty$ uniformly for $\lambda \in \mathbb{R}$, $\beta_1 \leq \beta \leq \beta_0$.

Putting expression (1.18) into (1.13) one arrives at the representation for the function $\Phi_{11}(\lambda)$

$$\Phi_{11,+}(\lambda) = 1 - \frac{\pi \gamma^2}{2i} \int_{-\infty}^{+\infty} \frac{\Phi_{11,+}(\mu) \, d\mu}{(1 - \pi \gamma + e^{\mu^2 - \beta})(1 + \pi \gamma + e^{\mu^2 - \beta})(\mu - \lambda - i0)}$$

$$- \frac{1}{2\pi i} \int_{-\infty}^{+\infty} \frac{R(\mu)}{\mu - \lambda - i0} \, d\mu \tag{1.21}$$

with

$$R(\mu) = \frac{\pi \gamma \alpha_-(\mu) e^{-2i\mu x}}{\beta_-(\mu)(1 - \pi \gamma + e^{\mu^2 - \beta})} S(\mu) \tag{1.22}$$

(one uses again scalar RP's (XV.2.8), (XV.2.9) for $\alpha(\lambda)$, $\beta(\lambda)$).

Equation (1.21) may be now regarded as a singular integral equation for the inhomogeneous scalar RP in the case when the scalar functions $\tilde{\Phi}_+(\lambda)$ are equal to $\Phi_{11,+}(\lambda)$ (see Appendix 1 and, especially, equation (A.1.12)). This inhomogeneous RP can be written as

$$\tilde{\Phi}_-(\lambda) = \tilde{\Phi}_+(\lambda) g(\lambda) + R(\lambda), \quad \lambda \in \mathbb{R}, \tag{1.23a}$$

$$\tilde{\Phi}_+(\lambda) \equiv \Phi_{11,+}(\lambda), \quad \lambda \in \mathbb{R};$$

$$\tilde{\Phi}(\infty) = 1. \tag{1.23b}$$

The function $g(\lambda)$ is calculated from the relation

$$1 - g(\lambda) = -\frac{\pi^2 \gamma^2}{(1 - \pi \gamma + e^{\lambda^2 - \beta})(1 + \pi \gamma + e^{\lambda^2 - \beta})}$$

to be equal to

$$g(\lambda) = \frac{(1 + e^{\lambda^2 - \beta})^2}{(1 - \pi \gamma + e^{\lambda^2 - \beta})(1 + \pi \gamma + e^{\lambda^2 - \beta})} \tag{1.24}$$

so that, due to RP's (XV.2.8), (XV.2.9) for $\alpha(\lambda)$, $\beta(\lambda)$ one represents it in the form (A.1.8):

$$g(\lambda) = \frac{\alpha_-^{-1}(\lambda)\beta_-^{-1}(\lambda)}{\alpha_+^{-1}(\lambda)\beta_+^{-1}(\lambda)}. \tag{1.25}$$

It means that the homogeneous RP corresponding to (1.23) is solved explicitly, which, in turn, gives the solution of the inhomogeneous RP, see (A.1.11),

$$\widetilde{\Phi}(\lambda) = \alpha^{-1}(\lambda)\beta^{-1}(\lambda) - \frac{\alpha^{-1}(\lambda)\beta^{-1}(\lambda)}{2\pi i} \int\limits_{-\infty}^{+\infty} \frac{\alpha_-(\mu)\beta_-(\mu)R(\mu)}{\mu - \lambda}\, d\mu, \tag{1.26}$$

$\mathrm{Im}\,\lambda \neq 0$.

Note now that $\widetilde{\Phi}_+(\lambda) = \Phi_{11,+}(\lambda)$ ($\mathrm{Im}\,\lambda = 0$) and $\widetilde{\Phi}(\infty) = \Phi_{11}(\infty) = 1$. As both $\widetilde{\Phi}(\lambda)$ and $\Phi_{11}(\lambda)$ are analytic in the upper half-plane $\mathrm{Im}\,\lambda > 0$, one concludes that $\widetilde{\Phi}(\lambda) = \Phi_{11}(\lambda)$ for $\mathrm{Im}\,\lambda \geq 0$. So, for $\mathrm{Im}\,\lambda \geq 0$, one obtains from (1.26), using the explicit form (1.22), (1.19) for $R(\mu)$, the following representation:

$$\Phi_{11}(\lambda) = \alpha^{-1}(\lambda)\beta^{-1}(\lambda) - \frac{\alpha^{-1}(\lambda)\beta^{-1}(\lambda)}{2\pi i}$$
$$\times \sum_{k=1}^{\infty} [A(\lambda_k, \lambda)\Phi_{11}(\lambda_k) + A(-\lambda_k^*, \lambda)\Phi_{11}(-\lambda_k^*)], \tag{1.27}$$

$\mathrm{Im}\,\lambda \geq 0$

where

$$A(\lambda_k, \lambda) = \left(\frac{\pi^2\gamma^2}{2(\pi\gamma - 1)\lambda_k}\right)\left(\frac{\beta(\lambda_k)}{\alpha(\lambda_k)}\right)e^{2i\lambda_k x}$$
$$\times \int\limits_{-\infty}^{+\infty} \frac{\alpha_-^2(\mu)e^{-2i\mu x}\, d\mu}{(1 - \pi\gamma + e^{\mu^2 - \beta})(\lambda_k - \mu)(\mu - \lambda)}. \tag{1.28}$$

Due to the possibility of shifting the integration contour in the integral (1.28) into the lower half-plane of μ, this integral can be estimated as

$$C\left|\frac{e^{2i\lambda_1 x}}{\lambda + \lambda_1}\right|$$

so that one obtains for the main term of the asymptotic behavior of $\Phi_{11}(\lambda)$ for $x \to +\infty$

$$\Phi_{11}(\lambda) = \alpha^{-1}(\lambda)\beta^{-1}(\lambda) + O\left(e^{-4r_1 x \sin\varphi_1}\right) \tag{1.29}$$

where the notation

$$r_k \equiv |\lambda_k|, \qquad \varphi_k \equiv \arg \lambda_k,$$
$$\lambda_k = [\ln(1 - \pi\gamma) + \beta + (2k-1)i\pi]^{1/2}, \qquad k = 1, 2, \ldots, \qquad (1.30)$$
$$\operatorname{Im} \lambda_k > 0, \quad \operatorname{Re} \lambda_k > 0$$

(see (1.17)) is introduced.

From (1.14) one obtains for $\Phi_{12}(\lambda)$

$$\Phi_{12}(\lambda) = \frac{\gamma}{2i} \int\limits_{-\infty}^{+\infty} \frac{\alpha_+^{-2}(\mu) e^{2i\mu x} \, d\mu}{(1 - \pi\gamma + e^{\mu^2 - \beta})(\mu - \lambda)}$$
$$+ O\left(e^{-4r_1 x \sin \varphi_1}\right), \qquad (1.31)$$

$$\operatorname{Im} \lambda > 0, \quad x \to +\infty.$$

These estimates are uniform for $\operatorname{Im} \lambda \geq 0$, $\beta_1 \leq \beta \leq \beta_0$. One can prove that the estimate is also valid for $-\infty < \beta \leq \beta_0$. Expanding in powers of $1/\lambda$ one gets for the potentials B (see (1.19), (1.10)):

$$B_{+-} = \frac{1}{\pi} \int\limits_{-\infty}^{+\infty} d\lambda \, \ln \frac{1 + e^{\lambda^2 - \beta}}{1 - \pi\gamma + e^{\lambda^2 - \beta}}$$
$$+ O\left(e^{-4r_1 x \sin \varphi_1}\right), \qquad x \to +\infty; \qquad (1.32)$$

$$B_{++} = \gamma \int\limits_{-\infty}^{+\infty} d\lambda \, \frac{\alpha_+^{-2}(\lambda) e^{2i\lambda x}}{1 - \pi\gamma + e^{\lambda^2 - \beta}}$$
$$+ O\left(e^{-4r_1 x \sin \varphi_1}\right), \qquad x \to +\infty. \qquad (1.33)$$

Shifting the integration contour in the last formula to the upper half-plane one obtains a more exact estimate for B_{++}:

$$B_{++} = \frac{2\pi\gamma}{1 - \pi\gamma} \sum_{k=1}^{K} \frac{|\alpha_+(\lambda_k)|^{-2}}{r_k} e^{-2x r_k \sin \varphi_k}$$
$$\times \sin\left\{2x r_k \cos \varphi_k - 2 \arg \alpha(\lambda_k) - \varphi_k\right\}$$
$$+ O\left(e^{-4r_1 x \sin \varphi_1}\right), \qquad (1.34)$$

where the integer $K = K(t, \gamma)$ (which is the number of terms in the sum) has to be defined from the conditions

$$r_K \sin \varphi_K \equiv \operatorname{Im} \lambda_K < 2\operatorname{Im} \lambda_1 \equiv 2r_1 \sin \varphi_1,$$
$$r_{K+1} \sin \varphi_{K+1} > 2r_1 \sin \varphi_1. \qquad (1.35)$$

By means of equation (XIV.2.23),

$$\partial_x B_{+-} = B_{++}^2 \qquad (1.36)$$

one obtains the next terms in the asymptotic behavior of B_{+-} using (1.34):

$$B_{+-} = C(\beta, \gamma) - b^2(8r_1)^{-1}e^{-4xr_1 \sin \varphi_1}$$
$$\times \left[\frac{1}{\sin \varphi_1} + \sin \{4xr_1 \cos \varphi_1 + 2\theta_1 - \varphi_1\} \right]$$
$$+ O\left(e^{-4r_1 x \sin \varphi_1} \right), \quad x \to +\infty \tag{1.37}$$

where the notation

$$C(\beta, \gamma) \equiv \frac{1}{\pi} \int\limits_{-\infty}^{+\infty} d\lambda \ln \frac{1 + e^{\lambda^2 - \beta}}{1 - \pi\gamma + e^{\lambda^2 - \beta}}, \tag{1.38}$$

$$C(\beta, \gamma) \to 0, \quad \beta \to -\infty$$

and

$$b \equiv \frac{2\pi\gamma}{1 - \pi\gamma} \frac{|\alpha(\lambda_1)|^{-2}}{r_1}, \quad \theta_1 = -2 \arg \alpha(\lambda_1) - \varphi_1 \tag{1.39}$$

has been introduced. So the first terms of the asymptotics of the potentials are obtained.

Let us now apply these results to get the asymptotics of the logarithm σ of the Fredholm determinant, which is just the logarithm of the correlator (1.1), see (1.3). Taking into account expressions (1.7), (1.8) for $\partial_x \sigma$ and $\partial_\beta \sigma$ as well as the condition $\sigma(x, \beta = -\infty, \gamma) = 0$ (1.6), and the uniformity of the estimates (1.29), (1.31) for $\beta \le \beta_0$ one obtains from (1.32), (1.34) for the main term of the asymptotics:

$$\sigma(x, \beta, \gamma) = -xC(\beta, \gamma) + \frac{1}{2} \int\limits_{-\infty}^{\beta} \left(\frac{\partial C(\beta', \gamma)}{\partial \beta'} \right)^2 d\beta'$$
$$+ O\left(e^{-4r_1 x \sin \varphi_1} \right). \tag{1.40}$$

Using (1.34), (1.37) the next term of the asymptotics can be calculated easily with the result

$$\sigma(x, \beta, \gamma) = -xC(\beta, \gamma) + \frac{1}{2} \int\limits_{-\infty}^{\beta} \left(\frac{\partial C(\beta', \gamma)}{\partial \beta'} \right)^2 d\beta'$$
$$- \frac{b^2}{32r_1^2} e^{-4xr_1 \sin \varphi_1} \left[\frac{1}{\sin^2 \varphi_1} + \cos \{4xr_1 \cos \varphi_1 + 2\theta_1 - 2\varphi_1\} \right]$$
$$+ o\left(e^{-4r_1 x \sin \varphi_1} \right). \tag{1.41}$$

The asymptotics of correlator (1.1) are now readily obtained:

$$\langle \exp\{\alpha Q_1(x_1, x_2)\}\rangle_T = \exp\{\sigma(x, \beta, \gamma)\}\Big|_{\gamma = (1 - \exp \alpha)/\pi}$$

$$= \exp\left\{\frac{1}{2}\int_{-\infty}^{\beta}\left(\frac{\partial C(\beta', \gamma)}{\partial \beta'}\right)^2 d\beta'\right\} e^{-xC(\beta, \gamma)}$$

$$\times \left(1 - \frac{b^2}{32r_1^2}e^{-4xr_1 \sin \varphi_1}\left[\frac{1}{\sin^2 \varphi_1} + \cos\{4xr_1 \cos \varphi_1 + 2\theta_1 - 2\varphi_1\}\right]\right.$$

$$\left. + o\left(e^{-4r_1 x \sin \varphi_1}\right)\right), \tag{1.42}$$

$$\equiv \frac{1}{2}|x_1 - x_2|\sqrt{T} \to x \to \infty, \qquad \beta \equiv \frac{h}{T}.$$

So the first terms of the asymptotics are given.

It is to be emphasized that, in principle, no difficulties occur in calculations of the further terms so that the complete asymptotic expansion can be obtained. It can be done in two ways. The first way is to substitute into the nonlinear differential equation (XIV.2.24),

$$\partial_\beta(B_{++})^2 = 1 + \frac{1}{2}\partial_x\frac{\partial_x\partial_\beta B_{++}}{B_{++}}, \tag{1.43}$$

the following Ansatz:

$$B_{++} = \sum_k A_k \exp\{-2xr_k \sin \varphi_k\} \sin\{2xr_k \cos \varphi_k + \theta_k\}$$

$$+ \sum_{\pm, k, l} A_{kl}^{\pm} \exp\{-2x(r_k \sin \varphi_k + r_l \sin \varphi_l)\}$$

$$\times \sin\{2(r_k \sin \varphi_k \pm r_l \sin \varphi_l)x + \theta_{kl}\} + \cdots. \tag{1.44}$$

We use this approach in the next section in the relatively simpler case of the emptiness formation probability ($\gamma = 1/\pi$).

Another way to obtain asymptotic expansions is to use the representation (1.27). Putting there

$$\lambda = \lambda_k, \quad \lambda = -\lambda_k^* \qquad (k = 1, 2, \ldots)$$

one gets an infinite-dimensional linear system for the quantities

$$\Phi(\lambda_k), \quad \Phi(-\lambda_k^*),$$

the coefficients being proportional to

$$\exp\{-2r_k \sin \varphi_k - 2r_1 \sin \varphi_1\}.$$

This system can be used to construct the asymptotic expansion.

XVI.2 The emptiness formation probability

The emptiness formation probability $P(x_1, x_2)$,

$$P(x_1, x_2) = \langle e^{\alpha Q_1(x_1, x_2)} \rangle_T \Big|_{\alpha=-\infty} = e^{\sigma(x, \beta, \gamma)} \Big|_{\gamma=1/\pi} \tag{2.1}$$

is the simplest two-point equal-time temperature-dependent correlator of impenetrable bosons. It illustrates for us the structure of the state of thermal equilibrium. It gives the probability that there are no particles of the gas possessing coordinates in the interval $[x_1, x_2]$. The correlator (2.1) is, of course, a particular case of the generating functional (1.1), (1.3) for current correlators. However, the calculations of the asymptotics are somewhat different. Below, the complete asymptotic expansion of the logarithm of this correlator, i.e., of

$$\sigma(x, \beta) \equiv \sigma(x, \beta, \gamma) \Big|_{\gamma=1/\pi} \tag{2.2}$$

is constructed.

As has already been explained in the previous section, the problem is reduced to the asymptotic analysis of the system of singular integral equations (1.13), (1.14) (it should be remembered, however, that the parameter γ is fixed, $\gamma = 1/\pi$). The method of analysis is, in principle, the same, but the estimates are essentially different. For example, representation (1.18) is now changed to

$$\Phi_{12,+} = \frac{\Phi_{11,+}(\lambda)\beta_+(\lambda)e^{2i\lambda x - \lambda^2 + \beta}}{\alpha_+(\lambda)} + e^{-x^2 + \beta} R(\lambda) \tag{2.3}$$

where $R(\lambda)$ is a function continuous in $\lambda \in \mathbb{R}$, holomorphic for $\operatorname{Im}\lambda < 0$ and satisfying the estimate

$$|R(\lambda)| < \frac{C}{\sqrt{|\lambda|^2 + x^2}} \tag{2.4}$$

uniformly for $\operatorname{Im}\lambda \leq 0$, $\beta \leq \beta_0$ and $x \geq x_0$. Repeating the corresponding considerations of section 1 (but taking into account the new estimate (2.3)) one obtains for Φ_{11} the following representation instead of (1.26):

$$\Phi_{11}(\lambda) = \alpha^{-1}(\lambda)\beta^{-1}(\lambda)$$
$$- \frac{\alpha^{-1}(\lambda)\beta^{-1}(\lambda)}{2\pi i} e^{-2x^2 + \beta} \int\limits_{-\infty}^{+\infty} \frac{R(\mu - ix)e^{-\mu^2}\alpha^2(\mu - ix)}{\mu - \lambda - ix} d\mu$$

$$\tag{2.5}$$

(the functions $\alpha(\lambda)$, $\beta(\lambda)$ are given by (1.11), (1.12) with $\gamma = 1/\pi$), so that the following estimate is valid:

$$\Phi_{11}(\lambda) = \alpha^{-1}(\lambda)\beta^{-1}(\lambda) + e^{-2x^2+\beta}L(\lambda) \qquad (2.6)$$

where the function $L(\lambda)$ is continuous for $\lambda \in \mathbb{R}$, holomorphic for $\mathrm{Im}\,\lambda < 0$ and satisfies the estimate

$$|L(\lambda)| < \frac{C}{x\sqrt{|\lambda|^2 + x^2}} \qquad (2.7)$$

uniformly for $\mathrm{Im}\,\lambda \le 0$, $\beta \le \beta_0$, $x \ge x_0$.

The asymptotic representation of $\Phi_{12}(\lambda)$ is obtained from (1.14) and (2.6). For $\mathrm{Im}\,\lambda > 0$ it is of the form

$$\Phi_{12}(\lambda) = \frac{1}{2\pi i} \int\limits_{-\infty}^{+\infty} \frac{e^{2i\mu x+\beta-\mu^2}}{\alpha_+^2(\mu)(\mu-\lambda)}\, d\mu + e^{-3x^2+3\beta}M(\lambda), \qquad (2.8)$$

where

$$|M(\lambda)| < \frac{C}{x^2\sqrt{|\lambda|^2 + x^2}}. \qquad (2.9)$$

Thus one has, due to (1.9), (1.10), the following main terms in the asymptotics for the potentials B:

$$B_{+-}(x,\beta) = \frac{1}{\pi} \int\limits_{-\infty}^{+\infty} d\lambda \ln\left(1 + e^{\beta-\lambda^2}\right)$$

$$+ O\left(e^{-2x^2+2\beta}/x\right), \qquad x \to +\infty \qquad (2.10)$$

$$B_{++}(x,\beta) = \frac{1}{\pi} \int\limits_{-\infty}^{+\infty} d\lambda\, \alpha_+^{-2}(\lambda)e^{2i\lambda x+\beta-\lambda^2}$$

$$+ O\left(e^{-3x^2+3\beta}/x^2\right), \qquad x \to +\infty. \qquad (2.11)$$

These estimates are uniform for $\beta \le \beta_0$. So the main terms in the asymptotics of the potentials are given by exactly the same formulæ ((1.32), (1.33) with $\gamma = 1/\pi$) as for $\gamma < 1/\pi$; the corrections are, however, different.

We now construct complete asymptotic expansions for the potentials B_{++}, B_{+-} using the nonlinear partial differential equation (XIV.2.24) for B_{++},

$$\partial_\beta B_{++}^2 = 1 + \frac{1}{2}\frac{\partial}{\partial x}\left(\frac{\partial_x \partial_\beta B_{++}}{B_{++}}\right). \qquad (2.12)$$

One begins with the representation of the integral in (2.11) as

$$\frac{1}{\pi} \int\limits_{-\infty}^{+\infty} d\lambda\, \alpha_+^{-2}(\lambda) e^{2i\lambda x + \beta - \lambda^2} = e^{-x^2 + \beta}$$

$$\times \left(\frac{1}{\sqrt{\pi}} + \frac{1}{\pi} \sum_{k,l=0}^{\infty} \frac{(-1)^l \alpha_k (2l+k)!}{x^{2l+k+1} i^{k+1} k!\,(2l)!} \Gamma\left(l + \frac{1}{2}\right) \right) \quad (2.13)$$

where the quantities α_k are introduced by means of the following expansion of $\alpha^{-2}(\lambda)$ (1.11) (it is to be recalled that $\gamma = 1/\pi$):

$$\alpha^{-2}(\lambda) = 1 + \sum_{k=0}^{\infty} \frac{\alpha_k}{\lambda^{k+1}},$$

$$\lambda \to \infty, \quad \mathrm{Im}\,\lambda > 0. \quad (2.14)$$

Due to the explicit form of $\alpha(\lambda)$, all the α_k's can be calculated from the identity

$$1 + \sum_{k=0}^{\infty} \frac{\alpha_k}{\lambda^{k+1}} \equiv \exp\left\{ -i \sum_{k=0}^{\infty} \frac{c_k}{\lambda^{2k+1}} \right\},$$

$$c_k = \frac{1}{\pi} \int\limits_{-\infty}^{+\infty} \lambda^{2k} \ln\left(1 + e^{\beta - \lambda^2}\right) d\lambda. \quad (2.15)$$

Thus one obtains

$$\alpha_0 = -ic_0, \quad \alpha_1 = -\frac{c_0^2}{2}, \quad \alpha_2 = -ic_1 + i\frac{c_0^3}{6},$$

$$\alpha_3 = -c_0 c_1 + \frac{c_0^4}{4!}, \quad \text{etc.} \quad (2.16)$$

Returning now to the asymptotics (2.11) for B_{++} one can write

$$B_{++}(x, \beta) = e^{-x^2 + \beta} \sum_{n=0}^{\infty} \frac{a_n}{x^n} + O\left(e^{-3x^2 + 3\beta}/x^2\right), \qquad x \to +\infty, \quad (2.17)$$

where the coefficients a_n (functions of β) are expressed in terms of the α_k's:

$$a_0 = \frac{1}{\sqrt{\pi}}, \quad a_1 = \frac{\alpha_0}{i\sqrt{\pi}}, \quad a_2 = \frac{\alpha_1}{i^2 \sqrt{\pi}},$$

$$a_n = \frac{1}{\pi} \sum_{\substack{k,l=0,\dots,\infty \\ k+2l=n-1}} \frac{(-i)^l \alpha_k (2l+k)!}{i^{k+1} k!\,(2l)!} \Gamma\left(l + \frac{1}{2}\right) \quad (2.18)$$

(it is to be noted that the sums here are taken over integers k, l).

Let us turn now to the nonlinear partial differential equation (2.12) for $B_{++}(x, \beta)$. The estimate (2.17) suggests the following Ansatz:

$$
B_{++}(x, \beta) = e^{-x^2 + \beta} \left(a_0 + \frac{a_1}{x} + \frac{a_2}{x^2} + \frac{a_3}{x^3} + \cdots \right)
$$

$$
+ e^{-3x^2} \left(\frac{b_2}{x^2} + \frac{b_3}{x^3} + \frac{b_4}{x^4} + \cdots \right)
$$

$$
+ e^{-5x^2} \left(\frac{d_4}{x^4} + \frac{d_5}{x^5} + \cdots \right) + \cdots
$$

$$
+ e^{-(2N+1)x^2} \left(\frac{d_{2N}^{(N)}}{x^{2N}} + \frac{d_{2N+1}^{(N)}}{x^{2N+1}} + \cdots \right) + \cdots . \quad (2.19)
$$

The quantities a_j, b_j, d_j, ..., d_j, ..., $d_j^{(N)}$, ..., are functions of β which we need to determine, apart from the a_j which are already known (2.18). Putting the formal series (2.19) into equation (2.12) (the relation $\partial_x B_{+-} = B_{++}^2$ (XIV.2.23) and estimate (2.10) are also useful) one obtains linear ordinary differential equations in β for functions b, d, ..., $d^{(N)}$, ... For example, for b_2 one gets

$$
\partial_\beta b_2 - \frac{1}{3} b_2 = \frac{1}{6\pi^{3/2}} e^{3\beta} \quad (2.20)
$$

(for $c_0 \equiv c_0(\beta)$ see (2.15)). The solutions of these equations are defined up to solutions of the corresponding homogeneous equations, these being

$$
b_2(\beta) = \text{const } e^{\beta/3}. \quad (2.21)
$$

All the constants here should, however, be equal to zero as the functions $b(\beta)$, $d(\beta)$, ..., $d^{(N)}(\beta)$ should not contain fractional powers of e^β in the expansion for $\beta \to -\infty$. This is the direct consequence of the uniformity at $\beta \le \beta_0$ of estimate (2.11) (see also Appendix XIV.2 where the explicit expansion for $\beta \to -\infty$ is given). So one has from (2.20):

$$
b_2(\beta) = \frac{e^{3\beta}}{16\pi^{3/2}}. \quad (2.22)
$$

Arguments of this kind are also valid for any higher terms in the expansion (2.19). It should be emphasized, however, that it is essential that the coefficients a_j have already been defined independently. The reason is that a solution of the corresponding homogeneous equation is e^β and arguments about the presence of only entire powers of e^β give nothing in this case. So one arrives at the following asymptotic expansion for B_{++}:

$$B_{++}(x, \beta) = e^{-x^2 + \beta} \sum_{n=0}^{\infty} \frac{a_n}{x^n}$$

$$+ e^{-3x^2 + 3\beta} \left(\frac{1}{16\pi^{3/2} x^2} + \cdots \right) + \cdots$$

$$(x \to +\infty) \tag{2.23}$$

where the functions $a_n = a_n(\beta)$ are given by (2.18) and function $c_0(\beta)$ is defined in (2.15). Using now (2.10) and equation (XIV.2.23), $\partial_x B_{+-} = B_{++}^2$, one has for the potential B_{+-}:

$$B_{+-}(x, \beta) = c_0(\beta) + e^{-2x^2 + 2\beta} \sum_{n=0}^{\infty} \frac{A_n}{x^n}$$

$$+ e^{-4x^2 + 4\beta} \left(-\frac{1}{2^6 \pi^2 x^3} + \cdots + \right) + \cdots$$

$$(x \to +\infty) \tag{2.24}$$

where the functions $A_n = A_n(\beta)$ are calculated from the identity

$$e^{-2x^2 + 2\beta} \sum_{n=1}^{\infty} \frac{A_n}{x^n} = -\int_x^{\infty} dy \, e^{-2y^2 + 2\beta} \sum_{n \geq 0, m \geq 0} \frac{a_n a_m}{y_{n+m}}, \tag{2.25}$$

$$A_1 = -\frac{1}{4\pi}. \tag{2.26}$$

So the asymptotical expansions for $x \to +\infty$ of the potentials are described completely.

Now equations (1.7), (1.8) and condition

$$\sigma(x, \beta) \to 0 \quad \text{for} \quad \beta \to -\infty,$$

can be used to calculate the asymptotics of the logarithm of the correlator $\sigma(x, \beta)$, (2.2) (one also takes into account the uniformity of all the estimates for $\beta \leq \beta_0$). The result is:

$$\sigma(x, \beta) = -xc_0(\beta) + \frac{1}{2} \int_{-\infty}^{\beta} \left(\frac{dc_0(\beta')}{d\beta'} \right)^2 d\beta'$$

$$+ e^{-2x^2 + 2\beta} \left(-\frac{1}{16\pi x^2} + \cdots \right)$$

$$+ e^{-4x^2 + 4\beta} \left(-\frac{1}{2^9 \pi^2 x^4} + \cdots \right) + \cdots. \tag{2.27}$$

Recall that the emptiness formation probability (2.1) is easily related to $\sigma(x, \beta)$,

$$P(x_1, x_2) = \left\langle \exp\left\{ \alpha \int_{x_2}^{x_1} \Psi^\dagger(z)\Psi(z)\, dz \right\} \right\rangle T \Bigg|_{\alpha=-\infty} = \exp\{\sigma(x, \beta)\}.$$

So one can easily calculate the asymptotic expansion of $P(x_1, x_2)$ using (2.26). We write explicitly only the most interesting terms:

$$P(x_1, x_2) = \exp\left\{ \frac{1}{2} \int_{-\infty}^{\beta} \left(\frac{dc_0(\beta')}{d\beta'} \right)^2 d\beta' \right\} e^{-xc_0(\beta)}$$

$$\times \left[1 + e^{-2x^2+2\beta}\left(-\frac{1}{16\pi x^2} + \cdots \right) \right.$$

$$\left. + O\left(e^{-2x^2+2\beta}/x^3 \right) \right]. \tag{2.28}$$

Here the scaled distance x and chemical potential β are used:

$$x \equiv \frac{1}{2}|x_1 - x_2|\sqrt{T}, \qquad \beta \equiv \frac{h}{T}.$$

The function $c_0(\beta)$ is given as

$$c_0(\beta) = \frac{1}{\pi} \int_{-\infty}^{+\infty} d\lambda \, \ln\left(1 + e^{\beta - \lambda^2} \right). \tag{2.29}$$

It is to be emphasized that $c_0(\beta)$ is just the value of the function $C(\beta, \gamma)$ (1.38) at $\gamma = 1/\pi$,

$$c_0(\beta) = C(\beta, \gamma = 1/\pi),$$

so that the main terms of the asymptotic expansion of the generating functional of currents for $\gamma < 1/\pi$ (1.42) and at $\gamma = 1/\pi$ (2.27) are given (they are, in fact, of the same form). The corrections are, of course, different. Finally, let us remark that the asymptotics obtained are uniform in x for $\beta \leq \beta_0$ (β_0 is an arbitrary but finite constant).

So we have shown how to construct the complete asymptotic decomposition of $P(x)$ at large distance using differential equations. Let us note that at zero temperature the asymptotics will turn into gaussian decay (see formula (9.11) in [13]).

XVI.3 Long distance asymptotics of the temperature-dependent equal-time correlator of two fields in the case of negative chemical potential

We shall begin constructing the asymptotics of the equal-time correlator (XIV.2.1), (XIV.2.6)

$$\langle \Psi^\dagger(x_1)\Psi(x_2)\rangle_T = \sqrt{T}\, g(x,\beta), \quad x > 0$$

$$g(x,\beta) = \frac{1}{4} B_{++}(x,\beta,\gamma) e^{\sigma(x,\beta,\gamma)}\Big|_{\gamma=2/\pi} \tag{3.1}$$

in the relatively simpler case of negative chemical potential. As usual, x and β denote the scaled distance and chemical potential

$$x \equiv \frac{1}{2}|x_1 - x_2|\sqrt{T}, \qquad \beta \equiv \frac{h}{T}. \tag{3.2}$$

The Fredholm determinant logarithm $\sigma(x,\beta)$,

$$\sigma(x,\beta) \equiv \sigma(x,\beta,\gamma)\Big|_{\gamma=2/\pi}, \tag{3.3}$$

is expressed in terms of potentials $B_{++}(x,\beta)$, $B_{+-}(x,\beta)$,

$$B_{lm}(x,\beta) \equiv B_{lm}(x,\beta,\gamma)\Big|_{\gamma=2/\pi} \qquad (l,m = +,-) \tag{3.4}$$

by means of the same formulæ (1.6)–(1.8) as in the case $\gamma < 1/\pi$:

$$\partial_x \sigma = -B_{+-}, \tag{3.5}$$

$$\partial_\beta \sigma = -x\partial_\beta B_{+-} + \frac{1}{2}(\partial_\beta B_{+-})^2 - \frac{1}{2}(\partial_\beta B_{++})^2, \tag{3.6}$$

$$\sigma(x,\beta=-\infty) = \sigma(x=0,\beta) = 0 \tag{3.7}$$

so that the problem is reduced to calculating the asymptotics of the potentials. The asymptotics will be considered for $x \to +\infty$ and fixed $\beta < 0$. Our starting point is again the matrix RP (XV.1.6) which is solvable uniquely in the region (XV.1.10),

$$\gamma = \frac{2}{\pi}, \quad \beta < 0, \quad x > 0 \tag{3.8}$$

considered in this section. As has been mentioned already in section XV.2, the RP (XV.2.11) is more convenient to use; it is exactly the matrix RP exploited in section 1 for the generating functional of current correlators (only $\gamma = 2/\pi$ now). The potentials are expressed by the same formulæ (1.9), (1.10):

$$B_{+-} = 2i \lim_{\lambda\to\infty} \lambda\left[\Phi_{11}(\lambda) - \beta^{-1}(\lambda)\right] \tag{3.9}$$

$$B_{++} = -2i \lim_{\lambda\to\infty} \lambda\Phi_{12}(\lambda) \tag{3.10}$$

and functions $\alpha(\lambda)$ and $\beta(\lambda)$ are defined as in (1.11), (1.12):

$$\alpha(\lambda) = \exp\left\{-\frac{1}{2\pi i}\int\limits_{-\infty}^{+\infty}\frac{d\mu}{\mu-\lambda}\ln\left(\frac{e^{\mu^2-\beta}-1}{e^{\mu^2-\beta}+1}\right)\right\} \qquad (3.11)$$

$$\beta(\lambda) = \exp\left\{-\frac{1}{2\pi i}\int\limits_{-\infty}^{+\infty}\frac{d\mu}{\mu-\lambda}\ln\left(\frac{e^{\mu^2-\beta}+3}{e^{\mu^2-\beta}+1}\right)\right\}; \qquad (3.12)$$

they are the solutions of scalar RP's (XV.2.8), (XV.2.9), respectively. It is to be remarked that for $\beta < 0$ the expressions under the logarithms are positive and the logarithms themselves are chosen to be real.

The system of singular integral equations (XV.2.11) is, of course, the same (1.13), (1.14)

$$\Phi_{11,+}(\lambda) = 1 - \frac{1}{\pi i}\int\limits_{-\infty}^{+\infty}\frac{\Phi_{12,+}(\mu)\alpha_-(\mu)e^{-2i\mu x}}{(\mu-\lambda-i0)\beta_-(\mu)(e^{\mu^2-\beta}-1)}\,d\mu; \qquad (3.13)$$

$$\Phi_{12,+}(\lambda) = \frac{1}{\pi i}\int\limits_{-\infty}^{+\infty}\frac{\Phi_{11,+}(\mu)\beta_+(\mu)e^{2i\mu x}}{(\mu-\lambda-i0)\alpha_+(\mu)(e^{\mu^2-\beta}-1)}\,d\mu, \qquad (3.14)$$

as well as the estimates (1.15), (1.16)

$$|\Phi_{11}(\lambda,x,\beta)-1| \le \frac{C}{\lambda}, \qquad (3.15)$$

$$|\Phi_{12}(\lambda,x,\beta)| \le \frac{D}{\lambda} \qquad (3.16)$$

$$(\operatorname{Im}\lambda \ge 0, \quad \beta \le \beta_0, \quad x \ge x_0).$$

The difference is in the situation of the first order poles in the integrands in (3.13), (3.14), i.e., of the zeros of the expression

$$1 - \pi\gamma + e^{\mu^2-\beta}\bigg|_{\gamma=2/\pi} = e^{\mu^2-\beta} - 1$$

which, in the upper half-plane, are now at the points λ_k, $-\lambda_k^*$ ($k = 1, 2, .:.$), where

$$\lambda_k = [\beta + 2i(k-1)\pi]^{1/2}, \quad k = 1, 2, \ldots, \quad \operatorname{Im}\lambda_k > 0, \quad \operatorname{Re}\lambda_k > 0,$$

$$(\operatorname{Im}\lambda_k)^2 = \frac{1}{2}\left(|\beta| + \sqrt{\beta^2 + 4(k-1)^2\pi^2}\right)$$

$$(\operatorname{Re}\lambda_k)^2 = \frac{1}{2}\left(\sqrt{\beta^2 + 4(k-1)^2\pi^2} - |\beta|\right), \quad \beta < 0.$$

$$(3.17)$$

The pole of the integrand nearest to the real axis in the upper half-plane is now alone; it is situated at the point λ_1,

$$\lambda_1 \equiv i\sqrt{|\beta|}, \quad \lambda_1 = -\lambda_1^*. \tag{3.18}$$

The following representation is derived quite similarly to the corresponding representation (1.27):

$$\Phi_{11}(\lambda) = \alpha^{-1}(\lambda)\beta^{-1}(\lambda) - \frac{\alpha^{-1}(\lambda)\beta^{-1}(\lambda)}{2\pi i} A(\lambda_1, \lambda)\Phi_{11}(\lambda_1)$$
$$- \frac{\alpha^{-1}(\lambda)\beta^{-1}(\lambda)}{2\pi i} \sum_{k=2}^{\infty} [A(\lambda_k, \lambda)\Phi_{11}(\lambda_k)$$
$$+ A(-\lambda_k^*, \lambda)\Phi_{11}(-\lambda_k^*)], \tag{3.19}$$

where

$$A(\lambda_k, \lambda) = \frac{2\beta(\lambda_k)}{\lambda_k \alpha(\lambda_k)} e^{2i\lambda_k x}$$
$$\times \int_{-\infty}^{+\infty} \frac{\alpha_-^2(\mu)e^{-2i\mu x}\,d\mu}{(e^{\mu^2-\beta}-1)(\lambda_k-\mu)(\mu-\lambda)}. \tag{3.20}$$

The integral here (after shifting the integration contour into the lower half-plane) is estimated by

$$\text{const}\,\frac{|e^{2i\lambda_0 x}|}{|\lambda+\lambda_0|} \sim e^{-2x\sqrt{|\beta|}}, \quad x \to +\infty.$$

Thus one obtains from (3.19), (3.14) that

$$\Phi_{11}(\lambda) = \alpha^{-1}(\lambda)\beta^{-1}(\lambda) + O\left(e^{-4\sqrt{|\beta|}\,x}\right) \tag{3.21}$$

$$\Phi_{12}(\lambda) = \frac{1}{\pi i}\int_{-\infty}^{+\infty} \frac{\alpha_+^{-2}(\mu)e^{2i\mu x}\,d\mu}{(e^{\mu^2-\beta}-1)(\mu-\lambda)} + O\left(e^{-4\sqrt{|\beta|}\,x}\right). \tag{3.22}$$

By virtue of (3.9), (3.10) one has then

$$B_{+-}(\beta) = C(\beta) + O\left(e^{-4\sqrt{|\beta|}\,x}\right), \quad x \to +\infty \tag{3.23}$$

where

$$C(\beta) \equiv \frac{1}{\pi}\int_{-\infty}^{+\infty} d\lambda \ln\left(\frac{e^{\lambda^2-\beta}+1}{e^{\lambda^2-\beta}-1}\right) \tag{3.24}$$

is exactly the value of $C(\beta,\gamma)$ (1.38) at $\gamma = 2/\pi$, and

$$B_{++} = \frac{2}{\pi}\int_{-\infty}^{+\infty} d\lambda \frac{\alpha_+^{-2}(\lambda)e^{2i\lambda x}}{e^{\lambda^2-\beta}-1} + O\left(e^{-4\sqrt{|\beta|}\,x}\right), \quad x \to +\infty. \tag{3.25}$$

Evaluating the integral in the last formula one gets a more precise estimate for B_{++}:

$$B_{++} = 2\alpha^{-1}(\lambda_1)|\beta|^{-1/2}e^{-2\sqrt{|\beta|}\,x}$$
$$- 4\sum_{k=2}^{J\geq 2} |\alpha(\lambda_k)|^{-2}|\lambda_k|^{-1}e^{-2x\mathrm{Im}\,\lambda_k}$$
$$\times \sin\left[2x\,\mathrm{Re}\,\lambda_k - \arg\lambda_k - 2\arg\alpha(\lambda_k)\right] + O\left(e^{-4x\sqrt{|\beta|}}\right),$$
$$x \to +\infty. \tag{3.26}$$

The integer $J \equiv J(\beta)$ here should be defined by the inequalities

$$\mathrm{Im}\,\lambda_J < 2\,\mathrm{Im}\,\lambda_1, \qquad \mathrm{Im}\,\lambda_{J+1} \geq 2\,\mathrm{Im}\,\lambda_1,$$

the solution being bounded by

$$\frac{2\sqrt{3}}{\pi}|\beta| \leq J < \frac{2\sqrt{3}}{\pi}|\beta| + 1. \tag{3.27}$$

So the first terms of the potentials are derived.

For the logarithm σ of the Fredholm determinant, using (3.5)–(3.7), (3.23), (3.27) (and also the uniformity of the asymptotics obtained) one gets:

$$\sigma(x,\beta) = -xC(\beta) + \frac{1}{2}\int_{-\infty}^{\beta} d\beta' \left(\frac{dC(\beta')}{d\beta'}\right)^2$$
$$+ O\left(e^{-4x\sqrt{|\beta|}}\right), \qquad x \to +\infty \tag{3.28}$$

with $C(\beta)$ given by (3.24). Of course, the further terms of this expansion can be calculated quite similarly to (1.41); we leave it to the interested reader.

Finally, let us write the answer for the asymptotics of the correlator (3.1). Combining expressions (3.26) and (3.28) and taking into account the explicit formula for $\alpha(\lambda_1)$,

$$\ln\alpha(\lambda_1) = \frac{1}{2}a(\beta) \quad (\lambda_1 = i\sqrt{|\beta|}), \tag{3.29}$$

$$a(\beta) \equiv \frac{\sqrt{|\beta|}}{\pi}\int_{-\infty}^{+\infty}\frac{d\mu}{\mu^2 + |\beta|}\ln\left(\frac{e^{\mu^2 + |\beta|} + 1}{e^{\mu^2 + |\beta|} - 1}\right) \tag{3.30}$$

one gets

$$\langle\Psi^\dagger(x_1)\Psi(x_2)\rangle_T = \frac{\sqrt{T}}{4}B_{++}(x,\beta)e^{\sigma(x,\beta)}$$

$$= \frac{1}{2}\sqrt{\frac{T}{|\beta|}}\exp\left\{\frac{1}{2}\int\limits_{-\infty}^{\beta}\left(\frac{dC(\beta')}{d\beta'}\right)^2 d\beta' - a(\beta)\right\}e^{-x[C(\beta)+2\sqrt{|\beta|}]}$$

$$\times\left[1 - 2\sum_{k=2}^{J\geq 1}\left(\frac{\alpha^2(\lambda_1)}{|\alpha(\lambda_k)|^2}\frac{\sqrt{|\beta|}}{|\lambda_k|}e^{-2x(\operatorname{Im}\lambda_k-\sqrt{|\beta|})}\right.\right.$$

$$\left.\times\sin\left(2x\operatorname{Re}\lambda_k - \arg\lambda_k - 2\arg\alpha(\lambda_k)\right)\right)$$

$$+ O\left(e^{-2x\sqrt{|\beta|}}\right)\Bigg]\tag{3.31}$$

where, as usual, $x \equiv \frac{1}{2}|x_1 - x_2|\sqrt{T} \to +\infty$, $\beta \equiv h/T < 0$ are the rescaled distance and chemical potential.

XVI.4 Long distance asymptotics of the temperature-dependent equal-time correlator of two fields in the case of positive chemical potential

In the case $\gamma = 2/\pi$, $\beta > 0$, the correlator (3.1) with positive chemical potential differs essentially from both the case where $\gamma < 1/\pi$ considered in section 1 and the case where $\gamma = 2/\pi$, $\beta < 0$ (correlator (3.1) with negative chemical potential) studied in the last section. The reason is that now the sufficient conditions (XV.0.5) for the solvability of the original matrix RP (XV.1.6) are not fulfilled and it can possess particular indices. The method of asymptotic analysis of the matrix RP used in sections 1 and 3 runs across a serious difficulty when $\beta > 0$, since in this case the function $g_\alpha(\lambda)$ (XV.2.19) possesses zeros on the real λ-axis at points $\pm\lambda_1 = \pm\sqrt{\beta}$, so that the scalar RP (XV.2.8) for the function $\alpha(\lambda)$ certainly possesses an index. As was explained in section XV.2, in this case one regularizes the matrix RP (and also the scalar RP for $\alpha(\lambda)$) by going to the matrix RP (XV.2.35) for the function $\overset{\circ}{\Phi}(\lambda)$ (XV.2.41) and also by changing the conjugation contour from the real axis to the contour Γ shown in Figure XVI.1.

Correspondingly, using the relations (XV.2.46), (XV.2.52) and (XV.2.53),

$$B_{++} = -2ed^{-1} + \overset{\circ}{B}_{++};\tag{4.1}$$

$$B_{+-} = -\partial_x\ln d + \overset{\circ}{B}_{+-};\tag{4.2}$$

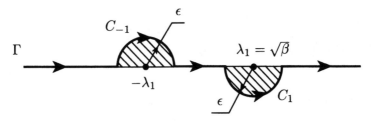

Figure XVI.1

$$\sigma = \sigma_0 + \ln d, \tag{4.3}$$

one represents the correlator (3.1) for $\beta > 0$ as

$$\langle \Psi^\dagger(x_1)\Psi(x_2)\rangle_T = \sqrt{T}\, g(x, \beta),$$

$$g(x, \beta) = \frac{1}{4}\left(-2e(x, \beta) + d(x, \beta)\mathring{B}_{++}(x, \beta)\right)\exp\{\sigma_0(x, \beta)\}, \tag{4.4}$$

$$\beta > 0, \quad x > 0$$

(as usual,

$$x \equiv \frac{1}{2}|x_1 - x_2|\sqrt{T}, \qquad \beta \equiv \frac{h}{T}).$$

The potentials $\mathring{B}_{+-}$ and $\mathring{B}_{++}$ and the functions e, d are expressed in terms of the matrix elements of the solution $\mathring{\Phi}(\lambda)$ of matrix RP (XV.2.35) by formulæ (XV.2.48), (XV.2.49), (XV.2.38), (XV.2.40):

$$\mathring{B}_{+-}(x, \beta) = 2i \lim_{\lambda \to \infty} \lambda\left(\mathring{\Phi}_{11}(\lambda) - \beta^{-1}(\lambda)\right) \tag{4.5}$$

$$\mathring{B}_{++}(x, \beta) = -2i \lim_{\lambda \to \infty} \lambda\mathring{\Phi}_{12}(\lambda) \tag{4.6}$$

$$d(x, \beta) = \mathring{\Phi}_{22}(\lambda_1)\mathring{\Phi}_{11}(-\lambda_1) - \mathring{\Phi}_{12}(\lambda_1)\mathring{\Phi}_{21}(-\lambda_1) \tag{4.7}$$

$$e(x, \beta) \equiv -2i\lambda_1\mathring{\Phi}_{11}(-\lambda_1)\mathring{\Phi}_{12}(\lambda_i) \tag{4.8}$$

(where $\lambda_1 \equiv \sqrt{\beta}$).

The x-derivative of $\sigma_0(x, \beta)$ is simply expressed in terms of $\mathring{B}_{+-}$, see (XV.2.53),

$$\partial_x \sigma_0(x, \beta) = -\mathring{B}_{+-}(x, \beta). \tag{4.9}$$

Expression (XV.2.3),

$$\partial_\beta \sigma = -x\partial_\beta B_{+-} + \frac{1}{2}(\partial_\beta B_{+-})^2 - \frac{1}{2}(\partial_\beta B_{++})^2 \tag{4.10}$$

is also very useful. So all the quantities entering representation (4.4) of the correlator are expressed in terms of the matrix $\mathring{\Phi}(\lambda)$. It is also to be emphasized that the scalar functions $\alpha(\lambda)$, $\beta(\lambda)$ are given now by (see (XV.2.23), (XV.2.24)):

$$\alpha(\lambda) = \exp\left\{-\frac{1}{2\pi i}\int_\Gamma \frac{d\mu}{\mu - \lambda}\ln\left(\frac{e^{\mu^2-\beta}-1}{e^{\mu^2-\beta}+1}\right)\right\} \tag{4.11}$$

$$\beta(\lambda) = \exp\left\{-\frac{1}{2\pi i}\int_\Gamma \frac{d\mu}{\mu - \lambda}\ln\left(\frac{e^{\mu^2-\beta}+3}{e^{\mu^2-\beta}+1}\right)\right\} \tag{4.12}$$

satisfying relations (XV.2.14)

$$\alpha^{-1}(\lambda) = \alpha(-\lambda), \quad \beta^{-1}(\lambda) = \beta(-\lambda) \tag{4.13}$$

so that (see (XV.2.36)):

$$\begin{aligned}
\mathring{\Phi}_{21}(\lambda) &= \mathring{\Phi}_{12}(-\lambda)\alpha^{-1}(\lambda)\beta^{-1}(\lambda) \\
\mathring{\Phi}_{22}(\lambda) &= \mathring{\Phi}_{11}(-\lambda)\alpha^{-1}(\lambda)\beta^{-1}(\lambda).
\end{aligned} \tag{4.14}$$

Thus it is sufficient to consider the matrix elements $\mathring{\Phi}_{11}(\lambda)$, $\mathring{\Phi}_{12}(\lambda)$.

The system of singular integral equations for these matrix elements is obtained quite similarly to (3.13), (3.14). Repeating formally the corresponding calculations of section 3 one arrives at the following representations for these matrix elements instead of (3.14), (3.19):

$$\mathring{\Phi}_{12}(\lambda) = \frac{1}{i\pi}\int_\Gamma \frac{\mathring{\Phi}_{11,+}(\mu)\beta_+(\mu)e^{2i\mu x}(\mu - \lambda_1)\,d\mu}{\alpha_+(\mu)(e^{\mu^2-\beta}-1)(\mu+\lambda_1)(\mu-\lambda)} \tag{4.15}$$

$$\begin{aligned}
\mathring{\Phi}_{11}(\lambda) &= \alpha^{-1}(\lambda)\beta^{-1}(\lambda) \\
&\quad - \frac{\alpha^{-1}(\lambda)\beta^{-1}(\lambda)}{2\pi i}\sum_{\substack{k=-\infty \\ k\neq 0,1}}^{\infty} A(\lambda_k,\lambda)\mathring{\Phi}_{11}(\lambda_k),
\end{aligned} \tag{4.16}$$

$$\lambda \in \Omega_+$$

where the λ_k are now (instead of (3.17)):

$$\begin{aligned}
\lambda_k &= [\beta + 2\pi i(k-1)]^{1/2}, \quad k = \pm 1, \pm 2, \dots, \quad \text{Im}\,\lambda_k \geq 0, \\
\lambda_{\pm 1} &= \pm\sqrt{\beta}, \quad \beta > 0
\end{aligned} \tag{4.17}$$

$$\begin{aligned}
A(\lambda_k,\lambda) &= \frac{(\lambda_k - \lambda_1)2\beta(\lambda_k)}{(\lambda_k + \lambda_1)\alpha(\lambda_k)\lambda_k}e^{2i\lambda_k x} \\
&\quad \times \int_\Gamma \frac{(\mu+\lambda_1)}{(\mu-\lambda_1)}\frac{\alpha_-^2(\mu)e^{-2i\mu x}\,d\mu}{(e^{\mu^2-\beta}-1)(\lambda_k-\mu)(\mu-\lambda)}.
\end{aligned} \tag{4.18}$$

Here Ω_+ is the part of the complex λ-plane lying to the left of contour Γ. It is worth mentioning that now the oscillatory term $A(\lambda_1,\lambda)\mathring{\Phi}_{11}(\lambda)$

nonvanishing as $x \to \infty$ is absent (compare with (3.19)) due to the fact that the matrix RP (XV.2.35) is used. Transforming via (XV.2.41) from the function $\chi(\lambda)$ to $\overset{\circ}{\Phi}(\lambda)$ effectively cancels the singularities at $\pm\lambda_1$ in all the integrals.

Strictly speaking, the reference to the case $\beta < 0$ is not quite correct when deriving representation (4.15), (4.16)—one cannot use now the results concerning the direct spectral problem (see, e.g., [4]) for the Dirac operator and hence estimates (3.15), (3.16) cannot be used. However, these representations can be justified independently. The result is that the series (4.16) converges uniformly in x, β, λ in the region

$$x \geq x_0, \quad \delta \leq \beta \leq \beta_0, \quad \lambda \in \Omega_\pm \tag{4.19}$$

where x_0, β_0, δ are arbitrary (but finite) positive constants and Ω_+ (Ω_-) is the part of the λ-plane to the left (right) of Γ. In addition, all the quantities entering (4.15), (4.16) possess limits when λ approaches Γ both from the left and from the right, and all the passages to the limit, summations and integrations can be done in any order. In other words, if one puts $\lambda = \lambda_k$ ($k = \pm 2, \pm 3, \ldots$) in (4.16), obtaining an infinite-dimensional system of linear equations for $\overset{\circ}{\Phi}_{11}(\lambda_k)$, then this system is uniquely solvable for $x \geq x_0$, $\delta \leq \beta \leq \beta_0$, and the functions $\overset{\circ}{\Phi}_{11}(\lambda)$, $\overset{\circ}{\Phi}_{12}(\lambda)$ defined by the right hand sides of (4.15), (4.16) give the solution of the matrix RP. The proof of these statements is given in paper [8].

So for x large enough one can use representations (4.15), (4.16) (and also relations (4.14)) to investigate the solution of the regularized matrix RP (XV.2.35). It is also worth mentioning that the following estimates are valid,

$$|\overset{\circ}{\Phi}_{11}(\lambda_k)| < C, \quad k = \pm 2, \pm 3, \ldots \tag{4.20}$$

$$|A(\lambda_k, \lambda)| < Ce^{-2x\mathrm{Im}\,\lambda_k}, \quad \forall \lambda, \ k = \pm 2, \pm 3, \ldots, \ x \geq 0 \tag{4.21}$$

$$|A(\lambda_j, \lambda_k)| < C\frac{e^{-2x\mathrm{Im}\,\lambda_2 - 2x\mathrm{Im}\,\lambda_k}}{|\lambda_k|^2|\lambda_j|}, \quad j, k = \pm 2, \pm 3, \ldots, \tag{4.22a}$$

and that the estimate (4.20) for the solution is uniform for $x \geq x_0$, $\delta \leq \beta \leq \beta_0$.

The representations (4.15), (4.16) can be used to calculate the asymptotics of $\overset{\circ}{\Phi}(\lambda)$ to within any accuracy given in advance, in the scale $e^{-2\mathrm{Im}\,\lambda_k}$. The scheme of calculations needed is given in paper [8]. Here we calculate explicitly the main term of each asymptotic, which is the most interesting one. It is evident from the above discussion that the following asymptotic formulæ are valid:

$$\overset{\circ}{\Phi}_{11}(\lambda) = \alpha^{-1}(\lambda)\beta^{-1}(\lambda) + O\left(e^{-2x\mathrm{Im}\,\lambda_2}\right), \quad x \to +\infty \tag{4.22b}$$

$$\mathring{\Phi}_{12}(\lambda) = \frac{1}{i\pi} \int_\Gamma \frac{e^{2i\mu x}(\mu - \lambda_1)\, d\mu}{\alpha_+^2(\mu)(e^{\mu^2 - \beta} - 1)(\mu + \lambda_1)(\mu - \lambda)}$$
$$+ O\left(e^{-4x \operatorname{Im} \lambda_2}\right), \quad x \to +\infty, \tag{4.23}$$

uniformly in $\lambda \in \Omega_\pm$, $\delta \le \beta \le \beta_0$. (In fact, these estimates can be made more precise. Namely, if $\lambda \in \Omega_+$, one can replace $O\left(e^{-2x \operatorname{Im} \lambda_2}\right)$ in (4.22) by $O\left(e^{-4x \operatorname{Im} \lambda_2}\right)$. Analogously, if $\lambda \in \Omega_-$, one can replace $O\left(e^{-4x \operatorname{Im} \lambda_2}\right)$ in (4.23) by $O\left(e^{-6x \operatorname{Im} \lambda_2}\right)$. This makes calculation of the higher order terms of the asymptotics easier.)

Using (4.5), (4.6) one may now easily calculate

$$\mathring{B}_{+-}(x, \beta) = \frac{1}{\pi} \int_\Gamma d\mu \, \ln\left(\frac{e^{\mu^2 - \beta} + 1}{e^{\mu^2 - \beta} - 1}\right) + O\left(e^{-2x \operatorname{Im} \lambda_2}\right)$$
$$= 2i\lambda_1 + C(\beta) + O\left(e^{-2x \operatorname{Im} \lambda_2}\right), \quad x \to +\infty \tag{4.24}$$

$$\mathring{B}_{++}(x, \beta) = O\left(e^{-2x \operatorname{Im} \lambda_2}\right) \tag{4.25}$$

where the notation

$$C(\beta) = \frac{1}{\pi} \int_{-\infty}^{+\infty} d\mu \, \ln\left|\frac{e^{\mu^2 - \beta} + 1}{e^{\mu^2 - \beta} - 1}\right| \tag{4.26}$$

is introduced.

Let us now calculate the asymptotics of the quantities d, e (4.7), (4.8) necessary to calculate the correlator (4.4). To this end one calculates first the matrix elements $\mathring{\Phi}_{11}(-\lambda_1)$, $\mathring{\Phi}_{12}(\lambda_1)$ with the result

$$\mathring{\Phi}_{11}(-\lambda_1) = \alpha(\lambda_1)\beta(\lambda_1) + O\left(e^{-2x \operatorname{Im} \lambda_2}\right), \quad x \to +\infty \tag{4.27}$$

$$\mathring{\Phi}_{12}(\lambda_1) = \frac{e^{2i\lambda_1 x}}{2\alpha^2(\lambda_1)\lambda_1^2} + O\left(e^{-2x \operatorname{Im} \lambda_2}\right). \quad x \to +\infty \tag{4.28}$$

It is shown in Appendix 2 that

$$2\lambda_1^2 \alpha^2(\lambda_1) = e^{-i\varphi(\beta)/2} \tag{4.29}$$

where the real phase $\varphi(\beta)$ is equal to

$$\varphi(\beta) = -\pi + \frac{2}{\pi} \fint_{-\infty}^{+\infty} \frac{d\mu}{\mu - \lambda_1} \ln\left|\frac{e^{\mu^2 - \beta} + 1}{e^{\mu^2 - \beta} - 1}\right|. \tag{4.30}$$

By virtue of relations (4.27), (4.28), (4.7), (4.8) and (4.14) one calculates

$$d(x, \beta) = -2i\alpha(\lambda_1)\beta(\lambda_1)e^{i\theta(x,\beta)} \sin\theta(x, \beta)$$

$$+ O\left(e^{-2x\operatorname{Im}\lambda_2}\right), \qquad x \to +\infty \qquad (4.31)$$

$$[-2e(x, \beta) + d(x, \beta)\mathring{B}_{++}(x, \beta)] = 4i\lambda_1\alpha(\lambda_1)\beta(\lambda_1)e^{i\theta(x,\beta)}$$

$$+ O\left(e^{-2x\operatorname{Im}\lambda_2}\right), \qquad x \to +\infty \qquad (4.32)$$

where the phase $\theta(x, \beta)$ is defined by

$$\theta(x, \beta) \equiv 2\lambda_1 x + \frac{\varphi(\beta)}{2} = 2x\sqrt{\beta} + \frac{\varphi(\beta)}{2}. \qquad (4.33)$$

Asymptotic formulæ (4.32), (4.24) together with the exact relation (4.9) result immediately in the following asymptotic representation for the correlator $g(x, \beta)$ (4.4):

$$g(x, \beta) = e^{c(\beta)}e^{-xC(\beta)}\left[1 + O\left(e^{-2x\operatorname{Im}\lambda_2}\right)\right] \qquad (4.34)$$

where the function $C(\beta)$ is given in (4.26) and $c(\beta)$ does not depend on x being equal to the sum of the expression

$$i\frac{\varphi}{2} + \ln\left[i\lambda_1\alpha(\lambda_1)\beta(\lambda_1)\right]$$

and the integration constant (β-dependent) in the relation equivalent to (4.9):

$$\sigma_0(x, \beta) = -\int dx\, \mathring{B}_{+-}(x, \beta).$$

Let us now obtain the dependence of $c(\beta)$ on β. To do this one uses relation (4.10) for $\partial_\beta\sigma$ rewriting it for the function $g(x, \beta)$ (which can also be represented as

$$g(x, \beta) = \frac{1}{4}B_{++}(x, \beta)e^{\sigma(x,\beta)})$$

as

$$\partial_\beta \ln g(x, \beta) = \partial_\beta \ln B_{++} - x\partial_\beta B_{+-}$$

$$+ \frac{1}{2}(\partial_\beta B_{+-})^2 - \frac{1}{2}(\partial_\beta B_{++})^2. \qquad (4.35)$$

Denote by $\{\mathring{x}_n\}$ the set of zeros of $d(x, \beta)$. It follows from the asymptotics (4.31) that

$$\mathring{x}_n = \frac{\pi n}{2\lambda_1} - \frac{\varphi}{4\lambda_1} + o(1), \qquad n \to \infty. \qquad (4.36)$$

So one can make compatible the following two conditions on the behavior of the variable x,

$$x \to +\infty, \qquad |x - \mathring{x}_n| \geq \rho_0 > 0 \qquad (4.37)$$

where ρ_0 is any number in the interval $0 < \rho_0 < \pi/4$.

It follows from (4.31), (4.32), (4.2) and (4.1) that under conditions (4.37) one gets the following asymptotics for the potentials B_{++}, B_{+-}:

$$B_{+-}(x, \beta) = -2\lambda_1 \cot \theta(x, \beta) + C(\beta) + O\left(e^{-2x \mathrm{Im}\, \lambda_2}\right) \qquad (4.38)$$

$$B_{++}(x, \beta) = -\frac{2\lambda_1}{\sin \theta(x, \beta)} + O\left(e^{-2x \mathrm{Im}\, \lambda_2}\right) \qquad (4.39)$$

$$(\lambda_1 \equiv \sqrt{\beta}).$$

Now let us put the asymptotics (4.34) of $g(x, \beta)$ into the left hand side of (4.35) and use (4.38), (4.39) to evaluate the right hand side. Taking into account that the function $c(\beta)$ does not depend on x, one comes after obvious calculations to the following identity:

$$\partial_\beta c(\beta) = -[\partial_\beta \theta(x, \beta)] \cot \theta(x, \beta) + 2x \partial_\beta[\lambda_1 \cot \theta(x, \beta)]$$

$$+ \frac{1}{2\beta} + \frac{1}{2}\left\{\partial_\beta C(\beta) - 2\partial_\beta[\lambda_1 \cot \theta(x, \beta)]\right\}^2$$

$$- \frac{1}{2}\left[2\partial_\beta \frac{\lambda_1}{\sin \theta(x, \beta)}\right]^2. \qquad (4.40)$$

Direct calculation using the equality established in Appendix 3,

$$\partial_\beta C(\beta) = -\frac{\lambda_1}{2} \partial_\beta \varphi(\beta), \qquad (4.41)$$

enables us to transform the right hand side in (4.40) to the explicitly x-independent form, obtaining

$$\partial_\beta c(\beta) = \frac{1}{2}[\partial_\beta C(\beta)]^2.$$

Thus one comes to the following representation for $c(\beta)$:

$$c(\beta) = -\frac{1}{2} \int\limits_\beta^\infty \left(\frac{dC(\beta')}{d\beta'}\right)^2 d\beta' + c_0 \qquad (4.42)$$

where c_0 is a numerical constant which does not depend on either x or β. The convergence of the integral at the upper limit is proved in Appendix 4.

Formulæ (4.32), (4.42) complete the analysis of the main terms in the asymptotic expansion of the correlator (4.4). Let us write down the answer including also the next terms (corresponding calculations are given in detail in paper [8]):

$$\langle \Psi^\dagger(x_1)\Psi(x_2)\rangle_T$$

$$= \rho_T \sqrt{\frac{T}{\pi}} \exp\left\{ -\frac{1}{2} \int_\beta^\infty \left(\frac{dC(\beta')}{d\beta'} \right)^2 d\beta' \right\}$$

$$\times e^{-xC(\beta)} \left[1 - 2\mathrm{Re}\left(\frac{ie^{2i\lambda_2 x}}{\lambda_1\lambda_2(\lambda_1 + \lambda_2)^2\alpha^2(\lambda_2)} \right. \right.$$

$$\left. \times \left[(\lambda_1^2 + \lambda_2^2) \sin\theta(x,\beta) + 2i\lambda_1\lambda_2 \cos\theta(x,\beta) \right] \right)$$

$$\left. + o\left(e^{-2x\mathrm{Im}\,\lambda_2} \right) \right], \tag{4.43}$$

$$x \equiv \frac{1}{2}|x_1 - x_2|\sqrt{T} \to +\infty, \quad \beta \equiv \frac{h}{T} > 0.$$

The function $C(\beta)$ here is given by (4.26), $\theta(x,\beta)$ by (4.33). The quantities λ_1, λ_2 (functions of β, see (4.17)) are

$$\lambda_1 \equiv \sqrt{\beta}; \quad \lambda_2 = (\beta + 2\pi i)^{1/2}, \quad \mathrm{Im}\,\lambda_1 > 0, \ \mathrm{Re}\,\lambda_1 > 0$$

$$\mathrm{Re}\,\lambda_2 = \sqrt{\frac{1}{2}\left(\beta + \sqrt{\beta^2 + 4\pi^2} \right)} \tag{4.44}$$

$$\mathrm{Im}\,\lambda_2 = \sqrt{\frac{1}{2}\left(\sqrt{\beta^2 + 4\pi^2} - \beta \right)}.$$

The function $\alpha(\lambda_2)$ is calculated (see Appendix 2) as

$$\alpha^{-1}(\lambda_2) = \sqrt{\frac{\lambda_2 + \lambda_1}{\lambda_2 - \lambda_1}} \exp\left\{ \frac{i}{2\pi} \int_{-\infty}^{+\infty} \frac{d\mu}{\mu - \lambda_2} \ln\left| \frac{e^{\mu^2 - \beta} + 1}{e^{\mu^2 - \beta} - 1} \right| \right\} \tag{4.45}$$

so that the right hand side in (4.43) is, in fact, independent of the contour Γ. Finally, the numerical constant ρ_T in (4.43) does not depend on either x or β. For numerical constant ρ_T see paper [8].

XVI.5 Explicit formulæ for asymptotics of the temperature-dependent equal-time field correlator for the impenetrable Bose gas

At zero temperature the main term in the asymptotics is a power [14], [19]:

$$\langle \Psi^\dagger(x_1)\Psi(x_2)\rangle_0 \simeq \frac{\mathrm{const}}{\sqrt{q|x_1 - x_2|}}, \tag{5.1}$$

$$|x_1 - x_2| \to +\infty, \quad h > 0, \quad q = \sqrt{h}.$$

The sign of the chemical potential h is very important for the asymptotics of correlation functions. We might expect this from the facts established in Part I (see (I.6.24), (I.6.25) and after (I.7.14)). The effect of changing the sign of the chemical potential is especially strong at zero temperature. The ground state of the Bose gas changes from zero density (at $h < 0$) to finite density (at $h > 0$). At positive temperatures the density of the ground state of the Bose gas is positive for any sign of the chemical potential, nevertheless the asymptotics of correlation functions do depend on the sign of h. For temperature correlations the asymptotics are given by (3.3) for $h < 0$ and by (4.43) in the case $h > 0$. The main term for $h < 0$ is

$$\langle \Psi^\dagger(x_1)\Psi(x_2)\rangle_T \simeq \frac{1}{2}\frac{T}{\sqrt{|h|}}\,\exp\left\{\frac{1}{2}\int\limits_{-\infty}^{h/T}\left(\frac{dC(\beta')}{d\beta'}\right)^2 d\beta' - a(h/T)\right\}$$

$$\times\, e^{-|x_1-x_2|/r_f}, \qquad |x_1-x_2| \to +\infty, \qquad h < 0 \qquad (5.2)$$

where the functions $C(\beta)$ and $a(\beta)$ are given by

$$C(\beta) \equiv \frac{1}{\pi}\int\limits_{-\infty}^{+\infty} d\mu\,\ln\left(\frac{e^{\mu^2-\beta}+1}{e^{\mu^2-\beta}-1}\right), \qquad \beta = \frac{h}{T} < 0 \qquad (5.3)$$

$$a(\beta) = \frac{\sqrt{|\beta|}}{\pi}\int\limits_{-\infty}^{+\infty}\frac{d\mu}{\mu^2-\beta}\,\ln\left(\frac{e^{\mu^2-\beta}+1}{e^{\mu^2-\beta}-1}\right), \qquad \beta = \frac{h}{T} < 0. \qquad (5.4)$$

The correlation radius (length) r_f is given by

$$\frac{1}{r_f} = \frac{1}{2}\sqrt{T}\,C\left(\frac{h}{T}\right) + \sqrt{|h|}, \qquad h < 0. \qquad (5.5)$$

For positive chemical potential the main term in the asymptotics is given by

$$\langle \Psi^\dagger(x_1)\Psi(x_2)\rangle_T \simeq \rho_T\sqrt{\frac{T}{\pi}}\,\exp\left\{-\frac{1}{2}\int\limits_{h/T}^{\infty}\left(\frac{dC(\beta')}{d\beta'}\right)^2 d\beta'\right\}$$

$$\times\, e^{-|x_1-x_2|/r_f}, \qquad |x_1-x_2| \to +\infty, \qquad h > 0 \qquad (5.6)$$

where now

$$C(\beta) \equiv \frac{1}{\pi}\int\limits_{-\infty}^{+\infty} d\mu\,\ln\left|\frac{e^{\mu^2-\beta}+1}{e^{\mu^2-\beta}-1}\right|, \qquad \beta = \frac{h}{T} > 0 \qquad (5.7)$$

and

$$\frac{1}{r_f} = \frac{1}{2}\sqrt{T}\,C\left(\frac{h}{T}\right), \qquad h > 0. \qquad (5.8)$$

For the numerical constant ρ_T see paper [8].

The behavior of the correlation radius (length) r_f for $T \to 0$ and $T \to \infty$ is as follows (see Appendix 4):

$$\frac{1}{r_f} = \frac{\pi T}{4\sqrt{h}} = \frac{T}{4D_0(h)} \qquad (h > 0, \ T \to 0)$$

$$\frac{1}{r_f} = \sqrt{|h|} \qquad (h < 0, \ T \to 0) \tag{5.9}$$

$$\frac{1}{r_f} = \sqrt{\frac{T}{\pi}} \left(1 - \frac{1}{2^{3/2}} \right) \zeta\left(\frac{3}{2}\right), \tag{5.10}$$
$$(T \to \infty, \ -\infty < h < +\infty)$$

where $\zeta(s)$ is the Riemann ζ-function,

$$\zeta(s) \equiv \sum_{n=1}^{\infty} \frac{1}{n^s}. \tag{5.11}$$

Let us also give the main term of the generating functional for currents (see (1.42)):

$$\langle \exp\{\alpha Q_1(x_1, x_2)\}\rangle_T \simeq \exp\left\{ \frac{1}{2} \int_{-\infty}^{h/T} \left(\frac{dC(\beta')}{d\beta'} \right)^2 d\beta' \right\} \exp\{-|x_1 - x_2|/r_q\}, \tag{5.12}$$

where the function $C(\beta, \gamma)$ is given by (1.38):

$$C(\beta, \gamma) \equiv \frac{1}{\pi} \int_{-\infty}^{+\infty} d\mu \ln\left(\frac{1 + e^{\mu^2 - \beta}}{1 - \pi\gamma + e^{\mu^2 - \beta}} \right) \tag{5.13}$$

and the parameter γ is related to α by

$$\gamma = \frac{1 - e^{\alpha}}{\pi}, \qquad 0 < \gamma < \frac{1}{\pi}, \qquad 0 > \alpha > -\infty. \tag{5.14}$$

The correlation length r_q is given by the formula

$$\frac{1}{r_q} = \frac{1}{2}\sqrt{T}\, C(\beta, \gamma). \tag{5.15}$$

This has a clear physical meaning and can be generalized to the case of finite coupling constant $(0 < c < \infty)$, see section XVII.5. Finally, the asymptotics of the "emptiness formation probability" (2.27) are:

$$P(x_1, x_2) \equiv \langle \exp\{\alpha Q_1(x_1, x_2)\}\rangle_T \Big|_{\alpha=-\infty}$$

$$\simeq \exp\left\{\frac{1}{2} \int_{-\infty}^{h/T} \left(\frac{dc_0(\beta')}{d\beta'}\right)^2 d\beta'\right\} \exp\{-|x_1 - x_2|/r_\infty\},$$

$$(5.16)$$

where (see (2.28))

$$c_0(\beta) \equiv \frac{1}{\pi} \int_{-\infty}^{+\infty} d\mu \, \ln\left(1 + e^{\beta - \mu^2}\right) \qquad (5.17)$$

and the correlation length r_∞ is given by

$$\frac{1}{r_\infty} = \frac{1}{2}\sqrt{T}\, c_0(\beta). \qquad (5.18)$$

At zero temperature this term (the right hand side of (5.16)) disappears from the asymptotics, and the asymptotics change to gaussian decay (see formula (9.11) from paper [13]).

XVI.6 Time-dependent correlator of two fields at finite temperature in the case of negative chemical potential

In sections 6–9 of this chapter the large time and distance asymptotics of the time-dependent correlator (XIV.6.2),

$$\langle \Psi(x_2, t_2)\Psi^\dagger(x_1, t_1)\rangle_T = \sqrt{T}g(x, t, \beta),$$

$$g(x, t, \beta) = -\frac{1}{2\pi} e^{2it\beta} b_{++}(x, t, \beta) e^{\sigma(x,t,\beta)}$$

$$(6.1)$$

are considered. Here x, t, β,

$$x \equiv \frac{1}{2}|x_1 - x_2|\sqrt{T} \to +\infty,$$

$$t \equiv \frac{1}{2}(t_2 - t_1)T \to +\infty, \qquad (6.2)$$

$$\beta \equiv \frac{h}{T} < 0$$

are the "scaled" distance, time and chemical potential, and $\sigma(x, t, \beta)$ (XIV.6.57) is the logarithm of the Fredholm determinant of "integrable linear integral operator" $\hat{V}_T$ (XIV.6.3). The potential $b_{++}(x, t, \beta)$ (XIV.6.8),

$$b_{++} = B_{++} - G \qquad (6.3)$$

and the derivatives of $\sigma(x, t, \beta)$ with respect to x, t, β (XIV.6.58)–(XIV.6.60),

$$\partial_x \sigma = -2iB_{+-}$$
$$\partial_t \sigma = -2i(C_{+-} + C_{-+} + GB_{--}) \tag{6.4}$$
$$\partial_\beta \sigma = -2it\partial_\beta(C_{+-} + C_{-+} + GB_{--}) - 2ix\partial_\beta B_{+-}$$
$$+ 2it(b_{++}\partial_\beta B_{--} - B_{--}\partial_\beta b_{++})$$
$$+ 2(\partial_\beta b_{++})(\partial_\beta B_{--}) - 2(\partial_\beta B_{+-})^2$$

are expressed in terms of the potentials B, C (XIV.6.9), (XIV.6.10) (the functions $G \equiv G(x, t)$ and $E(\lambda) \equiv E(\lambda | x, t)$,

$$G \equiv \int\limits_{-\infty}^{+\infty} d\mu\, e^{-2it\mu^2 - 2ix\mu}, \quad \partial_\beta G = 0$$

$$E(\lambda) \equiv \fint\limits_{-\infty}^{+\infty} \frac{d\mu}{\mu - \lambda} e^{-2it\mu^2 - 2ix\mu}, \quad \partial_\beta E(\lambda) = 0 \tag{6.5}$$

are defined in (XIV.6.5), (XIV.6.6)).

It was shown in sections XV.3, XV.4 that the potentials are expressed in terms of solutions of matrix RP (XV.3.3), so that the asymptotic analysis of the correlator is reduced to the analysis of the matrix RP. This asymptotic analysis is much more complicated than in the case of the equal-time correlators considered earlier in this chapter. The reason is that now, at the asymptotical condition

$$t \to +\infty, \quad x \to +\infty, \quad -\frac{x}{2t} = O(1) \tag{6.6}$$

the phase $\theta(x, t, \lambda)$,

$$\theta(x, t, \lambda) \equiv t\lambda^2 + x\lambda \tag{6.7}$$

(entering the conjugation matrix $G(\lambda)$ (XV.3.3c), see (XV.3.4)) possesses a stationary point λ_0,

$$\lambda_0 \equiv -\frac{x}{2t} < 0 \tag{6.8}$$

on the real axis. This fact is known in the classical theory of solitons to be critical for the asymptotic analysis (see, e.g., [1], [15], [16]. The corresponding difficulties were overcome, in principle, in paper [20]. Later on the scheme suggested there was developed in papers [2], [3], [10], [11], [17]. In the case considered the version of this method given in paper [10] is most suitable. It is based on the direct analysis of the matrix RP (and not of solutions of the corresponding nonlinear differential equations); and just the data of the RP (and not the initial data of the nonlinear evolution system (XIV.6.44)–(XIV.6.49)) are the original objects here. In the rest of this chapter we follow paper [9], where the asymptotics of correlator (6.1) were calculated. Here only the main points of the calculations and the results are given; for details see [9].

In this section the relatively simple case of negative chemical potential, $\beta < 0$, is considered. The simplicity here is due to the absence of zeros of the expression

$$e^{\lambda^2 - \beta} - 1$$

on the real λ-axis. So the original matrix RP (XV.3.3) as well as the matrix RP (XV.4.7) and scalar RP (XV.4.6) for the function, $\alpha(\lambda)$,

$$\alpha(\lambda) \equiv \exp\left\{ -\frac{1}{2\pi i} \int\limits_{-\infty}^{+\infty} \frac{d\mu}{\mu - \lambda} \ln \varphi(\mu^2, \beta) \right\}, \quad \beta < 0; \qquad (6.9)$$

$$\varphi(\mu^2, \beta) \equiv \frac{e^{\mu^2 - \beta} - 1}{e^{\mu^2 - \beta} + 1}, \qquad (6.10)$$

possess no indices and are uniquely solvable. It is convenient to use matrix RP (XV.4.7) for the function $\Phi(\lambda)$ (XV.4.3). It can be shown [9] that the boundary values $\Phi_{\pm}(\lambda)$ of $\Phi(\lambda)$ on the real axis are bounded uniformly in $\lambda \in \mathbb{R}$ for $t \to +\infty$.

The potentials B, C are expressed directly in terms of matrix elements of the coefficients $\Phi_1(x, t, \beta)$, $\Phi_2(x, t, \beta)$ in the expansion

$$\Phi(\lambda) = 1 + \frac{\Phi_1}{\lambda} + \frac{\Phi_2}{\lambda^2} + \cdots, \quad \lambda \to \infty$$

by means of the relations (XV.4.10),

$$\begin{aligned}
&B_{+-} = B_{-+} = -\alpha_0 - (\Phi_1)_{11}; \\
&b_{++} = B_{++} - G = (\Phi_1)_{12}; \qquad B_{--} = -(\Phi_1)_{21}; \qquad (6.11) \\
&C_{+-} + C_{-+} + B_{--}G = (\Phi_2)_{22} - (\Phi_2)_{11};
\end{aligned}$$

where α_0 is the coefficient of $1/\lambda$ in the expansion

$$\alpha(\lambda) = 1 + \frac{\alpha_0}{\lambda} + \cdots, \qquad \lambda \to \infty$$

$$\alpha_0 = \frac{1}{2\pi i} \int\limits_{-\infty}^{+\infty} d\mu \ln \varphi(\mu^2, \beta). \qquad (6.12)$$

In the asymptotic analysis of $\Phi(\lambda)$ one has to take into account properly the stationary point λ_0 (6.8). To this end, following the ideas of [10] and [15] one considers the piecewise analytic matrix function ("Manakov

Ansatz") $\Phi^m(\lambda)$ given as

$$\Phi^m(\lambda) = \begin{pmatrix} 1 & -I^p(\lambda) \\ -I^q(\lambda) & 1 \end{pmatrix} e^{\sigma_3 \ln \delta(\lambda)} \qquad (6.13)$$

where

$$I^p(\lambda) = \frac{1}{2\pi i} \int_{-\infty}^{+\infty} \frac{\delta_+(\mu)\delta_-(\mu)}{\mu - \lambda} p(\mu) e^{-2i\theta(x,t,\mu)} \, d\mu$$

$$\qquad (6.14)$$

$$I^q(\lambda) = \frac{1}{2\pi i} \int_{-\infty}^{+\infty} \frac{\delta_+^{-1}(\mu)\delta_-^{-1}(\mu)}{\mu - \lambda} q(\mu) e^{2i\theta(x,t,\mu)} \, d\mu.$$

The functions $p(\mu)$, $q(\mu)$ are given by (XV.4.8),

$$p(\lambda) = -2\pi i \alpha_-^2(\lambda) \frac{e^{\lambda^2 - \beta}}{e^{\lambda^2 - \beta} - 1}$$

$$\qquad (6.15)$$

$$q(\lambda) = -\frac{2i}{\pi} \alpha_+^{-2}(\lambda) \frac{1}{e^{\lambda^2 - \beta} - 1}.$$

These functions possess no singularities (or zeros) on the real λ-axis. The function $\delta(\lambda)$ solves the scalar RP

$$\delta_+(\lambda) = \delta_-(\lambda) \left[1 + p(\lambda)q(\lambda)\eta(\lambda_0 - \lambda)\right], \quad \text{Im } \lambda = 0 \qquad (6.16a)$$

$$\delta(\infty) = 1. \qquad (6.16b)$$

Due to the relation

$$1 + p(\lambda)q(\lambda) = \varphi^2(\lambda^2, \beta), \qquad (6.17)$$

this RP can be solved explicitly with the result

$$\delta(\lambda) = \exp\left\{ \frac{1}{\pi i} \int_{-\infty}^{\lambda_0} \frac{d\mu}{\mu - \lambda} \ln \varphi(\mu^2, \beta) \right\} \qquad (6.18)$$

(the notation $\eta(\lambda)$ for the step function,

$$\eta(\lambda) = \begin{cases} 1, & \text{if } \lambda > 0 \\ 0, & \text{if } \lambda < 0. \end{cases} \qquad (6.19)$$

is introduced here).

It is shown in paper [9] that the function $\Phi^m(\lambda)$ satisfies the relation

$$\Phi_-^m(\lambda) = \Phi_+^m(\lambda) \left[G_\Phi(\lambda, \{p, q\}) + r_0(\lambda)\right] \qquad (6.20)$$

where $G_\Phi(\lambda, \{p, q\})$ is just the conjugation matrix of RP (XV.4.7) for $\Phi(\lambda)$ and the matrix $r_0(\lambda)$ satisfies the estimates

$$r_0(\lambda) = \begin{cases} O\left(t^{-1/2}(\lambda - \lambda_0)^{-1}\right), & |\lambda - \lambda_0| \geq t^{-1/2+\varepsilon} \\ O(1), & |\lambda - \lambda_0| \leq t^{-1/2+\varepsilon} \end{cases} \tag{6.21}$$

$$t \to +\infty, \quad 0 < \varepsilon < \frac{1}{2}, \quad \lambda_0 = -\frac{x}{2t} = O(1).$$

This fact, together with the uniform boundedness in $\lambda \in \mathbb{R}$ of the exact solution $\Phi(\lambda)$, results in the following asymptotical representation of $\Phi(\lambda)$ for $\lambda \neq \lambda_0$:

$$\Phi(\lambda) = \left[1 + O\left(t^{-\varepsilon}\right)\right] \Phi^m(\lambda),$$

$$0 < \varepsilon < \frac{1}{2}, \quad t \to +\infty, \quad \lambda_0 = -\frac{x}{2t} = O(1) \tag{6.22}$$

which holds uniformly in λ for $|\mathrm{Im}\,\lambda| \geq c_0$, so that $\Phi^m(\lambda)$ is indeed the approximation to the solution of matrix RP (XV.4.7).

Though this estimate is not very accurate, one can use it to obtain the main terms in the asymptotics of the potentials (6.11) which are expressed in terms of the diagonal matrix elements of $\Phi(\lambda)$, i.e., of the potentials which are of order $O(1)$ for $t \to +\infty$. Namely, one obtains from (6.11), (6.22) the follwing asymptotic representations:

$$B_{+-} = -\alpha_0 - \delta_0 + o(1)$$

$$C_{+-} + C_{-+} + B_{--}G = -2\delta_1 + o(1) \tag{6.23}$$

where α_0 is given by (6.12) and δ_0, d_1 are extracted from the expansion at $\lambda = \infty$ of $\delta(\lambda)$ in the usual way:

$$\delta_0 \equiv \frac{i}{\pi} \int_{-\infty}^{\lambda_0} d\mu \, \ln \varphi(\mu^2, \beta)$$

$$\delta_1 \equiv \frac{i}{\pi} \int_{-\infty}^{\lambda_0} d\mu \, \mu \ln \varphi(\mu^2, \beta). \tag{6.24}$$

Due to (6.4), equalities (6.23) result in the following estimates for the derivatives with respect to x and t of $\sigma(x, t, \beta)$:

$$\partial_x \sigma = \frac{1}{\pi} \int_{-\infty}^{+\infty} \mathrm{sign}(\mu - \lambda_0) \ln \varphi(\mu^2, \beta) \, d\mu + o(1) \tag{6.25}$$

$$\partial_t \sigma = \frac{2}{\pi} \int_{-\infty}^{+\infty} \mathrm{sign}(\mu - \lambda_0) \mu \ln \varphi(\mu^2, \beta) \, d\mu + o(1) \tag{6.26}$$

where sign(λ) is the sign function,

$$\text{sign}(\lambda) = +1 \ \text{ if } \ \lambda > 0, \qquad \text{sign}(\lambda) = -1 \ \text{ if } \ \lambda < 0.$$

In writing (6.26) it was taken into account that $\varphi(\mu^2, \beta)$ is an even function of μ so that

$$\int\limits_{-\infty}^{+\infty} \mu \ln \varphi(\mu^2, \beta) \, d\mu = 0.$$

To obtain the leading terms of the asymptotics of the correlator $g(x, t, \beta)$ (6.1) estimates (6.25), (6.26) are not, however, sufficient. First, one has to evaluate the potential b_{++} for $t \to +\infty$. Second, it is necessary to have integrated estimates (6.25), (6.26) which requires making the next terms in these estimates more precise. For solving this problem it is convenient to use the nonlinear differential system (XIV.6.44)–(XIV.6.46) for the potentials b_{++}, B_{--}. Estimate (6.22) means, in particular, that the pair $b_{++}(x, t, \beta)$, $B_{--}(x, t, \beta)$ give a solution which decreases for $t \to +\infty$ as some power of t. It follows that necessarily

$$b_{++} = O\left(t^{-1/2}\right), \quad B_{--} = O\left(t^{-1/2}\right), \quad t \to +\infty \tag{6.27}$$

which is compatible, e.g., with equations (XIV.6.50),

$$\partial_x B_{+-} = 2i b_{++} B_{--} \tag{6.28}$$

and (6.23). What is more, the structure of the asymptotic expansion of the decreasing solutions of system (XIV.6.44)–(XIV.6.46) is known (see, e.g., [2]), being given by the following asymptotic series:

$$
\begin{aligned}
b_{++} &= t^{-1/2} \left(u_0 + \sum_{n=1}^{\infty} \sum_{k=0}^{2n} \frac{(\ln 4t)^k}{t^n} u_{nk} \right) \\
&\quad \times \exp\left\{ \frac{ix^2}{2t} - i\nu \ln 4t \right\}, \\
B_{--} &= t^{-1/2} \left(v_0 + \sum_{n=1}^{\infty} \sum_{k=0}^{2n} \frac{(\ln 4t)^k}{t^n} v_{nk} \right) \\
&\quad \times \exp\left\{ -\frac{ix^2}{2t} + i\nu \ln 4t \right\}.
\end{aligned}
\tag{6.29}
$$

The quantities u_0, v_0, u_{nk}, v_{nk} and ν here are functions of $\lambda_0 \equiv -x/2t$ (and of β), u_0 and v_0 being the determining functional parameters of the asymptotics (6.29). All the other coefficients ν, u_{nk}, v_{nk} are restored uniquely and in explicit form for given u_0, v_0. In particular, the following relations are valid (the prime in what follows denotes the derivative with respect to λ_0):

$$\nu = -4u_0 v_0$$
$$v_{12}u_0 + u_{12}v_0 = 0$$
$$v_{11}u_0 + u_{11}v_0 = \frac{1}{32}(\nu^2)'' \tag{6.30}$$
$$v_{10}u_0 + u_{10}v_0 = \frac{1}{16}(\nu\nu')' + \frac{i}{8}(v_0'u_0 - v_0 u_0')'.$$

The parameter ν can easily be calculated explicitly. Putting asymptotics (6.29) into equation (6.28) and taking equations (6.30) into account one has

$$\partial_x B_{+-} = -\frac{i\nu}{2t} + \frac{i(\nu^2)'' \ln 4t}{16t^2}$$
$$+ \frac{i}{t^2}\left[\frac{(\nu^2)''}{16} + \frac{i}{4}(v_0'' u_0 - v_0 u_0'')\right]$$
$$+ O\left(\frac{\ln^4 4t}{t^3}\right). \tag{6.31}$$

Comparing this with the main term of the asymptotics of $\partial_x B_{+-}$ as calculated from (6.23) one obtains for ν:

$$\nu = -\frac{1}{\pi}\ln\varphi(\lambda_0^2, \beta) > 0 \tag{6.32}$$

where the function $\varphi(\lambda^2, \beta)$ is given in (6.10).

Simultaneously, integrating estimate (6.31) over x one obtains the following improvement of the first equation in (6.23) and, equivalently, of (6.25):

$$\partial_x \sigma = -2iB_{+-}$$
$$= \frac{1}{\pi}\int_{-\infty}^{+\infty} \text{sign}(\mu - \lambda_0)\ln\varphi(\mu^2, \beta)\,d\mu$$
$$- \frac{(\nu^2)'\ln 4t}{4t} - \frac{(\nu^2)'}{4t} - \frac{i}{t}(v_0'u_0 - v_0 u_0')$$
$$+ O\left(\frac{\ln^4 4t}{t^2}\right), \qquad t \to +\infty. \tag{6.33}$$

Similar considerations based on identity (XIV.6.54),

$$\partial_x(C_{+-} + C_{-+} + B_{--}G) = b_{++}\partial_x B_{--} - B_{--}\partial_x b_{++} \tag{6.34}$$

result in an improvement of estimate (6.26),

$$\partial_t \sigma = -2i(C_{+-} + C_{-+} + B_{--}G)$$

$$= \frac{2}{\pi} \int\limits_{-\infty}^{+\infty} \text{sign}(\mu - \lambda_0)\mu \ln \varphi(\mu^2, \beta)\, d\mu$$

$$- \frac{\lambda_0(\nu^2)' \ln 4t}{2t} + \frac{\nu^2}{2t} - \frac{\lambda_0(\nu^2)'}{2t}$$

$$- \frac{2i\lambda_0}{t}(v_0' u_0 - v_0 u_0')$$

$$+ O\left(\frac{\ln^4 4t}{t^2}\right), \qquad t \to +\infty. \tag{6.35}$$

Integrating these estimates for $\partial_x \sigma$ and $\partial_t \sigma$ one obtains the following asymptotic formula for σ itself:

$$\sigma(x, t, \beta) = \frac{1}{\pi} \int\limits_{-\infty}^{+\infty} |x + 2t\mu| \ln \varphi(\mu^2, \beta)\, d\mu + \frac{\nu^2}{2} \ln 4t$$

$$+ \frac{\nu^2}{2} + 2i \int\limits_{-\infty}^{\lambda_0} [v_0'(\mu)u_0(\mu) - v_0(\mu)u_0'(\mu)]\, d\mu$$

$$+ c_0(\beta) + O\left(\frac{\ln^4 4t}{t}\right), \qquad \beta < 0 \tag{6.36}$$

where the function $c_0(\beta)$ (integration constant) depends on the chemical potential β only.

In the case $\beta < 0$ considered here it is possible to get (up to the as yet undefined functions u_0, v_0) the dependence of this function on β using the expression in (6.4) for $\partial_\beta \sigma$. It follows from (6.33), (6.35) that

$$-2it(C_{+-} - C_{-+} + GB_{--}) - 2ixB_{+-}$$

$$= \frac{1}{\pi} \int\limits_{-\infty}^{+\infty} |x + 2t\mu| \ln \varphi(\mu^2, \beta)\, d\mu + \frac{\nu^2}{2} + O\left(\frac{\ln^4 4t}{t}\right). \tag{6.37}$$

On the other hand, (6.29) (recall that $u_0 v_0 = -\nu/4$) results in

$$b_{++}\partial_\beta B_{--} - B_{--}\partial_\beta b_{++} = -\frac{i \ln 4t}{2t}\nu \partial_\beta \nu$$

$$+ \frac{1}{t}(u_0 \partial_\beta v_0 - v_0 \partial_\beta u_0) + O\left(\frac{\ln^3 4t}{t}\right) \tag{6.38}$$

so that using (6.4) one gets for $\partial_\beta \sigma$

$$\partial_\beta \sigma = \partial_\beta \left[\frac{1}{\pi} \int\limits_{-\infty}^{+\infty} |x + 2t\mu| \ln \varphi(\mu^2, \beta) \, d\mu + \frac{\nu^2}{2} \right]$$

$$+ (\ln 4t)\nu\partial_\beta\nu + 2i(u_0\partial_\beta v_0 - v_0\partial_\beta u_0)$$

$$+ \frac{1}{2\pi^2} \left(\partial_\beta \int\limits_{-\infty}^{+\infty} \text{sign}(\mu - \lambda_0) \ln \varphi(\mu^2, \beta) \, d\mu \right)^2$$

$$+ O\left(\frac{\ln^4 4t}{t} \right). \tag{6.39}$$

All the estimates in this section are easily seen to be uniform in β for $\beta \le \beta_0 < 0$. As also $\sigma \to 0$ as $\beta \to -\infty$, (6.36) and (6.39) mean that

$$\sigma(x, t, \beta) = \frac{1}{\pi} \int\limits_{-\infty}^{+\infty} |x + 2t\mu| \ln \varphi(\mu^2, \beta) \, d\mu + \frac{\nu^2}{2} \ln 4t + \frac{\nu^2}{2}$$

$$+ 2i \int\limits_{-\infty}^{\beta} (u_0\partial_\beta v_0 - v_0\partial_\beta u_0) \, d\beta$$

$$+ \frac{1}{2\pi^2} \int\limits_{-\infty}^{\beta} \left(\partial_\beta \int\limits_{-\infty}^{+\infty} \text{sign}(\mu - \lambda_0) \ln \varphi(\mu^2, \beta) \, d\mu \right)^2 d\beta$$

$$+ O\left(\frac{\ln^4 4t}{t} \right) \tag{6.40}$$

$t \to +\infty$,

so that the integration constant $c_0(\beta)$ in (6.36) is in fact fixed.

This asymptotic representation together with the first of the formulæ (6.29) results in the following asymptotics for correlator $g(x, t, \beta)$ (6.1):

$$g(x, t, \beta) = C_0(\lambda_0, \beta)(4t)^{(\nu-i)^2/2} \, e^{2it\beta + ix^2/2t}$$

$$\times \exp\left\{ \frac{1}{\pi} \int\limits_{-\infty}^{+\infty} |x + 2t\mu| \ln \varphi(\mu^2, \beta) \, d\mu \right\}$$

$$\times \left[1 + O\left(\frac{1}{\sqrt{t}} \right) \right], \tag{6.41}$$

$$t \to +\infty, \quad -\infty < -\frac{x}{2t} \equiv \lambda_0 < 0, \quad \beta < 0,$$

where the factor $C_0(\lambda_0, \beta)$ is defined in terms of the functions u_0, v_0 (6.29):

$$C_0(\lambda_0, \beta) = -\frac{1}{\pi} u_0(\lambda_0)$$

$$\times \exp\left\{\frac{\nu^2}{2} + 2i \int\limits_{-\infty}^{\beta} (u_0\partial_\beta v_0 - v_0\partial_\beta u_0)\, d\beta \right.$$

$$\left. + \frac{1}{2\pi^2} \int\limits_{-\infty}^{\beta} \left[\partial_\beta \int\limits_{-\infty}^{+\infty} \text{sign}(\mu - \lambda_0) \ln \varphi(\mu^2, \beta)\, d\mu\right]^2 d\beta\right\}.$$

(6.42)

The function $\nu = -4u_0v_0$ is given explicitly (6.32) as

$$\nu = -\frac{1}{\pi} \ln \varphi(\lambda_0^2, \beta) > 0$$

and the function φ is defined in (6.10):

$$\varphi(\lambda^2, \beta) = \frac{e^{\lambda^2 - \beta} - 1}{e^{\lambda^2 - \beta} + 1},$$

(6.43)

$$\varphi(\lambda^2, \beta) > 0, \quad \beta < 0, \quad -\infty < \lambda < +\infty.$$

To define the functions $u_0(\lambda_0, \beta)$, $v_0(\lambda_0, \beta)$ one has to consider corrections to the asymptotics given here, which demands many more calculations. It is done in paper [9] where the following answers are obtained:

$$u_0(\lambda_0) = \frac{\pi\sqrt{\nu}}{2} \exp\{(\lambda_0^2 - \beta)/2 + i\Psi_0\}$$

(6.44)

$$v_0(\lambda_0) = -\frac{\sqrt{\nu}}{2\pi} \exp\{-(\lambda_0^2 - \beta)/2 - i\Psi_0\}$$

where

$$\Psi_0 = -\frac{3\pi}{4} + \arg \Gamma(i\nu)$$

$$+ \frac{1}{\pi} \int\limits_{-\infty}^{+\infty} \text{sign}(\lambda_0 - \mu) \ln |\mu - \lambda_0|\, d\left(\ln \varphi(\mu^2, \beta)\right). \quad (6.45)$$

So the main term in the asymptotic expansion of the correlator (6.1) has been calculated. The explicit answer is given in section 9.

XVI.7 Time-dependent correlator of two fields at finite temperature in the case of positive chemical potential. Space-like region

As in the equal-time case, the matrix RP in terms of which the correlator is expressed should be regularized by going to contour Γ, in Figure XVI.1, if the chemical potential is positive ($\beta > 0$). This procedure for the time-dependent correlator was described in detail in section XV.4. As a result, the correlator (6.1) is represented in the form (XV.4.44)

$$\langle \Psi(x_2, t_2) \Psi^\dagger(x_1, t_1) \rangle_T = \sqrt{T} g(x, t, \beta), \tag{7.1}$$

$$g(x, t, \beta) = -\frac{1}{2\pi} e^{2it\beta} e^{\sigma_0(x,t,\beta)} \left[m(x, t, \beta) + d(x, t, \beta) \mathring{b}_{++}(x, t, \beta) \right].$$

In this section the space-like region (XV.4.19),

$$\lambda_0 < -\sqrt{\beta}, \quad \beta = \frac{h}{T} > 0 \tag{7.2}$$

is considered (due to (6.2), (6.8) this relation can be put into the form

$$|x_1 - x_2| > v(t_2 - t_1) > 0, \quad v = 2q, \quad q = \sqrt{h}, \ h > 0$$

hence the name "space-like region"). In this region one uses the regularized matrix RP (XV.4.22) for the function $\mathring{\Phi}(\lambda)$ (XV.4.21) instead of the original RP for $\Phi(\lambda)$. Correspondingly, the basic objects now are the potentials $\mathring{b}_{++}$, $\mathring{B}_{+-}$, $\mathring{B}_{--}$ and $\mathring{C}$, which are the matrix elements of the coefficients in the expansion in powers of $1/\lambda$ of $\mathring{\Phi}(\lambda)$ (see (XV.4.27), (XV.4.28)) and the functions m, d expressed (see (XV.4.45), (XV.4.25)) in terms of the values $\mathring{\Phi}(\pm\lambda_1)$ of the solution of the regularized RP at the "critical points" $\pm\lambda_1$,

$$\lambda_1 = \sqrt{\beta} > 0, \tag{7.3}$$

which are now real.

The derivatives of the function σ_0 (XV.4.46)

$$\sigma_0(x, t, \beta) = \sigma(x, t, \beta) - \ln d(x, t, \beta) \tag{7.4}$$

are expressed in terms of $\mathring{\Phi}(\lambda)$, see e.g., (XV.4.47), (XV.4.48),

$$\partial_x \sigma_0 = 2i\alpha_0 - 2i\mathring{B}_{+-},$$
$$\partial_t \sigma_0 = -2i \left(\mathring{C}_{+-} + \mathring{C}_{-+} \right). \tag{7.5}$$

The quantity α_0 here is given by (XV.4.26)

$$\alpha_0 = \frac{1}{2\pi i} \int\limits_{-\infty}^{+\infty} d\mu \, \ln|\varphi(\mu^2, \beta)| - \lambda_1, \tag{7.6}$$

being equal to the coefficient of $1/\lambda$ in the expansion

$$\alpha(\lambda) = 1 + \frac{\alpha_0}{\lambda} + \cdots$$

of the function

$$\alpha(\lambda) = \exp\left\{-\frac{1}{2\pi i}\int_\Gamma \frac{d\mu}{\mu - \lambda}\ln\varphi(\mu^2, \beta)\right\}, \qquad \beta > 0 \qquad (7.7)$$

$$\varphi(\mu^2, \beta) \equiv \frac{e^{\mu^2 - \beta} - 1}{e^{\mu^2 - \beta} + 1}. \qquad (7.8)$$

It is worth mentioning that the functions $p_1(\lambda)$, $q_1(\lambda)$ given a priori on the contour Γ by formulæ (XV.4.23),

$$p_1(\lambda) = p(\lambda)\frac{\lambda + \lambda_1}{\lambda - \lambda_1}$$

$$q_1(\lambda) = q(\lambda)\frac{\lambda - \lambda_1}{\lambda + \lambda_1} \qquad (7.9)$$

$$1 + p_1(\lambda)q_1(\lambda) = \varphi^2(\lambda^2, \beta)$$

can be continued by means of these formulæ into the semicircles Δ_-, Δ_+ (shown in Figure XVI.1) correspondingly. By equalities

$$p_1(\lambda) = -2\pi i[\alpha_+(\lambda)]^2\frac{e^{\lambda^2 - \beta}\,\varphi(\lambda^2, \beta)(\lambda + \lambda_1)}{(e^{\lambda^2 - \beta} + 1)(\lambda - \lambda_1)}$$

$$q_1(\lambda) = -\frac{2i}{\pi}\frac{[\alpha_-(\lambda)]^{-2}\varphi(\lambda^2, \beta)(\lambda - \lambda_1)}{(e^{\lambda^2 - \beta} + 1)(\lambda + \lambda_1)} \qquad (7.10)$$

(see (XV.4.8), (XV.4.6)) these functions can be also continued into semi-circles Δ_+, Δ_-, correspondingly. At this continuation they have no singularities there. Hence one can deform the conjugation contour Γ in the regularized RP (XV.4.22) back to the real axis. In what follows this reformulation is supposed to be done, so that, e.g., one has $\mathring{\Phi}(\pm\lambda_1) \equiv \mathring{\Phi}_\pm(\pm\lambda_1)$, etc.

Thus the asymptotic analysis of the correlator is reduced to the asymptotic analysis of the matrix function $\mathring{\Phi}(\lambda)$. This is done similarly to that for the function $\Phi(\lambda)$ in the previous section. The main difference is that now one has to consider not only the potentials (obtained from $\mathring{\Phi}(\lambda)$ for $\lambda \to \infty$) but also the values $\mathring{\Phi}(\pm\lambda_1)$ necessary to calculate the functions m and d. The detailed analysis of the problem is given in [9]; below we will give the main points and results.

The "zero order approximation" (Manakov's Ansatz) is now written for $\mathring{\Phi}(\lambda)$ in the form

$$\mathring{\Phi}^m(\lambda) = \begin{pmatrix} 1 & -I^p(\lambda) \\ -I^q(\lambda) & 1 \end{pmatrix} e^{\sigma_3 \ln \delta(\lambda)} \tag{7.11}$$

which differs from (6.13) only in the replacement of the functions $p(\mu)$, $q(\mu)$ in the expressions (6.14) by $p_1(\lambda)$, $q_1(\lambda)$:

$$I^p(\lambda) = \frac{1}{2\pi i} \int_{-\infty}^{+\infty} \frac{\delta_+(\mu)\delta_-(\mu)}{\mu - \lambda} p_1(\mu) e^{-2i\theta(x,t,\mu)} \, d\mu$$

$$I^q(\lambda) = \frac{1}{2\pi i} \int_{-\infty}^{+\infty} \frac{[\delta_+(\mu)\delta_-(\mu)]^{-1}}{\mu - \lambda} q_1(\mu) e^{+2i\theta(x,t,\mu)} \, d\mu. \tag{7.12}$$

The function $\delta(\lambda)$ is given by the same expression (6.18) as for $\beta < 0$:

$$\delta(\lambda) = \exp\left\{ \frac{1}{\pi i} \int_{-\infty}^{\lambda_0} \frac{d\mu}{\mu - \lambda} \ln \varphi(\mu^2, \beta) \right\}. \tag{7.13}$$

(It is to be emphasized that $\varphi(\mu^2, \beta) > 0$ for $\mu < -\sqrt{\beta}$, which is fulfilled in the space-like region as $\lambda_0 < -\sqrt{\beta}$ (7.2).) It is shown in paper [9] that $\mathring{\Phi}^m(\lambda)$ is indeed the approximation to the solution $\mathring{\Phi}(\lambda)$ of the regularized RP, i.e., that the following estimate (similar to (6.22)) holds true for any $\lambda \neq \lambda_0$:

$$\mathring{\Phi}(\lambda) = \left[1 + O\left(t^{-\varepsilon}\right)\right] \mathring{\Phi}^m(\lambda),$$

$$t \to +\infty, \quad -\frac{x}{2t} \equiv \lambda_0 < -\sqrt{\beta}, \quad \beta > 0 \tag{7.14}$$

with $0 < \varepsilon < 1/2$ (uniformly in λ for $|\text{Im }\lambda| \geq c_0$). This, similarly to (6.23), results in the following asymptotical formulæ for the potentials $\mathring{B}_{+-}$, $\mathring{C}$:

$$\mathring{B}_{+-} = -\delta_0 + o(1)$$

$$\mathring{C}_{+-} + \mathring{C}_{-+} = -2\delta_1 + o(1) \tag{7.15}$$

with δ_0, δ_1 given exactly by expressions (6.24), so that due to (7.5) one obtains the following asymptotics of the derivatives of $\sigma_0(x, t, \beta)$:

$$\partial_x \sigma_0 = 2i(\alpha_0 + \delta_0) + o(1)$$

$$= \frac{1}{\pi} \int_{-\infty}^{+\infty} \text{sign}(\mu - \lambda_0) \ln |\varphi(\mu^2, \beta)| \, d\mu - 2i\lambda_1 + o(1) \tag{7.16}$$

$$\partial_t \sigma_0 = 4i\delta_1$$

$$= \frac{2}{\pi} \int\limits_{-\infty}^{+\infty} \text{sign}(\mu - \lambda_0)\mu \ln |\varphi(\mu^2, \beta)| \, d\mu + o(1) \tag{7.17}$$

where expression (7.6) for α_0 should be used now.

As in the case $\beta < 0$, these estimates can be made more precise if one takes into account that a system of nonlinear partial differential equations in x, t quite similar to the system (XIV.6.44)–(XIV.6.46) can be written directly for the potentials $\mathring{b}_{++}$, $\mathring{B}_{--}$. It is known from (7.14) that $\mathring{b}_{++}$, $\mathring{B}_{--}$ decrease as some power of t for $t \to +\infty$. The same line of argument as in section 6 results in the validity of asymptotic expansions (6.29) for $\mathring{b}_{++}$, $\mathring{B}_{--}$, relations (6.13) being also valid. However, the function $\nu = \nu(\lambda_0, \beta)$ is now calculated to be

$$\nu = -\frac{1}{\pi} \ln |\varphi(\lambda_0^2, \beta)| > 0, \quad \beta > 0 \tag{7.18}$$

(the calculations repeat almost exactly the corresponding ones of section 6). As a result, one arrives at a formula for $\sigma_0(x, t, \beta)$ which differs only by an additional term $-2i\lambda_1 x$ from (6.36) for σ in the case $\beta < 0$:

$$\sigma_0(x, t, \beta) = -2i\lambda_1 x + \frac{1}{\pi} \int\limits_{-\infty}^{+\infty} |x + 2t\mu| \ln |\varphi(\mu^2, \beta)| \, d\mu + \frac{\nu^2}{2} \ln 4t$$

$$+ \frac{\nu^2}{2} + 2i \int\limits_{-\infty}^{\lambda_0} [v_0'(\mu)u_0(\mu) - v_0(\mu)u_0'(\mu)] \, d\mu$$

$$+ c(\beta) + O\left(\frac{\ln^4 4t}{t}\right), \quad t \to +\infty \tag{7.19}$$

where $c(\beta)$ (the integration constant) depends on chemical potential β only.

To obtain the asymptotics of the correlator $g(x, t, \beta)$ formulæ (7.19) and $\mathring{b}_{++} = t^{-1/2}u_0$ (6.29) are not now sufficient. One has also to calculate functions m and d entering representation (7.1). They are expressed in terms of the values $\mathring{\Phi}(\pm\lambda_1)$, see (XV.4.45), (XV.4.25). The explicit expression (7.11) for $\mathring{\Phi}^m(\lambda)$ results in the following asymptotic formulæ:

$$\mathring{\Phi}_+(\lambda_1) = \begin{pmatrix} \delta(\lambda_1) & 0 \\ -\delta^{-1}(\lambda_1)q_1(\lambda_1)e^{2i\theta(x,t,\lambda_1)} & \delta^{-1}(\lambda_1) \end{pmatrix} + o(1)$$

$$\mathring{\Phi}_-(-\lambda_1) = \begin{pmatrix} \delta(-\lambda_1) & \delta(-\lambda_1)p_1(-\lambda_1)e^{-2i\theta(x,t,-\lambda_1)} \\ 0 & \delta^{-1}(-\lambda_1) \end{pmatrix} + o(1) \tag{7.20}$$

$t \to +\infty$.

Taking into account the relations

$$q_1(\lambda_1) = -\frac{i}{\pi}[2\lambda_1^2\alpha^2(\lambda_1)]^{-1},$$

$$p_1(-\lambda_1) = -i\pi[2\lambda_1^2\alpha^2(\lambda_1)]^{-1}$$

(7.21)

(it is to be recalled that $\alpha(-\lambda) = \alpha^{-1}(\lambda)$), identity (A.2.8),

$$2\lambda_1^2\alpha^2(\lambda_1) = \exp\left\{\frac{i\pi}{2} + \frac{i}{\pi}\int_{-\infty}^{+\infty}\frac{d\mu}{\mu - \lambda_1}\ln|\varphi(\mu^2,\beta)|\right\}$$

(7.22)

and the explicit formula (7.13) for $\delta(\lambda)$ (as $\lambda_0 < -\lambda_1 \equiv \sqrt{\beta}$, $|\delta(\lambda_1)| = |\delta(-\lambda_1)| = 1$) one obtains for the functions d and m:

$$d = \delta^{-1}(-\lambda_1)\delta(\lambda_1)\left(1 + e^{4ix\lambda_1+i\Psi}\right) + o(1)$$

$$m = -2\pi\lambda_1\delta^{-1}(-\lambda_1)\delta(\lambda_1)e^{-2it\lambda_1^2-ix}e^{2ix\lambda_1+i\Psi/2} + o(1)$$

(7.23)

where the notation

$$\Psi \equiv \frac{2\lambda_1}{\pi}\int_{-\infty}^{+\infty}\text{sign}(\lambda_0 - \mu)\ln|\varphi(\mu^2,\beta)|\frac{d\mu}{\mu^2 - \lambda_1^2}$$

$$\chi \equiv \frac{1}{\pi}\int_{-\infty}^{+\infty}\text{sign}(\lambda_0 - \mu)\ln|\varphi(\mu^2,\beta)|\frac{\mu\,d\mu}{\mu^2 - \lambda_1^2}$$

(7.24)

is introduced.

From (7.19) and (7.23) one obtains for the main term in the asymptotics of correlator (7.1) in the space-like region (note that $d\mathring{b}_{++} \sim t^{-1/2}$ and does not contribute to this order):

$$g(x,t,\beta) = C_s(\lambda_0,\beta)(4t)^{\nu^2/2}$$

$$\times \exp\left\{\frac{1}{\pi}\int_{-\infty}^{+\infty}|x + 2t\mu|\ln|\varphi(\mu^2,\beta)|\,d\mu\right\}$$

$$\times \left[1 + O\left(t^{-1/2}\right)\right],$$

(7.25)

$$t \to +\infty, \quad -\frac{x}{2t} \equiv \lambda_0 < -\sqrt{\beta}$$

with functions φ and ν given in (7.8), (7.18).

The expression for function $C_s(\lambda_0,\beta)$ can easily be recovered from (7.19), (7.23). It contains, however, the as yet unknown functions u_0, v_0 and integration constant $c(\beta)$. To obtain the dependence of $C_s(\lambda_0,\beta)$ on λ_0, β one has to know functions u_0 and v_0. To do this, a more precise solution of matrix RP (XV.4.22) than one given by the Manakov

Ansatz is needed. The corresponding calculations are given in [9] where the following answers for u_0 and v_0 are obtained:

$$u_0(\lambda_0) = \frac{\pi\sqrt{\nu}}{2} \exp\{(\lambda_0^2 - \beta)/2 + i\Psi_0\}$$

$$v_0(\lambda_0) = -\frac{\sqrt{\nu}}{2\pi} \exp\{-(\lambda_0^2 - \beta)/2 - i\Psi_0\}$$

(7.26)

These are exactly the same formulæ as in (6.44), but the function ν is now given by (7.18) and

$$\Psi_0 = -\frac{3\pi}{4} + \arg\Gamma(i\nu)$$

$$+ \frac{1}{\pi} \fint_{-\infty}^{+\infty} \mathrm{sign}(\lambda_0 - \mu) \ln|\mu - \lambda_0| \, d\ln|\varphi(\mu^2, \beta)|.$$

(7.27)

The explicit answer for the correlator is given in section 9.

XVI.8 Time-dependent correlator of two fields at finite temperature in the case of positive chemical potential. Time-like region

In the time-like region,

$$-\lambda_1 \equiv -\sqrt{\beta} < \lambda_0 \equiv -\frac{x}{2t} < 0,$$

(8.1)

or

$$|x_1 - x_2| < v|t_2 - t_1|, \quad v = 2q, \quad q = \sqrt{h},$$

the correlator of two fields is expressed by the same formula (7.1) as in the space-like region but the regularized RP (XV.4.30) for $\mathring{\Phi}(\lambda)$, see (8.4), should be used. Correspondingly, the potentials $\mathring{b}, \mathring{B}, \mathring{C}$ are re-expressed now by means of formulæ (XV.4.35) and the functions m and d are given by (XV.4.45), (XV.4.33). Below only the main points of the calculations are given; for details see [9].

From the technical point of view the main difference from the space-like case is that the stationary point λ_0 lies between critical points $\pm\sqrt{\beta}$, so that the function $\delta(\lambda)$ (7.13) entering the zeroth approximation (Manakov's Ansatz) also requires regularization. We define now a piecewise analytic function $\delta(\lambda)$, changing the integration contour to the contour Γ_0 shown in Figure XVI.2 (contour Γ_0 is that part of contour Γ lying to the left of λ_0):

$$\delta(\lambda) = \exp\left\{ \frac{1}{\pi i} \int_{\Gamma_0} \frac{d\mu}{\mu - \lambda} \ln\varphi(\mu^2, \beta) \right\}.$$

(8.2)

The function $\delta(\lambda)$ solves the same scalar RP (6.16) where, however, the conjugation contour is now

$$\Gamma_0 \bigcup [\lambda_0, \infty)$$

and the function $\eta(\lambda_0 - \lambda)$ should be understood as follows:

$$\eta(\lambda_0 - \lambda) \equiv \begin{cases} 1, & \lambda \in \Gamma_0 \\ 0, & \lambda \in (\lambda_0, \infty). \end{cases} \tag{8.3}$$

Introducing now the new function $\overset{\circ}{\tilde{\Phi}}(\lambda)$ by the equality

$$\overset{\circ}{\Phi}(\lambda) = \overset{\circ}{\tilde{\Phi}}(\lambda) e^{\sigma_3 \ln \delta(\lambda)} \begin{pmatrix} \lambda - \lambda_0 & 0 \\ 0 & (\lambda - \lambda_0)^{-1} \end{pmatrix} \tag{8.4}$$

one rewrites matrix RP (XV.4.30) for $\overset{\circ}{\tilde{\Phi}}(\lambda)$ as

$$\overset{\circ}{\tilde{\Phi}}_-(\lambda) = \overset{\circ}{\tilde{\Phi}}_+(\lambda) \tilde{G}_t(\lambda), \quad \lambda \in \Gamma_0 \tag{8.5a}$$

$$\overset{\circ}{\tilde{\Phi}}(\infty) = 1 \tag{8.5b}$$

where the conjugation matrix is

$$\tilde{G}_t(\lambda) =$$
$$\begin{pmatrix} 1 + p_2 q_2 \eta(\lambda_0 - \lambda) & \delta_+ \delta_- p_2 (\lambda - \lambda_0)^2 e^{-2i\theta(x,t,\lambda)} \\ (\delta_+ \delta_-)^{-1} q_2 (\lambda - \lambda_0)^{-2} e^{2i\theta(x,t,\lambda)} & 1 + p_2 q_2 \eta(\lambda - \lambda_0) \end{pmatrix}. \tag{8.5c}$$

For the functions $p_2(\lambda)$, $q_2(\lambda)$ see (XV.4.31). It is shown in [9] that:
(1) The matrix $\tilde{G}_t(\lambda)$ can be continued analytically in semicircles Δ_+, Δ_- (shown in Figure XVI.1) without acquiring singularities at the points $\pm\lambda_1$;
(2) The explicit form of the functions $p_2(\lambda)$, $q_2(\lambda)$ and $\delta_\pm(\lambda)$ guarantees that there is no pole at $\lambda = \lambda_0$.

Results (1) and (2) mean that (as in the space-like case) one can change the conjugation contour Γ for the real λ-axis. In what follows this reformulation of RP (8.5) is supposed to have been done. Solution $\overset{\circ}{\Phi}(\lambda)$ of RP (XV.4.30) is obtained by means of the exact formula (8.4), putting

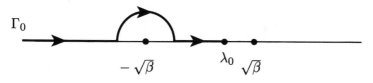

Figure XVI.2

into formulæ (XV.4.32), (XV.4.33) the expressions

$$\overset{\circ}{\Phi}(\pm\lambda_1) = \pm\overset{\circ}{\tilde{\Phi}}(\pm\lambda_1)e^{\sigma_3 \ln \delta(\pm\lambda_1)} \begin{pmatrix} \lambda_1 \mp \lambda_0 & 0 \\ 0 & (\lambda_1 \mp \lambda_0)^{-1} \end{pmatrix}. \tag{8.6}$$

Manakov's Ansatz for RP (8.5) is now

$$\overset{\circ}{\tilde{\Phi}}{}^{m}(\lambda) = \begin{pmatrix} 1 & -I^p(\lambda) \\ -I^q(\lambda) & 1 \end{pmatrix} \tag{8.7}$$

where

$$I^p(\lambda) = \frac{1}{2\pi i} \int\limits_{-\infty}^{+\infty} \frac{\delta_+(\mu)\delta_-(\mu)}{\mu - \lambda} p_2(\mu)(\mu - \lambda_0)^2 e^{-2i\theta(x,t,\mu)} \, d\mu$$

$$\tag{8.8}$$

$$I^q(\lambda) = \frac{1}{2\pi i} \int\limits_{-\infty}^{+\infty} \frac{[\delta_+(\mu)\delta_-(\mu)]^{-1}}{\mu - \lambda} q_2(\mu)(\mu - \lambda_0)^{-2} e^{+2i\theta(x,t,\mu)} \, d\mu.$$

It can be shown that, similarly to (6.20), the following estimate is valid:

$$\overset{\circ}{\tilde{\Phi}}{}^{m}_{-}(\lambda) = \overset{\circ}{\tilde{\Phi}}{}^{m}_{+}(\lambda) \left(\tilde{G}_t(\lambda) + r_0(\lambda) \right),$$

$$r_0(\lambda) = \begin{cases} O\left(t^{-1/2}(\lambda - \lambda_0)^{-1}\right), & |\lambda - \lambda_0| \geq t^{-1/2+\varepsilon} \\ O\left(1\right), & |\lambda - \lambda_0| \leq t^{-1/2+\varepsilon} \end{cases}$$

$$t \to +\infty, \quad 0 < \varepsilon < \frac{1}{2}, \quad \lambda_0 = -\frac{x}{2t} = O\left(1\right).$$

Hence for the function $\overset{\circ}{\Phi}{}^m(\lambda)$,

$$\overset{\circ}{\Phi}{}^m(\lambda) = \overset{\circ}{\tilde{\Phi}}{}^{m}(\lambda)e^{\sigma_3 \ln \delta(\lambda)} \begin{pmatrix} \lambda - \lambda_0 & 0 \\ 0 & (\lambda - \lambda_0)^{-1} \end{pmatrix}, \tag{8.9}$$

estimate (6.22), (7.14) is valid:

$$\overset{\circ}{\Phi}(\lambda) = \left(1 + O\left(t^{-\varepsilon}\right)\right) \overset{\circ}{\Phi}{}^m(\lambda),$$

$$0 < \varepsilon < \frac{1}{2}, \quad t \to +\infty, \quad \lambda \neq \lambda_0, \tag{8.10}$$

which permits one to calculate the quantities m, d and the potentials $\overset{\circ}{B}_{+-}, \overset{\circ}{C}_{+-}, \overset{\circ}{C}_{-+}$.

Using the equalities

$$\delta(\lambda_1) = \frac{2\lambda_1}{\lambda_1 - \lambda_0} \exp\left\{ \frac{1}{\pi i} \int\limits_{-\infty}^{\lambda_0} \frac{d\mu}{\mu - \lambda_1} \ln |\varphi(\mu^2, \beta)| \right\},$$

$$\tag{8.11}$$

$$\delta(-\lambda_1) = \frac{i}{\lambda_1(\lambda_1 + \lambda_0)} \exp\left\{ \frac{1}{\pi i} \fint\limits_{-\infty}^{\lambda_0} \frac{d\mu}{\mu + \lambda_1} \ln |\varphi(\mu^2, \beta)| \right\},$$

together with relation (A.2.8) one gets

$$m = -4i\lambda_1 e^{-i\chi} + o(1)$$

$$d = \frac{4i}{\pi} e^{2it\lambda_1^2} \cos\left(2x\lambda_1 + \frac{1}{2}\Psi\right) + o(1) \tag{8.12}$$

where the phases Ψ, χ are given by exactly the same expressions (7.24) as in the space-like case. For the potentials $\mathring{B}_{+-}$, $\mathring{C}_{+-} + \mathring{C}_{-+}$ one has instead of (7.15):

$$\mathring{B}_{+-} = -\delta_0 + \lambda_0 + o(1) \tag{8.13}$$

$$\mathring{C}_{+-} + \mathring{C}_{-+} = -2\delta_1 + \lambda_0^2 + o(1) \tag{8.14}$$

where now

$$\delta_0 \equiv \frac{i}{\pi} \int_{-\infty}^{\lambda_0} d\mu \, \ln|\varphi(\mu^2, \beta)| + \lambda_0 + \lambda_1$$

$$\delta_1 \equiv \frac{i}{\pi} \int_{-\infty}^{\lambda_0} d\mu \, \mu \ln|\varphi(\mu^2, \beta)| + \frac{\lambda_0^2}{2} - \frac{\lambda_1^2}{2}. \tag{8.15}$$

So one has for $\partial_x \sigma_0$ and $\partial_t \sigma_0$ (to be compared with (7.16), (7.17))

$$\partial_x \sigma_0 = \frac{1}{\pi} \int_{-\infty}^{+\infty} \operatorname{sign}(\mu - \lambda_0) \ln|\varphi(\mu^2, \beta)| \, d\mu + o(1) \tag{8.16}$$

$$\partial_t \sigma_0 = \frac{2}{\pi} \int_{-\infty}^{+\infty} \operatorname{sign}(\mu - \lambda_0)\mu \ln|\varphi(\mu^2, \beta)| \, d\mu - 2i\lambda_1^2 + o(1). \tag{8.17}$$

Making these estimates more precise is essentially the same as the improvement of the estimates (7.16), (7.17) in the space-like case. As a result, one obtains the following asymptotics of $\sigma_0(x, t, \beta)$:

$$\sigma_0(x, t, \beta) = -2i\lambda_1^2 t$$

$$+ \frac{1}{\pi} \int_{-\infty}^{+\infty} |x + 2t\mu| \ln|\varphi(\mu^2, \beta)| \, d\mu + \frac{\nu^2}{2} \ln 4t$$

$$+ \frac{\nu^2}{2} + 2i \int_0^{\lambda_0} [v_0'(\mu)u_0(\mu) - v_0(\mu)u_0'(\mu)] \, d\mu$$

$$+ c_0(\beta) + O\left(\frac{\ln^4 4t}{t}\right), \qquad t \to +\infty \tag{8.18}$$

where the functions u_0, v_0 are defined by equations (XIV.6.44)–(XIV.6.46) written for potentials $\mathring{b}_{++}$, $\mathring{B}_{--}$. The function ν is given by the same expression (7.18) as in the space-like case,

$$\nu = -\frac{1}{\pi} \ln |\varphi(\lambda_0^2, \beta)|,$$

and $c_0(\beta)$ is the integration constant.

For the correlator $g(x, t, \beta)$ (7.1) one obtains:

$$g(x, t, \beta) = C_t(\lambda_0, \beta)(4t)^{\nu^2/2}$$

$$\times \exp\left\{\frac{1}{\pi} \int_{-\infty}^{+\infty} |x + 2t\mu| \ln |\varphi(\mu^2, \beta)| \, d\mu\right\}$$

$$\times \left[1 + O\left(t^{-1/2}\right)\right], \tag{8.19}$$

$$t \to +\infty, \quad -\sqrt{\beta} < \lambda_0 < 0,$$

which can differ from the space-like case (7.25) only in the coefficient C_t. The expression for $C_t(\lambda_0, \beta)$ is again easily recovered but includes, however, the functions u_0, v_0 and integration constant $c_0(\beta)$. All these are calculated in [9], where it is shown also that the remainder in (8.19) is indeed of order $t^{-1/2}$. It should be mentioned that the functions u_0, v_0 are given by exactly the same formulæ (7.26), (7.27) as in the space-like case. The explicit results are given in the next section.

XVI.9 Results for the time-dependent correlator of two fields at finite temperature

First, let us recall that at zero temperature and zero density the correlation function of two fields is equal to

$$\langle \Psi(x_2, t_2) \Psi^\dagger(x_1, t_1) \rangle = \frac{1}{2\pi} \int_{-\infty}^{+\infty} d\lambda \, \exp\{it_{12}(\lambda^2 - h) - i\lambda x_{12}\},$$

$$T = 0, \quad h < 0.$$

Now let us present the results for the asymptotics of the field correlation function of the impenetrable Bose gas for finite temperature and density. As we have already mentioned the asymptotics depend essentially on the sign of the chemical potential h (at zero temperature the transition from negative to positive chemical potential means the transition of ground state density from zero to positive). First, we rescale the variables:

$$\langle \Psi(x_2, t_2) \Psi^\dagger(x_1, t_1) \rangle_T = \sqrt{T} \, g(x, t, \beta), \tag{9.1}$$

$$x \equiv \frac{1}{2}(x_1 - x_2)\sqrt{T}, \quad t = \frac{1}{2}(t_2 - t_1)T, \quad \beta = \frac{h}{T}.$$

In this section, the asymptotics of the correlation function (9.1) at

$$t \to +\infty, \quad x \to +\infty, \quad -\infty < \lambda_0 \equiv -\frac{x}{2t} < 0 \tag{9.2}$$

are given. It is to be mentioned that parity conservation

$$g(x, t, \beta) = g(-x, t, \beta), \tag{9.3}$$

and invariance under time inversion

$$g(x, t, \beta) = g^*(x, -t, \beta) \tag{9.4}$$

imply that the asymptotics in region (9.2) define all the asymptotics. The temperature-dependent field correlation function (9.1) decays exponentially with respect to both space x and time t separations. An important factor in the asymptotics is

$$\exp\left\{ \frac{1}{\pi} \int\limits_{-\infty}^{+\infty} d\mu \, |x + 2t\mu| \ln \left| \frac{e^{\mu^2 - \beta} - 1}{e^{\mu^2 - \beta} + 1} \right| \right\}.$$

One has to distinguish between the following situations:
(1) The case of negative chemical potential:

$$\beta < 0, \quad h < 0. \tag{9.5}$$

(2) Positive chemical potential, "space-like" region:

$$\beta > 0, \quad +\lambda_0 \equiv -\frac{x}{2t} < -\lambda_1 \equiv -\sqrt{\beta}; \tag{9.6}$$
$$h > 0, \quad |x_1 - x_2| > v(t_2 - t_1), \quad v \equiv 2\sqrt{h}.$$

(3) Positive chemical potential, "time-like" region:

$$\beta > 0, \quad -\lambda_1 \equiv -\sqrt{\beta} < -\frac{x}{2t} \equiv \lambda_0 < 0; \tag{9.7}$$
$$h > 0, \quad |x_1 - x_2| < v|t_2 - t_1|, \quad v \equiv 2\sqrt{h}.$$

(Here $v \equiv 2q$, $q = \sqrt{h}$, is the Fermi velocity.)
First let us give the result in the case $h < 0$ (9.5):

$$g(x, t, \beta) = C_0(\lambda_0, \beta) t^{(\nu - i)^2/2} \exp\left\{ 2it\beta + \frac{ix^2}{2t} \right\}$$

$$\times \exp\left\{ \frac{1}{\pi} \int\limits_{-\infty}^{+\infty} |x + 2t\mu| \ln \varphi(\mu^2, \beta) \, d\mu \right\}$$

$$\times \left[1 + O\left(\frac{1}{\sqrt{t}} \right) \right], \qquad \beta = \frac{h}{T} < 0 \tag{9.8}$$

where

$$\varphi(\mu^2, \beta) = \frac{e^{\mu^2 - \beta} - 1}{e^{\mu^2 - \beta} + 1}$$

$$C_0(\lambda_0, \beta) = -\frac{\sqrt{\nu}}{2} \exp\{(\lambda_0^2 - \beta)/2 + i\Psi_0\}$$

$$\times \exp\left\{\frac{\nu^2}{2} - \int_{-\infty}^{\beta} \left(\frac{i\nu}{2} + \nu\partial_\beta\Psi_0\right) d\beta \right.$$

$$+ \frac{1}{2\pi^2} \int_{-\infty}^{\beta} \left[\partial_\beta \int_{-\infty}^{+\infty} \text{sign}(\mu - \lambda_0) \ln \varphi(\mu^2, \beta) \, d\mu\right]^2 d\beta\Bigg\}.$$

$$(9.9)$$

Recall that $\lambda_0 \equiv -x/2t$ and

$$\nu = -\frac{1}{\pi} \ln |\varphi(\lambda_0^2, \beta)| > 0 \qquad (9.10)$$

$$\varphi(\lambda^2, \beta) = \frac{e^{\lambda^2 - \beta} - 1}{e^{\lambda^2 - \beta} + 1} \qquad (9.11)$$

$$\Psi_0 = -\frac{3\pi}{4} + \arg \Gamma(i\nu)$$

$$+ \frac{1}{\pi} \fint_{-\infty}^{+\infty} \text{sign}(\lambda_0 - \mu) \ln |\mu - \lambda_0| \, d \ln |\varphi(\mu^2, \beta)|. \qquad (9.12)$$

(It is to be recalled that in this case $\beta < 0$, $\varphi(\lambda^2, \beta) > 0$ and $|\varphi(\lambda^2, \beta)| = \varphi(\lambda^2, \beta)$; the principal value sign can be omitted.) It is essential that λ_0 is finite. If $t \to 0$, $\lambda_0 \to -\infty$ then the term (9.8) will drop out and the asymptotics will be given by (5.2)–(5.5). So the answer in the case $\beta < 0$ has been found.

Let us turn now to the case $\beta > 0$. The correlator is represented by the same formula in both the space-like (9.6) and time-like (9.7) regions:

$$g(x, t, \beta) = C(\lambda_0, \beta)(4t)^{\nu^2/2} \exp\left\{\frac{1}{\pi} \int_{-\infty}^{+\infty} |x + 2t\mu| \ln |\varphi(\mu^2, \beta)| \, d\mu\right\}$$

$$\times \left\{1 + g_0(\lambda_0, \beta)(4t)^{-1/2 - i\nu} e^{2it(\lambda_0^2 + \lambda_1^2)}\right.$$

$$\times \left[(\lambda_1^2 + \lambda_0^2) \cos\left(2x\lambda_1 + \frac{\Psi}{2}\right) + 2i\lambda_0\lambda_1 \sin\left(2x\lambda_1 + \frac{\Psi}{2}\right)\right]$$

$$+ o\left(t^{-1/2}\right)\Bigg\}$$

$$(9.13)$$

$$g_0(\lambda_0, \beta) = \frac{\sqrt{\nu} \exp\left\{\dfrac{\lambda_0^2 - \lambda_1^2}{2} + i(\chi + \Psi_0)\right\}}{\lambda_1(\lambda_1^2 - \lambda_0^2)}$$

$$\lambda_1 \equiv \sqrt{\beta}, \quad \lambda_0 = -\frac{x}{2t}, \quad \beta > 0.$$

The functions ν, φ and Ψ_0 are given by (9.10)–(9.12) (where now all the modulus signs and the principal value of the integral are essential). The functions $\Psi(\lambda_0, \beta)$ and $\chi(\lambda_0, \beta)$ are

$$\Psi = \frac{2\lambda_1}{\pi} \mathop{\rlap{\int}{\;\;-}}\limits_{-\infty}^{+\infty} \mathrm{sign}(\lambda_0 - \mu) \ln |\varphi(\mu^2, \beta)| \frac{d\mu}{\mu^2 - \lambda_1^2}$$

$$\chi = \frac{1}{\pi} \int_{-\infty}^{+\infty} \mathrm{sign}(\lambda_0 - \mu) \ln |\varphi(\mu^2, \beta)| \frac{\mu \, d\mu}{\mu^2 - \lambda_1^2}. \tag{9.14}$$

The description of the factor $C(\lambda_0, \beta)$ is now more complicated than in the case $\beta < 0$. We can represent it as an integral:

$$C(\lambda_0, \beta) = \exp\left\{\frac{\nu}{2} - i\chi + \int^{(\beta, \lambda_0)} \omega\right\} \tag{9.15}$$

where ω is a closed 1-form given by

$$\omega = \left(-i\frac{\nu}{2} - \nu\partial_\beta\Psi_0 + \frac{1}{2\pi^2}\left[\partial_\beta \int_{-\infty}^{+\infty} \mathrm{sign}(\lambda_0 - \mu) \ln \left|\frac{e^{\mu^2 - \beta} - 1}{e^{\mu^2 - \beta} + 1}\right| d\mu\right]^2\right) d\beta$$

$$+ (i\nu\lambda_0 - \nu\partial_{\lambda_0}\Psi) \, d\lambda_0. \tag{9.16}$$

These formulæ provide us with an explicit representation of the factor $C(\lambda_0, \beta)$ up to a numerical constant, which does not depend on x, t, λ_0 or β and might depend on whether a space-like or time-like region is considered. Its derivation can be found in [9].

Conclusion

We have evaluated the asymptotics of temperature- and time-dependent correlation functions for impenetrable bosons. We believe that a finite coupling constant will not change the results qualitatively. N.A. Slavnov has already obtained a differential equation which describes the field correlation function for finite coupling.

In this chapter the problem of the evaluation of correlation functions of the impenetrable Bose gas was completely solved. By means of the Riemann-Hilbert problem we have explicitly calculated the asymptotics of temperature quantum correlation functions, see sections 5 and 9. In this chapter we have followed the papers [6], [7], [8], [9], [12], [13].

Appendix XVI.1: Scalar Riemann-Hilbert problem

The scalar RP (in contrast with the matrix RP) has been investigated and completely solved, see, e.g., [5]. Here we give some elementary facts concerning the scalar RP.

The homogeneous scalar RP is formulated as a particular case ($N = 1$) of the matrix RP (XV.0.2)–(XV.0.4). One has to find a function $\alpha(\lambda)$, analytic in λ outside the conjugation contour Γ (which is supposed to be simply-connected), the limiting values on the contour from above ($\alpha_+(\lambda)$) and from below ($\alpha_-(\lambda)$) being related by a given conjugation function $g(\lambda)$:

$$\alpha_-(\lambda) = \alpha_+(\lambda)g(\lambda), \quad \lambda \in \Gamma. \tag{A.1.1}$$

The function $g(\lambda) \neq 0$ is usually supposed to satisfy, e.g., the Hölder condition,

$$|g(\lambda_1) - g(\lambda_2)| < C|\lambda_1 - \lambda_2|^k, \quad 0 < k \leq 1.$$

The index, χ, of function $g(\lambda)$ is defined by:

$$\chi \equiv \chi(g) = \frac{1}{2\pi} \operatorname{Var}_\Gamma \arg g(\lambda) = \frac{1}{2\pi i} \int_\Gamma d\ln g(\lambda). \tag{A.1.2}$$

The normalization condition

$$\alpha(\lambda = \infty) = 1 \tag{A.1.3}$$

is usually imposed.

If the index $\chi = 0$ then the problem (A.1.1) (with normalization (A.1.3)) admits a unique solution given by the obvious formula

$$\alpha(\lambda) = \exp\left\{-\frac{1}{2\pi i} \int_\Gamma \frac{\ln g(\mu)}{\mu - \lambda} d\mu\right\}, \quad \lambda \notin \Gamma. \tag{A.1.4}$$

The functions $\alpha_\pm(\lambda)$ ($\lambda \in \Gamma$) are given by the corresponding limiting values of (A.1.4).

If the index $\chi \le -1$, RP (A.1.1) is solvable but has more than one solution (if the normalization (A.1.3) is not imposed).

Finally, if $\chi \ge 1$ RP (A.1.1), (A.1.3) is not solvable.

It is not difficult to show that RP (A.1.1), (A.1.3) is equivalent to the following singular integral equation, which is the particular case ($N = 1$) of (XV.0.6) (for simplicity we take $\Gamma \equiv \mathbb{R}$)

$$\alpha_+(\lambda) = 1 + \frac{1}{2\pi i} \int\limits_{-\infty}^{+\infty} \frac{\alpha_+(\mu)[1 - g(\mu)]}{\mu - \lambda - i0} \, d\mu, \quad \lambda \in \mathbb{R}. \tag{A.1.5}$$

Turn now to the inhomogeneous RP,

$$\tilde{\alpha}_-(\lambda) = \tilde{\alpha}_+(\lambda)g(\lambda) + r(\lambda), \quad \lambda \in \Gamma \tag{A.1.6}$$

where $r(\lambda)$ is also supposed to satisfy Hölder's condition. The usual normalization is again

$$\tilde{\alpha}(\infty) = 1. \tag{A.1.7}$$

If the index χ of $g(\lambda)$ is equal to zero, $\chi = 0$ (which is the most important case for us) then RP (A.1.6), (A.1.7) is uniquely solvable. It is not difficult to see that the solution of the inhomogeneous problem can be given explicitly in terms of the solution of the corresponding homogeneous problem. Indeed, suppose that the solution $\alpha(\lambda)$ of (A.1.1), (A.1.3) is known; then one writes for $g(\lambda)$

$$g(\lambda) = \frac{\alpha_-(\lambda)}{\alpha_+(\lambda)}, \quad \lambda \in \Gamma \tag{A.1.8}$$

and (A.1.6) is rewritten as

$$\frac{\tilde{\alpha}_+(\lambda)}{\alpha_+(\lambda)} - \frac{\tilde{\alpha}_-(\lambda)}{\alpha_-(\lambda)} = -\frac{r(\lambda)}{\alpha_-(\lambda)}. \tag{A.1.9}$$

As the quantities $\tilde{\alpha}_\pm(\lambda)/\alpha_\pm(\lambda)$ are the boundary values of the function $\tilde{\alpha}(\lambda)/\alpha(\lambda)$, which is analytic at $\lambda \notin \Gamma$, and since $\tilde{\alpha}(\lambda)/\alpha(\lambda) \to 1$ for $\lambda \to \infty$, one represents this function as a Cauchy integral

$$\frac{\tilde{\alpha}(\lambda)}{\alpha(\lambda)} = 1 - \frac{1}{2\pi i} \int\limits_{\Gamma} \frac{r(\mu)}{\alpha_-(\mu)(\mu - \lambda)} \, d\mu, \quad \lambda \notin \Gamma \tag{A.1.10}$$

so that one has for $\tilde{\alpha}(\lambda)$:

$$\tilde{\alpha}(\lambda) = \alpha(\lambda) \left(1 - \frac{1}{2\pi i} \int\limits_{\Gamma} \frac{r(\mu)}{\alpha_-(\mu)(\mu - \lambda)} \, d\mu \right), \quad \lambda \notin \Gamma. \tag{A.1.11}$$

This is the solution of the inhomogeneous RP (A.1.6).

It is also not difficult to see that the inhomogeneous RP (A.1.8), (A.1.9) is equivalent to the following singular integral equation for $\widetilde{\alpha}_+(\lambda)$:

$$\widetilde{\alpha}_+(\lambda) = 1 + \frac{1}{2\pi i} \int\limits_{-\infty}^{+\infty} \frac{\widetilde{\alpha}_+(\mu)(1 - g(\mu))}{\mu - \lambda - i0} \, d\mu$$

$$- \frac{1}{2\pi i} \int\limits_{-\infty}^{+\infty} \frac{r(\mu)}{\mu - \lambda - i0} \, d\mu, \quad \lambda \in \mathbb{R}. \qquad (A.1.12)$$

(For simplicity of notation we again take conjugation contour Γ to be the real λ-axis, $\lambda \in \mathbb{R}$; the generalization to arbitrary Γ is, of course, evident.)

Appendix XVI.2:
Properties of $\alpha(\lambda_{1,2})$

In this Appendix properties of the quantities $\alpha(\lambda_1)$ and $\alpha(\lambda_2)$ are studied in more detail. Remember that the piecewise analytic function $\alpha(\lambda)$ is defined as the solution of the scalar Riemann problem (XV.2.23) and is given explicitly by the formula

$$\alpha(\lambda) \equiv \exp\left\{-\frac{1}{2\pi i}\int_\Gamma \frac{d\mu}{\mu - \lambda}\ln\left(\frac{e^{\mu^2 - \beta} - 1}{e^{\mu^2 - \beta} + 1}\right)\right\}. \qquad (A.2.1)$$

The contour Γ is shown in Figure XVI.1; we consider the case of positive chemical potential ($\beta > 0$); the points λ_k are given by (4.17):

$$\lambda_1 = +\sqrt{\beta}; \quad \lambda_2 = \sqrt{\beta + 2\pi i}, \quad \operatorname{Im}\lambda_2 > 0, \ \operatorname{Re}\lambda_2 > 0.$$

The radii ϵ of semicircles C_1 and C_{-1} on Figure XVI.1 are chosen so that the inequality

$$\operatorname{Im}\lambda_2 > \epsilon$$

is valid. The logarithm branch is fixed as

$$\ln\left(\frac{e^{\lambda^2 - \beta} + 1}{e^{\lambda^2 - \beta} - 1}\right) \to 0, \quad \lambda \to \infty. \qquad (A.2.2)$$

The main property of the contour Γ is

$$\operatorname{Var}_\Gamma \arg \frac{e^{\mu^2 - \beta} + 1}{e^{\mu^2 - \beta} - 1} = 0 \qquad (A.2.3)$$

which means that the index χ (see (A.1.2)) of the scalar RP for the function $\alpha(\lambda)$ is equal to zero and hence the RP is uniquely solvable. Equivalently, this guarantees the existence of the integral on the right hand side of (A.2.1). Under this condition, the values of $\alpha(\lambda)$ at the points $\lambda = \lambda_1$ and $\lambda = \lambda_2$ do not depend on the particular contour

Γ selected, that is, on the value of the parameter ϵ. Taking this into account, we transform the formula for $\alpha(\lambda_1)$ as follows:

$$\ln \alpha^{-1}(\lambda_1) = -\frac{1}{2\pi i} \int_{\Gamma} \frac{d\mu}{\mu - \lambda_1} \ln \left(\frac{e^{\mu^2 - \beta} + 1}{e^{\mu^2 - \beta} - 1} \right)$$

$$= -\frac{1}{2\pi i} \lim_{\epsilon \to 0} \left[\int_{-\infty}^{-\lambda_1 - \epsilon} + \int_{C_{-1}} + \int_{-\lambda_1 + \epsilon}^{\lambda_1 - \epsilon} + \int_{C_1} + \int_{\lambda_1 + \epsilon}^{+\infty} \right] \frac{d\mu}{\mu - \lambda_1}$$

$$\times \ln \left(\frac{e^{\mu^2 - \beta} + 1}{e^{\mu^2 - \beta} - 1} \right). \tag{A.2.4}$$

Due to (A.2.2) one obtains:

$$\ln \left(\frac{e^{\mu^2 - \beta} + 1}{e^{\mu^2 - \beta} - 1} \right) = \ln \left| \frac{e^{\mu^2 - \beta} + 1}{e^{\mu^2 - \beta} - 1} \right|,$$

$$\mu \in [\lambda_1 + \epsilon, +\infty), \quad \mu \in (-\infty, -\lambda_1 - \epsilon]; \tag{A.2.5}$$

$$\ln \left(\frac{e^{\mu^2 - \beta} + 1}{e^{\mu^2 - \beta} - 1} \right) = \ln \left| \frac{e^{\mu^2 - \beta} + 1}{e^{\mu^2 - \beta} - 1} \right| + i\pi,$$

$$\mu \in [-\lambda_1 + \epsilon, \lambda_1 - \epsilon]$$

so that (A.2.4) can be rewritten as

$$\ln \alpha^{-1}(\lambda_1) = -\frac{1}{2\pi i} \oint_{-\infty}^{+\infty} \frac{d\mu}{\mu - \lambda_1} \ln \left| \frac{e^{\mu^2 - \beta} + 1}{e^{\mu^2 - \beta} - 1} \right|$$

$$- \frac{1}{2\pi i} \lim_{\epsilon \to 0} \left(\pi i \ln \frac{\epsilon}{2\lambda_1 - \epsilon} + \int_{C_1} \frac{d\mu}{\mu - \lambda_1} \ln \left(\frac{e^{\mu^2 - \beta} + 1}{e^{\mu^2 - \beta} - 1} \right) \right).$$

$$\tag{A.2.6}$$

To evaluate the integral over C_1 one introduces the parametrization

$$\mu = \lambda_1 + e^{i\varphi}\epsilon, \quad -\pi \leq \varphi \leq 0.$$

Then

$$\int_{C_1} \frac{d\mu}{\mu - \lambda_1} \ln \left(\frac{e^{\mu^2 - \beta} + 1}{e^{\mu^2 - \beta} - 1} \right)$$

$$= -i \int_{-\pi}^{0} d\varphi \, (\ln \lambda_1 + \ln \epsilon + i\varphi) + O(\epsilon)$$

$$= -i\pi \ln \epsilon - i\pi \ln \lambda_1 - \frac{\pi^2}{2} + O(\epsilon), \quad \epsilon \to 0.$$

Putting this equation into (A.2.6) one obtains the following representation for $\alpha(\lambda_1)$:

$$\ln \alpha^{-1}(\lambda_1) = \frac{1}{2}\ln(2\lambda_1^2) - \frac{i\pi}{4}$$

$$- \frac{1}{2\pi i} \oint_{-\infty}^{+\infty} \frac{d\mu}{\mu - \lambda_1} \ln \left| \frac{e^{\mu^2-\beta}+1}{e^{\mu^2-\beta}-1} \right| \qquad (A.2.7)$$

which results in the relation (4.29)

$$2\lambda_1^2 \alpha^2(\lambda_1) = e^{-i\varphi/2}, \qquad (A.2.8)$$

$$\varphi = -\pi + \frac{2}{\pi} \oint_{-\infty}^{+\infty} \frac{d\mu}{\mu - \lambda_1} \ln \left| \frac{e^{\mu^2-\beta}+1}{e^{\mu^2-\beta}-1} \right|.$$

Next, consider the value $\alpha(\lambda_2)$. Here a considerable simplification appears: both at points λ_1 and $-\lambda_1$ the singularities of the integrand function are integrable. So one obtains:

$$\ln \alpha^{-1}(\lambda_2) = -\frac{1}{2\pi i} \lim_{\epsilon \to 0} \int_{\Gamma} \frac{d\mu}{\mu - \lambda_2} \ln \left| \frac{e^{\mu^2-\beta}+1}{e^{\mu^2-\beta}-1} \right|$$

$$= -\frac{1}{2\pi i} \int_{-\infty}^{+\infty} \frac{d\mu}{\mu - \lambda_2} \ln \left| \frac{e^{\mu^2-\beta}+1}{e^{\mu^2-\beta}-1} \right|$$

$$- \frac{1}{2} \ln \frac{\lambda_2 - \lambda_1}{\lambda_2 + \lambda_1}.$$

Thus

$$\alpha^{-1}(\lambda_2) = \sqrt{\frac{\lambda_2 + \lambda_1}{\lambda_2 - \lambda_1}} \exp \left\{ \frac{i}{2\pi} \int_{-\infty}^{+\infty} \frac{d\mu}{\mu - \lambda_2} \ln \left| \frac{e^{\mu^2-\beta}+1}{e^{\mu^2-\beta}-1} \right| \right\} \qquad (A.2.9)$$

which is exactly equation (4.45). Note that formula (A.2.9) in fact holds true for the value of function $\alpha(\lambda)$ at any point λ with a nonzero imaginary part. A direct consequence of (A.2.9) is the following symmetry property:

$$\alpha^2(-\lambda^*) = \left(\frac{\lambda^* + \lambda_1}{\lambda^* - \lambda_1} \right)^2 [\alpha^*(\lambda)]^2, \quad \operatorname{Im} \lambda \neq 0. \qquad (A.2.10)$$

Note also that

$$\beta^{-1}(\lambda_{\pm 2}) = \exp \left\{ \frac{i}{2\pi} \int_{-\infty}^{+\infty} \frac{d\mu}{\mu - \lambda_{\pm 2}} \ln \left(\frac{e^{\mu^2-\beta}+1}{e^{\mu^2-\beta}+3} \right) \right\}. \qquad (A.2.11)$$

Appendix XVI.3: Proof of (4.41)

In this Appendix relation (4.41) is proved. To do this let us represent the functions $\varphi(\beta)$ and $C(\beta)$ (4.30), (4.26) using equation (A.2.8) as

$$\varphi(\beta) = \frac{2}{\pi} \int_{\Gamma'} \frac{d\mu}{\mu} \ln\left(\frac{e^{\mu^2 + 2\mu\sqrt{\beta}} + 1}{e^{\mu^2 + 2\mu\sqrt{\beta}} - 1}\right) + 2i \ln 2\beta \qquad (\text{A.3.1})$$

$$C(\beta) = \frac{1}{\pi} \int_{\Gamma'} d\mu \, \ln\left(\frac{e^{\mu^2 + 2\mu\sqrt{\beta}} + 1}{e^{\mu^2 + 2\mu\sqrt{\beta}} - 1}\right) - 2i\sqrt{\beta} \qquad (\text{A.3.2})$$

where $\lambda_1 \equiv \sqrt{\beta} > 0$ and contour Γ' is shown in Figure XVI.3. This formulæ are convenient for carrying out the differentiation of functions $\varphi(\beta)$ and $C(\beta)$ with respect to β: one can differentiate inside the integrals, considering the contour Γ' to be independent of β. In other words, the following equations are obtained for $\partial_\beta C$ and $\partial_\beta \varphi$:

$$\partial_\beta C = -\frac{2}{\pi\sqrt{\beta}} \int_{\Gamma'} \frac{e^{\mu^2 + 2\mu\sqrt{\beta}}}{e^{2\mu^2 + 4\mu\sqrt{\beta}} - 1} \mu \, d\mu - \frac{i}{\sqrt{\beta}}$$

$$\partial_\beta \varphi = -\frac{4}{\pi\sqrt{\beta}} \int_{\Gamma'} \frac{e^{\mu^2 + 2\mu\sqrt{\beta}}}{e^{2\mu^2 + 4\mu\sqrt{\beta}} - 1} \, d\mu + \frac{2i}{\beta}.$$

From these equalities for the combination $\partial_\beta C + (\lambda_1/2)\partial_\beta \varphi$ one obtains

$$\partial_\beta C + \frac{\lambda_1}{2}\partial_\beta \varphi = \frac{1}{2\pi\sqrt{\beta}} \ln\left(\frac{e^{\mu^2 + 2\mu\sqrt{\beta}} + 1}{e^{\mu^2 + 2\mu\sqrt{\beta}} - 1}\right)\Bigg|_{\Gamma'}$$

$$= \frac{1}{2\pi\sqrt{\beta}} \ln\left(\frac{e^{\mu^2 - \beta} + 1}{e^{\mu^2 - \beta} - 1}\right)\Bigg|_{\Gamma} = 0.$$

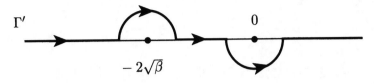

Figure XVI.3

So relation (4.41) is proved.

Appendix XVI.4: Asymptotics of correlation length at large β

Here the behavior for $\beta \to \pm\infty$ of the function $C(\beta)$,

$$C(\beta) = \frac{1}{\pi} \int\limits_{-\infty}^{+\infty} d\mu \, \ln \left| \frac{e^{\mu^2 - \beta} + 1}{e^{\mu^2 - \beta} - 1} \right| \qquad (A.4.1)$$

entering the asymptotic expansion (3.31) and (4.43) of the temperature-dependent equal-time correlator of two fields is given (for $\beta < 0$, the modulus sign in (A.4.1) can be, of course, omitted). It should be emphasized that this function defines in essence the exponent of the correlations (i.e., the correlation radius) as well as the preexponential factor.

Consider first the case $\beta = h/T < 0$. In this case one readily obtains:

$$C(\beta) = \frac{2}{\sqrt{\pi}} \sum_{k=1}^{\infty} \frac{e^{-(2k-1)|\beta|}}{(2k-1)^{3/2}} \qquad (\beta < 0) \qquad (A.4.2)$$

and hence

$$C(\beta) \simeq \frac{2}{\sqrt{\pi}} e^{-|\beta|} \qquad (\beta \to -\infty) \qquad (A.4.3)$$

$$C(\beta) \simeq \frac{2}{\sqrt{\pi}} \left(1 - 2^{-3/2}\right) \zeta(3/2) \qquad (\beta \to 0, \quad \beta < 0) \qquad (A.4.4)$$

(here ζ is the Riemann ζ-function (5.11)).

For $\beta > 0$, formula (A.4.4) also holds true,

$$C(\beta) \simeq \frac{2}{\sqrt{\pi}} \left(1 - 2^{-3/2}\right) \zeta(3/2) \qquad (\beta \to 0, \quad \beta > 0). \qquad (A.4.5)$$

Direct estimates of the integral in (A.4.1) result in the relation

$$C(\beta) \simeq \frac{\pi}{2\sqrt{\beta}} \qquad (\beta \to +\infty). \qquad (A.4.6)$$

The behavior of $C(\beta)$ as given by (A.4.3) and (A.4.6), in particular, ensures that the integrals

$$\int_{-\infty}^{\beta} \left(\frac{dC(\beta')}{d\beta'}\right)^2 d\beta', \quad \beta < 0; \qquad \int_{\beta}^{+\infty} \left(\frac{dC(\beta')}{d\beta'}\right)^2 d\beta', \quad \beta > 0$$

in (3.31) and (4.43) are convergent.

XVII

The Algebraic Bethe Ansatz and Asymptotics of Correlation Functions

Introduction

In the previous chapter we evaluated the asymptotics of the field correlation function using differential equations. In this chapter we present a completely different approach. Instead of determinant representations we shall write some special series (which emphasize the role of the R-matrix). This series is especially efficient for the current $(j(x) = \Psi^\dagger(x)\Psi(x))$ correlation function. It helps us to evaluate the asymptotics at zero temperature. The asymptotics of temperature correlation functions also can be obtained at any value of coupling constant for the Bose gas. The series for the correlation function

$$\langle j(x)j(0)\rangle$$

is based on the classification of all exactly solvable models (section VII.6) related to the fixed R-matrix. The series explicitly separates the contribution of the R-matrix and of the arbitrary functions $a(\lambda)$ and $d(\lambda)$. The Fourier coefficients of the irreducible part depend only on the R-matrix. Let us emphasize once more that in this chapter we shall consider the penetrable Bose gas $(0 < c < \infty)$.

In section 1 the algebraic foundation of the new approach to correlation functions is given. In section 2 the series representation for the current correlator $\langle j(x)j(0)\rangle$ at zero temperature (in the thermodynamic limit) is given. In section 3 temperature correlations (for the penetrable Bose gas, $0 < c < \infty$) are constructed. In section 4 explicit formulæ for asymptotics are presented. In section 5 the emptiness formation probability (the probability of absence of particles in some space interval due to thermal fluctuations) is evaluated. A similar series for the XXZ Heisenberg antiferromagnet can be found in [3], [8], [10].

XVII.1 Another approach to the calculation of correlators in the algebraic Bethe Ansatz

Here, a brief review of an alternative approach when calculating the correlators by means of the algebraic Bethe Ansatz is given. This approach is based on the further development of the ideas used in Chapter X for the calculation of the norms of Bethe wavefunctions. The details can be found in the original literature [7], [13] (see also the lectures in [2] and the book [3]).

We turn again to the generalized two-site model of section XI.1 and consider the mean value of the operator Q_1^2 (defined by (XI.1.11)) with respect to the Bethe eigenfunction $|\Psi_N\rangle$ (X.1.1), (X.1.2)

$$|\Psi_N(\{\lambda\})\rangle = \prod_{j=1}^{N} \mathbb{B}(\lambda_j)|0\rangle; \quad \mathbb{B}(\lambda) = \frac{B(\lambda)}{d(\lambda)}$$

$$\langle\Psi_N(\{\lambda\})| = \langle 0| \prod_{j=1}^{N} \mathbb{C}(\lambda_j); \quad \mathbb{C}(\lambda) = \frac{C(\lambda)}{d(\lambda)} \qquad (1.1)$$

$$\frac{a(\lambda_j)}{d(\lambda_j)} \prod_{\substack{k=1 \\ k \neq j}}^{N} \frac{f(\lambda_j, \lambda_k)}{f(\lambda_k, \lambda_j)} = 1$$

(the λ_j's are supposed to satisfy the system of Bethe equations (X.1.3)). It is convenient to define the quantities $\|Q_1^2\|_N$ ($N = 0, 1, \ldots$) similarly to (X.2.1),

$$\|Q_1^2\|_N \equiv c^{-N} \left(\prod_{j \neq k} f_{jk}\right)^{-1} \langle\Psi_N(\{\lambda\})|Q_1^2|\Psi_N(\{\lambda\})\rangle. \qquad (1.2)$$

These quantities can be investigated quite analogously to the norms of Bethe eigenfunctions in Chapter X. First of all, one establishes that in the generalized two-site model $\|Q_1^2\|_N$ is a function of $\lambda_1, \ldots, \lambda_N$ and a functional of the arbitrary functions $l(\lambda), m(\lambda)$

$$l(\lambda) = \frac{a_1(\lambda)}{d_1(\lambda)}, \quad m(\lambda) = \frac{a_2(\lambda)}{d_2(\lambda)},$$
$$r(\lambda) \equiv \frac{a(\lambda)}{d(\lambda)} = l(\lambda)m(\lambda) \qquad (1.3)$$

(It is worth mentioning that due to the Bethe equations $r(\lambda_j)$ can be expressed as an explicit function of λ_j's.) This functional is rather special. Namely, $\|Q_1^2\|_N$ depends only on the values of function $l(\lambda)$ at points λ_j and the values of the first derivatives of functions $l(\lambda), m(\lambda)$ at points λ_j, being thus a function of $4N$ complex independent variables,

$$\|Q_1^2\|_N = \|Q_1^2\|_N(\{\lambda_j\}_N, \{x_j\}_N, \{y_j\}_N, \{l_j\}_N) \qquad (1.4)$$

where

$$l_j \equiv l(\lambda_j), \quad x_j \equiv x(\lambda_j), \quad y_j \equiv y(\lambda_j); \qquad (1.5)$$

$$x(\lambda) \equiv i\frac{\partial \ln l(\lambda)}{\partial \lambda}, \quad y(\lambda) \equiv i\frac{\partial \ln m(\lambda)}{\partial \lambda}. \qquad (1.6)$$

This function $\|Q_1^2\|_N$ possesses the following properties which are quite similar to the properties of norms listed in section X.2 (see (X.2.3)–(X.2.9)):

(1) It is invariant under exchange of "quartets"

$$(\lambda_j, x_j, y_j, l_j) \leftrightarrow (\lambda_k, x_k, y_k, l_k). \qquad (1.7)$$

(2) It is a linear function of x_N and y_N (and hence, of course, of any x_j and any y_j).

(3) The coefficients at x_N and y_N are

$$\frac{\partial\|Q_1^2\|_N}{\partial y_N} = \|Q_1^2\|_{N-1}(\{\lambda_j\}_{N-1}, \{x_j\}_{N-1}, \{y_j + K_{jN}\}_{N-1}, \{l_j\}_{N-1}) \qquad (1.8)$$

$$\frac{\partial\|Q_1^2\|_N}{\partial x_N}$$
$$= \|(Q_1+1)^2\|_N(\{\lambda_j\}_{N-1}, \{x_j + K_{jN}\}_{N-1}, \{y_j\}_{N-1}, \{\tilde{l}_j\}_{N-1}). \qquad (1.9)$$

Here $K_{jN} \equiv K(\lambda_j, \lambda_N)$ and

$$K(\lambda_j, \lambda_N) = \frac{2c}{c^2 + (\lambda_j - \lambda_N)^2}$$

$$\tilde{l}_j = l_j \frac{f(\lambda_j, \lambda_N)}{f(\lambda_N, \lambda_j)}, \quad j = 1, \dots, N-1. \qquad (1.10)$$

Also

$$\|(Q_1+1)^2\|_N \equiv \|Q_1^2\|_N + 2\|Q_1\|_N + \|1\|_N. \qquad (1.11)$$

The value $\|Q_1\|_N$ is defined similarly to $\|Q_1^2\|_N$ but in (1.2) we should replace Q_1^2 by Q_1.

(4) The value of $\|Q_1^2\|_N$ at $x_1 = y_1 = x_2 = y_2 = \dots = x_N = y_N = 0$ is of primary importance. It is called the irreducible part I_N:

$$I_N(\{\lambda_j\}_N, \{l_j\}_N) \equiv \|Q_1^2\|_N(\{\lambda_j\}_N, \{x_j \equiv 0\}_N, \{y_j \equiv 0\}_N, \{l_j\}_N). \qquad (1.12)$$

The dependence of I_N on the variables can be extracted explicitly as follows:

$$I_N(\{\lambda_j\}_N, \{l_j\}_N)$$

$$= \sum \left(\prod_{(+)} l(\lambda^+) \prod_{(-)} l^{-1}(\lambda^-) \right) \mathcal{A}_N^n(\{\lambda^+\}_n, \{\lambda^-\}_n, \{\lambda_0\}_{N-2n}).$$

$$(1.13)$$

The sum here is taken over all the partitions of the set $\{\lambda\}_N$ into three disjoint subsets; $\mathrm{card}\{\lambda^+\} = \mathrm{card}\{\lambda^-\} = n$, $\mathrm{card}\{\lambda_0\} = N - 2n$; $0 \leq n \leq [N/2]$. The coefficients $\mathcal{A}_N^n$ are called the "Fourier coefficients" of the irreducible part I_N. They do not depend on l_j but only on λ_j being a rational function of the λ_j. The Fourier coefficients depend on the R-matrix only and do not depend on the specific model (which is obtained from the generalized model by choosing specific functions $l(\lambda)$, $m(\lambda)$). All the dependence on specific models enters through the "vacuum eigenvalues" $l(\lambda)$ and is written explicitly. In paper [7] the irreducible parts were studied in detail and the recursion procedure of their calculation was given. For instance, for $N = 0, 1, 2, 3$ one has:

$$I_0 = I_1 = 0; \tag{1.14}$$

$$I_2(\{\lambda_1, \lambda_2\}, \{l_1, l_2\}) = \mathcal{A}_2^1(\lambda_1, \lambda_2) \left(l_1 l_2^{-1} - 1 \right)$$

$$+ \mathcal{A}_2^1(\lambda_2, \lambda_1) \left(l_2 l_1^{-1} - 1 \right). \tag{1.15}$$

$$\mathcal{A}_2^1(\lambda_1, \lambda_2) = -\frac{2}{\lambda_{12}^2} \frac{\lambda_{12} + ic}{\lambda_{12} - ic} \qquad (\lambda_{jk} \equiv \lambda_j - \lambda_k).$$

$$I_3(\{\lambda_1, \lambda_2, \lambda_3\}, \{l_1, l_2, l_3\}) = \sum_{\mathcal{P}} \mathcal{A}_3^1(\lambda_{\mathcal{P}1}, \lambda_{\mathcal{P}2}, \lambda_{\mathcal{P}3}) \left(l_{\mathcal{P}1} l_{\mathcal{P}2}^{-1} - 1 \right)$$

$$(1.16)$$

$$\mathcal{A}_3^1(\lambda_1, \lambda_2, \lambda_3) = 8c\lambda_{12}^{-2}(\lambda_{12} + ic)(\lambda_{12} - ic)^{-1}$$

$$\times \left[\frac{\lambda_{32}}{\lambda_{31}} + \frac{\lambda_{31}}{\lambda_{32}} \right] (\lambda_{31} + ic)^{-1} (\lambda_{23} + ic)^{-1}.$$

The sum here is taken over all the permutations of λ_1, λ_2, λ_3. The irreducible part I_4 can be found in paper [14], and I_5 is calculated in [16]. The recurrence procedure was formulated in [7], [13], [14], [16]. It allows one to calculate the irreducible parts I_N for any N. The following properties can be established:

$$\begin{aligned} I_k \sim c^{k-2}, \quad &\mathcal{A}_k^n \sim c^{k-2}, \quad c \to 0 \quad (k \geq 2) \\ I_k \sim c^{2-k}, \quad &\mathcal{A}_k^n \sim c^{2-k}, \quad c \to \infty \quad (k \geq 2) \end{aligned} \tag{1.17}$$

which means that the irreducible parts are small at the limit of small coupling as well as that of large coupling.

(5) The last important property of $\|Q_1^2\|_N$ is that its value at $N = 1$ (which is easy to calculate) is

$$\|Q_1^2\|_{N=1} = x_1.$$

The five properties listed above allow one express the quantities $\|Q_1^2\|_N$ in terms of irreducible parts I_n $(n \leq N)$. That is why irreducible parts are of primary importance. The formulæ expressing $\|Q_1^2\|_N$ in terms of the irreducible parts I_N are rather bulky but they are somewhat simplified at the thermodynamic limit. The results for the current correlator at zero as well as nonzero temperature are given in sections 2 and 3, respectively.

It is also to be mentioned that the approach described in this section is quite general and can be applied not only to the correlator of currents but also to other equal-time correlators including multipoint ones (see paper [9]). It can be naturally generalized for models having the XXZ R- matrix. See paper [8] for the series representation of correlation functions for the Heisenberg antiferromagnet. In Russian it is published in the book [3].

XVII.2 The series for the correlator of currents in the nonrelativistic Bose gas at zero temperature

In this section the result of the analysis of the correlator of two currents at zero temperature by means of the approach described briefly in the previous section is presented. All the details are given in paper [13].

Consider first the normalized mean value with respect to the ground state $|\Omega\rangle$ of the gas of operator $Q_1^2(x)$,

$$\langle Q_1^2(x)\rangle \equiv \frac{\langle\Omega|Q_1^2(x)|\Omega\rangle}{\langle\Omega|\Omega\rangle} \tag{2.1}$$

where

$$Q_1(x) = \int_0^x j(y)\,dy, \qquad j(x) = \Psi^\dagger(x)\Psi(x) \tag{2.2}$$

is the operator for the number of gas particles in the interval $[0, x]$. As has already been mentioned, its mean value with respect to $|\Omega\rangle$ is given by a simple formula:

$$\langle Q_1(x)\rangle = Dx \tag{2.3}$$

where

$$D = \frac{N}{L} = \int_{-q}^{q} \rho(\lambda)\,d\lambda$$

is the density of gas particles in the ground state. The most interesting is the mean value (2.1) of $Q_1^2(x)$ in terms of which the correlator of two currents is expressed, see (XI.1.3). For this quantity the following representation can be obtained. First, one writes

$$\langle Q_1^2(x)\rangle = \langle Q_1^2(x)\rangle^0 + \langle\langle Q_1^2(x)\rangle\rangle. \tag{2.4}$$

The first term in the right hand side is a quadratic function of x:

$$\langle Q_1^2(x)\rangle^0 = x^2 D^2 + x \int_{-q}^{q} P''(t)\,dt \tag{2.5}$$

where the function $P''(t)$ is defined by means of the linear integral equation

$$2\pi P''(t) - \int_{-q}^{q} P''(s)K(t,s)\,ds = [2\pi\rho(s)]^2 \tag{2.6}$$

with the standard kernel $K(\lambda,\mu)$

$$K(\lambda,\mu) = \frac{2c}{c^2 + (\lambda-\mu)^2} \tag{2.7}$$

(q is the bare Fermi momentum).

The second, nontrivial, term on the right hand side of (2.4) is represented as a series

$$\langle\langle Q_1^2(x)\rangle\rangle = \sum_{k=2}^{\infty} \Gamma_k(x). \tag{2.8}$$

The quantity $\Gamma_k(x)$ (corresponding to the contribution of k-particle processes to the correlator) is given by the k-multiple integral (with weight $\omega(\lambda)$). We integrate the k-particle "dressed" irreducible part $I_k^d(x,\{\lambda\})$:

$$\Gamma_k(x) = \frac{1}{k!} \int_{-q}^{q} \left(\prod_{j=1}^{k} \frac{\omega(\lambda_j)\,d\lambda_j}{2\pi}\right) I_k^d(x,\{\lambda\}) \tag{2.9}$$

($\{\lambda\} \equiv \{\lambda_1, \ldots, \lambda_k\}$). The dependence on distance x here enters only through the dressed irreducible parts, which are defined by a formula (similar to (1.13) defining the "bare" irreducible parts I_k):

$$I_k^d(x,\{\lambda\}) = \sum \exp[xp_n(\{\lambda^+\}, \{\lambda^-\})]\mathcal{A}_k^n(\{\lambda^+\}, \{\lambda^-\}, \{\lambda_0\}). \tag{2.10}$$

The sum here is taken exactly as in (1.13), i.e., over all the partitions of the set $\{\lambda_j\}$ ($\mathrm{card}\{\lambda_j\} = k$), into three disjoint subsets $\{\lambda^+\}_n$, $\{\lambda^-\}_n$ and $\{\lambda_0\}_{k-2n}$ ($\mathrm{card}\{\lambda^+\} = \mathrm{card}\{\lambda^-\} = n$, $\mathrm{card}\{\lambda_0\} = k - 2n$). The Fourier coefficients $\mathcal{A}_k^n$ are the same as in the bare irreducible part I_k (1.13). They depend only on the R-matrix being the same for any specific model obtained from the two-site generalized model at specific values of the arbitrary functional parameters $l(\lambda)$, $m(\lambda)$. In particular, the nonrelativistic Bose gas is obtained by the choice

$$l(\lambda) = e^{-ix\lambda} \qquad (2.11)$$

so that in this case the dressed irreducible part (2.10) differs from the bare one (1.13) by the "dressing", i.e., the replacement of bare exponents by dressed ones:

$$\left(\prod_{(+)} l(\lambda^+) \right) \left(\prod_{(-)} l^{-1}(\lambda^-) \right) = \exp\left\{ -ix \left(\sum_{(+)} \lambda^+ - \sum_{(-)} \lambda^- \right) \right\}$$

$$\longrightarrow \exp\left\{ x p_n(\{\lambda^+\}, \{\lambda^-\}) \right\}. \qquad (2.12a)$$

The function p_n is defined by

$$p_n(\{\lambda^+\}, \{\lambda^-\}) = -i \sum_{j=1}^{n} (\lambda_j^+ - \lambda_j^-)$$

$$+ \int_{-q}^{q} dt \, P_n(t, \{\lambda^+\}, \{\lambda^-\}). \qquad (2.12b)$$

The "dressing" function P_n here is the solution of the nonlinear integral equation

$$1 + 2\pi P_n(t) = S(t) \exp\left\{ \int_{-q}^{q} K(t, s) P_n(s) \, ds \right\}, \qquad (2.13)$$

$$\operatorname{Re} P_n(t) \le 0$$

where $S(t)$ is a product of the corresponding "bare" S-matrices

$$S(t) = \prod_{j=1}^{n} \frac{f(\lambda_j^+, t) \, f(t, \lambda_j^-)}{f(t, \lambda_j^+) \, f(\lambda_j^-, t)} = \prod_{j=1}^{n} \frac{(\lambda_j^+ - t + ic)(t - \lambda_j^- + ic)}{(\lambda_j^+ - t - ic)(t - \lambda_j^- - ic)} \qquad (2.14)$$

(see (VI.3.18)). In [2], [13] the theorem that equation (2.13) has unique solution is proved. Finally, the weight $\omega(\lambda)$ is

$$\omega(\lambda) = \exp\left\{ -\frac{1}{2\pi} \int_{-q}^{q} K(\lambda, \mu) \, d\mu \right\} \qquad (2.15)$$

satisfying the obvious conditions

$$0 < e^{-1} < \omega(\lambda) \le \omega_0 < 1 \qquad (2.16)$$

$$\omega_0 \equiv \exp\left\{ -\frac{2cq}{\pi(c^2 + 4q^2)} \right\}.$$

The k-th term Γ_k of series (2.8) contains a factor $\omega^k \leq \omega_0^k$ ($\omega_0 < 1$); this improves the convergence of the series. It should also be mentioned that for $c \to \infty$ the quantities Γ_k, $(k > 2)$ are small:

$$\Gamma_k \sim c^{2-k}, \qquad c \to \infty \tag{2.17}$$

so that expansion (2.8) is analogous to the $1/c$-expansion (however, Γ_k is not a monomial in $1/c$).

We turn now to the correlator of two currents

$$j(x) = \Psi^\dagger(x)\Psi(x)$$

which is related to the mean value (2.1) of the operator $Q_1^2(x)$ by relation (XI.1.3),

$$\langle :j(x)j(0): \rangle = \frac{1}{2}\frac{\partial^2}{\partial x^2}\langle Q_1^2(x)\rangle, \tag{2.18}$$

so that (2.4), (2.8) result in the following representation:

$$\langle :j(x)j(0): \rangle = \langle j(0)\rangle^2 + \langle\langle j(x)j(0)\rangle\rangle \tag{2.19}$$

where the first term on the right hand side corresponds to the contribution of $\langle Q_1^2(x)\rangle^0$,

$$\langle j(0)\rangle^2 = D^2. \tag{2.20}$$

The nontrivial part of the correlator is given as

$$\langle\langle j(x)j(0)\rangle\rangle = \sum_{k=2}^{\infty}\frac{1}{2}\frac{\partial^2}{\partial x^2}\Gamma_k(x). \tag{2.21}$$

It is to be emphasized that differentiation with respect to x of the function $\Gamma_k(x)$ is reduced to differentiation of the exponent in (2.10).

At the strong coupling limit $c \to +\infty$, as has already been mentioned $\Gamma_k \sim c^{2-k}$ $(k > 2)$ and so to obtain the conventional $1/c$-expansion from series (2.21) we used a finite number of terms of the series. In particular, at $c = \infty$ ("impenetrable bosons") only the term with Γ_2 contributes. As $K(\lambda, \mu) = 0$ at $c = \infty$, one obtains the following explicit answer for the correlator:

$$\langle :j(x)j(0): \rangle = \left(\frac{q}{\pi}\right)^2 - \left(\frac{\sin qx}{\pi x}\right)^2 \tag{2.22}$$

$(c = +\infty)$.

Let us also give the first correction to this expansion (the calculations needed are quite obvious):

$$\langle:j(x)j(0):\rangle = \left(\frac{q}{\pi}\right)^2 + \frac{4}{c}\left(\frac{q}{\pi}\right)^3$$
$$- \left(1 + \frac{2}{c}\frac{\partial}{\partial x_r}\right)\left(\frac{\sin qx_r}{\pi x_r}\right)^2$$
$$+ \frac{2}{\pi^2 c}\frac{\partial}{\partial x}\left[\frac{\sin qx}{\pi x}\int_{-q}^{q} d\lambda \sin \lambda x \ln \frac{q+\lambda}{q-\lambda}\right]$$
$$- \frac{8q}{\pi c}\left(\frac{\sin qx}{\pi x}\right)^2 + O\left(\frac{1}{c^2}\right). \tag{2.23}$$

Here

$$x_r = \left(1 + \frac{2q}{\pi c}\right)x.$$

The correction to (2.23) is of order $1/c^2$ for any x.

Here we have considered the equal-time correlation function of two currents. The time-dependent correlation function of currents was considered by N.A. Slavnov. He obtained a representation similar to (2.21). In paper [8] correlation functions for the Heisenberg antiferromagnet were evaluated. It was shown that the correlation functions can be described by elementary functions of distance (on the lattice) only in the free fermion case ($\Delta = 0$, XXO model) and that they depend on the magnetization of the ground state very strongly.

XVII.3 The series for the correlator of currents in the nonrelativistic Bose gas at finite temperature

To evaluate the correlator of two currents,

$$\langle:j(x)j(0):\rangle_T = \frac{\text{tr}\left(e^{-H/T}:j(x)j(0):\right)}{\text{tr}\,e^{-H/T}}$$

one should evaluate the mean value

$$\frac{\langle\Omega_T|:j(x)j(0):|\Omega_T\rangle}{\langle\Omega_T|\Omega_T\rangle} \tag{3.1}$$

where Ω_T is one of the eigenfunctions of the Hamiltonian, which is present in the state of thermal equilibrium (see section I.8). The evaluation of this mean value is very similar to the zero temperature case (see section 2). The answer differs only in the measure of integration

$$\int_{-q}^{q} d\lambda \longrightarrow \int_{-\infty}^{+\infty} d\lambda\,\vartheta(\lambda) \tag{3.2}$$

as in section I.8 (see formula I.8.5). Here

$$\vartheta(\lambda) = \frac{1}{1 + e^{\varepsilon(\lambda)/T}} = \frac{\rho_p(\lambda)}{\rho_t(\lambda)} \tag{3.3}$$

is the Fermi weight (see (I.5.28)) and $\varepsilon(\lambda)$ is the solution of the Yang-Yang equation (I.5.19),

$$\varepsilon(\lambda) = \lambda^2 - h - \frac{T}{2\pi} \int\limits_{-\infty}^{+\infty} K(\lambda, \mu) \ln\left(1 + e^{-\varepsilon(\mu)/T}\right) d\mu. \tag{3.4}$$

The particle density ρ_p and the density of vacancies $\rho_t = \rho_p + \rho_h$ are related by equation (I.5.24)

$$2\pi\rho_t(\lambda) = 1 + \int\limits_{-\infty}^{+\infty} K(\lambda, \mu)\rho_p(\mu)\, d\mu \tag{3.5}$$

and the density D of the gas is given by

$$D = \int\limits_{-\infty}^{+\infty} \rho_p(\lambda)\, d\lambda. \tag{3.6}$$

So the mean value (3.1) is the same for all eigenstates present in the state of thermal equilibrium, because all of them have the same $\varepsilon(\lambda)$ and $\vartheta(\lambda)$. Now according to section I.8 (formula (I.8.20)) this means that the mean value (3.1) is equal to the temperature correlation function

$$\frac{\text{tr}\left(e^{-H/T} :j(x)j(0):\right)}{\text{tr}\, e^{-H/T}} = \frac{\langle\Omega_T|:j(x)j(0):|\Omega_T\rangle}{\langle\Omega_T|\Omega_T\rangle}.$$

Correspondingly, the following representation for the correlator was obtained in papers [4], [5]

$$\langle :j(x)j(0): \rangle_T = \langle j(0)\rangle_T^2 + \langle\langle j(x)j(0)\rangle\rangle_T \tag{3.7}$$

with

$$\langle j(0)\rangle_T = D \tag{3.8}$$

and the nontrivial part given as a series,

$$\langle\langle j(x)j(0)\rangle\rangle_T = \sum_{k=2}^{\infty} \frac{1}{2}\frac{\partial^2}{\partial x^2}\Gamma_k^{(T)}(x), \tag{3.9}$$

where $\Gamma_k^{(T)}$ is represented as

$$\Gamma_k^{(T)}(x) = \frac{1}{k!} \int\limits_{-\infty}^{+\infty} \left(\prod_{j=1}^{k} \frac{\vartheta(\lambda_j)\omega(\lambda_j)d\lambda_j}{2\pi}\right) I_k^d(x, \{\lambda\}, T). \tag{3.10}$$

The "statistical weight" $\omega(\lambda)$ is now

$$\omega(\lambda) = \exp\left\{-\frac{1}{2\pi}\int_{-\infty}^{+\infty} K(\lambda, \mu)\vartheta(\mu)\,d\mu\right\}, \tag{3.11}$$

$$0 < \omega(\lambda) < 1, \quad \omega(\lambda) = \omega(-\lambda).$$

The dressed irreducible part is constructed by means of the same Fourier coefficients $\mathcal{A}_k^n$ as in the zero temperature case (2.10),

$$I_k^d(x, \{\lambda\}, T) = \sum e^{xp_n^T(\{\lambda^+\}, \{\lambda^-\})}\mathcal{A}_k^n(\{\lambda^+\}, \{\lambda^-\}, \{\lambda_0\}). \tag{3.12}$$

(The sum here is taken exactly as in (2.10).) The dressed function p_n^T is different from the zero temperature case:

$$p_n^T(\{\lambda^+\}, \{\lambda^-\}) = -i\sum_{j=1}^{n}(\lambda_j^+ - \lambda_j^-)$$

$$+ \int_{-\infty}^{+\infty} dt\,\vartheta(t)P_n^T(t, \{\lambda^+\}, \{\lambda^-\}) \tag{3.13}$$

where the dressing function P_n^T is the solution of the integral nonlinear equation

$$1 + 2\pi P_n^T(t) = S(t)\exp\left\{\int_{-\infty}^{+\infty} K(t, s)P_n^T(s)\vartheta(s)\,ds\right\}, \tag{3.14}$$

$$S(t) = \prod_{j=1}^{n}\frac{(\lambda_j^+ - t + ic)(t - \lambda_j^- + ic)}{(\lambda_j^+ - t - ic)(t - \lambda_j^- - ic)}$$

$$\operatorname{Re} P_n^T(t) \leq 0.$$

The factor $S(t)$ (the product of "bare" S-matrices) is exactly the same (2.14) as in the zero temperature case. In papers [4], [5] it is proved that equation (3.14) has a unique solution. And again, since

$$\Gamma_k^{(T)} \sim c^{2-k}, \quad c \to +\infty \tag{3.15}$$

one easily reproduces, e.g., the conventional strong coupling limit expansion in powers of $1/c$ from series (3.9). In the strong coupling limit this series is especially effective and the results for the correlator are simple.

In the "free fermion" point $c = +\infty$ (impenetrable bosons) only $\Gamma_2^{(T)}$ contributes. As the kernel $K(\lambda, \mu) = 0$, and the Fermi weight $\vartheta(\lambda)$ is given explicitly by

$$\vartheta(\lambda) = \frac{1}{1 + e^{(\lambda^2 - h)/T}} \quad (c = +\infty) \tag{3.16}$$

one easily obtains

$$\langle :j(x)j(0): \rangle_T = D_\infty^2(h) + \langle\!\langle j(x)j(0) \rangle\!\rangle_T, \qquad (3.17)$$

where the gas density is given by

$$D_\infty(h) = \frac{1}{2\pi} \int\limits_{-\infty}^{+\infty} \frac{d\lambda}{1 + e^{(\lambda^2 - h)/T}} \qquad (c = +\infty) \qquad (3.18)$$

and the nontrivial part of the correlator is given as the squared modulus of the Fourier transform of the Fermi weight:

$$\langle\!\langle j(x)j(0) \rangle\!\rangle_T = -\frac{1}{4\pi^2} \left(\int\limits_{-\infty}^{+\infty} e^{i\lambda x}\, \vartheta(\lambda)\, d\lambda \right)^2. \qquad (3.19)$$

Corrections (of order $1/c$) related to interactions of fermions are obtained in [2] and [4].

The approach in this section has also been applied to the time- and temperature-dependent correlation function of currents. In papers [8] and [10] the correlation functions of the XXZ Heisenberg antiferromagnet were evaluated. It was shown that they depend essentially on the magnetic field applied to the model. Many current correlation functions of impenetrable bosons depending on space, time and temperature can be found in [6].

XVII.4 Asymptotics of the correlator of currents at finite temperature

The long distance asymptotics of correlators for integrable quantum systems at zero temperature constitute the main interest from the physical point of view. For systems in one space dimension zero temperature is the only possible phase transition point. Such models as the nonrelativistic Bose gas (the quantum NS model) or the Heisenberg spin chain possess no gap in the energy spectrum ($-1 < \Delta < 1$). The exponential decrease of correlation with distance at finite temperature is changed to a power law decrease at zero temperature for these models. The corresponding power is called the critical exponent. Its calculation is one of the main problems in phase transition theory. In papers [1], [11], [12] the asymptotics of correlation functions at zero temperature were studied on the basis of the series representation (2.21). The critical exponents were evaluated for both the Bose gas and the Heisenberg antiferromagnet. Explicit formulæ can be found in Chapter XVIII. In papers [1], [11], [12] it was proved that the critical exponents depend only on the R-matrix of the theory. It is interesting to mention that the series representation (3.9) can be used to evaluate the asymptotics of correlation functions at finite temperature as well.

Let us consider the long distance asymptotics of the equal-time temperature correlator of currents (3.9). In this section, following papers [4], [5], we present the asymptotics for the case of finite coupling constant c, calculating the correlator asymptotics by taking into account the analytical properties of the coefficients Γ_k in (3.9). Properties of the Yang-Yang energy $\varepsilon(\lambda)$ and of the Fermi weight $\vartheta(\lambda)$ for $0 < c < +\infty$ are investigated in detail in the Appendix. It is shown that the zeros of the function $\vartheta^{-1}(\lambda)$ do exist and are situated in the complex λ-plane forming a rectangle

$$\alpha, \quad \alpha^*, \quad -\alpha, \quad -\alpha^*; \quad \mathrm{Im}\,\alpha > 0, \quad \mathrm{Re}\,\alpha > 0. \tag{4.1}$$

The singularities of the function $\varepsilon(\lambda)$ nearest to the real λ-axis (simple poles) are situated at the distance $\mathrm{Im}\,(\alpha+ic)$ from the real axis. The "statistical weight" $\omega(\lambda)$ can also be continued into the complex λ-plane, the nearest singularities being at the same distance from the real axis. The analytic continuations of $\varepsilon(\lambda)$ and $\omega(\lambda)$ possess the following properties:

$$\begin{aligned}
\varepsilon^*(\lambda) &= \varepsilon(\lambda^*), \quad \omega^*(\lambda) = \omega(\lambda^*), \\
\varepsilon(-\lambda) &= \varepsilon(\lambda), \quad \omega(-\lambda) = \omega(\lambda).
\end{aligned} \tag{4.2}$$

Consider first the current correlator taking into account only the first term in the expansion (3.9):

$$\langle\!\langle j(x)j(0)\rangle\!\rangle_T \simeq \frac{1}{2}\frac{\partial^2}{\partial x^2}\Gamma_2^{(T)}(x)$$

$$= -\frac{1}{4\pi^2}\int\limits_{-\infty}^{+\infty} d\lambda_1\,\omega(\lambda_1)\vartheta(\lambda_1)\int\limits_{-\infty}^{+\infty} d\lambda_2\,\omega(\lambda_2)\vartheta(\lambda_2)$$

$$\times\left(\frac{\lambda_1-\lambda_2+ic}{\lambda_1-\lambda_2-ic}\right)\left[\frac{p_1^T(\lambda_1,\lambda_2)}{\lambda_1-\lambda_2}\right]^2 \exp\{xp_1^T(\lambda_1,\lambda_2)\}, \tag{4.3}$$

$$x > 0.$$

Let us study this expression in the limit $x \to \infty$. We move the integration contour in the variable λ_1 into the lower half-plane and the integration contour in variable λ_2 into the upper half-plane. The nearest obstacles are the poles of the Fermi weight

$$\vartheta(\lambda) = \frac{1}{1 + \exp\{\varepsilon(\lambda)/T\}}$$

which near the pole behave as follows:

$$\vartheta(\lambda) \longrightarrow -\frac{T}{\varepsilon'(\alpha)(\lambda - \alpha)},$$

$$e^{\varepsilon(\alpha)/T} = -1, \quad \varepsilon(\alpha) = i\pi T. \tag{4.4}$$

The contribution of these poles to $\frac{1}{2}\partial^2\Gamma_2^{(T)}(x)/\partial x^2$ is $(x \to +\infty)$

$$2T^2 \left|\frac{\omega(\alpha)}{\varepsilon'(\alpha)}\right|^2 \left(\frac{2\operatorname{Im}\alpha - c}{2\operatorname{Im}\alpha + c}\right) \left[\frac{p_1^T(\alpha^*,\alpha)}{2\operatorname{Im}\alpha}\right]^2 \exp\{xp_1^T(\alpha^*,\alpha)\} \qquad (4.5)$$

$$+ 2T^2\operatorname{Re}\left\{\left(\frac{\omega(\alpha)}{\varepsilon'(\alpha)}\right)^2 \left(\frac{2\alpha - ic}{2\alpha + ic}\right)\left[\frac{p_1^T(-\alpha,\alpha)}{2\alpha}\right]^2\right\}\exp\{xp_1^T(-\alpha,\alpha)\}$$

$$(4.6)$$

To obtain other contributions to (4.3) one should move the integration contours further from the real axis obtaining thus expressions decreasing faster with distance than (4.5), (4.6). The arguments based on perturbation theory show that the term (4.6) is decreasing faster than (4.5), so that the long distance asymptotics of the correlator are of the form

$$\langle\!\langle j(x)j(0)\rangle\!\rangle_T \sim e^{-x/r_c}. \qquad (4.7)$$

The correlation radius (length) r_c is given as

$$\frac{1}{r_c} = -p_1^T(\alpha^*,\alpha) = 2\operatorname{Im}\alpha - \int_{-\infty}^{+\infty} dt\,\vartheta(t)P_1^T(t,\alpha^*,\alpha) \geq 2\operatorname{Im}\alpha \geq 0; \quad (4.8)$$

it should be stated that function $p_1(\alpha^*,\alpha)$ is real and

$$-\int_{-\infty}^{+\infty} dt\,\vartheta(t)P_1^T(t,\alpha^*,\alpha) \geq 0.$$

The higher order terms of expansion (3.9) were analyzed in [14], where the following formula for the correlation radius (length) at finite temperature was proposed:

$$\frac{1}{r_c} = -p_1^T(\alpha^*,\alpha). \qquad (4.9)$$

It coincides with (4.8). This formula was studied in the papers [2], [4], [5]. It was shown that for small T it agrees with the conformal approach (XVIII.2.11) and at large coupling constant the correlation length becomes

$$\frac{1}{r_c} = \sqrt{2\left[(h^2 + \pi^2 T^2)^{1/2} - h\right]}.$$

A similar approach is applicable to the time-dependent correlation functions of currents

$$\langle j(x,t)j(0,0)\rangle_T.$$

Formula (4.9) is valid at finite temperature but at infinite temperature the asymptotics turn into gaussian decay [14]. The large time asymptotics were calculated (for finite coupling) in paper [15]. It was shown that the correlations decay slowly with time:

$$\langle j(x,t)j(0,0)\rangle - \langle j(0,0)\rangle^2 \xrightarrow[t\to\infty]{} \frac{\text{const}}{t}. \tag{4.10}$$

XVII.5 Asymptotics of the emptiness formation probability

In this section we evaluate the probability $P(x_1, x_2)$ of the absence of particles from some space interval $[x_1, x_2]$ due to thermal fluctuations. We argue that the long distance asymptotics of $P(x_1, x_2)$ for positive temperature, $\infty > T > 0$, are given by

$$P(x_1, x_2) \simeq \exp\left\{-\frac{|x_1 - x_2|}{T}\mathcal{P}\right\}, \quad |x_1 - x_2| \to \infty. \tag{5.1}$$

Here $\mathcal{P}$ is the pressure (see (I.5.22)). To show this let us remind the reader that $P(x_1, x_2)$ is a special value of the generating functional of current correlators:

$$\langle e^{\alpha Q_1(x_1,x_2)}\rangle_T = \frac{\text{tr}\left(e^{-H/T}e^{\alpha Q_1(x_1,x_2)}\right)}{\text{tr}\, e^{-H/T}} \tag{5.2}$$

$$Q_1(x_1, x_2) = \int_{x_2}^{x_1} \Psi^\dagger(y)\Psi(y)\, dy, \quad x_1 > x_2$$

$$P(x_1, x_2) = \langle e^{\alpha Q_1(x_1,x_2)}\rangle_T \Big|_{e^\alpha = 0}. \tag{5.3}$$

Let us rewrite the results of section I.5. Now our Hamiltonian includes the chemical potential h

$$H = \int dx \left[\partial_x \Psi^\dagger(x)\, \partial_x \Psi(x) + c\Psi^\dagger(x)\Psi^\dagger(x)\Psi(x)\Psi(x) - h\Psi^\dagger(x)\Psi(x)\right]. \tag{5.4}$$

This changes formulæ (I.5.1), (I.5.21) a little:

$$Z = \text{tr}\, e^{-H/T} = e^{-F/T} \tag{5.5}$$

$$F = -\frac{TL}{2\pi} \int_{-\infty}^{+\infty} d\lambda \ln\left(1 + e^{-\varepsilon(\lambda)/T}\right) \tag{5.6}$$

$$\varepsilon(\lambda) = \lambda^2 - h - \frac{T}{2\pi} \int_{-\infty}^{+\infty} K(\lambda, \mu) \ln\left(1 + e^{-\varepsilon(\mu)/T}\right) d\mu. \tag{5.7}$$

Here we consider the grand canonical ensemble (the chemical potential h is fixed but the number of particles is not), and L is the length of the box. Equation (5.7) coincides with (I.5.25). The pressure $\mathcal{P}$ is equal to (I.5.22)

$$\mathcal{P} = \frac{T}{2\pi} \int\limits_{-\infty}^{+\infty} d\lambda \, \ln\left(1 + e^{-\varepsilon(\lambda)/T}\right). \tag{5.8}$$

Now let us consider the asymptotics of (5.2). If $|x_1 - x_2|$ is very large we can say that on the interval $[x_2, x_1]$ we have thermodynamic equilibrium with a different value of the chemical potential

$$h \longrightarrow h + \alpha T. \tag{5.9}$$

So on the interval $[x_2, x_1]$ $\varepsilon(\lambda)$ from (5.7) is replaced by $\varepsilon_\alpha(\lambda)$ defined by

$$\varepsilon_\alpha(\lambda) = \lambda^2 - h - \alpha T - \frac{T}{2\pi} \int\limits_{-\infty}^{+\infty} K(\lambda, \mu) \ln\left(1 + e^{-\varepsilon_\alpha(\mu)/T}\right) d\mu. \tag{5.10}$$

Now we can replace (5.2) by the following relation:

$$\langle e^{\alpha Q_1(x_1, x_2)} \rangle_T = \frac{\operatorname{tr}\left(e^{-H/T} e^{\alpha Q_1(x_1, x_2)}\right)}{\operatorname{tr} e^{-H/T}}$$

$$= \exp\left\{\frac{|x_1 - x_2|}{2\pi} \int\limits_{-\infty}^{+\infty} d\lambda \, \ln\left(\frac{1 + e^{-\varepsilon_\alpha(\lambda)/T}}{1 + e^{-\varepsilon(\lambda)/T}}\right)\right\}. \tag{5.11}$$

This asymptotic formula is interesting in itself because it gives the generating functional of current correlators for finite coupling c. It is in agreement with the impenetrable case ($c = \infty$), see (XVI.5.12). Now let us turn to the emptiness formation probability. We need to let $\alpha \to -\infty$ in (5.11). This means that $\varepsilon_\alpha(\lambda) \to +\infty$, and (5.11) becomes

$$P(x_1, x_2) = \exp\left\{-\frac{|x_1 - x_2|}{2\pi} \int\limits_{-\infty}^{+\infty} d\lambda \, \ln\left(1 + e^{-\varepsilon(\lambda)/T}\right)\right\} \tag{5.12}$$

which, by virtue of (5.8), gives (5.1). It might be that the analytic prolongation of (5.11) to $\alpha = i\pi$ is related to the field correlator, as it was in the impenetrable case (see the text after formula (XIII.3.12), and also (XVI.5.6)–(XVI.5.8)). Using this method, we have managed to get only the leading term in the asymptotics; to get the complete asymptotic expansion for large distance one needs differential equations like those in Chapters XIV–XVI.

Let us now discuss the zero temperature case. We would expect, for the long-distance asymptotics,

$$\langle e^{\alpha Q_1(x_1,x_2)} \rangle \xrightarrow[|x_1-x_2|\to\infty]{} e^{\alpha D\,|x_1-x_2|}$$

where $D = N/L$.

Conclusion

In this chapter we have constructed a very special series for quantum correlation functions. It is based on the algebraic Bethe Ansatz and emphasizes the role of the R-matrix. It is especially effective for current correlation functions and gives explicit formulæ for asymptotics. Here we have followed the papers [2], [4], [5], [7], [13], [14].

Appendix XVII.1: Analytic properties of solutions of the Yang-Yang equation

Let us study the analytic properties of the function $\varepsilon(\lambda)$, which is the solution of the Yang-Yang equation:

$$\varepsilon(\lambda) = \lambda^2 - h - \frac{T}{2\pi} \int\limits_{-\infty}^{+\infty} d\mu \, K(\lambda, \mu) \ln\left(1 + e^{-\varepsilon(\mu)/T}\right) \quad (A.1.1)$$

$$K(\lambda, \mu) = \frac{2c}{c^2 + (\lambda - \mu)^2}. \quad (A.1.2)$$

In section I.6 it was proved that the solution of (A.1.1) exists and is unique on the real axis $\text{Im } \lambda = 0$. Let us try to make an analytic continuation of $\varepsilon(\lambda)$ to the complex λ-plane. At first sight it seems that such a continuation can be made up to $\text{Im } \lambda = c$, due to the pole of $K(\lambda, \mu)$ (A.1.2). The continued function $\varepsilon(\lambda)$ is given by the right hand side of (A.1.1) for complex λ. This pole is not an essential obstacle and we can continue $\varepsilon(\lambda)$ deeper into the complex plane. Indeed, we can shift the integration on the right hand side of (A.1.1) from the real axis into the complex plane (because $\varepsilon(\lambda)$, as we showed, is analytic in the vicinity of the real axis). Let us move the contour for example into the upper half-plane. This way of analytically continuing $\varepsilon(\lambda)$ can be prevented only by zeros of the function

$$1 + e^{\varepsilon(\mu)/T}.$$

One can prove that these zeros must exist, otherwise we would get a contradiction (we follow paper [2]). If such zeros do not exist then this means that $\varepsilon(\lambda)$ can be analytically continued by equation (A.1.1) into the whole complex plane without singularities. In this case $\varepsilon(\lambda)$ would be an entire function of λ with the asymptotic behavior

$$\varepsilon(\lambda) - (\lambda^2 - h) \xrightarrow[\lambda \to \infty]{} 0.$$

500

Thus we would have

$$\varepsilon(\lambda) = \lambda^2 - h.$$

But this is a contradiction since $\varepsilon(\lambda) = \lambda^2 - h$ does not satisfy equation (A.1.1) (with a nonzero kernel) and, moreover, the function

$$1 + e^{(\lambda^2 - h)/T}$$

has zeros in the complex plane. This proves that zeros of

$$1 + e^{\varepsilon(\lambda)/T}$$

must necessarily exist somewhere in the complex plane.

The zeros nearest to the real axis are the most important. They play the role of the Fermi momentum for finite temperature. After analytic continuation of $\varepsilon(\lambda)$ into the complex plane it will have the properties:

$$\varepsilon(\lambda) = \varepsilon(-\lambda), \quad \varepsilon^*(\lambda) = \varepsilon(\lambda^*). \tag{A.1.3}$$

This means that there will be four zeros of $1 + e^{\varepsilon(\lambda)/T} = 1/\vartheta(\lambda)$ nearest to the real axis,

$$\{\lambda\} = \{\alpha, -\alpha, \alpha^*, -\alpha^*\}. \tag{A.1.4}$$

Here α satisfies the equation

$$\varepsilon(\alpha) = i\pi T, \quad \text{Im}\,\alpha \geq 0, \ \text{Re}\,\alpha \geq 0. \tag{A.1.5}$$

The singularities of $\varepsilon(\lambda)$ lie deeper in the complex plane. They are at the points

$$\{\lambda\} = \{\alpha + ic, -\alpha - ic, \alpha^* - ic, -\alpha^* + ic\}. \tag{A.1.6}$$

If we consider the Fermi weight

$$\vartheta(\lambda) = \frac{1}{1 + e^{\varepsilon(\lambda)/T}}$$

then one can also prove that it can be analytically continued from the real axis into the complex plane. The simple poles at (A.1.4) will be the singularities that are closest to the real axis.

XVIII

Asymptotics of Correlation Functions and the Conformal Approach

Introduction

The long distance asymptotics of correlation functions in integrable models have attracted long-standing interest. For the gapless one-dimensional systems considered in this book, zero temperature is a critical point. At zero temperature, $(T = 0)$, correlation functions decay as a power of the distance, but for $T > 0$, correlation functions decay exponentially. The powers of the distance by which correlation functions decay at zero temperature are called the critical exponents and they are the subject of this chapter.

A recent development in the understanding of critical phenomena in (1+1)-dimensional systems is connected with conformal field theory [4] which provides a powerful method for calculating critical exponents. However, many important results in (1+1)-dimensional systems have been obtained from perturbation calculations and renormalization group treatments [52], [53], [45], [23], [20] and [22]. It should also be noted that the quantum inverse scattering method (QISM) approach to correlation functions [42], [37] confirms the predictions of conformal field theory (CFT), but more importantly QISM techniques show that the critical exponents of integrable models depend only on the underlying R-matrix.

The Luttinger liquid approach is very powerful; it is also close to CFT. This approach ([28], [29], [30], [31]) is essentially based on the representation of a critical one–dimensional model in terms of a gaussian model [40]. In the framework of this approach, the critical exponent was calculated first [30] for Bethe Ansatz solvable models with a single degree of freedom. One should also mention the bosonization technique [21].

Conformal invariance constrains the behavior of (1+1)-dimensional quantum systems at a critical point. A conformal theory is characterized by the central charge, c, of the underlying Virasoro algebra. The critical exponents in a conformally invariant theory are related to the scaling (conformal) dimensions of the operators within this theory. Bethe Ansatz solvable models (which we consider here) are related to a Virasoro algebra with $c = 1$. In this case, the critical exponents may demonstrate continuous dependence on the parameters of the model [4], [19], [26]. We shall evaluate the critical exponents with the help of finite scaling which connects the critical behavior of the model with the behavior of the model in a finite volume [1], [7]. For unitary conformal field theory, the central charge, c, is the coefficient of the $1/L$ term (L is the size of the system) in the expansion of the ground state energy for $L \to \infty$

$$E = L\epsilon - c\frac{\pi v}{6L} \quad + \quad \text{h.o.c.} \tag{0.1}$$

while the conformal dimensions are determined from the spectrum of low-lying excitations [14]. Here v is the Fermi velocity.

Putting the system in a periodic box of length L will cause the correlation functions to decay exponentially. (This is due to the conformal mapping explained in section 1.) In addition, the spectrum of the Hamiltonian will also be changed; a gap of order $1/L$ will arise. Comparing this gap with the rate of exponential decay permits us to evaluate the conformal dimensions of the model. These gaps can be explicitly evaluated by means of the Bethe Ansatz. This approach has been used successfully in various exactly solvable models [16], [17], [18] and in models which cannot be exactly solved as well [48].

Fluctuations near the Fermi surface are important for the asymptotics of correlation functions. In the models under consideration (the Bose gas, the Heisenberg magnet and the Hubbard chain), the dispersion curve is linear near the Fermi surface (k close to k_F):

$$\varepsilon(k) = v|k - k_F|.$$

This allows the application of the conformal description to the asymptotics of correlation functions in these models. We restrict this approach to asymptotics because the models themselves are not conformally invariant.

Comparing formula (0.1) and the results of finite size analysis from Part I ((I.9.13) and (II.5.1)), we see that both the Bose gas and the Heisenberg magnet are related to a Virasoro algebra with $c = 1$. Below we shall express the whole spectrum of conformal dimensions in terms of the dressed charge $\mathcal{Z}$ which was defined in section 9 of Chapter I and section 5 of Chapter II (see (I.9.20) and (II.5.6)). By $\mathcal{Z}$ we mean the value of $Z(\lambda)$ on the Fermi surface.

The explicit method for evaluating the conformal dimensions will be shown in section 1 where the general form for the asymptotics of the correlation functions will be derived. In sections 2 and 3, the asymptotics of the Bose gas and the Heisenberg magnet, respectively, will be considered. Section 4 will be devoted to the Hubbard model. It should also be mentioned that in this approach, it is much more convenient to use Euclidean time. We shall not give here a complete description of conformal field theory. Instead we recommend the book [36].

XVIII.1 Finite size effects

To describe the long distance asymptotics of correlation functions at zero temperature, it is convenient to introduce the complex variable

$$z = ix + vt$$

where v is the Fermi velocity and t is the Euclidean time. The conformal fields $\varphi(z, \overline{z})$ depend on z and $\overline{z}$. Under the conformal transformation

$$z = z(w); \qquad \overline{z} = \overline{z}(\overline{w}),$$

primary fields transform in the following way:

$$\varphi(w, \overline{w}) = \left(\frac{\partial z}{\partial w}\right)^{\Delta^+} \left(\frac{\partial \overline{z}}{\partial \overline{w}}\right)^{\Delta^-} \varphi(z(w), \overline{z}(\overline{w})). \tag{1.1}$$

Here Δ^+ and Δ^- are the conformal dimensions of the primary field φ. Conformal fields with different conformal dimensions are orthogonal (their correlation function is equal to zero). The correlation function of conformal fields with the same conformal dimensions is given by

$$\langle \varphi(z_1, \overline{z}_1)\varphi(z_2, \overline{z}_2) \rangle = (z_1 - z_2)^{-2\Delta^+} \ (\overline{z}_1 - \overline{z}_2)^{-2\Delta^-}. \tag{1.2}$$

This formula is valid for the whole complex plane without the origin ($z_1 \neq z_2$). To evaluate $\Delta^\pm$, we shall make a conformal mapping of this plane (without the origin) to a periodic strip (cylinder) using

$$z = e^{2\pi w/L}. \tag{1.3}$$

The complex variable in the strip will be denoted by

$$w = ix + vt \quad \text{with} \quad 0 < x \le L. \tag{1.4}$$

By means of (1.1), the correlation function (1.2) can be evaluated in the strip to give

$$\langle \varphi(w_1, \overline{w}_1)\varphi(w_2, \overline{w}_2) \rangle$$

$$= \left\{ \frac{\pi/L}{\sinh[\frac{\pi}{L}(w_1 - w_2)]} \right\}^{2\Delta^+} \left\{ \frac{\pi/L}{\sinh[\frac{\pi}{L}(\overline{w}_1 - \overline{w}_2)]} \right\}^{2\Delta^-}. \tag{1.5}$$

Thus the asymptotics change to exponential decay

$$\langle \varphi(w_1, \overline{w}_1) \varphi(w_2, \overline{w}_2) \rangle_L \sim \exp \left\{ \frac{-2\pi \Delta^+}{L} [i(x_1 - x_2) + v(t_1 - t_2)] \right\}$$

$$\times \exp \left\{ \frac{-2\pi \Delta^-}{L} [-i(x_1 - x_2) + v(t_1 - t_2)] \right\}. \tag{1.6}$$

This can be compared with the spectral decomposition of the correlation function $(t_1 > t_2)$

$$\langle \varphi(w_1, \overline{w}_1) \varphi(w_2, \overline{w}_2) \rangle_L$$

$$= \sum_Q |\langle 0 | \varphi(0,0) | Q \rangle|^2 \times \exp\{-(t_1 - t_2)(E_Q - E_0) - i(x_1 - x_2)(P_Q - P_0)\} \tag{1.7}$$

where a complete set of intermediate states $|Q\rangle$ has been inserted between the fields $\varphi(w_1, \overline{w}_1)$ and $\varphi(w_2, \overline{w}_2)$. ($|0\rangle$ represents the ground state.) The leading term corresponds to the smallest gap.

Comparison of (1.6) and the leading term of (1.7) leads to

$$E_Q - E_0 = \frac{2\pi v}{L} (\Delta^+ + \Delta^-) \tag{1.8}$$

and

$$P_Q - P_0 = \frac{2\pi}{L} (\Delta^+ - \Delta^-). \tag{1.9}$$

A detailed presentation can be found in [14]. We have assumed that both gaps are small (of order $1/L$). (Below, we shall discuss the case when the gap in the momentum spectrum is of order 1.) Formulæ (1.8) and (1.9) are very important since they provide an opportunity to calculate the conformal dimensions, $\Delta^\pm$, in terms of energy and momentum gaps which arise when the system is placed in a large periodic box of length L. These gaps have already been calculated in Part I (see section 9 of Chapter I, sections 5 and 6 of Chapter II and section 2 of Chapter IV).

Bethe Ansatz solvable models have excitation spectra which are linear (in momentum) in the vicinity of the Fermi surface. This provides an opportunity to apply the ideas of conformal field theory. To do this carefully, one should represent the local field of the Bethe Ansatz solvable model, φ, as a combination of conformal fields, φ_Q,

$$\varphi(x, t) = \sum_Q \widetilde{A}(Q) \varphi_Q(z, \overline{z}). \tag{1.10}$$

See papers [11], [6]. Here the $\widetilde{A}(Q)$ are coefficients and the conformal field φ_Q is related to excitations with quantum numbers $Q = \{\Delta N, d, N^\pm\}$. These excitations have been classified previously in sections I.9, II.5 and

IV.2. The quantum number ΔN is characteristic of the local field $\varphi(x,t)$. In the Bose gas, ΔN is the change in the number of particles produced by $\varphi(x,t)$; in the Heisenberg magnet, it is the number of spins overturned by $\varphi(x,t)$. This quantum number is the same for all conformal fields φ_Q in decomposition (1.10). The quantum number d represents the number of particles backscattered and can be different for different terms in (1.10). $N^{\pm}$ shows the change of integer numbers n_j (from the periodic boundary conditions of the Bethe Ansatz) in the vicinity of the Fermi surface (see I.9.4) (in Chapter I). In the conformal language, $N^{\pm}$ describes the level of the descendents.

If the gap in the momentum spectrum, $P_Q - P_0$, is of order 1, the coefficients $\tilde{A}(Q)$ will depend on x in the following way [11]

$$\tilde{A}(Q) = A(Q)e^{-2ixk_Fd} \tag{1.11}$$

where $2k_Fd$ is the gap in the momentum spectrum; in (I.9.18), the following results were presented where the gap in the momentum spectrum is

$$P_Q - P_0 = 2k_Fd + \frac{2\pi}{L}(N^+ - N^- + d\Delta N) \tag{1.12}$$

and the gap in the energy spectrum is

$$E_Q - E_0 = \frac{2\pi v}{L}\left[\left(\frac{\Delta N}{2\mathcal{Z}}\right)^2 + (\mathcal{Z}d)^2 + N^+ + N^-\right]. \tag{1.13}$$

$\mathcal{Z}$ is the dressed charge defined by the following expression

$$\mathcal{Z} = Z(\Lambda) \tag{1.14}$$

where $Z(\lambda)$ is defined by the following integral equation:

$$Z(\lambda) - \frac{1}{2\pi}\int\limits_{-\Lambda}^{\Lambda} K(\lambda,\mu)Z(\mu)d\mu = 1. \tag{1.15}$$

For more information, see sections I.9 and II.5. Combining formulæ (1.8) and (1.9) with (1.12) and (1.13) gives the expression for the conformal dimensions as

$$2\Delta^{\pm} = 2N^{\pm} + \left(\frac{\Delta N}{2\mathcal{Z}} \pm \mathcal{Z}d\right)^2. \tag{1.16}$$

Combining formulæ (1.2), (1.10) and (1.11), we obtain the generic formula for the asymptotics of correlation functions at zero temperature:

$$\langle\varphi(x,t)\varphi(0,0)\rangle = \sum_Q C(Q)e^{-2ixk_Fd}(x-ivt)^{-2\Delta^+}(x+ivt)^{-2\Delta^-}. \tag{1.17}$$

The leading terms here correspond to minimal $\Delta^{\pm}$. It should also be mentioned that since the dressed charge $\mathcal{Z}$ depends only on the R-matrix (and Fermi momentum), the conformal dimensions demonstrate universality properties.

So far we have only discussed the zero temperature case. At nonzero temperatures, all the correlation functions decay exponentially. We shall find the low temperature asymptotics of the Euclidean time correlation functions (with period $1/T$ along the imaginary time axis) using conformal invariance. These correlation functions will be calculated in the strip geometry and determined in terms of infinite-plane correlation functions at $T = 0$ (1.2). The calculation is very similar to the evaluation of correlation functions in the finite box (1.5). However, the role of space and time should be interchanged. Instead of the conformal mapping (1.3), we shall now use

$$z = e^{2\pi T w/v} \quad \text{where} \quad z = x - ivt.$$

This gives the finite temperature correlator as

$$\langle \varphi(x, t)\varphi(0, 0)\rangle_T = \sum_Q B(Q)e^{-2ixk_Fd}$$

$$\times \left\{ \frac{\pi T/v}{\sinh[\frac{\pi T}{v}(x-ivt)]} \right\}^{2\Delta^+} \left\{ \frac{\pi T/v}{\sinh[\frac{\pi T}{v}(x+ivt)]} \right\}^{2\Delta^-}$$

$$\text{(1.18)}$$

which is only valid for small temperatures in the vicinity of the phase transition. The leading term corresponds to the smallest $\Delta^{\pm}$.

For equal-time correlators ($t = 0$) in the limit $x \to \infty$, expression (1.18) becomes

$$\langle \varphi(x, 0)\varphi(0, 0)\rangle_T = \sum_Q B(Q)e^{-2ixk_Fd}\left(\frac{2\pi T}{v}\right)^{2(\Delta^+ + \Delta^-)}e^{-x/R_c}. \quad \text{(1.19)}$$

with correlation length R_c given by

$$R_c^{-1} = \frac{2\pi T}{v}(\Delta^+ + \Delta^-), \quad \text{(1.20)}$$

To conclude this section, it should be mentioned that deviations from conformality of the models under consideration can lead to corrections in the conformal predictions of the correlation functions. Concrete examples of correlation functions for the models under consideration will be given in the following sections. In this chapter, we shall not pay attention to the coefficients in (1.17) or (1.18).

XVIII.2 Asymptotics of correlation functions for the Bose gas

The Bose gas was considered in detail in Chapter I. Here we shall use the results from section I.9. The conformal dimensions are given by (1.16)

$$2\Delta^{\pm} = 2N^{\pm} + \left(\frac{\Delta N}{2\mathcal{Z}} \pm \mathcal{Z}d\right)^2 \qquad (2.1)$$

where $\mathcal{Z}$ is the value of the dressed charge, $Z(\lambda)$, on the Fermi surface. Remember, $Z(\lambda)$ is defined by equation (I.9.20),

$$Z(\lambda) - \frac{1}{2\pi}\int_{-q}^{q} K(\lambda, \mu)Z(\mu)d\mu = 1; \qquad (2.2)$$

thus,

$$\mathcal{Z} = Z(q). \qquad (2.3)$$

Below, it will prove more convenient to use

$$\theta = 2\mathcal{Z}^2 \qquad (2.4)$$

instead of the dressed charge. In Chapter I, it was shown that θ can be expressed in terms of the density D and the Fermi velocity (I.9.21),

$$\theta = \frac{4\pi D}{v}. \qquad (2.5)$$

It is probably useful to remind the reader that the density and the Fermi momentum are related by $k_F = \pi D$.

First, let us consider the asymptotic behavior of the field correlator $\langle\Psi(x,t)\Psi^{\dagger}(0,0)\rangle$. In this case $\Delta N = 1$. To obtain the leading term, we should set $N^{\pm} = d = 0$. Combining (1.2) and (2.1) gives

$$\langle\Psi(x,t)\Psi^{\dagger}(0,0)\rangle \longrightarrow A|x + ivt|^{-1/\theta}. \qquad (2.6)$$

To obtain higher corrections to the asymptotics, one should use (1.17):

$$\langle\Psi(x,t)\Psi^{\dagger}(0,0)\rangle = \sum_{d,N^{\pm}} \frac{A(d, N^{\pm})e^{-2idk_F x}}{(x - ivt)^{2\Delta^+}(x + ivt)^{2\Delta^-}}$$

where $\Delta^{\pm}$ are given by (2.1), $\Delta N = 1$, and d and $N^{\pm}$ are integers. The leading term of the asymptotics corresponds to the minimum value of $\Delta^{\pm}$; the other terms are corrections. This is the purely conformal contribution. To obtain an exact formula for the correlation functions, one should take into account the deviations from conformality. One of the consequences of this is that the higher order coefficients A start to depend on distance in a logarithmic way.

For the correlator of densities, $\langle j(x,t)j(0,0)\rangle$, we set $\Delta N = 0$. Again using the above formulæ, we obtain

$$\langle\langle j(x,t)j(0,0)\rangle\rangle = \langle j(x,t)j(0,0)\rangle - \langle j(0,0)\rangle^2$$

$$= \frac{A}{(x+ivt)^2} + \frac{A}{(x-ivt)^2} + A_3\frac{\cos 2k_F x}{|x+ivt|^\theta}. \qquad (2.7)$$

The first term corresponds to $d = 0, N^+ = 0, N^- = 1$ and the second term corresponds to $d = 0, N^+ = 1, N^- = 0$. The quantum numbers of the last term are $d = \pm 1, N^+ = 0, N^- = 0$. At finite coupling constant, $c < \infty$, the last term has $\theta > 2$ and thus decays faster than the other terms. The asymptotic expressions (2.6) and (2.7) coincide with results from perturbation theory ([52], [53]; a special version of the renormalization group was used in these papers), the Luttinger liquid approach [29], [30] and QISM calculations [10], [42].

It is interesting to note that the asymptotics of many-point correlation functions can be reduced to a product of pair correlators. It was found in [52] that

$$\left\langle \prod_{j=1}^{m} \Psi(x_j,t_j) \prod_{j=m+1}^{2m} \Psi^\dagger(x_j,t_j) \right\rangle$$

is of the conformal form

$$\left\langle \prod_{j=1}^{2m} \Psi^{l_j}(x_j,t_j) \right\rangle = \prod_{1\le i<j\le 2m} [G(x_j - x_i, t_j - t_i)]^{-l_i l_j} \qquad (2.8)$$

with

$$l_j = \pm 1; \quad \sum l_j = 0; \quad (x_j - x_i)^2 + v^2(t_j - t_i)^2 \to \infty$$

where

$$G(x,t) = \langle\Psi(x,t)\Psi^\dagger(0,0)\rangle \simeq A|x + ivt|^{-1/\theta}$$

and we define $\Psi^1 = \Psi$ and $\Psi^{-1} = \Psi^\dagger$. In (2.8), it should be noted that all $\Psi^\dagger$ stand to the right of all Ψ and G is raised to the power ± 1.

To obtain the finite-temperature correlation functions, one should apply formula (1.18) from section 1. Thus the asymptotics of the field correlator and the density correlator are

$$\langle\Psi(x,t)\Psi^\dagger(0,0)\rangle_T = B_1 \left| \frac{\pi T/v}{\sinh\frac{\pi T}{v}(x+ivt)} \right|^{1/\theta} + \cdots \qquad (2.9)$$

and

$$\langle\!\langle j(x,t)j(0,0)\rangle\!\rangle_T = \mathrm{Re}\, B_2 \left\{ \frac{\pi T/v}{\sinh\dfrac{\pi T}{v}(x+ivt)} \right\}^2$$

$$+ B_3 \cos(2xk_F) \left| \frac{\pi T/v}{\sinh\dfrac{\pi T}{v}(x+ivt)} \right|^\theta + \cdots . \quad (2.10)$$

The equal-time correlator of densities is given by

$$\langle\!\langle j(x,0)j(0,0)\rangle\!\rangle_T = B_2 \left(\frac{2\pi T}{v}\right)^2 \exp\left\{-\frac{2\pi T}{v}x\right\}$$

$$+ B_3 \left(\frac{2\pi T}{v}\right)^\theta \exp\left\{-\frac{\pi T\theta}{v}x\right\} \cos 2k_F x \quad (2.11)$$

and coincides with one obtained from an exact expression [8], [43] and one obtained by means of functional integration [53].

It was mentioned in section 1 that these formulæ are only applicable for small temperatures (in the vicinity of the phase transition). At larger temperatures, different methods can be used to analyze the asymptotic behavior of correlation functions. Another comment should be made about correlation functions which depend on Minkowski time (real time as opposed to Euclidean time). To obtain these asymptotics, one should make an analytic continuation with respect to time in formulæ (2.6)–(2.10). However, one should be careful since the asymptotics obtained in such a way will only be valid in the space-like region, $x^2 > v^2 t^2$, where t is Minkowski time. Let us present some of the results for time and temperature dependent correlation functions (for finite temperature which is not necessarily close to zero). In the paper [44], it was shown that large time asymptotics of density correlation functions are universal:

$$\langle j(x,t)j(0,0)\rangle - \langle j(0,0)\rangle^2 \longrightarrow \frac{A}{t}; \qquad t \to \infty \quad (2.12)$$

where t is real (not Euclidean) time. The time- and temperature-dependent field correlator $\langle \Psi(x,t)\Psi^\dagger(0,0)\rangle$ has completely different asymptotics. It decays exponentially as $t \to \infty$. Detailed formulæ are presented in [34]. As functions of the distance, correlation functions (including $\langle j(x,t)j(0,0)\rangle$ and $\langle \Psi(x,t)\Psi^\dagger(0,0)\rangle$) at finite temperature always decay exponentially; details are given in [33], [8] and [9].

XVIII.3 Asymptotics of correlation functions for the Heisenberg magnet

The asymptotics of correlation functions of the XXZ and XXX Heisenberg magnets are quite similar to those of the Bose gas. We shall use the results of sections II.5 and II.6 where finite size corrections for the Heisenberg magnet were evaluated. We shall also use

$$\theta = 2\mathcal{Z}^2$$

defined in (II.6.6) and (II.6.12); here $\mathcal{Z}$ is the value of $Z(\lambda)$ on the Fermi surface,

$$\mathcal{Z} = Z(\lambda)\Big|_{\lambda=\Lambda}.$$

The function $Z(\lambda)$ is defined by the integral equation

$$Z(\lambda) - \frac{1}{2\pi} \int_{-\Lambda}^{\Lambda} K(\lambda, \mu) Z(\mu) \, d\mu = 1$$

where

$$K(\lambda, \mu) = \frac{\sin 4\eta}{\sinh(\lambda - \mu + 2i\eta)\sinh(\lambda - \mu - 2i\eta)}$$

$$\Delta = \cos 2\eta.$$

The quality Λ (value of the spectral parameter on the Fermi surface) depends on the magnetization of the ground state. Explicit formulæ are given in Chapter II. θ can also be expressed in terms of the Fermi velocity and the magnetic susceptibiliy, χ, (II.6.7)

$$\theta = \frac{\pi}{2} v_F \chi. \tag{3.1}$$

We shall list here the main results of section II.6 for convenience. For the XXZ magnet, we have:

$$\theta = 2 + 4(\pi \tan \eta \tan 2\eta)^{-1} \sqrt{h_c - h}$$

$$h \to h_c; \quad h \le h_c. \tag{3.2}$$

For $h \to 0$, we have different formulæ:

$$\theta = \frac{\pi}{2\eta}(1 + \alpha_1 h^2); \qquad 0 < 2\eta < \frac{2\pi}{3}$$

$$\theta = \frac{\pi}{2\eta}(1 + \alpha_2 h^{(2\pi/\eta)-4}); \qquad \frac{2\pi}{3} < 2\eta < \pi. \tag{3.3}$$

Here $\alpha_{1,2}$ are independent of the magnetic field; they are presented in section II.6 between formulæ (II.6.16) and (II.6.17). For $h = 0$, we obtain

$$\theta = \frac{\pi}{2\eta}; \quad v = \frac{2\pi \sin 2\eta}{\pi - 2\eta}. \tag{3.4}$$

Thus, we see that the critical exponent in the XXZ model is a function of the coupling constant, η, and the external field, h.

For the XXX magnet with the external field near h_c and $h \leq h_c$, we have

$$\theta = 2 \left(1 - \frac{1}{\pi} \sqrt{h_c - h} \right); \quad h \to h_c \tag{3.5}$$

while for a weak magnetic field we have

$$\theta = 1 + [2 \ln(h_o/h)]^{-1}; \quad h \to 0. \tag{3.6}$$

For zero magnetic field we obtain

$$\theta = 1; \quad v = 2\pi. \tag{3.7}$$

The magnetization σ (it should be recalled that $\chi = \partial\sigma/\partial h$) is connected with the Fermi momentum (II.3.4) as follows:

$$k_F = \frac{\pi}{2}(1 - \sigma) = \pi D \tag{3.8}$$

and if $h = 0$ then

$$k_F = \frac{\pi}{2}. \tag{3.9}$$

Now let us consider the asymptotics of correlation functions. First, we will evaluate the correlator $\langle \sigma_- \sigma_\pm \rangle$. In this case, $\Delta N = 1$ and the leading term is given by (see (1.16) and (1.17))

$$\langle \sigma_-(x,t)\sigma_+(1,0) \rangle \longrightarrow A|x + ivt|^{-1/\theta} \tag{3.10}$$

(for $d = 0$ and $N^\pm = 0$) where x is the distance on the lattice and $\sigma_\pm(x,t)$ is the Pauli matrix at the site number x of the lattice at time t. This formula is true if the magnetic field varies between zero and the critical value, $0 \leq h \leq h_c$. h_c corresponds to the ferromagnetic state (see II.2.10).

The asymptotics of the correlator for the z component of spin are similar to the correlator of densities in the Bose gas (2.7):

$$\langle\langle \sigma_z(x,t)\sigma_z(1,0) \rangle\rangle = A_1 \frac{x^2 - (vt)^2}{[x^2 + (vt)^2]^2}$$

$$+ A_2 \frac{\cos 2k_F x}{|x + ivt|^\theta} + \cdots; \quad 0 \leq h \leq h_c \tag{3.11}$$

Now let us make some comments on the XXX Heisenberg magnet. The XXX model is a special case of the XXZ model. It was explained in section 4 of Chapter II how to deform XXZ formulæ in order to obtain the corresponding XXX formulæ. For example, in the definition of the fractional charge (1.15), one should use the kernel (II.4.4). The value of θ can still be defined as $\theta = 2\mathcal{Z}^2$ and formula (3.1) for θ is still valid. However, the dependence of θ on the magnetic field is now different. This is described by formulæ (3.5)–(3.7). For correlation functions of the XXX magnet, formulæ (3.10) and (3.11) are still valid. At zero magnetic field, the coefficient A_1 of the leading term begins to depend on the distance as follows [2]:

$$\langle \sigma_z(x)\sigma_z(1)\rangle \longrightarrow a\frac{(-1)^x}{x}\sqrt{\ln x}, \quad h = 0. \tag{3.12}$$

To obtain the leading term of the temperature dependent correlation function, one should use the substitution

$$x + ivt \longrightarrow \frac{v}{\pi T}\sinh\left[\frac{\pi T}{v}(x + ivt)\right] \tag{3.13}$$

in formulæ (3.10) and (3.11) according to the general formula (1.18). The asymptotics are similar to those for the Bose gas (2.9) and (2.10). The asymptotics of multipoint correlation functions are given by formulæ similar to (2.8). Let us emphasize once more that we are not concerned with the coefficients (i.e., A) here.

The first papers to evaluate the asymptotics of correlation functions in the Heisenberg model used perturbative techniques [45] and [23]. The Luttinger liquid approach was developed in [31] and the QISM method was developed in [39], [37]. Formula (3.12) was obtained in [2] where the results from the calculations of finite size corrections in [56] were used.

XVIII.4 Asymptotics of correlation functions for the one-dimensional Hubbard model

In this section we shall use the results of Chapter IV where the Hubbard model was introduced. The Bethe equations for the Hubbard model are a matrix generalization of those for the Bose gas. The dressed charge is now a 2×2 matrix (see (IV.2.4)). We will consider the repulsive interaction at less than half filling of the states. Now there are two interacting degrees of freedom: charge density waves and spin density waves. There are also two Fermi velocities (IV.2.2). It is instructive to compare the finite size correction to the ground state energy (IV.2.1)

$$E = N\epsilon_0 - \frac{\pi v_c}{6N} - \frac{\pi v_s}{6N} + o\left(\frac{1}{N}\right) \tag{4.1}$$

with the conformal formula (0.1). This shows that the long distance asymptotics of correlation functions will be driven by two Virasoro algebras each having a central charge equal to 1. Typical contributions to the asymptotics of the correlation functions will be of the form

$$\frac{\exp\{-2id_cP_{F\uparrow}x\}\exp\{-2i(d_c + d_s)P_{F\downarrow}x\}}{(x - iv_ct)^{2\Delta_c^+}(x + iv_ct)^{2\Delta_c^-}(x - iv_st)^{2\Delta_s^+}(x + iv_st)^{2\Delta_s^-}} \tag{4.2}$$

which is a generalization of (1.17). The definition of $P_{F\uparrow,\downarrow}$ was given in (IV.2.11). Here $\Delta_c^\pm$ are the conformal dimensions related to the Virasoro algebra describing charge degrees of freedom and $\Delta_s^\pm$ are the conformal dimensions related to the other Virasoro algebra describing the spin degrees of freedom. The formula for these conformal dimensions is the matrix generalization of (1.16):

$$2\Delta_c^\pm = \left(\mathcal{Z}_{cc} d_c + \mathcal{Z}_{sc} d_s \pm \frac{\mathcal{Z}_{ss} \Delta N_c - \mathcal{Z}_{cs} \Delta N_s}{2 \det \mathcal{Z}} \right)^2 + 2N_c^\pm \qquad (4.3)$$

$$2\Delta_s^\pm = \left(\mathcal{Z}_{cs} d_c + \mathcal{Z}_{ss} d_s \pm \frac{\mathcal{Z}_{cc} \Delta N_s - \mathcal{Z}_{sc} \Delta N_c}{2 \det \mathcal{Z}} \right)^2 + 2N_s^\pm \qquad (4.4)$$

where $\mathcal{Z}$ is the matrix version of the dressed charge given by formulæ (2.6), (2.5) of Chapter IV;

$$\mathcal{Z} = \begin{pmatrix} \mathcal{Z}_{cc} & \mathcal{Z}_{cs} \\ \mathcal{Z}_{sc} & \mathcal{Z}_{ss} \end{pmatrix} = \begin{pmatrix} Z_{cc}(k_0) & Z_{cs}(\lambda_0) \\ Z_{sc}(k_0) & Z_{ss}(\lambda_0) \end{pmatrix}. \qquad (4.5)$$

The quantum numbers d_c, d_s, ΔN_s and ΔN_c are defined in section 2 of Chapter IV. To obtain this formula, one should study the finite size corrections from (IV.2.7) and (IV.2.10) and compare them with the conformal expressions (1.8) and (1.9).

Formulæ (4.2)–(4.4) provide an opportunity to evaluate the asymptotics of correlation functions in this model. Examples considered in the paper [24] are

$$\langle \psi_\downarrow^\dagger(x,t)\psi_\downarrow(0,0) \rangle$$

$$\langle \psi_\uparrow^\dagger(x,t)\psi_\uparrow(x,t)\psi_\uparrow^\dagger(0,0)\psi_\uparrow(0,0) \rangle$$

$$\langle \psi_\uparrow^\dagger(x,t)\psi_\downarrow(x,t)\psi_\downarrow(0,0)\psi_\uparrow^\dagger(0,0) \rangle \qquad (4.6)$$

$$\langle \psi_\uparrow^\dagger(x,t)\psi_\downarrow^\dagger(x,t)\psi_\downarrow(0,0)\psi_\uparrow(0,0) \rangle$$

$$\langle \psi_\uparrow^\dagger(x,t)\psi_\uparrow(x+1,t)\psi_\uparrow^\dagger(1,0)\psi_\uparrow(0,0) \rangle.$$

To illustrate the approach, consider the field correlator

$$G(x,t) = \langle \psi_\uparrow(x,t)\psi_\uparrow^\dagger(0,0) \rangle. \qquad (4.7)$$

In this case $\Delta N_c = 1$, $\Delta N_s = 0$ and d runs through half-integer values

$$d_c = \pm\frac{1}{2}, \pm\frac{3}{2}, \ldots \qquad d_s = \pm\frac{1}{2}, \pm\frac{3}{2}, \ldots \qquad (4.8)$$

Thus, we get the asymptotics

$$G(x,t) = \langle \psi_\uparrow(x,t)\psi_\uparrow^\dagger(0,0) \rangle$$

$$= \sum_{d_c,d_s,N_{s,c}^\pm} \frac{A(d,N^\pm)\exp\{-2id_c P_{F\uparrow}x\}\exp\{-2i(d_c+d_s)P_{F\downarrow}x\}}{(x-iv_ct)^{2\Delta_c^+}(x+iv_ct)^{2\Delta_c^-}(x-iv_st)^{2\Delta_s^+}(x+iv_st)^{2\Delta_s^-}}$$

$$(4.9)$$

where $\Delta_{c,s}^{\pm}$ are given by (4.3) and (4.4). The dressed charge was derived in Chapter IV. The leading term of the above series corresponds to the following values of d:

$$d_c = -\frac{1}{2}; \qquad d_s = \frac{1}{2} \qquad (4.10)$$

and the conjugated part.

The critical exponents $\Delta_{s,c}^{\pm}$ depend on all the parameters of the model: the density of the ground state, the magnetization of the ground state and the coupling constant. The dressed charge defines this dependence; it was studied explicitly in [24], [25].

Now we shall present the asymptotics in the the simplest case: infinite coupling constant, $u = \infty$, zero magnetic field and less than half filling of the states. The leading critical exponents are given by

$$2\Delta_c^+ = \frac{1}{16}; \qquad 2\Delta_c^- = \frac{9}{16}; \qquad 2\Delta_s^+ = 0; \qquad 2\Delta_s^- = \frac{1}{2} \qquad (4.11)$$

which means that the leading term for the field correlator (4.7) is

$$\langle \psi_\uparrow(x,t)\psi_\uparrow^\dagger(0,0)\rangle \to A \, \mathrm{Re} \, \frac{\exp\{\frac{i\pi}{2}n_c x\}}{(x - iv_c t)^{\frac{1}{16}}(x + iv_c t)^{\frac{9}{16}}(x + iv_s t)^{\frac{1}{2}}} \qquad (4.12)$$

$$\text{with} \qquad u = \infty \qquad h = 0.$$

Here n_c is the concentration of electrons $n_c = N_c/N$ (see (IV.1.8)). Logarithmic corrections to the power law decay (which appear at zero magnetic field) are discussed in [27]. The temperature-dependent correlation functions are given by formulæ similar to (1.18), details of which can be found in [24]. The correlation functions for the half–filled repulsive Hubbard model are very different because a gap arises in the spectrum of charge excitations [55]. Correlation functions for the attractive Hubbard model were evaluated in [13]. They are also quite different from the case we considered above since there is a gap in the spin excitations for the attractive Hubbard model. More details about correlation functions in the Hubbard model can be found in [3], [24], [25], [49], [50], [54]. The most important formulæ are (4.3) and (4.4) for the conformal dimensions. They were obtained in [24]. They are quite remarkable because they describe the conformal dimensions in any model solvable by the nested Bethe Ansatz. In the paper [41] formulæ (4.3) and (4.4) were applied to the description of the correlation functions in the t-J model.

Conclusion

In this chapter we have seen that the asymptotic behavior of correlation functions for integrable models is defined by the R-matrix (and Fermi momentum). This was noted in [10], [43], [39], [42], [37]. The calculation of the conformal dimensions presented in this chapter was based on the papers [11], [12]. In relation to this calculation, the following papers should be mentioned: [56], [55], [5], [6], [51], [13], [16], [17] and [18]. Finite size corrections in Bethe Ansatz solvable models with more than one gapless excitation were obtained in [38], [57]. The generalization of the relationship between critical exponents and finite size corrections for models with more than one gapless excitation was made in [24].

Conclusion to Part IV

We have solved the problem of the calculation of correlation functions for completely integrable models; we have considered different approaches to the evaluation of quantum correlation functions. The most interesting result is that quantum correlation functions satisfy classical completely integrable differential equations. This makes it possible to evaluate the asymptotics of correlation functions (even for time-dependent temperature correlation functions). This result is very general and not restricted to the nonlinear Schrödinger equation. One can compare our approach with another approach to quantum correlation functions based on the quantum Gel'fand-Levitan equation. There are similarities. The Fredholm integral operator which appears in the determinant representation for quantum correlation functions (see Part III) we consider as a Gel'fand-Levitan operator for the differential equation which drives the quantum correlation function. For the penetrable Bose gas ($0 < c < \infty$, finite coupling constant) quantum dual fields enter our Gel'fand-Levitan operator. Quantum dual fields are canonical Bose fields. So in this case we should compare our Fredholm integral operator with the quantum Gel'fand-Levitan operator. They are similar, with one essential difference. In our case quantum fields are arranged in commuting combinations (all quantum operators belong to a commuting subalgebra). This permits us to avoid perturbation theory when considering our Fredholm determinant, which leads to differential equations and explicit formulæ for the asymptotics of correlation functions. We believe that similar results for quantum correlation functions can be obtained for any (1+1)-dimensional quantum completely integrable system (exactly solvable). Our approach has already been succesfully applied to the sine-Gordon model and to nonrelativistic fermions (references ([2] and ([1] in the Introduction to Part IV) and to the Heisenberg antiferromagnet with special anisotropy. We would like to mention that differential equations for correlation functions were first obtained in [1] .

Final Conclusion

We have explained the theory of exactly solvable models on both the quantum and classical levels. We hope that the reader appreciates the beauty and perfection of the Bethe Ansatz. The theory of completely integrable nonlinear partial differential equations plays an equally important role in the book. It is deeply related to the Bethe Ansatz. We hope that our book can be used for further development of the theory of exactly solvable models (we think that the interested reader can find problems to work out in our book). Our book does not close the theory of the Bethe Ansatz; it opens new possibilities for further development.

We have explained how to construct eigenfunctions of the Hamiltonian and how to describe the ground state and its excitations. We have also evaluated dispersion curves and scattering matrices for the excitations. The thermodynamics of exactly solvable models is explained in this book in all the details, including the evaluation of temperature correlation functions (even if they depend on real time). The problem of the evaluation of correlation functions (for Bethe Ansatz solvable models) is solved in our book, using the example of the one-dimensional Bose gas (quantum nonlinear Schrödinger equation). It was done using the ideas of the inverse scattering method (the Lax representation, the Riemann-Hilbert problem, the Gel'fand-Levitan-Marchenko equation, etc.). We would like also to emphasize that all other Bethe Ansatz solvable models are constructed in a similar way.

It is difficult to overestimate the role of exactly solvable models in modern physics and mathematics. Some solvable models have direct application to physical problems, especially to condensed matter, solid state and statistical physics (such as the Kondo problem). Exactly solvable models are also important in more abstract physics. They are related to string theory, matrix models, two-dimensional gravitation, conformal field theory and the theory of factorizable S-matrices. To recall the role of the theory of exactly solvable models in mathematics it is enough to mention the theory of braids and knots or quantum groups. This is the reason why we hope that the book will be interesting and useful for the reader. The Bethe Ansatz was created in 1931; it has been successfully developed for sixty years. We are sure that more interesting discoveries await us in the years ahead.

References

Preface

[1] M.J. Ablowitz and H. Segur, *Solitons and the Inverse Scattering Transform*, Philadelphia: SIAM, (1981).

[2] A. Andrei, K. Furuya and J.H. Lowenstein, Solution of the Kondo problem, *Rev. Mod. Phys.* **55**, 331–402, (1983).

[3] M.F. Atiyah, N.J. Hitchin, V.G. Drinfel'd and Yu.I. Manin, Construction of instantons, *Phys. Lett.* **65A**, 185–7, (1978).

[4] E. Barouch, B.M. McCoy and T.T. Wu, Zero-field susceptibility of the two-dimensional Ising model near T_c, *Phys. Rev. Lett.* **31**, 1409–11, (1973).

[5] R.J. Baxter, *Exactly Solved Models in Statistical Mechanics*, New York: Academic Press, (1982).

[6] A.A. Belavin, A.M. Polyakov, and A.B. Zamolodchikov, Infinite conformal symmetry in two-dimensional quantum field theory, *Nucl. Phys.* **B241**, 333–80, (1984).

[7] H. Bethe, Zur Theorie der Metalle I. Eigenwerte und Eigenfunktionen der Linearen Atomkette, *Zeitschrift für Physik* **71**, 205–26, (1931).

[8] E. Brezin and V. Kazakov, Exactly solvable field theories of closed strings, *Phys. Lett.* **236B**, 144–50, (1990).

[9] R.K. Bullough and S. Olafsson, Complete integrability of the integrable models: Quick review, in *Swansea 1988, Proceedings of Mathematical Physics*, pp. 329–34, Bristol: Adam Hilger, (1988)

[10] R.K. Bullough and P.J. Cadrey (eds.), *Solitons*, New York: Springer-Verlag, (1980).

[11] F. Calogero and A. Degasperis, *Spectral Transforms and Solitons*, vol. 1, Amsterdam: North Holland, (1982).

[12] R.F. Dashen, B. Hasslacher and A. Neveu, Nonperturbative methods and extended-hadron models in field theory. I. Semiclassical functional methods, *Phys. Rev.* **D10**, 4114–29, (1974).

[13] R.F. Dashen, B. Hasslacher and A. Neveu, Nonperturbative methods and extended-hadron models in field theory. II. Two-dimensional models and extended-hadrons, *Phys. Rev.* **D10**, 4130–38, (1974).

[14] R.F. Dashen, B. Hasslacher and A. Neveu, Particle spectrum in model field theories from semiclassical integral techniques, *Phys. Rev.* **D11**, 3424–50, (1975).

[15] A. Davey and K. Stewartson, On three-dimensional packets of surface waves, *Proc. Roy. Soc. London* **A338**, 101–10, (1974).

[16] M. Douglas and S. Shenker, Strings in less than one dimension, *Nucl. Phys.* **B335**, 635–54, (1990).

[17] V.G. Drinfel'd, Quantum groups, in *Proceedings of the International Congress of Mathematicians*, pp. 798–820, Berkeley: AMS, (1987).

[18] L.D. Faddeev and L.A. Takhtajan, *Hamiltonian Methods in the Theory of Solitons*, Berlin: Springer-Verlag, (1987).

[19] L.D. Faddeev, Quantum inverse scattering method, *Sov. Sci. Rev. Math.* **C1**, 107–60, (1980).

[20] L.D. Faddeev and V.E. Korepin, Quantum theory of solitons, *Phys. Rep.* **42**, 1–87, (1978).

[21] H. Flaschka, The Toda Lattice, I. Existence of integrals, *Phys. Rev.* **B9**, 1924–5, (1974).

[22] H. Flaschka, The Toda Lattice, II. Inverse scattering solution, *Prog. Theor. Phys.* **51**, 703–6, (1974).

[23] H. Flaschka and D.W. McLaughlin, Canonically conjugate variables for KdV and Toda lattice under periodic boundary conditions, *Prog. Theor. Phys.* **55**, 438–56, (1976).

[24] C.S. Gardner, J.M. Greene, M.D. Kruskal and R.M. Miura, Method for solving the Korteweg-deVries equation, *Phys. Rev. Lett.* **19**, 1095–7, (1967).

[25] M. Gaudin, *La Fonction d'Onde de Bethe*, Paris: Masson, (1983).

[26] J. Goldstone and R. Jackiw, Quantization of nonlinear waves, *Phys. Rev.* **D11**, 1486–98, (1975).

[27] D.J. Gross and A. Migdal, Non perturbative two-dimensional quantum gravity, *Phys. Rev. Lett.* **64**, 127–30, (1990).

[28] A.R. Its, A.G. Izergin, V.E. Korepin and N.A. Slavnov, Differential equations for quantum correlation functions, *Int. J. Mod. Phys.* **B4**, 1003–37, (1990).

[29] A.R. Its and V.Yu. Novokshenov, *The Isomonodromic Deformation Method in the Theory of Painlevé Equations,* Lecture Notes in Mathematics, **1191**, Berlin: Springer-Verlag, (1986).

[30] C. Itzykson, H. Saleur and J.B. Zuber, *Conformal Invariance and Applications to Statistical Mechanics*, Singapore: World Scientific, (1986).

[31] A.G. Izergin and V.E. Korepin, A lattice model related to the nonlinear Schrödinger equation, *Sov. Phys. Dokl.* **26**, 653–4, (1981).

[32] Yu.A. Izyumov and Yu.N. Skryabin, *Statistical Mechanics of Magnetically Ordered Systems*, English translation; New York: Plenum, (1990).

[33] M. Jimbo, T. Miwa, Y. Mori and M. Sato, Density matrix of an impenetrable Bose gas and the fifth Painlevé transcendent, Physica **1D**, 80–158, (1980).

[34] M. Jimbo, *Yang Baxter Equation in Integrable Systems*, Advanced Series in Mathematical Physics **10**, Singapore: World Scientific, (1990).

[35] B.B. Kadomtsev and V.I. Petviashvilli, On the stability of solitary waves in weakly dispersing media, *Sov. Phys. Dokl.* **15**, 539–41, (1970).

[36] P.P. Kulish and E.K. Sklyanin, *Integrable Quantum Field Theories*, Lecture Notes in Physics **151**, pp. 61–119, Berlin: Springer-Verlag, (1982).

[37] G.L. Lamb, *Elements of Soliton Theory*, New York: Wiley, (1986).

[38] P.D. Lax, Integrals of nonlinear equations of evolution and solitary waves, *Comm. Pure Appl. Math.* **21**, 467–90, (1968).

[39] E.H. Lieb and F.Y. Wu, Absence of Mott transition in an exact solution of the short-range one-band model in one-dimension, *Phys. Rev. Lett.* **20**, 1445–9, (1968).

[40] E.H. Lieb and W. Liniger, Exact analysis of an interacting Bose gas I, *Phys. Rev.* **130**, 1605–16, (1963).

[41] E.H. Lieb, Exact analysis of an interacting Bose gas II. The excitation spectrum, *Phys. Rev.* **130**, 1616–24, (1963).

[42] E.H. Lieb and D.C. Mattis (eds.), *Mathematical Physics in One-Dimension*, New York: Academic Press, (1966).

[43] B.M. McCoy and T.T. Wu, *The Two-Dimensional Ising Model*, Cambridge, MA: Harvard University Press, (1973).

[44] J.B. McGuire, Study of exactly solvable one-dimensional N-body problems, *J. Math. Phys.* **5**, 622–36, (1964).

[45] A.C. Newell, *Solitons in Mathematics and Physics*, Philadelphia: SIAM, (1985).

[46] A.M. Polyakov and P.B. Wiegmann, Theory of nonabelian Goldstone bosons in two dimensions, *Phys. Lett.* **131B**, 121–6, (1983).

[47] A.M. Polyakov and P.B. Wiegmann, Goldstone fields in two dimensions with multivalued action, *Phys. Lett.* **141B**, 223–8, (1984).

[48] V.I. Rupasov and V.I. Yudson, Rigorous theory of cooperative spontaneous emission of radiation from a lumped system of two-level atoms; Bethe ansatz method, *Sov. Phys. JETP* **60**, 927–34, (1984).

[49] B.S. Shastry (ed.), *Exactly Solvable Problems in Condensed Matter and Relativistic Field Theory*, Lecture Notes in Physics, **242**, Berlin: Springer-Verlag, (1985).

[50] E.K. Sklyanin, L.A. Takhtajan and L.D. Faddeev, Quantum inverse problem method I, *Theor. Math. Phys.* **40**, 688–706, (1980).

[51] E.K. Sklyanin and L.D. Faddeev, Quantum mechanical approach to completely integrable field theory models, *Sov. Phys. Dokl.* **23**, 902–4, (1978).

[52] E.K. Sklyanin, Quantum version of the method of inverse scattering problem, *J. Sov. Math.* **19**, 1546–96, (1982).

[53] F.A. Smirnov, *Form Factors in Completely Integrable Models in Quantum Field Theory*, Singapore: World Scientific, (1992).

[54] H.B. Thacker, Exact integrability in quantum field theory and statistical systems, *Rev. Mod. Phys.* **53**, 253–85, (1981).

[55] C.A. Tracy and B.M. McCoy, Neutron scattering and correlation functions of the Ising model near T_c, *Phys. Rev. Lett.* **31**, 1500–4, (1973).

[56] A.M. Tsvelick and P.B. Wiegmann, Exact results in the theory of magnetic alloys, *Adv. Phys.* **32**, 453–713, (1983).

[57] C.N. Yang, Some exact results for the many-body problem in one dimension with repulsive delta-function interaction, *Phys. Rev. Lett.* **19**, 1312–14, (1967).

[58] C.N. Yang and M.L. Ge, *Braid Group, Knot Theory, and Statistical Mechanics*, Singapore: World Scientific, (1989).

[59] V.E. Zakharov and A.B. Shabat, Exact theory of two-dimensional selffocussing and one-dimensional self-modulation of waves in nonlinear media, *Sov. Phys. JETP* **34**, 62–9, (1972).

[60] V.E. Zakharov and L.D. Faddeev, Korteweg-deVries equation is completely integrable Hamiltonian system, *Funct. Anal. Appl.* **5**, 280–7, (1971).

[61] V.E. Zakharov, S.V. Manakov, S.P. Novikov and L.P. Pitaievski, *Theory of Solitons. The Inverse Scattering Method*, Russian; Moscow: Nauka (1980), English translation; New York: Plenum, (1984).

[62] A.B. Zamolodchikov and Al.B. Zamolodchikov, Factorized S-matrices in two dimensions as the exact solutions of certain relativistic quantum fields theory models, *Ann. Phys.* **120**, 253–91, (1979).

Introduction to Part I

[1] H. Bethe, Zur Theorie der Metalle I. Eigenwerte und Eigenfunktionen der Linearen Atomkette, *Zeitschrift für Physik* **71**, 205–26, (1931).

[2] M. Gaudin, *La Fonction d'Onde de Bethe*, Paris: Masson, (1983).

[3] E.H. Lieb and D.C. Mattis (eds.), *Mathematical Physics in One Dimension*, New York–London: Academic Press, (1966).

[4] C.N. Yang, *Selected Papers 1945–1980 with Commentary*, New York: W.H. Freeman and Co., (1983).

Chapter I

[1] F.A. Berezin, C.P. Pokhil and V.M. Finkelberg, The Schrödinger equation for a system of one-dimensional particles with point interactions, *Vestn. Mosk. Gos. Univ.* **1**, 21–8, (1964) (in Russian).

[2] A. Berkovich and G. Murthy, Time dependent multipoint correlation functions of the nonlinear Schrödinger model, *Phys. Lett.* **142A**, 121–7, (1989).

[3] N.M. Bogoliubov and V.E. Korepin, Correlation length of one-dimensional Bose gas, *Nucl. Phys.* **B257**, 766–78, (1985).

[4] N.M. Bogoliubov, A.G. Izergin, and N.Yu. Reshetikhin, Finite-size corrections and infrared asymptotics of the correlation functions in two dimensions, *J. Phys.* **A20**, 5361–9, (1987).

[5] E. Brezin and J. Zinn-Justin, Un problemé d'N corps soluble, *Compt. Rend. Acad. Sci.* **265B**, 670–3, (1966).

[6] D.B. Creamer, H.B. Thacker, and D. Wilkinson, Gelfand-Levitan, method for operator fields, *Phys. Rev.* **D21**, 1523–6, (1980).

[7] M. Flicker and E.H. Lieb, Delta-function fermi gas with two-spin deviates, *Phys. Rev.* **161**, 179–88, (1967).

[8] M.K. Fung, Validity of the Bethe–Young hypothesis in the delta-function interaction problem, *J. Math. Phys.* **22**, 2017–19, (1981).

[9] M. Gaudin, *La Fonction d'Onde de Bethe*, Paris: Masson, (1983).

[10] M. Gaudin, Un systeme à une dimension de fermions en interaction, *Phys. Lett.* **24A**, 55–6, (1967).

[11] M. Girardeau, Relationship between systems of impenetrable bosons and fermions in one-dimension, *J. Math. Phys.* **1**, 516–23, (1960).

[12] A.R. Its, A.G. Izergin, V.E. Korepin and N.A. Slavnov, Differential equations for quantum correlation functions, *Int. J. Mod. Phys.* **B4**, 1003–37, (1990).

[13] V.E. Korepin, Direct calculation of the scattering matrix in the massive Thirring model, *Theor. Math. Phys.* **41**, 953–67, (1979).

[14] E.H. Lieb and W. Liniger, Exact analysis of an interacting Bose gas I. The general solution and the ground state, *Phys. Rev.* **130**, 1605–16, (1963).

[15] E.H. Lieb, Exact analysis of an interacting Bose gas. II. The excitation spectrum, *Phys. Rev.* **130**, 1616–34, (1963).

[16] E.H. Lieb and D.C. Mattis (eds.), *Mathematical Physics in one dimension*, New York–London: Academic Press, (1966).

[17] J.B. McGuire, Study of exactly soluble one-dimensional N-body problems, *J. Math. Phys.* **5**, 622–9, (1964).

[18] J.B. McGuire, Interacting fermions in one-dimension: I Repulsive potential, *J. Math. Phys.* **6**, 432–9, (1965).

[19] J.B. McGuire, Interacting fermions in one-dimension: II Attractive potential, *J. Math. Phys.* **7**, 123–32, (1966).

[20] C.N. Yang, *Selected Papers 1945–1980 with Commentary*, New York: W.H. Freeman and Co., (1983).

[21] C.N. Yang, Some exact results for the many-body problem in one dimension with repulsive delta-function interaction, *Phys. Rev. Lett.* **19**, 1312–14, (1967).

[22] C.N. Yang and C.P. Yang, Thermodynamics of a one-dimensional system of bosons with repulsive delta-function interactions, *J. Math. Phys.* **10**, 1115–22, (1969).

Chapter II

[1] D. Babbitt and L. Thomas, Ground state representation of the infinite one-dimensional Heisenberg Ferromagnet, *Comm. Math. Phys.* **54**, 255–78, (1977).

[2] R.J. Baxter, *Exactly Solved Models in Statistical Mechanics*, New York: Academic Press, (1982).

[3] H. Bethe, Zur Theorie der Metalle I. Eigenwerte und Eigenfunktionen Atomkete, *Zeitschrift für Physik* **71** (3–4), 205–26, (1931).

[4] N.M. Bogoliubov, A.G. Izergin and V.E. Korepin, Critical exponents for integrable models, *Nucl. Phys.* **B275**, 687–705, (1986).

[5] N.M. Bogoliubov, A.G. Izergin and N.Yu. Reshetikhin, Finite size corrections and infrared asymptotics of the correlation functions in two dimensions, *J. Phys.* **A20**, 5361–9, (1987).

[6] J. Bonner and M. Fisher, Linear magnetic chains with anisotropic coupling, *Phys. Rev.* **A135**, 640–58, (1964).

[7] R.K. Bullough, Y.Z. Chen, J. Timonen, V. Tognetti and R. Vain, Chemical thermodynamics of the Heisenberg chain in a field by generalized Bethe Ansatz method, *Phys. Lett.* **145A**, 154, (1990).

[8] H.J. de Vega, Finite-size corrections for nested Bethe Ansatz models and conformal invariance, *J. Phys.* **A20**, 6023–36, (1987).

[9] L.D. Faddeev and L.A. Takhtajan, Spectrum and scattering of excitations in the one-dimensional isotropic Heisenberg model, *J. Sov. Math.* **24**, 241, (1984).

[10] L.D. Faddeev and L.H. Takhtajan, What is the spin of a spin wave?, *Phys. Lett.* **85A**, 375–7, (1981).

[11] M. Fowler and X. Zotos, *Phys. Rev.* **B24**, 2634, (1981).

[12] M. Gaudin, *La Fonction d'Onde de Bethe*, Paris: Masson, (1983)

[13] W. Heisenberg, Zur Theorie der Ferromagnetismus, *Zeitschrift für Physik* **49** (9–10), 619–36, (1928).

[14] L. Hulthén, Über has Austauschproblem eines Kristalls, *Arkiv. Mat. Astron. Fys.* **26A**(11), 1–106, (1939).

[15] A.N. Kirillov, Combinatorial identities and completeness of eigenstates of the Heisenberg magnet, *J. Sov. Math.* **30**, 2298, (1985).

[16] A.N. Kirillov, Completeness of states of the generalized Heisenberg magnet, *J. Sov. Math.* **36**, 115, (1987).

[17] V.E. Korepin, Calculation of norms of Bethe wave functions, *Comm. Math. Phys.* **86**, 391–418, (1982).

[18] S.V. Pokrovskii and A.M. Tsvelick, Conformal dimension spectrum for lattice integrable models of magnets, *Sov. Phys. JETP* **66**, 1275–82, (1987).

[19] M. Takahashi, Low-temperature specific heat of spin 1/2 anisotropic Heisenberg ring, *Prog. Theor. Phys.* **50**, 1519–36, (1973).

[20] M. Takahashi and M. Suzuki, One-dimensional anisotropic Heisenberg model at finite temperatures, *Prog. Theor. Phys.* **48**, 2187–209, (1972).

[21] M. Takahashi, *Prog. Theor. Phys.* **46**, 401, (1971).

[22] L.A. Takhtajan and L.D. Faddeev, The quantum inverse problem method and the Heisenberg XYZ model, *Russ. Math. Survey* **34**(5), 11–68, (1979).

[23] A.A. Vladimirov, Proof of the invariance of the Bethe-Ansatz solutions under complex conjugation, *Theor. Math. Phys.* **66**, 102–5, (1986).

[24] F. Woynarovich, Finite-size effects in a non-half-filled Hubbard chain, *J. Phys.* **A22**, 4243, (1989).

[25] C.N. Yang and C.P. Yang, One-dimensional chain of anisotropic spin-spin interactions. I. Proof of Bethe's hypothesis for ground state in finite system, *Phys. Rev.* **150**, 321–7, (1966).

[26] C.N. Yang and C.P. Yang, One-dimensional chain of anisotropic spin-spin interactions. II. Properties of the ground state energy per lattice site for an infinite system, *Phys. Rev.* **150**, 327–39, (1966).

[27] C.N. Yang and C.P. Yang, One-dimensional chain of anisotropic spin-spin interactions. III. Applications, *Phys. Rev.* **151**, 258–64, (1966).

[28] C.N. Yang, *Selected Papers 1945–1980 with Commentary*, New York: W.H. Freeman and Co., (1983).

Chapter III

[1] I.Yu. Areef'eva and V.E. Korepin, Scattering in two-dimensional model with Lagrangian $L = \frac{1}{\gamma}[\frac{1}{2}(\partial_\mu u)^2 + m^2(\cos u - 1)]$, *JETP Lett.* **20**, 312–14, (1974).

[2] F.A. Beresin and V.N. Sushko, Relativistic two-dimensional model of a self-interacting fermion field with non-vanishing rest mass, *Sov. Phys. JETP* **21**, 865–73, (1965).

[3] H. Bergknoff and H.B. Thacker, Structure and solution of the massive Thirring model, *Phys. Rev.* **D19**, 3666–81, (1979).

[4] N.M. Bogoliubov, A.G. Izergin and V.E. Korepin, Structure of the vacuum in the quantum sine-Gordon model, *Phys. Lett.* **159B**, 345–7, (1985).

[5] R.K. Bullough, D.J. Pilling, and J. Timonen, Statistical mechanics of the sine-Gordon field, in *Nonlinear Phenomena in Physics* (ed. F. Claro), pp. 70–128, Heidelberg: Springer-Verlag, (1984).

[6] R.K. Bullough, D.J. Pilling and J.T. Timonen, Quantum and classical statistical mechanics of the nonlinear Schrödinger, sinh-Gordon, and sine-Gordon equations, in *Dynamical Problems in Soliton Systems* (ed. S. Takeno), p. 105, Heidelberg: Springer-Verlag, (1985).

[7] N.N. Chen, M.D. Johnson and M. Fowler, Classical limit of Bethe Ansatz thermodynamics for the sine-Gordon system, *Phys. Rev. Lett.* **56**, 904–7, (1986).

[8] S. Coleman, Quantum sine-Gordon equation as the massive Thirring model, *Phys. Rev.* **D11**, 2088–97, (1975).

[9] R.F. Dashen, B. Hasslacher, and A. Neveu, Particle spectrum in model field theories from semiclassical functional integral techniques, *Phys. Rev.* **D11**, 3424–50, (1975).

[10] L.D. Faddeev and V.E. Korepin, Quantum theory of solitons, *Phys. Rep.* **42C**, 1–87, (1978).

[11] R. Haag, Quantum field theories with composite particles and asymptotic conditions, *Phys. Rev.* **112**, 669–73, (1958).

[12] H. Itoyama and P. Moxhay, Neutral excitations and the massless limit of the sine–Gordon or massive Thirring theory, *Phys. Rev. Lett.* **65**, 2102–5, (1990).

[13] M.D. Johnson, N.N. Chen and M. Fowler, Classical limit of sine-Gordon thermodynamics using the Bethe Ansatz, *Phys. Rev.* **B34**, 7851–65, (1986).

[14] M.D. Johnson and M. Fowler, Finite temperature excitations of the quantum sine-Gordon massive Thirring model: variation of the soliton mass with coupling constant and temperature, *Phys. Rev.* **B31**, 536–45, (1985).

[15] M. Karowskii, H.-J. Thun, T.T. Truong, and P.H. Weisz, On the uniqueness of a purely elastic S-matrix in $(1+1)$ dimensions, *Phys. Lett.* **67B**, 321–2, (1977).

[16] V.E. Korepin, Above-barrier soliton reflection, *JETP Lett.* **23**, 201–4, (1976).

[17] V.E. Korepin, Direct calculation of the scattering matrix in the massive Thirring model, *Theor. Math. Phys.* **41**, 953–67, (1979).

[18] V.E. Korepin, The mass spectrum and the S-matrix of the massive Thirring model in the repulsive case, *Comm. Math. Phys.* **76**, 165–76, (1980).

[19] V.E. Korepin and L.D. Faddeev, Quantization of solitons, *Theor. Math. Phys.* **25**, 1039–49, (1975).

[20] P.P. Kulish and N.Yu. Reshetikhin, Generalized Heisenberg ferromagnet and the Gross-Neveu model, *Sov. Phys. JETP* **53**, 108–14, (1981).

[21] M. Lüsher, *Nucl. Phys.* **B117**, 475 (1976).

[22] A. Luther, Eigenvalue spectrum of interacting massive fermions in one dimension, *Phys. Rev.* **B14**, 2153–9, (1976).

[23] S. Mandelstam, Soliton operators for the quantized sine-Gordon equation, *Phys. Rev.* **D11**, 3026–30, (1975).

[24] A.K. Pogrebkov and V.N. Sushko, Quantum solitons and their connection with fermion fields for the $(\sin\varphi)_2$ interaction, *Theor. Math. Phys.* **26**, 286–9, (1976).

[25] W.E. Thirring, A solvable relativistic field theory, *Ann. Phys.* **3**, 91–112, (1958).

[26] A.B. Zamolodchikov, Exact two-particle S-matrix of quantum solitons of the sine-Gordon model, *JETP Lett.* **25**, 468–71, (1977).

[27] W. Zimmerman, On the bound state problem in quantum field theory, *Nuovo Cimento* **10**, 567–614, (1958).

Chapter IV

[1] N.M. Bogoliubov, A.G. Izergin and N.Yu. Reshetikhin, Finite size effects and infrared asymptotics of the correlation functions in two dimensions, *J. Phys.* **A20**, 5361–9, (1987).

[2]   N.M. Bogoliubov and V.E. Korepin, Formation of Cooper pairs in the Hubbard model, *Mod. Phys. Lett.* **B1**, 349–52, (1988).

[3]   N.M. Bogoliubov and V.E. Korepin, The role of quasi one-dimensional structures in high-T_c superconductivity, *Int. J. Mod. Phys.* **B3**, 427–39, (1989).

[4]   F.H.L. Essler, V.E. Korepin and K. Schoutens, New eigenstates of the 1-dimensional Hubbard model, *Nucl. Phys.* **B372**, 559, (1992); *PRL* **67**, 3848 (1991); *Nucl. Phys.* **B384**, 431 (1992).

[5]   A.G. Izergin, V.E. Korepin and N.Yu Reshetikhin, Conformal dimensions in Bethe Ansatz solvable models, *J. Phys.* **A22**, 2615–21, (1989).

[6]   E.H. Lieb and F.Y. Wu, Absence of Mott transition in an exact solution of the short range one band model in one dimension, *Phys. Rev. Lett.* **20**, 1445–8, (1968).

[7]   B.S. Shastry, Exact integrability of the one-dimensional Hubbard model, *Phys. Rev. Lett.* **56**, 2453–5, (1986).

[8]   B. Sutherland, in *Exactly Solvable Problems in Condensed Matter and Relativistic Field Theory*, Lecture Notes in Physics **242** (eds. B.S. Shastry, S.S. Jha and V. Singh), Berlin: Springer-Verlag, (1985).

[9]   M. Takahashi, One-dimensional Hubbard model at finite temperature, *Prog. Theor. Phys.* **47**, 69–82, (1972).

[10] F. Woynarovich, Excitations with complex wave numbers in a Hubbard chain. I: States with one pair of complex wave numbers; II: States with several pairs of complex wave numbers, *J. Phys.* **C15**, 85–109, (1982).

[11] F. Woynarovich and H.P. Eckle, Finite size corrections for the low lying states of a half-filled Hubbard chain, *J. Phys.* **A20**, L449, (1987).

[12] F. Woynarovich, Finite-size effects in a non-half-filled Hubbard chain, *J. Phys.* **A22**, 4243–56, (1989).

[13] F. Woynarovich, Spin excitations in a Hubbard chain, *J. Phys.* **C16**, 5293–304, (1983).

[14] C.N. Yang, Some exact results for the many-body problem in one dimension with repulsive delta-function interaction, *Phys. Rev. Lett.* **19**, 1312-15, (1967).

Introduction to Part II

[1]   V.G. Drinfel'd, Quantum groups, in *Proceedings of the International Congress of Mathematicians*, pp. 798–820, Berkeley: AMS, (1987).

[2]   L.D. Faddeev and L.A. Takhtajan, *Hamiltonian Methods in the Theory of Solitons*, Springer-Verlag, (1987).

[3]   I.Z. Gokhberg and M.P. Krein, Systems of integral equations on a half-axis with kernels depending on difference of variables, *Usp. Matem. Nauk*, **13**, 3–72, (1958).

[4]   S.V. Manakov, Nonlinear Fraunhofer diffraction, *Sov. Phys. JETP* **38**, 693, (1974).

[5]   C.N. Yang and M.L. Ge, *Braid Group, Knot Theory, and Statistical Mechanics*, Singapore: World Scientific, (1989).

Chapter V

[1]   M.J. Ablowitz, D.J. Kaup, A.C. Newell and H. Segur, Method for solving the sine-Gordon equation, *Phys. Rev. Lett.* **30**, 1262–4 (1973).

[2]   A.A. Belavin and V.G. Drinfel'd, Solutions of the classical Yang-Baxter equation for simple Lie algebras, *Funct. Anal. Appl.* **16**, 159–80, (1983).

[3]   R.K. Bullough and R.K. Dodd, Polynomial conserved densities for the sine-Gordon equations, *Proc. Roy. Soc. London* **352A**, 481–503, (1977).

[4]   V.G. Drinfel'd, Hamiltonian structures on Lie groups, Lie bialgebras and the geometric meaning of the classical Yang-Baxter equations, *Sov. Math. Dokl.* **27**, 68–71, (1983).

[5]   L.D. Faddeev and L.A. Takhtajan, *Hamiltonian Methods in the Theory of Solitons*, Berlin: Springer-Verlag, (1987).

[6]   A.G. Izergin and V.E. Korepin, The inverse scattering method approach to the quantum Shabat-Mikhailov model, *Comm. Math. Phys.* **79**, 303–16, (1981).

[7]   A.G. Izergin and V.E. Korepin, Quantum Inverse Scattering Method, *Fizika Ehlementarynykh Chastits i Atomnogo Yadra* **13**, 501–41, (1982) (in Russian); English translation: *Sov. J. Part. Nucl.* **13**, 207–23.

[8]   D.J. Kaup and A.C. Newell, The Goursat and Cauchy problems for the sine-Gordon equation, *SIAM J. Appl. Math.* **34**, 37–54, (1978).

[9]   V.E. Korepin, Quantization of non-abelian Toda chain, *J. Sov. Math.* **23**, 2429 (1983).

[10] P.P. Kulish and E.K. Sklyanin, Solutions of the Yang-Baxter equation, *J. Sov. Math.* **19**, 1596, (1982).

[11] A.V. Mikhailov, Two-dimensional generalization of Toda chain, *JETP Lett.* **30**, 414–18, (1979).

[12] M.A. Semenov-Tyan-Shanskii, What is a classical R-matrix?, *Funct. Anal. Appl.* **17**, 259–72, (1983).

[13] E.K. Sklyanin, Quantum version of the method of the Inverse Scattering Problem, *J. Sov. Math.* **19**, 1546, (1982).

[14] E.K. Sklyanin, L.A. Takhtajan and L.D. Faddeev, Quantum Inverse Problem. I. *Theor. Math. Phys.* **40**, 688–706, (1979).

[15] L.A. Takhtajan, Exact theory of propagation of ultrashort optical pulses in two-level media, *Sov. Phys. JETP* **39**, 228–33, (1974).

[16] L.A. Takhtajan and L.D. Faddeev, Hamiltonian system connected with the equation $u_{\xi\eta} + \sin u = 0$. *Proc. Steklov Inst. Math,* **3**, 277–89, (1979).

[17] L.A. Takhtajan and L.D. Faddeev, Essentially nonlinear one-dimensional model of classical field theory, *Theor. Math. Phys.* **21**, 1046–57, (1975).

[18] V.E. Zakharov and S.V. Manakov, On complete integrability of a nonlinear Schrödinger equation, *Theor. Math. Phys.* **19**, 551–9, (1974).

[19] V.E. Zakharov and A.B. Shabat, Exact theory of two-dimensional self-focusing and one-dimensional self-modulation of waves in nonlinear media, *Sov. Phys. JETP* **34**, 62–9; *Theor. Math. Phys.* **34**, 62, (1972).

[20] V.E. Zakharov, L.A. Takhtajan and L.D. Faddeev, Complete description of solutions of the sine-Gordon equation, *Sov. Phys. Dokl.* **19**, 824–6, (1975).

[21] A.V. Zhiber and A.B. Shabat, Klein-Gordon equation with a nontrivial group, *Sov. Phys. Dokl.* **24**, 607–9, (1979).

Chapter VI

[1]   R.J. Baxter, One-dimensional anisotropic Heisenberg chain, *Ann. Phys. (NY)* **70**, 323–37, (1972).

[2]   R.J. Baxter, Partition function of the eight-vertex lattice model, *Ann. Phys.* **70**, 193–228, (1972).

[3]   R.J. Baxter, Eight-vertex model in lattice statistics and one-dimensional anisotropic Heisenberg chain I, II, III, *Ann. Phys.* **76**, 1–24, 25–47, 48–71, (1973).

[4]   R.J. Baxter, Solvable eight-vertex model on an arbitrary planar lattice, *Phil. Trans. Royal Soc. London* **289**, 315–46, (1978).

[5]   R.J. Baxter, *Exactly Solved Models in Statistical Mechanics*, London–New York: Academic Press, (1982).

[6]   R.J. Baxter, q colourings of the triangular lattice, *J. Phys.* **A19**, 2821–39, (1986).

[7]   V.V. Bazhanov and Ya.G. Stroganov, A new class of factorized S-matrices and triangle equations, *Phys. Lett.* **105B**, 278–80, (1981).

[8]   V.V. Bazhanov and Ya.G. Stroganov, Trigonometric and S_n symmetric solutions of triangle equations with variables on the faces, *Nucl. Phys.* **B205**, 505–26, (1982).

[9]   V.V. Bazhanov, Integrable quantum systems and classical Lie algebras, *Comm. Math. Phys.* **113**, 471–503, (1987).

[10] A.A. Belavin, Dynamical symmetry of integrable quantum systems, *Nucl. Phys.* **B180**, 189–200, (1981).

[11] A.A. Belavin, Discrete groups and the integrability of quantum systems, *Funct. Anal. Appl.* **14**, 260–7, (1981).

[12] F.A. Berezin, C.P. Pokhil and V.M. Finkelberg, The Schrödinger equation for a system of one-dimensional particles with point interactions, *Vestn. Mosk. Gos. Univ.* **1**, 21–8, (1964) (in Russian).

[13] F.A. Berezin and V.N. Sushko, Relativistic two-dimensional model of a self-interacting fermion field with nonvanishing rest mass, *Sov. Phys. JETP* **21**, 865–73, (1965).

[14] B. Berg, M. Karowski, P. Weisz and V. Kurak, Factorized $U(n)$ symmetric S-matrices in two dimensions, *Nucl. Phys.* **B134**, 125–32, (1978).

[15] N.M. Bogoliubov, A.G. Izergin and V.E. Korepin, *Quantum Inverse Scattering Method and Correlation Functions*, Lecture Notes in Physics **242**, pp. 220–316, Berlin: Springer, (1985).

[16] E. Brezin and J. Zinn-Justin, Un problème à N corps soluble, *C. R. Acad. Sci. Paris* **B263**, 670–3, (1966).

[17] K.M. Case, Polynomial constants for the quantized NLS equation, *J. Math. Phys.* **25**, 2306–14, (1984).

[18] I.V. Cherednik, On 'quantum' deformations of irreducible finite-dimensional representations of gl$_N$, *Sov. Math. Dokl.* **33**, 507–10, (1986).

[19] E. Date, M. Jimbo, A. Kuniba, T. Miwa and M. Okado, Exactly solvable SOS models. II: Proof of the star-triangle relation and combinatorial identities, *Adv. Stud. Pure Math.* **16**, 17–122, (1988).

[20] B. Davies and T.D. Kieu, *Inverse Problems* **2**, 141, (1986).

[21] B. Davies, The quantum Gelfand–Levitan equation and the non-linear Schrödinger equation, *Inverse Problems* **4**, 47, (1988).

[22] B. Davies, Higher conservation laws for the quantum non-linear Schrödinger equation, *Physica A* **167**, 433–56, (1990).

[23] L.D. Faddeev and E.K. Sklyanin, Quantum-mechanical approach to completely integrable field theory models, *Sov. Phys. Dokl.* **23**, 902–4, (1978).

[24] L.D. Faddeev, Quantum inverse scattering method, *Sov. Sci. Rev. Math. Phys.* **C1**, 107–60, (1981).

[25] L.D. Faddeev and N.Yu. Reshetikhin, Integrability of the principal chiral field model in 1+1 dimension, *Ann. Phys. (NY)* **167**, 227–56, (1986).

[26] L.D. Faddeev and L.A. Takhtajan, *Liouville Model on the Lattice*, Lecture Notes in Physics **246**, Berlin: Springer, 166–79, (1986).

[27] V.A. Fateev and A.B. Zamolodchikov, Self-dual solutions of the star-triangle relations in **Z**$_N$ models, *Phys. Lett.* **92A**, 37–9, (1982).

[28] M. Fowler, The Quantum Inverse Scattering Method and the Jordan-Wigner transformation for the Heisenberg-Ising spin chain, *Phys. Lett.* **94B**, 189–91, (1980).

[29] A.G. Izergin and V.E. Korepin, The inverse scattering method approach to the quantum Shabat-Mikhailov model, *Comm. Math. Phys.* **79**, 303–16, (1981).

[30] A.G. Izergin and V.E. Korepin, Lattice model connected with non-linear Schrödinger equation, *Sov. Phys. Dokl.* **26**, 653–4, (1981).

[31] A.G. Izergin and V.E. Korepin, Quantum Inverse Scattering Method, *Fizika Ehlementarynykh Chastits i Atomnogo Yadra* **13**, 501–41, (1982) (in Russian); English translation: *Sov. J. Part. Nucl.* **13**, 207–23.

[32] A.G. Izergin, V.E. Korepin and F.A. Smirnov, Trace formulas for quantum nonlinear Schrödinger equation, *Theor. Math. Phys.* **48**, 773–6, (1982).

[33] A.G. Izergin and V.E. Korepin, *J. Sov. Math.* **30**, 2292, (1985).

[34] M. Jimbo, Quantum R-matrix for the generalized Toda system, *Comm. Math. Phys.* **102**, 537–47, (1986).

[35] M. Jimbo, T. Miwa and M. Okado, Solvable lattice models related to the vector representation of classical simple Lie algebras, *Comm. Math. Phys.* **116**, 507–25, (1988).

[36] M. Jimbo (ed.), *Yang-Baxter Equation in Integrable Models*, Singapore: World Scientific, (1989).

[37] A.N. Kirillov and N. Yu. Reshetikhin, Exact solution of the integrable XXZ Heisenberg model with arbitrary spin: I. The ground state and the excitation spectrum, *J. Phys.* **A20**, 1565–85, (1987).

[38] V.E. Korepin, Calculations of norms of Bethe wave functions, *Comm. Math. Phys.* **86**, 391–418, (1982).

[39] V.E. Korepin, Analysis of bilinear relation for 6-vertex model, *Sov. Phys. Dokl.* **27**, 612–13, (1982).

[40] V.E. Korepin, Quantisation of Toda lattice in nonabelian case, *J. Sov. Math.* **23**, 2429, (1983).

[41] P.P. Kulish, S.V. Manakov and L.D. Faddeev, Comparison of the exact quantum and quasiclassical results for the nonlinear Schrödinger equation, *Theor. Math. Phys.* **28**, 615–20, (1976).

[42] P.P. Kulish and N. Yu. Reshetikhin, *J. Phys.* **A16**, L591–6, (1983).

[43] P.P. Kulish, N. Yu. Reshetikhin and E.K. Sklyanin, Yang-Baxter equation and representation theory I, *Lett. Math. Phys.* **5**, 393–403, (1981).

[44] P.P. Kulish and N.Yu. Reshetikhin, Generalized Heisenberg ferromagnet and the Gross-Neveu model, *Sov. Phys. JETP* **53**, 108–14, (1981).

[45] P.P. Kulish and E.K. Sklyanin, *Quantum Spectral Transform Method. Recent Developments*, Lecture Notes in Physics **151**, pp. 61–119, Berlin–New York: Springer, (1982).

[46] P.P. Kulish and E.K. Sklyanin, Solutions of the Yang-Baxter equation, *J. Sov. Math.* **19**, 1596–620, (1982).

[47] I.M. Krichever, Baxter's equations and algebraic geometry, *Funct. Anal. Appl.* **15**, 92, (1981).

[48] J.B. McGuire, Study of exactly solvable one-dimensional N-body problem, *J. Math. Phys.* **5**, 622-36, (1964).

[49] V. Pasquier, Etiology of IRF models, *Comm. Math. Phys.* **118**, 335–64, (1988).

[50] J.H.H. Perk and C.L. Schultz, Families of commuting transfer matrices in q-state vertex models, in *Nonlinear Integrable Systems—Classical Theory and Quantum Theory* (eds. M. Jimbo and T. Miwa), pp. 135–52, Singapore: World Scientific, (1983).

[51] N.Yu. Reshetikhin, The functional equation method in the theory of exactly soluble quantum systems, *Sov. Phys. JETP* **57**, 691–6, (1983).

[52] N.Yu. Reshetikhin and P. Wiegmann, Towards the classification of completely integrable quantum field theories (the Bethe Ansatz associated with Dynkin diagrams and their automorphisms), *Phys. Lett.* **189B**, 125–31, (1987).

[53] R. Shankar and E. Witten, S-matrix of the supersymmetric σ model, *Phys. Rev.* **D17**, 2134, (1978).

[54] E.K. Sklyanin, L.A. Takhtajan and L.D. Faddeev, Quantum Inverse Problem Method I, *Theor. Math. Phys.* **40**, 688–706, (1979).

[55] E.K. Sklyanin, Method of the inverse scattering problem and the nonlinear quantum Schrödinger equation, *Sov. Phys. Dokl.* **24**, 107–9, (1979).

[56] E.K. Sklyanin, Quantum version of the method of inverse scattering problem, *J. Sov. Math.* **19**, 1546–96, (1982).

[57] E.K. Sklyanin, Quantization of integrable models in the infinite volume. Nonlinear Schrödinger equation, *Algebra and Analysis* **1**, 189–206, (1989) (in Russian).

[58] L.A. Takhtajan and L.D. Faddeev, The quantum inverse problem method and the Heisenberg XYZ model, *Russ. Math. Survey* **34**, 11–68, (1979).

[59] L.A. Takhtajan, The picture of low-lying excitations in the isotropic Heisenberg chain of arbitrary spins, *Phys. Lett.* **87A**, 479–82 (1982).

[60] L.A. Takhtajan, in *Proceedings of the International Congress of Mathematicians*, Warsaw: North–Holland, (1983).

[61] L.A. Takhtajan, *Introduction to Bethe Ansatz*, Lecture Notes in Physics **242**, Berlin: Springer, (1985).

[62] H.B. Thacker, Polynomial conservation laws in $(1+1)$-dimensional classical and quantum field theory, *Phys. Rev.* **D17**, 1031–40 (1978).

[63] H.B. Thacker, Exact integrability in quantum field theory and statistical systems, *Rev. Mod. Phys.* **53**, 253–85, (1981).

[64] V.I. Vichirko and N.Yu. Reshetikhin, Excitation spectrum of the anisotropic generalization of an SU_3 magnet, *Theor. Math. Phys.* **56**, 805–12, (1984).

[65] C.N. Yang, Some exact results for the many-body problem in one-dimension with repulsive delta function interaction, *Phys. Rev. Lett.* **19**, 1312–14, (1967).

[66] C.N. Yang, S-matrix for the one-dimensional N-body problem with repulsive or attractive δ-function interaction, *Phys. Rev.* **168**, 1920–3, (1968).

[67] A.B. Zamolodchikov, Exact two-particle S-matrix of quantum solitons of the sine-Gordon model, *JETP Lett.* **25**, 468–71, (1977).

[68] A.B. Zamolodchikov and Al.B. Zamolodchikov, Realistic factorized S-matrix in two-dimensional space-time with isotropic symmetry, *JETP Lett.* **26**, 457–60, (1977).

[69] A.B. Zamolodchikov and Al.B. Zamolodchikov, Exact S-matrix of Gross-Neveu "elementary" fermions, *Phys. Lett.* **72B**, (1978).

[70] A.B. Zamolodchikov and Al.B. Zamolodchikov, Factorized S-matrices in two dimensions as the exact solutions of certain relativistic quantum field theories, *Ann. Phys.* **120**, 253–91, (1979).

Chapter VII

[1] V.G. Drinfel'd, Hamiltonian structures on Lie groups, Lie bialgebras and the geometric meaning of the classical Yang-Baxter equations, *Sov. Math. Dokl.* **27**, 68–71, (1983).

[2] L.D. Faddeev and L.A. Takhtajan, Quantum inverse scattering method, *Sov. Sci. Rev. Math.* **C1**, 107, (1981).

[3] A.G. Izergin and V.E. Korepin, Lattice model connected with nonlinear Schrödinger equation, *Sov. Phys. Dokl.* **26**, 653–4, (1981).

[4] A.G. Izergin and V.E. Korepin, Lattice versions of quantum field theory models in two dimensions, *Nucl. Phys.* **A20**, 401–13, (1982).

[5]   A.G. Izergin and V.E. Korepin, Pauli principle for one-dimensional bosons and Algebraic Bethe Ansatz, *Lett. Math. Phys.* **6**, 283–9, (1982).

[6]   A.G. Izergin, Partition function of the six-vertex model in a finite volume, *Sov. Phys. Dokl.* **32**, 878–9, (1987).

[7]   A.G. Izergin, V.E. Korepin and N.Yu. Reshetikhin, *J. Phys.* **A20**, 4799–822, (1987).

[8]   V.E. Korepin, Analysis of bilinear relation for the six-vertex model, *Sov. Phys. Dokl.* **27**, 612–13, (1982).

[9]   V.E. Korepin, Calculations of norms of Bethe wave functions, *Comm. Math. Phys.* **86**, 391–418, (1982).

[10] V.E. Korepin, *Dokl. Akad. Nauk. SSSR* **262**, 78 (1982).

[11] P.P. Kulish, Classical and quantum inverse problem method and generalized Bethe Ansatz, *Physica* **3D**, 246–57, (1981).

[12] P.P. Kulish and N.Yu. Reshetikhin, Generalized Heisenberg ferromagnet and the Gross-Neveu model, *Sov. Phys. JETP* **53**, 108–14, (1981).

[13] P.P. Kulish, N.Yu. Reshetikhin and E.K. Sklyanin, Yang–Baxter equation and representation theory: 1, *Lett. Math. Phys.* **5**, 393–403, (1981).

[14] P.P. Kulish and E.K. Sklyanin, *Quantum Spectral Transform Method. Recent Developments*, Lecture Notes in Physics **151**, pp. 61–119, Berlin–New York: Springer, (1982).

[15] E.H. Lieb, Residual entropy of square ice, *Phys. Rev.* **162**, 162–72, (1967).

[16] E.H. Lieb, Exact solution of the *F* model of an antiferromagnet, *Phys. Rev. Lett.* **18**, 1046–8, (1967).

[17] E.H. Lieb, Exact solution of the two-dimensional Slater KDP model of a ferroelectric, *Phys. Rev. Lett.* **19**, 108–10, (1967).

[18] N.Yu. Reshetikhin, The functional equation method in the theory of exactly soluble quantum systems, *Sov. Phys. JETP* **57**, 691–6, (1983).

[19] B.S. Shastry, Infinite conservation laws in the one-dimensional Hubbard model, *Phys. Rev. Lett.* **56**, 1529–34, (1986).

[20] B.S. Shastry, Exact integrability of the one-dimensional Hubbard model, *Phys. Rev. Lett.* **56**, 2453–5, (1986).

[21] B.S. Shastry, Decorated star-triangle relations and exact integrability of one-dimensional Hubbard model, *J. Stat. Phys.* **50**, 57–79, (1988).

[22] E.K. Sklyanin, L.A. Takhtajan and L.D. Faddeev, Quantum Inverse Problem Method I, *Theor. Math. Phys.* **40**, 688–706, (1979).

[23] E.K. Sklyanin, Some algebraic structures connected to the Yang-Baxter equation, *Funct. Anal. Appl.* **16**, 263, (1983).

[24] E.K. Sklyanin, Some algebraic structures connected with the Yang-Baxter equation. Representation of quantum algebras, *Funct. Anal. Appl.* **17**, 273, (1983).

[25] F.A. Smirnov, Gel'fand-Levitan equations for quantum nonlinear Schrödinger equation with interaction, *Sov. Phys. Dokl.* **27**, 34–6, (1982).

[26] B. Sutherland, Exact solution of a two-dimensional model for hydrogen-bonded crystal, *Phys. Rev. Lett.* **19**, 103–4, (1967).

[27] L.A. Takhtajan, *J. Sov. Math.* **23** (4), 2470, (1983).

[28] L.A. Takhtajan and L.D. Faddeev, The quantum inverse problem method and the Heisenberg XYZ model, *Russ. Math. Survey* **34**(5), 11–68, (1979).

[29] V.O. Tarasov, Structure of quantum L-operators for the R-matrix of the XXZ model, *Theor. Math. Phys.* **61**, 1065–72, (1984).

[30] V.O. Tarasov, L.A. Takhtajan and L.D. Faddeev, Local Hamiltonian for integrable quantum models on a lattice, *Theor. Math. Phys.* **57**, 1059–73, (1983).

Chapter VIII

[1] M.J. Ablowitz and J.F. Ladik, A nonlinear difference scheme and inverse scattering, *Stud. Appl. Math.* **55**, 213, (1976).

[2] N.M. Bogoliubov, A.G. Izergin and V.E. Korepin, Structure of the vacuum in the quantum sine-Gordon model, *Phys. Lett.* **159B**, 345–7, (1985).

[3] N.M. Bogoliubov, On the physical sector of the lattice sine-Gordon model, *Theor. Math. Phys.* **51**, 540–7, (1982).

[4] N.M. Bogoliubov, Thermodynamics of one-dimensional lattice Bose-gas, *Theor. Math. Phys.* **67**, 614–22, (1986).

[5] N.M. Bogoliubov and A.G. Izergin, Lattice completely integrable regularization of sine-Gordon model, *Theor. Math. Phys.* **59**, 441–52, (1984).

[6]   N.M. Bogoliubov and A.G. Izergin, Lattice sine-Gordon model with local Hamiltonian, *Theor. Math. Phys.* **61**, 1195–204, (1985).

[7]   N.M. Bogoliubov and V.E. Korepin, Quantum nonlinear Schrödinger equation on the lattice, *Theor. Math. Phys.* **66**, 300–5, (1986).

[8]   H.W. Capel, G.L. Wiersma, G.R.W. Quispel and F.W. Nijhoff, Linearizing integral transforms and partial difference equations, *Phys. Lett.* **103A**, 293–7, (1984).

[9]   H.W. Capel and G.L. Wiersma, Lattice equations, hierarchies and Hamiltonian structures. II. KP-type of hierarchies and 2D lattices, *Physica*, **149A**, 49, (1988).

[10] R. Hiroto, Nonlinear partial difference equations III: Discrete sine-Gordon equation, *J. Phys. Soc. Jap.* **43**, 2079, (1977).

[11] T. Holstein and H. Primakoff, Field dependence of the intrinsic domain magnetization of a ferromagnet, *Phys. Rev.* **58**, 1098, (1940).

[12] A.G. Izergin and V.E. Korepin, The lattice quantum sine-Gordon model, *Lett. Math. Phys.* **5**, 199–205, (1981).

[13] A.G. Izergin and V.E. Korepin, The most general L-operator for the R-matrix of the XXX model, *Lett. Math. Phys.* **8**, 259–65, (1984).

[14] A.G. Izergin and V.E. Korepin, Lattice versions of quantum field theory models in two dimensions, *Nucl. Phys.* **B205**, 401–13 (1982).

[15] A.G. Izergin and V.E. Korepin, Lattice model connected with the nonlinear Schrödinger equation, *Sov. Phys. Dokl.* **26**, 653–4 (1981).

[16] A.G. Izergin and V.E. Korepin, *Vestnik LGU* **22**, 84, (1981) (in Russian).

[17] A.G. Izergin and V.E. Korepin, *J. Sov. Math.* **34**, 1937, (1982).

[18] A.G. Izergin and V.E. Korepin, Problem of describing all L-operators of the XXX and XXZ models, *J. Sov. Math.* **30**, 2292, (1985).

[19] P.P. Kulish, *Lett. Math. Phys.* **5**, 191, (1981).

[20] P.P. Kulish and N.Yu. Reshetikhin, Diagonalization of $GL(N)$ invariant transfer matrices and quantum N-wave system (Lee model), *J. Phys.* **A16**, L591–6, (1983).

[21] S.J. Orfanidis, Group-theoretical aspects of the discrete sine-Gordon equation, *Phys. Rev.* **D21**, 1507–22, (1980).

[22] E.K. Sklyanin, Some algebraic structures connected to the Yang-Baxter equation, *Funct. Anal. Appl.* **16**, 263, (1983).

[23] E.K. Sklyanin, Some algebraic structures connected to the Yang-Baxter equation. Representation of quantum algebras, *Funct. Anal. Appl.* **17**, 273, (1983).

[24] F.A. Smirnov, Gel'fand-Levitan equations for the quantum nonlinear Schrödinger equation with attraction, *Sov. Phys. Dokl.* **27**, 34–6, (1982).

[25] V.O. Tarasov, *J. Sov. Math.* **34**, 2011, (1986).

[26] V.O. Tarasov, Structure of quantum L-operators for the R-matrix of XXZ model, *Theor. Math. Phys.* **61**, 1065–72, (1984).

[27] V.O. Tarasov, L.A. Takhtajan and L.D. Faddeev, Local Hamiltonians for integrable quantum models on a lattice, *Theor. Math. Phys.* **57**, 1059–73, (1984).

[28] A.Yu. Volkov, *J. Sov. Math.* **46**, 1576, (1989).

Chapter IX

[1] A.G. Izergin, Partition function of the six-vertex model in a finite volume, *Sov. Phys. Dokl.* **32**, 878, (1987).

[2] V.E. Korepin, Calculations of norms of Bethe wave functions, *Comm. Math. Phys.* **86**, 391–418, (1982).

[3] V.E. Korepin, Dual field formulation of quantum integrable models, *Comm. Math. Phys.* **113**, 177–90, (1987).

Chapter X

[1] M. Gaudin, *La Fonction d'Onde de Bethe*, Paris: Masson, (1983).

[2] A.N. Kirillov and F.A. Smirnov, Solution of some combinatorial problems which arise in calculating correlators in exactly solvable models, *J. Sov. Math.* **47**, 2413–23, (1989).

[3] V.E. Korepin, Calculations of norms of Bethe wave functions, *Comm. Math. Phys.* **86**, 391–418, (1982).

[4] N.A. Slavnov, Calculation of scalar products of the wave functions and form factors in the framework of the Algebraic Bethe Ansatz, *Theor. Math. Phys.* **79**, 502–8, (1989).

[5] F.A. Smirnov, *Form Factors in Completely Integrable Models in Quantum Field Theory*, Advanced Series in Mathematics and Physics, Singapore: World Scientific, (1992).

Chapter XI

[1]   N.M. Bogoliubov and V.E. Korepin, Correlation length of the one-dimensional Bose gas, *Nucl. Phys.* **B257**, 766–78, (1985).

[2]   N.M. Bogoliubov, V.E. Korepin and A.G. Izergin, *Quantum Inverse Scattering Method and Correlation Functions*, Lecture Notes in Physics **242**, pp. 220–316, Berlin: Springer-Verlag, (1985).

[3]   A.G. Izergin and V.E. Korepin, The quantum inverse scattering method approach to correlation functions, *Comm. Math. Phys.* **94**, 67–92, (1984).

[4]   A.G. Izergin and V.E. Korepin, Correlation functions for the Heisenberg XXZ antiferromagnet, *Comm. Math. Phys.* **99**, 271–302 (1985).

[5]   V.E. Korepin, Dual field formulation of quantum integral models, *Comm. Math. Phys.* **113**, 177–90, (1987).

[6]   V.E. Korepin, Generating functional of correlation functions for the nonlinear Schrödinger equation, *Funk. Analiz. i jego Prilozh.* **23**, 15–23, (1989) (in Russian).

Chapter XII

[1]   A.G. Izergin, V.E. Korepin and N.Yu. Reshetikin, Correlation functions in a one-dimensional Bose gas, *J. Phys.* **A20**, 4799–822 (1987).

[2]   V.E. Korepin and N.A. Slavnov, Correlation functions of fields in a one-dimensional Bose gas, *Comm. Math. Phys.* **136**, 633–44, (1991).

Conclusion to Part III

[1]   F.C. Pu and B.H. Zhao, *Phys. Rev.* **D30**, 2253, (1984).

[2]   F.A. Smirnov, Gel'fand-Levitan equation for quantum nonlinear Schrödinger equation in the attractive case, *Doklady Akad. Nauka USSR* **262**, 78–83, (1981).

[3]   H.B. Thacker, Integrability, duality, monodromy, and the structure of Bethe Ansatz, in *Les Houches Lectures in Theoretical Physics*, (eds. J.-B. Zuber and R. Stora), Amsterdam: Elsevier, (1982).

Introduction to Part IV

[1]   A. Berkovich, Temperature and magnetic field dependent correlators of the exactly integrable (1+1)-dimensional gas of impenetrable fermions, *J. Phys.* **A24**, 1543–56, (1991).

[2]   H. Itoyama, H. Thacker and V.E. Korepin, Correlation functions of sine-Gordon at free fermion point as Fredholm determinant, *Int. J. Modern Phys.* **B**, in press.

[3]   A.R. Its, A.G. Izergin and V.E. Korepin, Long-distance asymptotics of temperature correlators of the impenetrable Bose gas, *Comm. Math. Phys.* **130**, 471–86, (1990).

[4]   A.R. Its, A.G. Izergin and V.E. Korepin, Temperature correlators of the impenetrable Bose gas as an integrable system, *Comm. Math. Phys.* **129**, 209–22, (1990).

[5]   A.R. Its, A.G. Izergin, V.E. Korepin and N.A. Slavonov, Differential equations for quantum correlation functions, *Int. J. Mod. Phys.* **B4**, 1003–37, (1990).

[6]   A.R. Its, A.G. Izergin, N.Yu. Novokshenov, V.E. Korepin, Temperature autocorrelators of the transverse Ising chain at the critical magnetic field, *Nucl. Phys.* **B340**, 752, (1990).

[7]   M. Jimbo, T. Miwa, Y. Mori and M. Sato, *Physica* **1D**, 80–158, (1980).

Chapter XIII

[1]   N.M. Bogoliubov, A.G. Izergin and V.E. Korepin, *Quantum Inverse Scattering Method and Correlation Functions*, Lecture Notes in Physics **242**, pp. 221–316, Berlin: Springer-Verlag, (1985).

[2]   N.M. Bogoliubov and V.E. Korepin, Correlation length of the one-dimensional Bose gas, *Nucl. Phys.* **B257**, 766–78, (1985).

[3]   N.M. Bogoliubov and V.E. Korepin, Correlation functions of the one-dimensional Bose gas in thermodynamic equilibrium, *Theor. Math. Phys.* **60**, 808, (1984).

[4]   A.R. Its, A.G. Izergin, V.E. Korepin and N.A. Slavnov, Differential equations for quantum correlation functions, *Int. J. Mod. Phys.* **B4**, 1003–37, (1990).

[5]   A.R. Its, A.G. Izergin and V.E. Korepin, Temperature correlators of the impenetrable Bose gas as an integrable system, *Comm. Math. Phys.* **129**, 205–22, (1990).

[6]   V.E. Korepin and N.A. Slavnov, The time-dependent correlation function of an impenetrable Bose gas as a Fredholm minor, *Comm. Math. Phys.* **129**, 103–13, (1990).

[7]   A. Lenard, One-dimensional impenetrable bosons in thermal equilibrium, *J. Math. Phys.* **7**, 1268–72, (1966).

[8]   A. Lenard, Momentum distribution in the ground state of the one-dimensional system of impenetrable bosons, *J. Math. Phys.* **5**, 930–43, (1964).

[9]   J.H.H. Perk, H.W. Capel, G.R.W. Quispel and F.W. Nijhoff, Finite-temperature correlations for the Ising chain in a transverse field, *Physica* **A123**, 1–49, (1984).

Chapter XIV

[1]   M.J. Ablowitz and A.S. Fokas, Linearization of a class of nonlinear evolution equations, *J. Math. Phys.* **27**, 2614–19, (1964).

[2]   S.P. Burtsev, I.R. Galitov and V.E. Zakharov, Exact theory of Maxwell-Bloch equation with pumping, in *Plasma Theory and Nonlinear and Turbulent Processes in Physics*, p. 897, World Scientific: Singapore, (1988).

[3]   A.S. Fokas and M.J. Ablowitz, Linearization of the Korteweg-de Vries and Painlevé II equations, *Phys. Rev. Lett.* **47**, 1096, (1981).

[4]   A.R. Its, A.G. Izergin and V.E. Korepin, Correlation radius for one-dimensional impenetrable bosons, *Phys. Lett.* **141A**, 121–5, (1989).

[5]   A.R. Its, A.G. Izergin and V.E. Korepin, Temperature correlators of the impenetrable Bose gas as an integrable system, *Comm. Math. Phys.* **129**, 205–22, (1990).

[6]   A.R. Its, A.G. Izergin, V.E. Korepin and N.A. Slavnov, Differential equations for quantum correlation functions, *Int. J. Mod. Phys.* **B4**, 1003–37, (1990).

[7]   A.R. Its, A.G. Izergin, V.E. Korepin and N.Yu. Novokshenov, Temperature autocorrelations of the transverse Ising chain at the critical magnetic field, *Nucl. Phys.* **B340**, 752, (1990).

[8]   M. Jimbo, T. Miwa, Y. Mori and M. Sato, Density matrix of an impenetrable Bose gas and the fifth Painlevè transcendent, *Physica* **1D**, 80–158, (1980).

[9]   M.G. Krein, *Doklady of Academy of Sciences of the USSR* (in Russian), **97**, 21–4, (1954).

[10] M.G. Krein, *Doklady of Academy of Sciences of the USSR* (in Russian), **105**, 433–6, (1955).

[11] B.M. McCoy, T.T. Wu, *The two-dimensional Ising model*, Harvard Univ. Press, Cambridge, MA, (1973).

[12] A.Yu. Orlov and E.I. Schulman, On the additional symmetries of the nonlinear Schrödinger equation, *Theor. Math. Phys.* **64**, 323–8, (1985).

[13] A.Yu. Orlov and E.I. Schulman, Additional symmetries for integrable equations and conformal algebra representation, *Lett. Math. Phys.* **12**, 171–9, (1986).

[14] G.R.W. Quispel, F.W. Nijhoff and H.W. Capel, *Phys. Lett.* **91A**, 143, (1982).

[15] N.A. Slavnov, Differential equation for space- and time-dependent correlation functions of fields in quantum Nonlinear Schrödinger equation at finite coupling constant, Dubna preprint, (1991).

Chapter XV

[1] L.D. Faddeev and L.A. Takhtajan, *Hamiltonian Methods in the Theory of Solitons*, Berlin: Springer-Verlag, (1987).

[2] I.Z. Gokhberg and M.P. Krein, Systems of integral equations on a half-axis with kernels depending on difference of variables, *Usp. Matem. Nauk*, **13**, 3–72, (1958).

[3] A.R. Its, A.G. Izergin, V.E. Korepin and N.A. Slavnov, Differential equations for quantum correlation functions, *Int. J. Mod. Phys.* **B4**, 1003–37, (1990).

[4] A.R. Its, A.G. Izergin and V.E. Korepin, Space correlations in the one-dimensional impenetrable Bose gas at finite temperature, *Physica* **D53**, 187, (1991).

[5] A.R. Its, A.G. Izergin, V.E. Korepin and G.G. Varzugin, Large time and distance asymptotics of the correlator of the impenetrable bosons at finite temperature, preprint ITP-SB-91-27, SUNY Stony Brook, (1991).

[6] A.R. Its and G.G. Varzugin, Long distance and large time asymptotics of the field correlation functions of impenetrable bosons for negative chemical potential, *Vestnik LGU* **137**, (1991).

[7] A.R. Its, A.G. Izergin and V.E. Korepin, Long-distance asymptotics of temperature correlators of the impenetrable Bose gas, *Comm. Math. Phys.* **130**, 471–88, (1990).

[8] A.R. Its, A.G. Izergin and V.E. Korepin, Temperature correlations of impenetrable Bose gas as an integrable system, *Comm. Math. Phys.* **129**, 205–22, (1990).

[9]   A. Jimbo, T. Miwa, Y. Mori and M. Sato, Density matrix of an impenetrable Bose gas and the fifth Painlevè transcendent, *Physica* **1D**, 80–158, (1980).

[10] A.V. Mikhailov, The reduction problem and the inverse scattering method, *Physica* **3D**, 73–117, (1981).

[11] A.B. Shabat, Inverse scattering problem for a system of differential equations, *Funkz. anal. i jego prilozh.* **9**, 75–8, (1975).

[12] V.E. Zakharov and S.V. Manakov, The theory of resonance interactions of wave packets in the nonlinear media, *Sov. Phys. JETP* **42**, 842, (1975).

[13] V.E. Zakharov and A.B. Shabat, Integration of nonlinear equations by means of the inverse scattering method, *Funkz. Anal. i Jego Prilozh.* **13**, 13–22, (1979).

Chapter XVI

[1]   M.J. Ablowitz and H. Segur, Asymptotic solutions of the Korteweg-deVries equation, *Stud. Appl. Math.* **57**, 13–44, (1977).

[2]   M.J. Ablowitz and H. Segur, *Solitons and the Inverse Scattering Transform*, Philadelphia: SIAM, (1981).

[3]   V.S. Buslaev and V.V. Sukhanov, Asymptotical behavior of solutions of the Korteweg-deVries equation at large time, *Zap. Nauk. Semin. LOMI* **120**, 35–59, (1982).

[4]   L.D. Faddeev and L.A. Takhtajan, *Hamiltonian Methods in the Theory of Solitons*, Berlin: Springer-Verlag, (1987).

[5]   F.D. Gakhov, *Boundary problems*, Moscow: Nauka, (1977).

[6]   A.R. Its, A.G. Izergin and V.E. Korepin, Correlation radius for one-dimensional impenetrable bosons, Phys. Lett. **A141**, 121–25, (1989).

[7]   A.R. Its, A.G. Izergin and V.E. Korepin, Long-distance asymptotics of temperature correlators of the impenetrable Bose gas, *Comm. Math. Phys.* **130**, 471–86, (1990).

[8]   A.R. Its, A.G. Izergin and V.E. Korepin, Space correlations in the one-dimensional impenetrable Bose gas at finite temperature, *Physica* **D53**, 187, (1991).

[9]   A.R. Its, A.G. Izergin, V.E. Korepin and G.G. Varzugin, Large time and distance asymptotics of the correlator of the impenetrable bosons at finite temperature, *Physica* **D54**, 351, (1991).

[10] A.R. Its, Asymptotics of solutions of the nonlinear Schrödinger equation and isomonodromic deformations of systems of linear differential equations, *Sov. Math. Dokl.* **24**, 452–6, (1981).

[11] A.R. Its and V.Yu. Novokshenov, The isomonodromic deformation method in the theory of Painlevé Equations, Lecture Notes in Mathematics **1191**, Berlin: Springer-Verlag, (1986).

[12] A.R. Its, A.G. Izergin and V.E. Korepin, Large time and distance asymptotics of the temperature field correlator in the impenetrable Bose gas, *Nucl. Phys.* **B348**, 757–65, (1991).

[13] A.R. Its, A.G. Izergin, V.E. Korepin and N.A. Slavnov, Differential equations for quantum correlation functions, *Int. Jour. Mod. Phys.* **B4**, 1003–37, (1990).

[14] A. Jimbo, T. Miwa, Y. Mori and M. Sato, Density matrix of an impenetrable Bose gas and the fifth Painlevè transcendent, *Physica.* **1D**, 80–158, (1980).

[15] S.V. Manakov, Nonlinear Fraunhofer diffraction, *Sov. Phys. JETP* **38**, 693, (1974).

[16] J.W. Miles, The asymptotic solutions of the Korteweg-deVries equation, *Stud. Appl. Math.* **60**, 59–72, (1979).

[17] V.Yu. Novokshenov, Asymptotics at $t \to \infty$ of the solutions of the Cauchy problem for the nonlinear Schrödinger equation, *Sov. Math. Dokl.* **21**, 529, (1980).

[18] J.H.H. Perk, H.W. Capel, G.R.W. Quispel and F.W. Nijhoff, Finite-temperature correlations for the Ising chain in a transverse field, *Physica* **A123** 1–49, (1984).

[19] H.G. Vaidiya and C.A. Tracy, One-particle reduced density matrix of impenetrable bosons in one dimension at zero temperature, *J. Math. Phys.* **20**, 2291–303, (1979).

[20] V.E. Zakharov and S.V. Manakov, Asymptotical behavior of nonlinear wave systems integrated by the inverse scattering method, *Sov. Phys. JETP*, **44**, 106, (1976).

Chapter XVII

[1] N.M. Bogoliubov, A.G. Izergin and V.E. Korepin, Critical exponents for integrable models, *Nucl. Phys.* **B275**, 687–705, (1986).

[2] N.M. Bogoliubov, A.G. Izergin and V.E. Korepin, Quantum inverse scattering method and correlation functions, in *Exactly Solvable*

Problems in Condensed Matter and Relativistic Field Theory, (eds. B.S. Shastry, S.S. Jha and V. Singh), Lecture Notes in Physics **242**, pp. 220–316, Berlin–Heidelberg–New York–Tokyo: Springer-Verlag, (1985).

[3] N.M. Bogoliubov, A.G. Izergin and V.E. Korepin, *Correlation Functions of Integrable Systems and Quantum Inverse Scattering Method*, Nauka, Moscow, (1991) (in Russian).

[4] N.M. Bogoliubov and V.E. Korepin, Correlation functions of the one-dimensional Bose gas, *Nucl. Phys.* **B257**, 766–78, (1985).

[5] N.M. Bogoliubov, and V.E. Korepin, Correlation functions of the one-dimensional Bose gas in thermodynamic equilibrium, *Theor. Math. Phys.* **60**, 808, (1984).

[6] A.R. Its, A.G. Izergin, V.E. Korepin and N.A. Slavnov, Differential equations for quantum correlation functions, *Int. J. Mod. Phys.* **B4**, 1003–37, (1990).

[7] A.G. Izergin and V.E. Korepin, The quantum inverse scattering method approach to correlation functions, *Comm. Math. Phys.* **94**, 67–92, (1984).

[8] A.G. Izergin and V.E. Korepin, Correlation functions for the Heisenberg XXZ antiferromagnet, *Comm. Math. Phys.* **99**, 271–307 (1985).

[9] A.G. Izergin, V.E. Korepin and N.Yu. Reshetikhin, Correlation functions in a one-dimensional Bose gas, *J. Phys.* **A20**, 4799–822 (1987).

[10] A.G. Izergin, V.E. Korepin and N.A. Slavnov, Temperature correlation functions of Heisenberg antiferromagnet, *Theor. Math. Phys.* **72**, 878–84, (1988).

[11] A.G. Izergin and V.E. Korepin, Phase transition in the one-dimensional Heisenberg model, *Vestnik Leningrad Univ.* **4**, 3–8, (1986).

[12] A.G. Izergin and V.E. Korepin, Critical exponents in the Heisenberg magnet, *JETP Lett.* **42**, 414–16, (1985).

[13] V.E. Korepin, Correlation functions of the one-dimensional Bose gas in the repulsive case, *Comm. Math. Phys.* **94**, 93–113, (1984).

[14] V.E. Korepin and N.A. Slavnov, Correlation function in the one-dimensional Bose gas, *Theor. Math. Phys.* **68**, 955–60, (1986).

[15] V.E. Korepin and N.A. Slavnov, Time dependence of the density-density temperature correlation function of a one-dimensional Bose gas, *Nucl. Phys.* **B340**, 759, (1990).

[16] A.V. Zabrodin, On the irreducible part of the correlator of currents in quantum integrable models, *Theor. Math. Phys.* **71**, 485–91, (1987).

Chapter XVIII

[1] I. Affleck, Universal term in the free energy at a critical point and the conformal anomaly, *Phys. Rev. Lett.* **56**, 746–8, (1986).

[2] I. Affleck, D. Gepner, H. Schulz and T. Ziman, Critical behavior of spin-s Heisenberg chains: analytical and numerical results, *J. Phys.* **A22**, 511–29, (1989).

[3] P.W. Anderson and Y. Ren, Princeton University preprint, 1990.

[4] A.A. Belavin, A.M. Polyakov and A.B. Zamolodchikov, Infinite conformal symmetry in two-dimensional quantum field theory, *Nucl. Phys.* **B241**, 333–80, (1984).

[5] A. Berkovich and G. Murthy, Operator dimensions and surface exponents for the nonlinear Schrödinger model at $T = 0$, *J. Phys.* **A21**, 3703–21, (1988).

[6] A. Berkovich and G. Murthy, Time-dependent multipoint correlation functions of the nonlinear Schrödinger model, *Phys. Lett.* **142A**, 121–7, (1989).

[7] H.W. Blöte, J.L. Cardy and M.P. Nightingale, Conformal invariance, the central charge and universal finite-size amplitudes at criticality, *Phys. Rev. Lett.* **56**, 742–5, (1986).

[8] N.M. Bogoliubov and V.E. Korepin, Correlation length of the one-dimensional Bose-gas, *Nucl. Phys.* **B257**, 766–78, (1985).

[9] N.M. Bogoliubov and V.E. Korepin, Correlation length of the one-dimensional Bose-gas, *Phys. Lett.* **111A**, 419–22, (1985).

[10] N.M. Bogoliubov, A.G. Izergin and V.E. Korepin, Critical exponents for integrable models, *Nucl. Phys.* **B275**, 687–705, (1986).

[11] N.M. Bogoliubov, A.G. Izergin and N.Yu. Reshetikhin, Finite-size effects and critical indices of $1D$ quantum models, *JETP Lett.* **44**, 521–3, (1986).

[12] N.M. Bogoliubov, A.G. Izergin and N.Yu. Reshetikhin, Finite-size effects and infrared asymptotics of the correlation functions in two dimensions, *J. Phys.* **A20**, 5361–9, (1987).

[13] N.M. Bogoliubov and V.E. Korepin, The role of quasi-one-dimensional structures in high-T_c superconductivity, *Int. J. Mod. Phys.* **B3** 427–39 (1989).

[14] J.L. Cardy, Operator content of two-dimensional conformally invariant theories, *Nucl. Phys.* **B270**, 186–204, (1986).

[15] J.L. Cardy, Logarithmic corrections to finite-size scaling in strips, *J. Phys.* **A19**, L1093–8, (1986).

[16] H.J. de Vega and F. Woynarovich, Method for calculating finite size corrections in Bethe Ansatz systems: Heisenberg chain and six-vertex model, *Nucl. Phys.* **B251**, 439–56, (1985).

[17] H.J. de Vega, Finite-size corrections for nested Bethe ansatz models and conformal invariance, *J. Phys.* **A20**, 6023–36, (1987).

[18] H.J. de Vega, Integrable vertex models and extended conformal invariance, *J. Phys.* **A21**, L1089–95, (1988).

[19] R. Dijkgraaf, E. Verlinde and H. Verlinde, $c = 1$ conformal field theories on Riemann surfaces, *Comm. Math. Phys.* **115**, 649–90, (1988).

[20] K.B. Efetov and A.I. Larkin, Effect of fluctuations on the transition temperature in quasi-one-dimensional superconductors, *Sov. Phys. JETP* **39**, 1129–34, (1974).

[21] V. Emery, in *Highly Conducting One-Dimensional Solids* (ed. J.T. Devereese, *et al.*) New York: Plenum, (1979).

[22] A.M. Finkelshtein, Correlation functions in one-dimensional Hubbard model, *JETP Lett.* **25**, 73–6, (1977).

[23] H.C. Fogedby, Correlation functions for the Heisenberg-Ising chain at $T = 0$, *J. Phys.* **C22**, 4767–91, (1978).

[24] H. Frahm and V.E. Korepin, Critical exponents for the one-dimensional Hubbard model, *Phys. Rev.* **B42**, 10553–65, (1990).

[25] H. Frahm and V.E. Korepin, Correlation functions of the one-dimensional Hubbard model in a magnetic field, *Phys. Rev.* **B43**, 5653–62, (1991).

[26] D. Friedan, Z. Qui and S. Shenker, Conformal invariance, unitarity and critical exponents in two dimensions, *Phys. Rev. Lett.* **52**, 1575–8, (1984).

[27] T. Giamarchi and H.J. Schultz, Correlation functions of the one-dimensional quantum systems, *Phys. Rev.* **B39**, 4620–9, (1989).

[28] F.D.M. Haldane, Luttinger liquid theory of one-dimensional quantum fluids, *J. Phys.* **C14**, 2589–610, (1981).

[29] F.D.M. Haldane, Effective harmonic-fluid approach to low energy properties of one-dimensional quantum fluids, *Phys. Rev. Lett.* **47**, 1840–3, (1981).

[30] F.D.M. Haldane, Demonstration of the "Luttinger Liquid" character of Bethe-Ansatz-Soluble models of 1-D quantum fluids, *Phys. Lett.* **81A**, 153–5, (1981).

[31] F.D.M. Haldane, in *Electron Correlation and Magnetism in Narrow-band Systems* (ed. T. Moriya), p. 150, Berlin: Springer-Verlag, (1981).

[32] C.J. Hamer, M.T. Batchelor and M.N. Barber, Logarithmic corrections to finite-size scaling in the four-state Potts models, *J. Stat. Phys.* **B52**, 679–710, (1988).

[33] A.R. Its, A.G. Izergin and V.E. Korepin, Long-distance asymptotics of temperature correlations of the impenetrable Bose gas, *Comm. Math. Phys.* **130**, 471–86, (1990).

[34] A.R. Its, A.G. Izergin and V.E. Korepin, Large time and distance asymptotics of the temperature field correlator in the impenetrable Bose gas, *Nucl. Phys.* **B348**, 757–65, (1991).

[35] A.R. Its, A.G. Izergin, V.E. Korepin and V.Yu. Novokshenov, Temperature autocorrelations of the transverse Ising chain at the critical magnetic field, *Nucl. Phys.* **B340**, 752–8, (1990).

[36] C. Itzykson, M. Saleur and J.B. Zuber, *Conformal Invariance and Applications to Statistical Mechanics*, Singapore: World Scientific, (1986).

[37] A.G. Izergin and V.E. Korepin, Correlation functions for the Heisenberg XXZ-antiferromagnet, *Comm. Math. Phys.* **99**, 271–307 (1985).

[38] A.G. Izergin, V.E. Korepin and N.Yu. Reshetikhin, Conformal dimensions in the Bethe Ansatz solvable models, *J. Phys.* **A22**, 2615–20, (1989).

[39] A.G. Izergin and V.E. Korepin, Critical exponents in the Heisenberg magnet, *JETP Lett.* **42**, 414–16, (1985).

[40] L. Kadanoff and A.C. Brown, Correlation functions on a critical lines of a Baxter and Ashkin-Teller models, *Ann. Phys. (NY)* **121**, 318–42, (1979).

[41] N. Kawakami and S.K. Yang, Luttinger liquid properties of highly correlated electron systems in one dimension, *J. Phys.* **C3**, 5983–6008, (1991).

[42] V.E. Korepin, Correlation functions of the one-dimensional Bose gas in the repulsive case, *Comm. Math. Phys.* **94**, 93–113, (1984).

[43] V.E. Korepin and N.A. Slavnov, Correlation functions in one-dimensional Bose gas, *Theor. Math. Phys.* **68**, 955–60, (1986).

[44] V.E. Korepin and N.A. Slavnov, Time dependence of the density-density temperature correlation function of a one-dimensional Bose gas, *Nucl. Phys.* **B340**, 759–66, (1990).

[45] A. Luther and I. Peschel, Calculations of critical exponents in two dimensions from quantum field theory in one dimension, *Phys. Rev.* **B12**, 3908–17, (1975).

[46] B.M. McCoy, J.H.H. Perk and R.E. Shrock, Time-dependent correlation functions of the transverse Ising chain at the critical magnetic field, *Nucl. Phys.* **B220**, 35–47, (1983).

[47] B.M. McCoy, J.H.H. Perk and R.E. Shrock, Correlation functions of the transverse Ising chain at the critical field for large temporal and spatial separations, *Nucl. Phys.* **B220**, 269–82, (1983).

[48] A.D. Mironov and A.V. Zabrodin, Critical exponents in mutiparticle $1D$ quantum systems from finite-size effects in conformal field theories, *Phys. Rev. Lett.* **56**, 534–7, (1991).

[49] M. Ogata and H. Shiba, Bethe-ansatz wave-function, momentum distribution, and spin correlation in the one-dimensional strongly correlated Hubbard model, *Phys. Rev.* **B41**, 2326–38, (1990).

[50] A. Parola and S. Sorella, Asymptotic spin-spin correlations of the $U \to \infty$ one-dimensional Hubbard model, *Phys. Rev. Lett.* **64**, 1831–4, (1990).

[51] S.V. Pokrovsky and A.M. Tsvelick, Conformal dimension spectrum for lattice integrable models of magnets, *Sov. Phys. JETP* **66**, 1275–82, (1987).

[52] V.N. Popov, Long-wave asymptotics form of the many-body Green's functions of a one-dimensional Bose gas, *JETP Lett.* **31**, 526–9, (1980).

[53] V.N. Popov, *Functional Integrals in Quantum Field Theory and Statistical Physics*, Dordrecht: Reidel, (1983).

[54] H.J. Schulz, Correlation exponents and the metal-insulator transition in the one-dimensional Hubbard model, *Phys. Rev. Lett.* **64**, 2831–4, (1990).

[55] F. Woynarovich and H.P. Eckle, Finite-size corrections for the low lying states of a half-filled Hubbard chain, *J. Phys.* **A20**, L443–9, (1987).

[56] F. Woynarovich, H.P. Eckle and T.T. Truong, Non-analytic finite-size corrections in the one-dimensional Bose gas and Heisenberg chain, *J. Phys.* **A22**, 4027–43, (1989).

[57] F. Woynarovich, Finite-size effects in a non-half-filled Hubbard chain, *J. Phys.* **A22**, 4243–56, (1989).

Conclusion to Part IV

[1] B.M. McCoy, E. Barouch and D.B. Abraham, *Phys. Rev.* **A4**, 2331, (1971).

Index

Printed in the United States
By Bookmasters